Digital Signal Processing

# Digital Signal Processing

Theory and Practice

Tenth Edition

*Maurice Bellanger*
CNAM, Paris
France

Translated by
*Benjamin A. Engel*

WILEY

This edition first published 2024

Originally published in France as: Traitement numérique du signal 10th edition By Maurice BELLANGER


*Registered Offices*
John Wiley & Sons, Inc., 111 River Street, Hoboken, NJ 07030, USA
John Wiley & Sons Ltd, The Atrium, Southern Gate, Chichester, West Sussex, PO19 8SQ, UK

For details of our global editorial offices, customer services, and more information about Wiley products visit us at www.wiley.com.

Wiley also publishes its books in a variety of electronic formats and by print-on-demand. Some content that appears in standard print versions of this book may not be available in other formats.



*Library of Congress Cataloging-in-Publication Data*

Names: Bellanger, Maurice, author. | Engel, Benjamin A., translator.
Title: Digital signal processing : theory and practice / Maurice Bellanger, CNAM; translated by Benjamin A. Engel.
Other titles: Traitement numérique du signal. English
Description: Tenth edition. | Hoboken, NJ, USA : Wiley, 2024. | Translation of: Traitement numérique du signal.
Identifiers: LCCN 2023049303 (print) | LCCN 2023049304 (ebook) | ISBN 9781394182664 (cloth) | ISBN 9781394182671 (adobe pdf) | ISBN 9781394182688 (epub)
Subjects: LCSH: Signal processing–Digital techniques.
Classification: LCC TK5102.9 .B4513 2024 (print) | LCC TK5102.9 (ebook) | DDC 621.382/2–dc23/eng/20240110
LC record available at https://lccn.loc.gov/2023049303
LC ebook record available at https://lccn.loc.gov/2023049340

Cover Design: Wiley
Cover Image: © MR.Cole_Photographer/Getty Images

Set in 9.5/12.5pt STIXTwoText by Straive, Chennai, India
Printed and bound by CPI Group (UK) Ltd, Croydon, CR0 4YY

C9781394182664_150324

# Contents

# Foreword (Historical Perspective)

The most important and most impactful technical revolutions are not always those that are most evident to a product's end user. Modern digital signalprocessing methods fall into the category of impactful technical revolutions whose consequences are not immediately perceptible, and which do not make the front page.

It is interesting to reflect, for a moment, on the way in which such techniques emerge. Digital computation, applied to a signal in the broadest sense, is certainly not a new idea in itself. When Kepler derived the laws of motion of the planets from the series of observations made by his father-in-law Tycho Brahe, his was a truly numerical computation of the signal – in this case, the signal being Brahe's observations of the planets' positions over time. In recent decades, though, digital signal processing has become a discipline in its own right. What has changed is the way it can now process electrical signals in real time, using digital technologies.

This leap forward is the cumulative result of technical progress in numerous fields – starting, of course, with the capability of recording the data we wish to process in the form of an electrical signal. This has been contingent on the gradual development of what are known as information sensors, which can range in complexity from a simple stress gage (which, in itself, took a great deal of research in solid mechanics to make possible) to a radar system.

In addition, with the marvelous progress in micro-electronics came the necessary technological tools, capable, at the extremely fast rates required for real-time processing, of performing the arithmetical operations that the earliest computers (the ENIAC was built in 1945, not long ago in the grand scheme of things) took hours to do, often being interrupted by repeated breakdowns. Today, these operations can be carried out by microprocessors weighing only a few grams and consuming only a few milliwatts of power, capable of functioning for over a decade without breakdown.

Finally, we have had to wait for progress in programming techniques – i.e. the optimal use of these technologies – because though the computational capacities of modern microprocessors are vast, it is unwise to waste those capacities on performing unnecessary operations. The invention of the fast Fourier transform algorithms is one of the most striking examples of the importance of programming methods. This convergence of technical progress in fields ranging from physics to electronics to mathematics has not been unintentional. To a certain extent, every step forward has created a new problem, which was then solved by new progress in a different field. It would undoubtedly be helpful, from the standpoint of the history and epistemology of science and technology, to have an in-depth study of this lengthy and complicated process.

Indeed, the consequences are already considerable. Indisputably, analog processing of electrical signals came before digital processing, and analog processing will surely continue to have an important role to play in certain applications, but the benefits of digital processing can be expressed in two words: accuracy and reliability. Certain applications have only been made possible by the accuracy and reliability offered by digital technologies, which go far beyond the sectors of electronics and telecommunications in which these techniques first emerged. As one example among many, in X-ray tomodensitography, scanners are based on the application of a theorem developed by Johann Radon in 1917. Only the developments mentioned above have enabled the practical implementation of this new medical diagnostic tool. It is a safe bet that, in tomorrow's world, digital signal processing techniques will be used in increasingly varied products, including consumer electronics. However, it is an equally safe bet that the general public, while benefitting from the lower prices and higher performance and reliability offered by these techniques, will remain blissfully unaware of the phenomenal and complex combination of research, technology, and invention represented by this progress. This shift has already begun in the case of television receivers.

However, when these technical revolutions take place, another problem almost inevitably arises. We need to train users to get to grips not just with a new tool, but often, an entirely new way of thinking. If we are not careful, such training can easily become a bottleneck, delaying the introduction of new techniques. Therefore, this book is a particularly important addition to the field. Its author, Maurice Bellanger, has been teaching for many years at the École Nationale Supérieure des Télécommunications and the Institut Supérieur d'Électronique de Paris. It is a highly didactic book, containing relevant exercises as well as in-depth explanations and multiple programs, which certain people will often be able to make use of exactly as they are. Without a doubt, it will help open the door to desirable and necessary evolution.

P. Aigrain, 1981

# Preface

In signal processing, digital techniques offer a fantastic range of possibilities: rigorous system design, flexibility, reproducibility of equipment, stability of operating features, and ease of supervision and monitoring. However, there is a certain amount of abstractness in these techniques, and, in order to apply them to real-world cases, we need a set of theoretical knowledge, which may represent an obstacle to their use. This book aims to break down the barriers and make digital techniques accessible to readers by drawing the connection between theory and practice and providing users with the most widely used results in the domain, at their fingertips.

The foundation upon which this book is built is the author's teaching at engineering schools – first the École nationale supérieure des télécommunications and the Institut supérieur d'électronique de Paris, and later, Supélec and CNAM. The book offers a clear and concise presentation of the main techniques used in digital processing, comparing them on their merits and giving the most useful results in a form that is directly usable, both for the design and for the concrete implementation of systems. Theoretical explanations have been condensed to what is absolutely necessary for a thorough understanding and a correct application of the results. Bibliographic references are provided, where interested readers will find further information about the topics discussed herein. At the end of each chapter are a few exercises, often drawn from real-world examples, to allow readers to test their absorption of the material in the chapter and familiarize themselves with its application. Answers to these exercises and guidelines are given at the end of the book.

With respect to previous editions, this new edition offers additional information, simplifications, and also a new chapter about one of the most important tools in the field of artificial intelligence – neural networks, as they relate to adaptive systems.

As with the previous editions, this one owes a great deal to the author's students and colleagues. Thanks to them all for their contributions and assistance.

# Introduction

A signal is the medium carrying information, transmitted by a source to a receiver. In other words, a signal is the vehicle of intelligence in systems. It transports commands in control and remote-control equipment; it carries data such as information, spoken words, or images across networks. It is particularly fragile and needs to be handled with a great deal of care. Signal processing is applied in order to extract information, alter the message being carried, or adapt the signal to the transmission techniques being used. It is here that digital techniques come into play. Indeed, if we imagine substituting the signal with a set of numbers, representing its value or amplitude at carefully chosen times, then its processing, even in the most elaborate of forms, boils down to a sequence of logical and arithmetical operations on that set of numbers, committing the results to memory.

A continuous analog signal is converted into a digital signal by sensors which act on readings, or directly in the devices producing or receiving the signal. The operations taking place in the wake of that conversion are carried out by digital computers, tasked or programmed to perform the sequence of operations by which the desired processing is defined.

Before introducing the content of each chapter of this book, it is wise to precisely define the processing of which we speak here.

Digital signal processing refers to the set of operations, arithmetic calculations, and number manipulations, which are applied to a signal to be processed, represented by a series or a set of numbers, to produce another series or set of numbers, which represent the processed signal. In this way, an immense variety of functions can be performed, such as spectral analysis, linear or nonlinear filtration, transcoding, modulation, detection, estimation, and parameter extraction. The machines used are digital computers.

The systems corresponding to this processing obey the laws of discrete systems. In certain cases, the numbers to which the processing is applied may be derived from a discrete process. However, they often represent the amplitude of samples taken from a continuous signal, and, in that case, the computer must be downstream of an analog-to-digital converter and possibly upstream of a digital-to-analog converter. In designing such systems, and in studying how they work, signal digitization is fundamentally important, and the operations of sampling and encoding must be analyzed in terms of their principles and their consequences. The theory of distributions is a concise, simple, and effective means of such analysis. Following the presentation of certain fundamental aspects concerning Fourier analysis, distributions, and signal representation, Chapter 1 contains the most important and most useful results for sampling and encoding of a signal.

The advent of digital processing dates from the discovery of fast computational algorithms of the discrete Fourier transform. Indeed, this transform is the basis for the study of discrete systems. In digital processing, it is the equivalent of the Fourier transform in analog processing, enabling us to transition from the discrete-time domain to the discretefrequency domain. It lends itself very well to spectral analysis, with a frequency step dividing the sampling frequency of the signals being analyzed.

Fast computation algorithms offer gains, as they enable operations to be performed in real time, provided certain elementary conditions are met. Thus, the discrete Fourier transform is not only a fundamental tool in determining the processing characteristics and in the study of the impacts of those characteristics on the signal, but it is also used in the production of popular devices, such as mobile radio and digital television. Chapters 2 and 3 are dedicated to these algorithms. To begin with, they present the elementary properties and the mechanism of fast computation algorithms and their applications before moving on to a set of variants associated with practical situations.

A significant portion of this book is devoted to the study of one-dimensional invariant linear time-discrete systems, which are easily accessible and highly useful. Multi-dimensional systems, and, in particular, two- and three-dimensional systems, are experiencing significant development. For example, they are applied to images. However, their properties are generally deduced from those of one-dimensional systems, of which they are often merely simplified extensions. Nonlinear or time-variable systems either contain a significant subset, retaining the properties of linearity and time-invariance, or can be analyzed with the same techniques as systems that have those properties.

Linearity and time-invariance lead to the existence of a convolution relation, which governs the operation of the system or filter having those properties. This convolution relation is defined on the basis of the system's response to the elementary signal which represents a pulse – the impulse response – by an integral in the case of analog signals. Thus, if $x(t)$ denotes the signal to be filtered, and $h(t)$ is the filter impulse response, the filtered signal $y(t)$ is given by the equation:

$$y(t) = \int_{-\infty}^{\infty} h(\tau)\, x(t-\tau)\, d\tau$$

In these conditions, such a relation, which directly expresses the filter's real operation, offers limited practical interest. To begin with, it is not particularly easy to determine the impulse response on the basis of criteria that define the filter's intended operation. In addition, an equation that contains an integral cannot easily be used to recognize and check the filter's behavior. Design is much easier to address in the frequency domain because the Laplace transform or Fourier transform can be used to move to a transformed plane where the convolution relations from the amplitude–time plane become simple products of functions. The Fourier transform matches the system's frequency response to the impulse response, and the filtration is then the product of that frequency response by the Fourier transform, or spectrum, of the signal to be filtered.

In discrete digital systems, the convolution is expressed by a sum. The filter is defined by a series of numbers, representing its impulse response. Thus, if the series to be filtered is written as $x(n)$, the filtered series $y(n)$ is expressed by the following sum, where $n$ and $m$ are integers:

$$y(n) = \sum_{m} h(m)\, x(n-m)$$

Two scenarios then arise. Firstly, the sum may pertain to a finite number of terms – i.e. the $h(m)$ values are zero, except for a finite number of values of the integer variable $m$. The filter is known as a finite impulse-response filter. In reference to its realization, it is also referred to as non-recursive, because it does not require a feedback loop from output to input in its implementation. It occupies finite memory space because it only retains the memory of an elementary signal – an impulse, for example – for a limited time. The numbers $h(m)$ are called the coefficients of the filter, which they define completely. They can be calculated directly, in a very simple way – for instance, by means of a Fourier series development of the frequency response. This type of filter exhibits highly interesting original features (for example, the possibility of a rigorously linear phase response – i.e., a constant group delay); the signals whose components are within the filter's passband are not deformed as they pass through the filter. This possibility is exploited in data transmission systems, or spectral analysis, for example.

Alternatively, the sum may pertain to an infinite number of terms, and the $h(m)$ may have an infinite number of nonzero values; the filter is called an infinite impulse-response filter, or recursive, because its memory must be set up as a feedback loop from output to input. Its operation is governed by an equation whereby an element in the output series $y(n)$ is calculated by the weighted sum of a number of elements of the input series $x(n)$, and a certain number of elements of the previous output series. For example, if $L$ and $K$ are integers, the filter's operation may be defined by the following equation:

$$y(n) = \sum_{l=0}^{L} a_l x(n-l) - \sum_{k=1}^{K} b_k y(n-k)$$

The $al(l = 0, 1, \ldots, L)$ and $bk(k = 1, 2, \ldots, K)$ are the coefficients. As is the case with analog filters, this type of filter generally cannot easily be studied directly; it is necessary to go through a transformed plane. The Laplace or Fourier transforms could be used for this purpose. However, there is a transform that is much more suitable – the $Z$ transform, which is the equivalent for discrete systems. A filter is characterized by its $Z$-transfer function, generally written as $H(Z)$, which involves the coefficients in the following equation:

$$H(Z) = \frac{\sum_{l=0}^{L} a_l \mathbf{Z}^{-l}}{1 + \sum_{k=1}^{K} b_k \mathbf{Z}^{-k}}$$

To obtain the filter's frequency response, in $H(Z)$, we simply need to replace the variable $Z$ with the following expression, where $f$ denotes the frequency variable, and $T$ the time step between the signal samples:

$$Z = e^{j2\pi f}T$$

In this operation, the imaginary axis in the Laplacian plane corresponds to the circle with unit radius, centered at the origin in the plane of the variable $Z$. It is plain that the frequency response of the filter defined by $H(Z)$ is a periodic function whose period is the sampling frequency. Another representation of the function $H(Z)$, which is useful in the design of filters and the study of a number of properties, explicitly includes the roots of the numerator, also known as the zeroes of the filter, $Zl(l = 1, 2, \ldots, L)$, and the roots of the denominator, also known as the poles, $Pk(k = 1, 2, \ldots, K)$:

$$H(Z) = a_0 \frac{\prod_{l=1}^{L} (1 - Z_l Z^{-1})}{\prod_{k=1}^{K} (1 - P_k Z^{-1})}$$

The term $a_0$ is a scaling factor which defines the gain of the filter. The filter stability condition is expressed very simply by the following constraint: all the poles must be within the unit circle. The position of the poles and zeroes with respect to the unit circle offers a very simple way of determining the characteristics of the filter; this technique is very widely used in practice.

Four chapters are devoted to the study of the characteristics of these digital filters. Chapter 4 presents the properties of time-invariant discrete linear systems, recaps the main properties of the $Z$-transform, and lays down the fundamental groundwork necessary for the study of filters. Chapter 5 discusses finite impulse-response filters – their properties are studied, the techniques for calculating the coefficients are described, and the structures of real-world filters are examined. Infinite impulse-response filters are generally produced by cascading first- and second-order elementary cells, or sections, so Chapter 6 describes these sections and their properties. To begin with, this makes the study of this type of system considerably easier; in addition, the chapter provides a set of results that are highly useful in practice. Chapter 7 outlines the methods for calculating the coefficients for infinite impulse-response filters and discusses the problems posed by their real-world implementation, with the limitations that are encountered and the consequences of those limitations – in particular, computational noise.

As the properties of infinite impulse-response filters are comparable to those of continuous analog filters, it is natural to envisage similar structures for these filters to those generally employed in analog filtering. This is the subject of Chapter 8, which presents ladder structures. We then take a diversion to look at switched-capacitor filters, which are not digital in the strictest sense of the word, but which are sampled, and are highly useful additions to digital filters. To guide users, a summary of the respective merits of the structures described is given at the end of the chapter.

Certain devices – for example, in instrumentation or telecommunications – work on signals represented by a series of complex numbers. Out of all signals of this type, one category is of particular practical interest: analytic signals. Their properties are studied in Chapter 9, as is the design of devices apt for the generation or processing of such signals. Additional concepts relating to filtering are also explained in this chapter, which, in a unified manner, presents the main interpolation techniques. Signal restoration is also discussed.

Digital processing machines, when operating in real time, operate at a rate that is closely linked to the signal sampling frequency. Their complexity depends on the volume of operations being carried out, and the length of time available in which to perform this processing. The signal sampling frequency is generally imposed either at system input or at output, but within the system itself, it is possible to vary this rate in order to adapt it to the characteristics of the signal and the processing, and thereby reduce the volume of operations and the computation rate. The machines may be simplified – potentially very significantly – if, over the course of the processing, the sampling frequency is adapted to suit the usable bandwidth of the signal; this is multirate filtering, which is presented in Chapter 10. The impacts on the processing characteristics are described, along with realization methods. Rules are provided on usage and assessment. This technique produces particularly interesting results for narrow passband filters or the implementation of sets known as filter banks. In this case, the system associates, with a set of phase-shifting circuits, a discrete Fourier transform calculator.

Filter banks for the breakdown and reconstruction of signals have become a fundamental tool for compression. The way in which they work is described in Chapters 11 and 12 with design methods and realization structures.

The filters can be determined on the basis of time-domain specifications; such is the case, for example, with the modeling of a system, as described in Chapter 13. If the characteristics vary, it may be interesting to adapt the coefficients as a function of changes occurring in the system. This adaptation may depend on an approximation criterion and take place at a rate that may come to equal the system's sampling rate; then, the filter is said to be adaptive. Chapter 14 is devoted to adaptive filtering, in the simplest of cases, but also the most common and the most useful – where the approximation criterion chosen is the minimization of the mean squared error, and where the coefficients vary depending on the gradient algorithm. After recapping details of random signals and their properties in Chapter 13 – in particular, the autocorrelation function and matrix, whose eigenvalues play an important role – the gradient algorithm is presented in Chapter 14, and its convergence conditions are studied. Then, the two main adaptation parameters, the time constant and the residual error, are analyzed along with the arithmetic complexity. Different structures are proposed for concrete implementation.

Chapter 15 can be viewed as an extension of Chapters 13 and 14 to the domain of neural networks in artificial intelligence. These devices are characterized by the systematic use of nonlinear circuits for the functions of modeling, classification, or shape recognition. Adaptive techniques are used during the learning phases.

Chapter 16 discusses a very specific application: error-correction coding. Indeed, information processing and transmission systems include error-correction coding techniques, which are generally introduced by a mathematical approach, though some of the most widely used types of coding are actually direct applications of the fundamental signal processing techniques. Thus, the chapter puts forward a signalprocessing vision of certain types of coding, to facilitate readers' access to and use of these techniques.

Finally, Chapter 17 briefly describes some applications, showing how the fundamental methods and techniques are put to use.

# 1

# Signal Digitizing – Sampling and Coding

The conversion of an analog signal to digital form involves a twofold approximation. Firstly, in the time domain, the signal function $s(t)$ is replaced by its values at integral time increments $T$ and is thus converted to $s(nT)$. This process is called sampling. Secondly, in the amplitude domain, each value of $s(nT)$ is approximated by a whole multiple of an elementary quantity. This process is called quantization. The approximate value thus obtained is then associated with a number. This process is called coding – a term often used to describe the whole process by which the value of $s(nT)$ is transformed into the number representing it.

The effect of these two approximations on the signal will be analyzed in this chapter. To achieve this, two basic tools will be used: Fourier analysis and distribution theory.

## 1.1 Fourier Analysis

Fourier analysis is a method of decomposing a signal into a sum of individual components which can easily be produced and observed. The importance of this decomposition is that a system's response to the signal can be deduced from these individual components using the superposition principle. These elementary component signals are periodic and complex, so both the amplitude and phase of the systems can be studied. They are represented by a function $s_e(t)$ such that:

$$S_e(t) = e^{j2\pi ft} = \cos(2\pi ft) + j\sin(2\pi ft) \tag{1.1}$$

where $f$ is the inverse of the period – that is, the frequency of the elementary signal.

Since the elementary signals are periodic, clearly, the analysis is simplified when the signal itself is periodic. This case will be examined first, although it is not the most interesting, since a periodic signal is completely determinate and carries practically no information.

### 1.1.1 Fourier Series Expansion of a Periodic Function

Let $s(t)$ be a function of a periodic variable $t$ with period $T$ – that is, satisfying the relation:

$$s(t + T) = s(t) \tag{1.2}$$

Under certain conditions, this function can be expanded in a Fourier series as:

$$s(t) = \sum_{n=-\infty}^{\infty} C_n e^{j2\pi nt/T} \tag{1.3}$$

*Digital Signal Processing: Theory and Practice*, Tenth Edition. Maurice Bellanger.

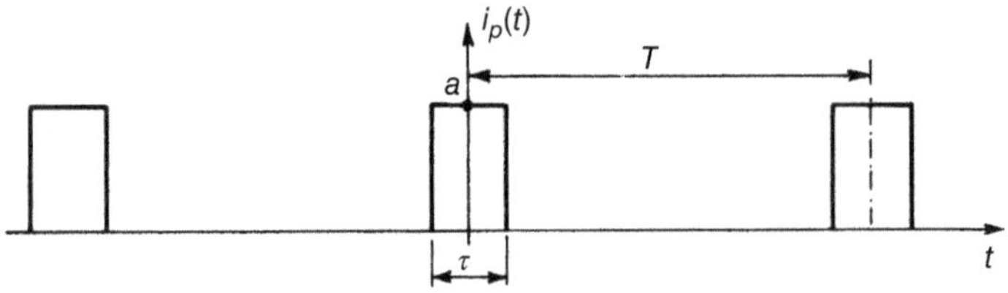


**Figure 1.1** Impulse train.

The index $n$ is an integer, and the $C_n$, called the Fourier coefficients, are defined by:

$$C_n = \frac{1}{T}\int_0^T s(t)e^{-j2\pi nt/T}dt \tag{1.4}$$

In fact, the Fourier coefficients minimize the square of the difference between the function $s(t)$ and the series (1.3). Expression (1.4) is obtained by taking the derivative with respect to the index $n$ coefficient of the quantity:

$$\int_0^T \left(s(t) - \sum_{m=-\infty}^{\infty} C_m e^{j2\pi mt/T}\right)^2 dt$$

and setting that derivative to zero.

*Example:* Figure 1.1 shows an example of a Fourier expansion of a function $i_p(t)$ composed of a train of impulses, each of width $\tau$ and amplitude $a$, occurring at time intervals $T$. The time origin is taken as being at the center of an impulse.

The coefficients $C_n$ are given by:

$$C_n = \frac{1}{T}\int_{-\tau/2}^{\tau/2} ae^{-j2\pi nt/T}dt = \frac{a\tau}{T}\frac{\sin(\pi n\tau/T)}{\pi n\tau/T} \tag{1.5}$$

and the Fourier expansion is:

$$i_p(t) = \frac{a\tau}{T}\sum_{n=-\infty}^{\infty}\frac{\sin(\pi n\tau/T)}{\pi n\tau/T}e^{j2\pi nt/T} \tag{1.6}$$

The importance of this example for the study of sampled systems is readily apparent.

The properties of Fourier series expansions are given in Ref. [1]. One important property, expressed by the Bessel–Parseval equation, is that power is conserved in the expansion of the signal:

$$\sum_{n=-\infty}^{\infty} |C_n|^2 = \frac{1}{T}\int_0^T |s(t)|^2\, dt \tag{1.7}$$

The constituent elements resulting from the expansion of a periodic signal have frequencies which are integer multiples of $1/T$ (the inverse of the period). They form a discrete set in the space of all frequencies. In contrast, if the signal is not periodic, the Fourier components form a continuous domain in the frequency space.

### 1.1.2 Fourier Transform of a Function

Let $s(t)$ be a function of $t$. Under certain conditions, one can write:

$$s(t) = \int_{-\infty}^{\infty} S(f)e^{j2\pi ft}\, df \tag{1.8}$$

where

$$S(f) = \int_{-\infty}^{\infty} S(f)e^{j2\pi ft}\, dt \tag{1.9}$$

The function $S(f)$ is the Fourier transform of $s(t)$. More commonly, $S(f)$ is called the spectrum of signal $s(t)$.

*Example:* To calculate the Fourier transform $I(f)$ of an isolated pulse $i(t)$ of width $\tau$ and amplitude $a$, centered on the time origin (Figure 1.2):

$$I(f) = \int_{-\infty}^{\infty} i(t)e^{-j2\pi ft}\, dt = a\int_{-\tau/2}^{\tau/2} e^{-j2\pi ft}\, dt$$

$$I(f) = a\tau \frac{\sin(\pi f\tau)}{\pi f\tau} \tag{1.10}$$

Figure 1.3 represents the function $I(f)$. This will be used frequently in this book. It is important to note that it will be zero for nonzero frequencies which are whole multiples of the inverse of the impulse width. A table of this function is given in Appendix 1.

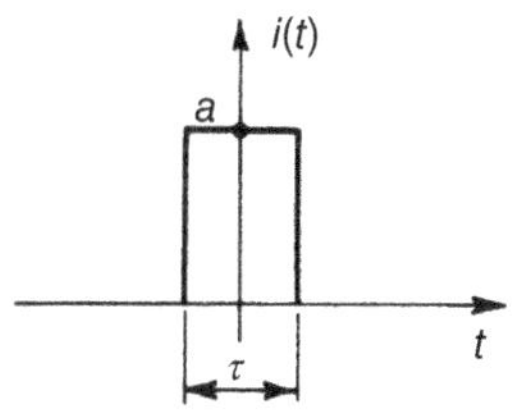

**Figure 1.2** Isolated impulse.

This example clearly shows the correspondence between the Fourier coefficients and the spectrum. In effect, by comparing equations (1.6) and (1.10), it can be verified that, apart from the factor $1/T$, the coefficients of the Fourier series expansion of an impulse train correspond to the values of the spectrum of the isolated impulse at frequencies which are whole multiples of the inverse of the period of the impulses.

In the case of a nonperiodic function, there is an expression similar to the Bessel–Parseval relation, but this time the energy in the signal is conserved, instead of the power:

$$\int_{-\infty}^{\infty} |S(f)|^2 df = \int_{-\infty}^{\infty} |s(t)|^2 dt \tag{1.11}$$

Let $s'(t)$ be the derivative of the function $s(t)$; its Fourier transform $S_d(f)$ is given by:

$$S_d(f) = \int_{-\infty}^{\infty} e^{-j2\pi ft} s'(t) dt = j2\pi f S(f) \tag{1.12}$$

Thus, taking the derivative of a signal leads to multiplying its spectrum by $j2\pi f$.

One essential property of the Fourier transform (in fact, the main reason for its use) is that it transforms a convolution into a simple product. Consider two time functions, $x(t)$ and $h(t)$, with Fourier transforms $X(f)$ and $H(f)$, respectively. The convolution $y(t)$ is defined by:

$$y(t) = x(t) * h(t) = \int_{-\infty}^{\infty} x(t-\tau)h(\tau)d\tau \tag{1.13}$$

**Figure 1.3** Spectrum of an isolated impulse.

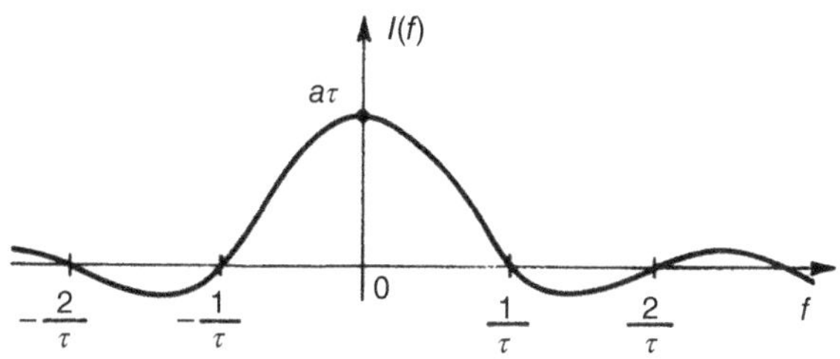

The Fourier transform of this product is:

$$Y(f) = \int_{-\infty}^{\infty} \left( \int_{-\infty}^{\infty} x(t-\tau)h(\tau)d\tau \right) e^{-j2\pi ft}\mathrm{d}t$$

$$Y(f) \int_{-\infty}^{\infty} h(\tau)e^{-j2\pi f\tau}\mathrm{d}\tau \int_{-\infty}^{\infty} x(u)e^{-j2\pi fu}\mathrm{d}u = H(f)X(f)$$

Conversely, it can be shown that the Fourier transform of a simple product is a convolution product.

An interesting result can be derived from the abovementioned properties. Let us consider the Fourier transform $II(f)$ of the function $i^2(t)$. Because of equations (1.10) and (1.13), we obtain:

$$II(f) = I(f) * I(f) = aI(f) \tag{1.14}$$

and therefore,

$$\int_{-\infty}^{\infty} \frac{\sin(\pi\phi\tau)}{\pi\phi\tau} \frac{\sin(\pi(f-\phi)\tau)}{\pi(f-\phi)\tau} \mathrm{d}\phi = \frac{1}{\tau} \frac{\sin(\pi f\tau)}{\pi f\tau}$$

Taking $f = n/\tau$, for any integer $n$,

$$\int_{-\infty}^{\infty} \frac{\sin(\pi\phi\tau)}{\pi\phi\tau} \cdot \frac{\sin(\pi(\phi\tau - n))}{\pi(\phi\tau - n)} \mathrm{d}\phi = 0 \tag{1.15}$$

Thus, the functions $\sin \pi(x - n)/[\pi(x - n)]$, with $n$ being an integer, forms a set of orthogonal functions.

The definition and properties of the Fourier transform can be extended to multivariate functions. Let $s(x_1, x_2, \ldots, x_n)$ be a function of $n$ real variables. Its Fourier transform is a function $S(\lambda_1, \lambda_2, \ldots, \lambda_n)$ defined by:

$$S(\lambda_1, \lambda_2, \ldots, \lambda_n) = \iint_{\mathbb{R}^n} \cdots \int s(x_1, x_2, \ldots, x_n) \times \boldsymbol{e}^{-\mathbf{j}2\pi(\lambda_1 x_1 + \lambda_2 x_2 + \cdots + \lambda_n x_n)} \mathbf{d}x_1 \mathbf{d}x_2 \cdots \mathbf{d}x_n \tag{1.16}$$

If the function $s(x_1, x_2, \ldots, x_n)$ is separable – that is, if:

$$s(x_1, x_2, \ldots, x_n) = s(x_1)s(x_2) \cdots s(x_n)$$

then:

$$S(\lambda_1, \lambda_2, \ldots, \lambda_n) = S(\lambda_1)S(\lambda_2) \cdots S(\lambda_n)$$

The variables $x_i$ $(1 \leqslant i \leqslant n)$ often represent distances (for example, for two dimensions), and in that case, $\lambda_i$ are called spatial frequencies.

## 1.2 Distributions

Mathematical distributions constitute a formal mathematical representation of the physical distributions found in experiment [1].

### 1.2.1 Definition

A distribution $D$ is defined as a continuous linear function in the vector space $\mathscr{D}$ of functions defined in $\mathbb{R}^n$, indefinitely differentiated, and having a bounded support.

With each function $\varphi$ belonging to $\mathscr{D}$, the distribution $D$ associates a complex number $D(\varphi)$ which will also be denoted by $\langle D, \varphi\rangle$, with the properties:

(1) $D(\varphi_1 + \varphi_2) = D(\varphi_1) + D(\varphi_2)$.
(2) $D(\lambda\varphi) = \lambda D(\varphi)$ where $\lambda$ is a scalar.
(3) If $\varphi_j$ converges to $\varphi$ when $j$ tends toward infinity, the sequence $D(\varphi_j)$ converges to $D(\varphi)$.

*Examples:*

(i) If $f(t)$ is a function which is summable over any bounded ensemble, it defines a distribution $D_f$ by:

$$\langle D_f, \phi\rangle = \int_{-\infty}^{\infty} f(t)\phi(t)\mathrm{d}t \tag{1.17}$$

(ii) If $\varphi'$ denotes the derivative of $\varphi$, the function:

$$\langle D, \phi\rangle = \int_{-\infty}^{\infty} f(t)\phi'(t)\mathrm{d}t = \langle f, \phi\rangle \tag{1.18}$$

is also a distribution.

(iii) The Dirac distribution $\delta$ is defined by:

$$\langle \delta, \phi\rangle = \phi(0) \tag{1.19}$$

The Dirac distribution $\delta$ at a real point $x$ is defined by:

$$\langle \delta(t-x), \phi\rangle = \phi(x) \tag{1.20}$$

This distribution is said to represent a mass of $+1$ at the point $x$.

(iv) Consider a pulse $i(t)$ of duration $\tau$, with amplitude $a = 1/\tau$, centered on the origin. It defines a distribution $D_i$:

$$\langle D_i, \phi\rangle = \frac{1}{\tau}\int_{-\tau/2}^{\tau/2} \phi(t)\mathrm{d}t$$

For very small values of $\tau$, this becomes:

$$\langle D_i, \phi\rangle \simeq \phi(0)$$

that is, the Dirac distribution can be regarded as the limit of the distribution $D_i$ when $\tau$ tends toward 0.

### 1.2.2 Differentiation of Distributions

The derivative $\partial D/\partial t$ of a distribution $D$ is defined by the relation:

$$\left\langle \frac{\partial D}{\partial t}, \phi\right\rangle = -\left\langle D, \frac{\partial \phi}{\partial t}\right\rangle \tag{1.21}$$

To illustrate this, consider the Heaviside function $Y$, or single-step function, which is zero when $t < 0$ and $+1$ if $t \geqslant 0$:

$$\left\langle \frac{\partial Y}{\partial t}, \phi\right\rangle = -\left\langle Y, \frac{\partial \phi}{\partial t}\right\rangle = -\int_{0}^{\infty} \phi'(t)\mathrm{d}t = \phi(0) = \langle \delta, \phi\rangle \tag{1.22}$$

As a result, the discontinuity in $Y$ appears in the derivative as a point of unit mass.

This example illustrates the considerable practical interest of the notion of a distribution, which means that a number of the concepts and properties of continuous functions can be extended to discontinuous functions.

#### 1.2.2.1 The Fourier Transform of a Distribution

By definition, the Fourier transform of a distribution $D$ is a distribution denoted by $FD$ such that:

$$\langle FD, \phi \rangle = \langle D, F\phi \rangle \tag{1.23}$$

By applying this definition to distributions with a point support, we obtain:

$$\langle F\delta, \phi \rangle = \langle \delta, F\phi \rangle = \int_{-\infty}^{\infty} \phi(t)\mathrm{d}t = \langle 1, \phi \rangle \tag{1.24}$$

Consequently, $F\delta = 1$. Similarly, $F\delta(t - a) = e^{-\mathrm{j}2\pi fa}$.

A case which is fundamental to the study of sampling is that of the set $(u)$ of Dirac distributions shifted by $T$ and such that:

$$u(t) = \sum_{n=-\infty}^{\infty} \delta(t - nT) \tag{1.25}$$

This set is a distribution of unit mass points separated on the abscissa by whole multiples of $T$. Its Fourier transform is:

$$Fu = \sum_{n=-\infty}^{\infty} e^{-\mathrm{j}2\pi fnT} = U(f) \tag{1.26}$$

and it can be shown that this sum is a point distribution.

A simple demonstration can be obtained from the Fourier series development of the function $i_\mathrm{p}(t)$, formed by the set of separate pulses of duration $T$, with width $\tau$, and amplitude $1/\tau$, centered on the time origin.

One can consider $u(t)$ as the limit of $i_\mathrm{p}(t)$ as $\tau$ tends toward zero:

$$u(t) = \lim_{\tau \to 0} i_p(t)$$

and by referring to relation (1.6), we find that:

$$\lim_{\tau \to 0} i_\mathrm{p}(t) = \frac{1}{T} \sum_{n=-\infty}^{\infty} e^{\mathrm{j}2\pi nt/T}$$

The following fundamental property is demonstrated in Ref. [2].

*The Fourier transform of the time distribution, represented by unit mass points separated by whole multiples of* T *is a frequency distribution of points of mass* $1/T$ *separated by whole multiples of* $1/T$.

That is:

$$U(f) = \sum_{n=-\infty}^{\infty} e^{\mathrm{j}\pi fnT} = \frac{1}{T} \sum_{n=-\infty}^{\infty} \delta\left(f - \frac{n}{T}\right) \tag{1.27}$$

This result will be used when studying signal sampling. The property of the Fourier transform whereby it exchanges convolution and multiplication applies equally to distributions.

Before considering the influence of the sampling and quantizing operations on the signal, it is useful to discuss the characteristics of those signals which are most often studied.

## 1.3 Some Commonly Studied Signals

A signal is defined as a function of time $s(t)$. This function can be either an analytic expression or the solution of a differential equation, in which case the signal is said to be deterministic.

### 1.3.1 Deterministic Signals

Sine waves are the most frequently used signals of this type. For example,

$$s(t) = A\ \cos(\omega t + \alpha)$$

where $A$ is the amplitude, $\omega = 2\pi f$ is the angular frequency, and $\alpha$ is the phase of the signal.

Signals of this type are easy to reproduce and recognize at different points of a system. They allow the various characteristics to be visualized in a simple way. Moreover, as mentioned above, they serve as the basis for the decomposition of any deterministic signal through the Fourier transform.

If the system is linear and invariant in time, it can be characterized by its frequency response $H(\omega)$. For each value of the frequency, $H(\omega)$ is a complex number whose modulus is the amplitude of the response. By convention, the function $\varphi(\omega)$ such that:

$$H(\omega) = |H(\omega)|e^{-j\phi(\omega)} \tag{1.28}$$

is defined as the phase. This convention allows the group delay $\tau(\omega)$, a positive function for real systems, to be expressed as:

$$\tau(\omega) = \frac{d\phi}{d\omega} \tag{1.29}$$

The group delay refers to transmission lines on which different frequencies of the signal propagate at different speeds, which leads to energy dispersion in time. As an illustration of the concept, let us consider two close frequencies $\omega \pm \Delta\omega$ and the corresponding phases per unit length $\varphi \pm \Delta\varphi$. The sum signal is expressed by:

$$s(t) = \cos[(\omega + \Delta\omega)t - (\varphi + \Delta\varphi)] + \cos[(\omega - \Delta\omega)t - (\varphi - \Delta\varphi)]$$

or

$$s(t) = 2\cos(\omega t - \varphi)\cos(\Delta\omega t - \Delta\varphi)$$

This is a modulated signal, and there is no dispersion if the two factors in the above expression undergo the same delay per unit length – that is, $\Delta\varphi/\Delta\omega$ is constant. Thus, the group delay characterizes the dispersion imposed on the signal energy by a transmission line or any equivalent system.

If the sinusoidal signal $s(t)$ is applied to the system, then an output signal $s_r(t)$ is obtained such that:

$$s_r(t) = A|H(\omega)|\cos[\omega t + \alpha - \phi(\omega)] \tag{1.30}$$

Once again, this is a sinusoidal signal, and comparison with the applied signal reveals the response of the system. The importance of this procedure (for example, for test operations) can readily be appreciated.

Deterministic signals, meanwhile, do not give a good representation of real signals because they do not carry any information except by the fact of their presence. Real signals are generally represented by a random function $s(t)$. For testing and analyzing systems, random signals are also used, but they must have particular characteristics which do not present undue complications for their generation and use. A study of such signals is given in Vol. 2 of Ref. [2].

### 1.3.2 Random Signals

A random signal is defined at each time, $t$, by a probability law for its amplitude. This law can be expressed by a probability density $p(x, t)$ defined by:

$$p(x,t) = \lim_{\Delta x \to 0} \frac{\text{Prob}[x \le s(t) \le x + \Delta x]}{\Delta x} \tag{1.31}$$

It is stationary if its statistical properties are independent of time – that is, if the probability density is independent of time:

$$p(x,t) = p(x)$$

It is of second-order if it possesses a first-order moment called the mean value, which is the mathematical expectation of $s(t)$, denoted by $E[s(t)]$ and defined by:

$$m_1(t) = E[s(t)] = \int_{-\infty}^{\infty} xp(x,t)\mathrm{d}x \tag{1.32}$$

and a second-order moment called the covariance:

$$E[s(t_1)s(t_2)] = m_2(t_1,t_2) = \int_{-\infty}^{\infty}\int_{-\infty}^{\infty} x_1x_2p(x_1,x_2;t_1,t_2)\mathrm{d}x_1\mathrm{d}x_2$$

where $p(x_1,x_2;t_1,t_2)$ is the probability density of a pair of random variables $[s(t_1),s(t_2)]$.

The stationarity can be limited to the moments of first and second order; then, the signal is said to be stationary of order 2 or stationary in the wider sense. For such a signal:

$$E[s(t)] = \int_{-\infty}^{\infty} xp(x)\mathrm{d}x = m_1$$

The independence of time is translated for the probability density $p(x_1, x_{2;}t_1,t_2)$ as follows:

$$p(x_1,x_2;t_1,t_2) = p(x_1,x_2;0,t_2-t_1) = p(x_1,x_2;\tau)$$

where,

$$\tau = t_2 - t_1$$

Only the difference between the two observation times is involved:

$$E[s(t_1)s(t_2)] = m_2(\tau) \tag{1.33}$$

The function $r_{xx}(\tau)$, such that:

$$r_{xx}(\tau) = E[s(t)s(t-\tau)] \tag{1.34}$$

is called the autocorrelation function of the signal.

A random signal $s(t)$ also has a time mean $m_{\mathrm{T}}$. This is a random variable defined by:

$$m_{\mathrm{T}} = \lim_{T\to\infty}\frac{1}{T}\int_{-T/2}^{T/2} s(t)\mathrm{d}t \tag{1.35}$$

The ergodicity of this average illustrates that it takes a particular value $k$ with probability 1. For a stationary signal, the ergodicity of the time average implies equality with the average of the amplitudes at a given instant. In effect, we use the expectation of the variable $m_{\mathrm{T}}$:

$$E[m_{\mathrm{T}}] = k = E\left[\lim_{T\to\infty}\frac{1}{T}\int_{-T/2}^{T/2} s(t)\mathrm{d}t\right] = \lim_{T\to\infty}\frac{1}{T}\int_{-T/2}^{T/2} E[s(t)]\mathrm{d}t = m_1$$

This result has important consequences in practice, as it provides a means of finding the statistical properties of the signal at a given instant from observation over the time period. The ergodicity of the covariance in the stationary case is also very interesting because it leads to the relation:

$$r_{xx}(\tau) = \lim_{T\to\infty} \frac{1}{T} \int_{-T/2}^{T/2} s(t)s(t-\tau)\mathrm{d}t \tag{1.36}$$

The autocorrelation function $r_{xx}(\tau)$ of the signal $s(t)$ is fundamental for the study of ergodic second-order stationary signals. Its principal properties are:

(1) It is an even function:

$$r_{xx}(\tau) = r_{xx}(-\tau)$$

(2) Its maximum is at the origin and corresponds to the power $P$ of the signal:

$$r_{xx}(0) = E[s^2(t)] = P$$

(3) The power spectral density is the Fourier transform of the autocorrelation function:

$$\Phi_{xx}(f) = \int_{-\infty}^{\infty} r_{xx}(\tau)e^{j2\pi f\tau}\mathrm{d}\tau = 2\int_{0}^{\infty} r_{xx}(\tau)\cos(2\pi f\tau)\mathrm{d}\tau$$

In effect, $r_{xx}(\tau) = s(\tau)^* s(-\tau)$, and if $S(f)$ denotes the Fourier transform of $s(t)$, we obtain:

$$\Phi_{xx}(f) = S(f)\overline{S(f)} = |S(f)|^2 \tag{1.37}$$

This last property is physically translated by the fact that the more rapidly the signal varies (that is, the more its spectrum tends toward high frequencies), the narrower its autocorrelation function becomes. In the limit case, the signal is purely random, and the function becomes zero for $\tau \neq 0$. This signal is called white noise and is described by:

$$r_{xx}(\tau) = P\delta$$

The spectral density is then constant:

$$\Phi_{xx}(f) = P$$

In fact, such a signal has no physical reality since its power is infinite, but it does offer a useful mathematical model for signals with a spectral density that is virtually constant over a wide frequency band.

### 1.3.3 Gaussian Signals

Among the probability distributions that can be considered for a signal $s(t)$, one particular category is of special interest – normal or Gaussian distributions. In effect, normal random distributions retain their normal character under any linear operation, such as convolution through a certain distribution, filtering, differentiation, or integration. These random distributions are also frequently used for modeling real signals and for testing systems.

A random variable $x$ is said to be Gaussian if its probability law has a density $p(x)$ which follows the normal or Gaussian law:

$$p(x) = \frac{1}{\sigma\sqrt{(2\pi)}} e^{-(x-m)^2/(2\sigma^2)} \tag{1.38}$$

The parameter $m$ is the mean of the variable $x$; the variance $\sigma^2$ is the second-order moment of the centered random variable $(x-m)$; and $\sigma$ is the standard deviation. The variable $(x-m)/\sigma$ has a zero average and a unit standard deviation. A tabulation is given in Appendix 2.

A random variable is characterized by its amplitude probability law and also by its moments $m_n$, such that:

$$m_n = \int_{-\infty}^{\infty} x^n p(x)dx \tag{1.39}$$

These moments are the coefficients of the series expansion of the function $F(u)$, called the characteristic function of the random variable $x$ and defined by:

$$F(u) = \int_{-\infty}^{\infty} e^{jux} p(x)dx \tag{1.40}$$

It is the inverse Fourier transform of the probability density $p(x)$ which is expressed by:

$$p(x) = \frac{1}{2\pi}\int_{-\infty}^{\infty} e^{-jux} F(u)du \tag{1.41}$$

On the basis of equation (1.40), the following series expansion is obtained:

$$F(u) = \sum_{n=0}^{\infty} \frac{(ju)^n}{n!} m_n \tag{1.42}$$

and for a centered Gaussian variable:

$$F(u) = e^{-\sigma^2 u^2/2} \tag{1.43}$$

The normal law can be generalized to multidimensional random variables [3]. The characteristic function of a $k$-dimensional Gaussian variable $x(x_1,x_2, \dots ,x_k)$ is expressed by:

$$F(u_1, u_2, \dots , u_k) = \exp\left(-\frac{1}{2}\sum_{i=1}^{k}\sum_{j=1}^{k} r_{ij}u_iu_j\right) \tag{1.44}$$

where

$$r_{ij} = E(x_ix_j)$$

The probability density is obtained through Fourier transformation. For two dimensions, we get:

$$p(x_1, x_2) = \frac{1}{2\pi\sigma_1\sigma_2\sqrt{(1-r^2)}} \exp\left\{-\frac{1}{2(1-r^2)}\left[\frac{x_1^2}{\sigma_1^2} - \frac{2rx_1x_2}{\sigma_1\sigma_2} + \frac{x_2^2}{\sigma_2^2}\right]\right\} \tag{1.45}$$

where $r$ is the correlation coefficient:

$$r = \frac{E(x_1x_2)}{\sigma_1\sigma_2}$$

A random signal $s(t)$ is said to be Gaussian if, for any set of $k$ times $t_i$ $(1 \leqslant i \leqslant k)$, the $k$-dimensional random variable $s = [s(t_1),\dots, s(t_k)]$ is Gaussian. According to equation (1.44), the probability law of that variable is completely defined by the autocorrelation function $r_{xx}(\tau)$ of the signal $s(t)$.

*Example:* The signal defined by the equations:

$$r_{xx}(\tau) = \sigma^2 e^{-|\tau|/(RC)} \tag{1.46}$$

$$p(x) = \frac{1}{\sigma\sqrt{(2\pi)}} e^{-x^2/(2\sigma^2)} \tag{1.47}$$

is an approximation to white Gaussian noise which is used in the analysis of systems or for modeling signals. It is a stationary signal with a zero average and a spectral density which is not strictly constant, but which corresponds to a uniform distribution filtered by an *RC*-type low-pass filter. It is obtained by amplifying the thermal noise generated at the terminals of a resistor.

The Gaussian distribution can be obtained from a uniform distribution. Let $p(y)$ be the Rayleigh distribution:

$$p(y) = \frac{y}{\sigma^2} e^{-(y^2/2\sigma^2)}; \quad y \geqslant 0 \tag{1.48}$$

and $p(x)$ the uniform distribution over the interval [0,1]. Changing variables so that:

$$p(x)\mathrm{d}x = p(y)\mathrm{d}y$$

one gets:

$$p(y) = p(x)\frac{\mathrm{d}x}{\mathrm{d}y} = \frac{\mathrm{d}x}{\mathrm{d}y} = \frac{y}{\sigma^2}\mathrm{e}^{-(y^2/2\sigma^2)}$$

and therefore,

$$x = e^{-(y^2/2\sigma^2)}; \quad y = \sigma\sqrt{\left[21\mathrm{n}\left(\frac{1}{x}\right)\right]} \tag{1.49}$$

The Gaussian distribution is obtained by considering two independent variables, $x$ and $y$, and the variable $z$, given by:

$$z = y \cos 2\pi x \tag{1.50}$$

With the help of the variable:

$$z' = y \sin 2\pi x$$

the following equation is obtained:

$$p(z, z')\mathrm{d}z\,\mathrm{d}z' = p(z)p(z')\mathrm{d}z\,\mathrm{d}z' = p(y)p(x)\mathrm{d}x\,\mathrm{d}y = p(z)p(z')y\,\mathrm{d}y2\pi\mathrm{d}x$$

Hence:

$$p(z)p(z') = \frac{1}{2\pi}\frac{1}{\sigma^2}e^{-(y^2/2\sigma^2)} = \frac{1}{2\pi\sigma^2}e^{-(z^2+z'^2/2\sigma^2)}$$

and finally,

$$p(z) = \frac{1}{\sigma\sqrt{2\pi}}e^{-(z^2/2\sigma^2)}$$

The method is often used to generate digital Gaussian signals.

#### 1.3.3.1 Peak Factor of a Random Signal

A random signal is defined by a probability law of its amplitude, such that this amplitude is often not bounded. This is the case for Gaussian signals, as can be seen from equation (1.38).

Signal processing can only be performed for a limited range of amplitudes, and restrictions are necessary. An important parameter is the peak factor defined as the ratio of a certain amplitude $A_m$ to the standard deviation $\sigma$. By convention, the amplitude $A_m$ is that value which is exceeded for not more than $10^{-5}$ of the total time. This relationship is often expressed in decibels (dB) by $f_c$, where:

$$f_c = 20\log_{10}(A_m/\sigma) \tag{1.51}$$

Following that convention, the peak factor for a Gaussian signal is 12.9 dB, and when this definition is applied to a sinusoidal signal, a peak factor of 3 dB results.

A stationary model used to represent telephone signals is formed by the random signal whose amplitude probability density obeys the following exponential, or Laplace, distribution:

$$p(x)\frac{1}{\sigma\sqrt{2}}e^{-\sqrt{2}|x|/\sigma} \tag{1.52}$$

The peak factor rises, in this case, to 17.8 dB.

In conclusion, ergodic second-order stationary functions characterized by an amplitude probability distribution and an autocorrelation function can be used to model the majority of signals of interest. They are widely used in system analysis.

In addition to the other possibilities for representing signals, it is important to have some sort of global measure, so as to be able, for example, to follow a signal through the processing system. Such a measure can be obtained by defining norms on the function which represents the signal.

## 1.4 The Norms of a Function

A norm is a positive real function which satisfies the relations:

$$||x|| \geq 0; \quad k||x|| = ||kx||$$

where $k$ is real and positive.

One category of norms that is frequently employed is the set called $L_p$ norms [4]. The $L_p$ norm of a continuous function $s(t)$ defined over the interval [0,1] is denoted by $||s||_p$ and defined by:

$$||s||_p = \left[\int_0^1 |s(t)|^p \mathrm{d}t\right]^{1/p} \tag{1.53}$$

Three values of $p$ are of particular interest:

(1) $p = 1$:

$$||s||_1 = \int_0^1 |s(t)|\mathrm{d}t \tag{1.53a}$$

(2) $p = 2$:

$$||s||_2^2 = \int_0^1 |s(t)|^2 \, \mathrm{d}t \tag{1.53b}$$

which is the expression for the energy of the signal $s(t)$.

(3) $p = \infty$:

$$||s||_\infty = \max_{0\leqslant t\leqslant 1} |s(t)| \tag{1.53c}$$

This is Chebyshev's norm.

Norms are also used in approximation techniques to measure the discrepancy between a function $f(x)$ and the function $F(x)$ being approximated. The approximation is made by the least squares method if the norm $L_2$ is used and, in Chebyshev's sense, if the norm $L_\infty$ is used.

The $L_p$ norms can be generalized by introducing a real, positive weighting function of $p(x)$. The weighted $L_p$ norm of the difference function $f(x) - F(x)$ is then written:

$$\|f(x) - F(x)\|_p = \left[\int_0^1 |f(x) - F(x)|^p p(x) \mathrm{d}x\right]^{1/p} \tag{1.53d}$$

These expressions are applied when calculating filter coefficients and in realization problems – in particular for the scaling of internal data in memories and for noise estimation.

## 1.5 Sampling

Sampling consists of representing a signal function $s(t)$ by its value $s(nT)$ taken at whole multiples of the time interval $T$, called the sampling period. Such an operation can be analyzed in a simple and concise way by using distribution theory. In fact, by definition, the distribution of unit masses at whole multiple points on the axis, with period $T$, associates with the function $s(t)$ the ensemble of its values $s(nT)$, where $n$ is a whole number. Conforming to the notation given earlier, this distribution is denoted by $u(t)$ and is written:

$$u(t) = \sum_{n=-\infty}^{\infty} \delta(t - nT)$$

The sampling operation affects the spectrum $S(f)$ of the signal. Considering the fundamental relation (1.27), it appears that the spectrum $U(f)$ of the distribution $u(t)$ is formed of lines of amplitude $1/T$ at frequencies which are whole multiples of the sampling frequency $f_s = 1/T$. Thus, $u(t)$ is expressed as a sum of elementary sinusoidal signals, collectively called carriers:

$$u(t) = \frac{1}{T} \sum_{n=-\infty}^{\infty} e^{j2\pi nt/T} \tag{1.54}$$

Hence, the set of values for the signal $s(nT)$ corresponds to the product with the signal $s(t)$ of the ensemble of component signals which make up $u(t)$. That is, the operation of sampling is an amplitude modulation of the signal by an infinite number of carriers with frequencies which are whole multiples of the sampling frequency $f_s = 1/T$. Consequently, the sampled signal spectrum includes the function $S(f)$, called the baseband, and other images or sidebands, which correspond to the translation of the baseband by whole multiples of the sampling frequency. The operation of sampling and its influence on the signal spectrum are represented in Figure 1.4.

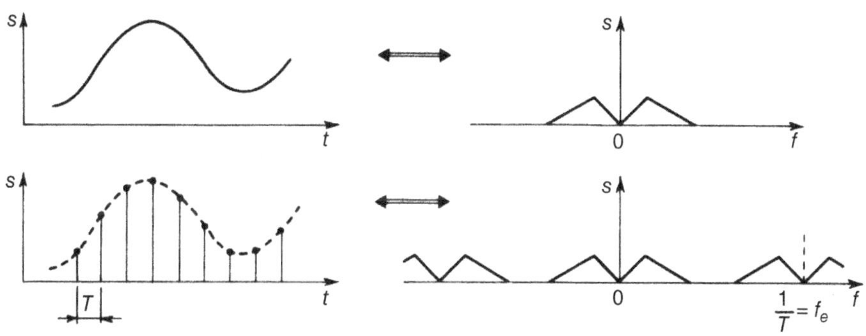


**Figure 1.4** The spectral incidence of sampling.

The sampled signal spectrum $S_s(f)$ is expressed as the convolution of $S(f)$ with $U(_f)$, so that:

$$S_s(f) = 1/T \sum_{n=-\infty}^{\infty} S(f - n/T) \tag{1.55}$$

It is important to note that the function $S_s(f)$ is periodic – that is, the sampling has introduced a periodicity into the frequency space, which constitutes a fundamental characteristic of the sampled signals.

The sampling operation, as described above, which is called ideal sampling, may seem rather unrealistic in that it would be difficult in practice to manipulate or reconstitute an instantaneous signal value. Real samplers, or circuits which reconstitute samples, all possess a certain aperture time. In fact, it can be shown that sampling, or the reconstitution of samples by pulses having a given width, simply introduces a modification of the signal spectrum.

In effect, when sampling a signal $x(t)$ by a set of pulses separated by period $T$, with width $\tau$ and amplitude $a$, it is quite possible for a quantity $\sigma_n$ to be collected in the $n$th period, and this is written:

$$\sigma_n = a \int_{nT-\tau/2}^{nT+\tau/2} s(t)\mathrm{d}t$$

This quantity expresses the result of the convolution of the signal $s(t)$ by the elementary pulse $i(t)$. The function which results in this case at the sampling time $nT$ is the function $s*i$. That is, the sampled signal does not have $S(f)$ for its spectrum, but the product:

$$S_p(f) = U(f)S(f) = \frac{1}{T} \sum_{n=-\infty}^{\infty} S\left(\frac{n}{T}\right) \delta\left(f - \frac{n}{T}\right)$$

Similar reasoning applies in the case of reconstitution of samples with a duration $\tau$. In fact, it is the convolution product of the samples $s(nT)$ with the elementary pulse $i(t)$ which is reconstituted. Thus, we have the proposition:

*Sampling or the reconstitution of samples by pulses with width $\tau$ can be treated as ideal sampling or ideal reconstitution, provided that the signal spectrum is multiplied by the spectrum of the elementary pulse train.*

In practice, however, whenever $\tau$ is small in comparison with the period $T$, the correction is negligible.

## 1.6 Frequency Sampling

The type of sampling considered above is carried out on a temporal basis. Nevertheless, the general characteristics are also applicable to frequency sampling.

Let us calculate the spectrum of a periodic function $s_p(t)$ with period $T$. Such a function can be regarded as resulting from the convolution product of the function $s(t)$ which takes values of $s_p(t)$ over one period and is zero elsewhere, and the point distribution $u(t)$. This results in the following relation between the Fourier transforms:

$$S_p(f) = U(f)S(f) = \frac{1}{T} \sum_{n=-\infty}^{\infty} S\left(\frac{n}{T}\right) \delta\left(f - \frac{n}{T}\right) \tag{1.56}$$

where we again meet the coefficients of the Fourier series expansion of the function $s_p(t)$. The case where $s(t)$ is a square pulse is represented in Figure 1.5.

It is apparent that the spectrum of the periodic function $s_p(t)$ is a set of lines forming a sample of the spectrum of one period of the function. Sampling in frequency space corresponds to a periodicity in the time space. This is a useful interpretation for digital analysis of spectra.

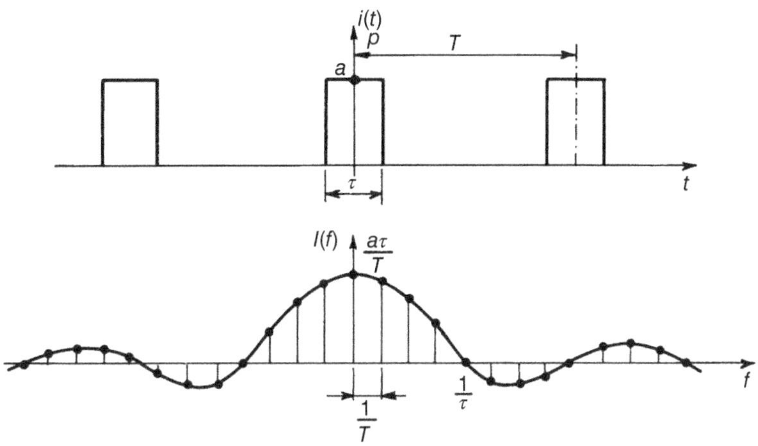


**Figure 1.5** The spectrum of an impulse train.

## 1.7 The Sampling Theorem

This theorem establishes the conditions under which the set of samples of a signal correctly represents that signal. A signal is said to be correctly represented by the set of its samples taken with the periodicity $T$, if it is possible, from this set of values, to completely reconstitute the original signal.

The sampling has introduced a periodicity of the spectrum in frequency space. To restore the original signal, this periodicity must be suppressed – that is, the image bands must be eliminated. This can be achieved using a low-pass filter with a transfer function $H(f)$ which is equal to $1/f_s$ up to the frequency $f_s/2$ and is zero at higher frequencies. At the output of such a filter, a continuous signal appears, which can be expressed as a function of the values $s(nT)$. The square wave impulse response $h(t)$ of the filter is written, using equation (1.10), as:

$$h(t) = \frac{\sin(\pi t/T)}{\pi t/T}$$

The signal at the output of the filter, $s(f)$, corresponds to the convolution product of the set $s(nT)$ with the function $h(t)$. It is written as:

$$s(t) = \int_{-\infty}^{\infty} \left[ \sum_{n=-\infty}^{\infty} s(\theta)\delta(\theta - nT) \right] \frac{\sin \pi(t-\theta)/T}{\pi(t-\theta)/T} \mathrm{d}\theta$$

and hence,

$$s(t) = \sum_{n=-\infty}^{\infty} s(nT) \frac{\sin \pi(t/T - n)}{\pi(t/T - n)} \tag{1.57}$$

This is the formula for interpolating signal values at points sited between the samples. It can be verified that it reproduces $s(nT)$ for multiples of the period $T$, and the reconstitution process is represented in Figure 1.6.

In order for the calculated signal $s(t)$ to be identical to the original signal, the spectrum $S(f)$ has to be identical to the spectrum of the original signal. As shown in Figure 1.6, this condition is satisfied if and only if the original spectrum does not contain any components with frequencies greater than or equal to $f_s/2$.

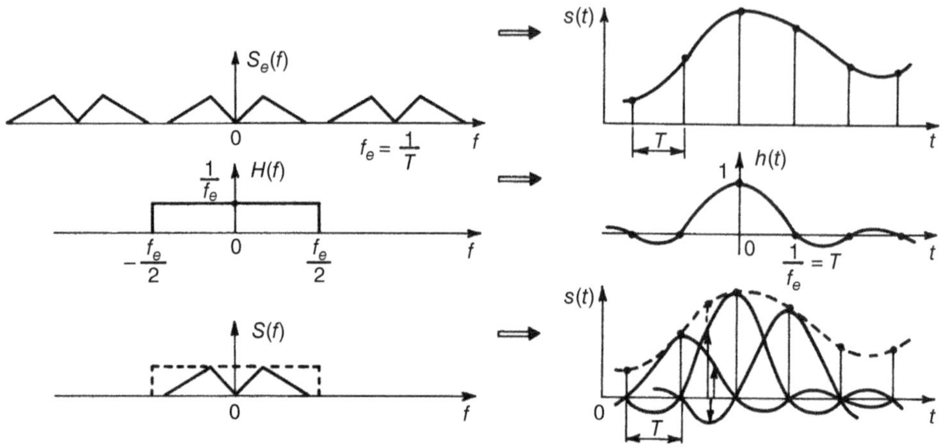


**Figure 1.6** Reconstruction of the signal after sampling.

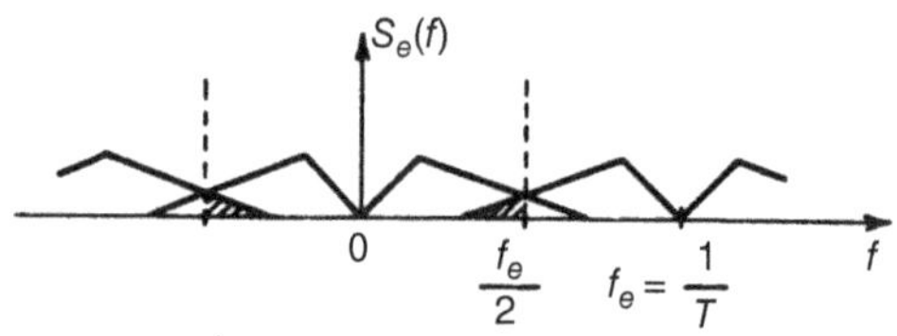


**Figure 1.7** Spectrum aliasing.

If this is not the case, and the image bands overlap the baseband, as shown in Figure 1.7, a foldover distortion or aliasing is introduced and the restoring filter produces a signal that is different from the original one. From this, the sampling theorem or Shannon's theorem is derived:

*A signal which does not contain any component with a frequency equal to or greater than a value $f_m$ is completely determined by the set of its values at regularly spaced intervals of period $T = 1/(2f_m)$.*

Thus, the sampling frequency of a signal is determined by the upper limit of its frequency band. In practice, the signal band is generally limited by filtering to a value below $f_s/2$ before sampling at frequency $f_s$, in order that the restoring filter can be practically realizable.

It is interesting to note that the sampling frequency is determined by the bandwidth of the signal. The reconstitution illustrated in Figure 1.6 was for a low-frequency signal with which a low-pass filter was associated. It is apparent that the same reasoning can also be applied to a signal occupying a restricted region in frequency space by using an associated band-pass filter. In particular, this property is applicable to modulated signals and is used in certain types of digital filters.

The result given at the end of Section 1.1.2 enables sampling to be presented from another viewpoint. Equation (1.57) shows that sampling corresponds to decomposition of the signal $s(t)$ in accordance with the set of orthogonal functions $\frac{\sin \pi(t/T-n)}{\pi(t/T-n)}$ and Shannon's theorem simply expresses the condition for this set to form a signal decomposition basis (several decomposition bases may exist).

## 1.8 Sampling of Sinusoidal and Random Signals

The properties given above can be clearly illustrated by sampling sinusoidal signals, whose features are of use in numerous applications.

### 1.8.1 Sinusoidal Signals

Let the signal $s(t) = \cos(2\pi ft + \varphi)$, with $0 \leqslant \varphi \leqslant \pi/2$, be sampled with the period $T = 1/f_s = 1$. The samples are given by the set $s(n)$ such that:

$$s(n) = \cos(2\pi fn + \phi)$$

If the ratio $f/f_s = f$ is a rational number, this becomes:

$$f = N_1/N_2 \quad \text{with } N_1 \text{ and } N_2 \text{ as integers}$$

Then,

$$s(n + N_2) = \cos[2\pi f(n + N_2) + \phi] = s(n)$$

The set exhibits the periodicity $N_2$ and comprises at most $N_2$ different numbers. On the other hand, since the sampling frequency is more than twice the signal frequency, $N_1/N_2 < \frac{1}{2}$. The ensemble of $N_2$ different samples permits the representation of a number of sinusoidal signals equal to the largest whole number less than $N_2/2$. For example, if $N_2 = 8$, with the ensemble of numbers – $2\cos(2\pi n/8 + \varphi)$, $(n = 0,1, \ldots,7)$, it is possible to represent samples of three sinusoidal signals:

$$2\cos\left(2\pi\frac{N_1}{8}t + \phi\right) \quad \text{with} \quad N_1 = 1,2,3$$

Figure 1.8 represents this sampling for $\varphi = 0$, and in this particular case, four numbers are sufficient: $\pm 2$ and $\pm\sqrt{2}$.

If we then add to the three sinusoidal signals in Figure 1.8 the continuous signal with the value 1, and the oscillating signal $\cos(\pi t)$ with frequency $\frac{1}{2}$, and amplitude 1, the result is that when the composite signal is sampled, zero values appear. This is true except for points which are multiples of 8, where the value 8 is obtained as shown in Figure 1.9(a). The spectrum of this sum is obtained directly by applying the relation:

$$\cos x = \frac{1}{2}(e^{jx} + e^{-jx})$$

It is formed of lines with an amplitude of 1 at frequencies which are multiples of $\frac{1}{8}$ (Figure 1.9(b)). This spectrum has already been studied in Section 1.2, and equation (1.27) again applies.

Spectrum analyzers and digital frequency synthesizers use the fact that it is possible to produce a range of sinusoidal signals from a limited set of numbers which are stored, for example, in a computer memory.

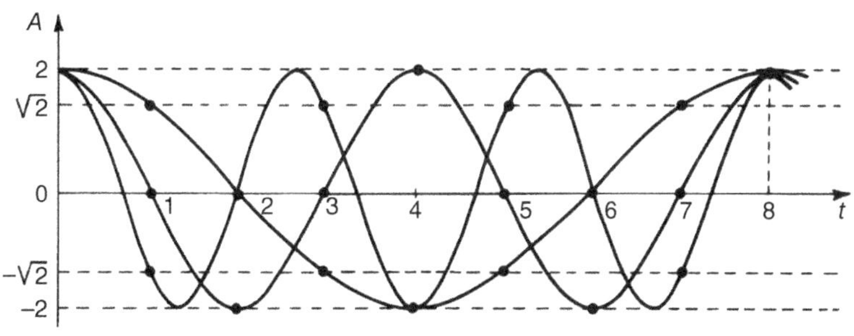


**Figure 1.8** Sampling the signals $\cos(2\pi N/8t)$.

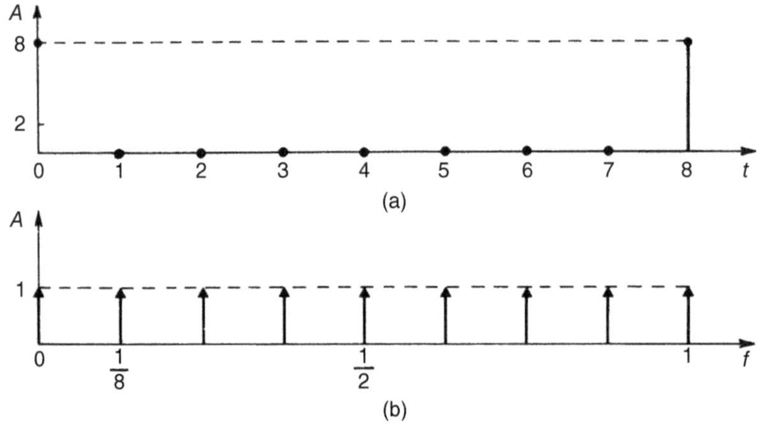


**Figure 1.9** (a) Sampling the signal: $s(t) = 1 + 2\sum_{n=1}^{3} \cos(2\pi n/8t) + \cos(\pi t)$. (b) corresponding spectrum.

### 1.8.2 Discrete Random Signals

If the random signal $s(t)$ is sampled with the unit period $T = 1$, a discrete random signal results, which, by definition, has the same probability distribution of amplitude. The results obtained for the continuous case can be applied to the discrete one, particularly for the random signals which are second-order stationary and ergodic [5].

Thus, the autocorrelation function of the discrete signal $s(n)$ is the sequence $r(n)$, such that:

$$r(n) = E[s(i)s(i-n)] \tag{1.58}$$

This is the sampling of the autocorrelation function $r_{xx}(\tau)$ of the continuous random signal defined by expression (1.34). Its Fourier transform gives the power spectrum density $\Phi_d(f)$ of the discrete signal, which is related to the spectral density $\Phi_{xx}(f)$ of the continuous signal by a relation similar to equation (1.55); that is,

$$\Phi_d(f) = \frac{1}{T}\sum_{n=-\infty}^{\infty} \Phi_{xx}\left(f - \frac{n}{T}\right) \tag{1.59}$$

If the sampling frequency is not high enough, or if the spectrum $\Phi_{xx}(f)$ spans an infinite domain, aliasing takes place.

The hypothesis of ergodicity for the discrete signal $s(n)$ leads to the relation:

$$r(n) = \lim_{N\to\infty} \frac{1}{2N+1}\sum_{i=-N}^{N} s(i)s(i-n) \tag{1.60}$$

This relation gives the opportunity to extend the concept of autocorrelation function to deterministic signals. Then, for a periodic signal with period $N_0$, the autocorrelation function is the sequence $r(n)$ given by:

$$r(n) = \frac{1}{N_0}\sum_{i=0}^{N_0-1} s(i)s(i-n) \tag{1.61}$$

It is a periodic sequence with the same period.

*Example:*

$$s(n) = A \sin\left(2\pi\frac{n}{N_0}\right)$$

$$r(n) = \frac{1}{N_0}\sum_{i=0}^{N_0} A^2 \sin\left(2\pi\frac{i}{N_0}\right) \sin 2\pi\left(\frac{i-n}{N_0}\right)$$

$$r(n) = \frac{A^2}{2}\cos\left(2\pi\frac{n}{N_0}\right)$$

The period of $r(n)$ is $N_0$ and $r(0)$ is the signal power.

A discrete random signal can also be defined directly. For example, if $r(n) = 0$ for $n \neq 0$, the signal $s(n)$ is a discrete white noise, and the spectral density is constant over the frequency interval $\left[-\frac{1}{2}, \frac{1}{2}\right]$. This signal represents a physical reality – it is a sequence of noncorrelated random variables, and it can be obtained through an algorithm which produces statistically independent numbers.

### 1.8.3 Discrete Noise Generation

It has been shown in Section 1.8.1 that the generation of digital sinusoidal signals is particularly simple. Such signals can be used to simulate noise, for example, by addition of a large number of sine waves of different frequencies and constant amplitudes, having random or pseudorandom phases. This approach can lead to particularly efficient realizations, like that which has been standardized for measuring equipment in digital telephone transmission and is as follows:

A pseudorandom sequence is created, which is a periodic sequence of $2^N - 1$ bits comprising approximately as many "zeros" as "ones" and which simulates a random sequence in which the bits are independent and have the probability $\frac{1}{2}$ of having a value of 0 or 1 (or in order to center the variables, a value of $\pm\frac{1}{2}$).

If such a set is filtered (filtering in fact consists of weighted summation), the numbers obtained after filtering obey a probability distribution which approximates the normal distribution.

Pseudorandom sequences are studied in Ref. [6] and can be obtained easily using a suitably looped $N$-bit shift register. Figure 1.10 gives an example used in measuring equipment where $N = 17$. The generator polynomial is written:

$$g(x) = 1 + x^3 + x^{17} \tag{1.62}$$

The sequence contains $2^N - 1 = 131071$ bits, and it is periodic with a period $T = (2^N - 1)\tau$, where $\tau = 1/f_H$ denotes the clock period. The spectrum is formed of lines $1/T$ apart. For $f_H = 370$ kHz, the distance between the lines is 2.8 Hz and there are 36 lines per 100 Hz.

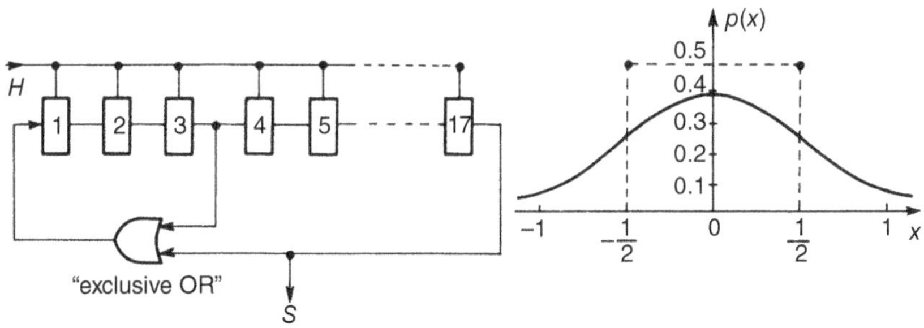


**Figure 1.10** Pseudorandom generator and the probability distribution after filtering.

By applying to this set a narrowband filter which passes only the band 450–550 Hz, an approximately Gaussian signal is obtained. The signal has a peak factor of 10.5 dB and is an excellent test signal for digital transmission equipment. If the filtering is performed numerically, the set of numbers obtained can be used to test digital processing equipment.

## 1.9 Quantization

Quantization is the approximation of each signal value $s(t)$ by a whole multiple of an elementary quantity $q$ which is called the quantizing step. If $q$ is constant for all signal amplitudes, the quantization is said to be uniform. This operation is carried out by passing the signal through a device which has a staircase characteristic and produces the signal $s_q(t)$, as shown in Figure 1.11 for $q = 1$.

The way in which the approximation is made defines the centering of this characteristic. For example, the diagram represents the case (called rounding), where each value of the signal between $\left(n - \frac{1}{2}\right) q$ and $\left(n + \frac{1}{2}\right) q$ is rounded to $nq$. This approximation minimizes the power of the error signal. It is also possible to have approximation by default, which is obtained by truncation and consists of approximating every value between $nq$ and $(n + 1)q$ by $nq$; the characteristic is therefore displaced by $q/2$ toward the right on the abscissa.

The effect of this approximation is to superimpose on the original signal an error signal $e(t)$ called the quantizing distortion or, more commonly, the quantizing noise. Thus:

$$s(t) = s_q(t) + e(t) \tag{1.63}$$

The case of rounding is illustrated in Figure 1.12. The amplitudes at odd multiples of $q/2$ are called the decision amplitudes. The amplitude of the error signal lies between $-q/2$ and $q/2$ and power measures the degradation undergone by the signal.

When the variations in the signal are large relative to the quantizing step – that is, when quantization has been carried out sufficiently finely – the error signal is equivalent to a set of elementary signals which are each formed from a straight-line segment (Figure 1.13). The power of such an

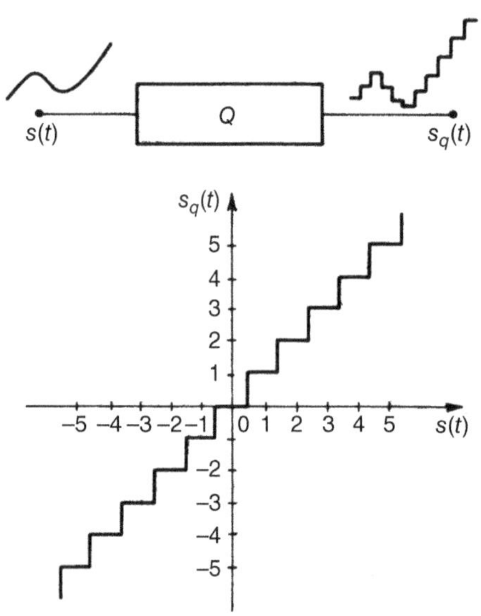


**Figure 1.11** The quantization operation.

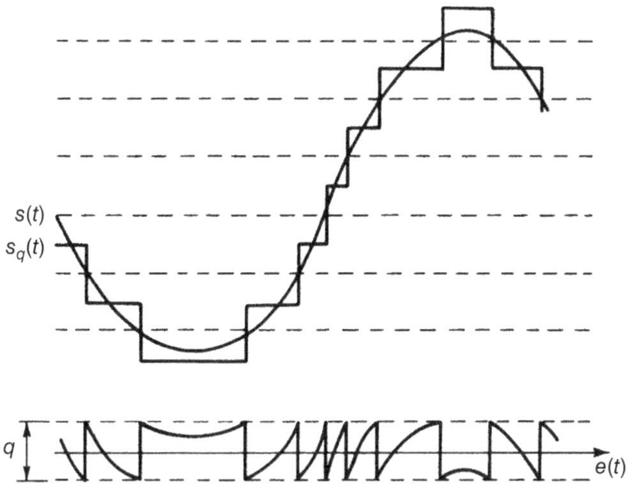


**Figure 1.12** Quantization error.

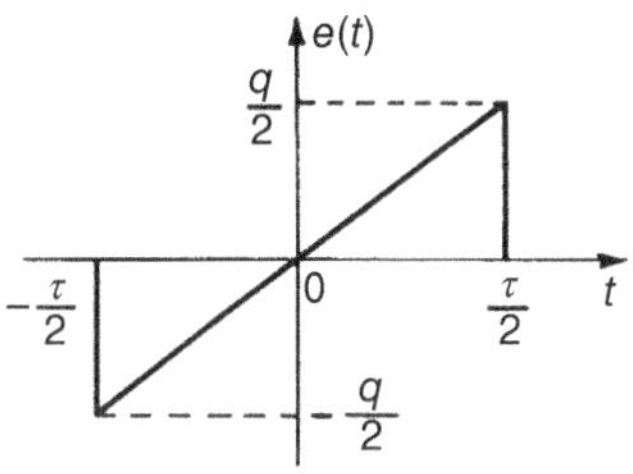


**Figure 1.13** Elementary error signal.

elementary signal of width $\tau$ is written:

$$B = \frac{1}{\tau}\int_{-\tau/2}^{\tau/2} e^2(t)\mathrm{d}t = \frac{1}{\tau}\left(\frac{q}{\tau}\right)^2 \int_{-\tau/2}^{\tau/2} t^2\mathrm{d}t = \frac{q^2}{12} \tag{1.64}$$

The value obtained in this way, $B = q^2/12$, is a satisfactory estimate of the power of the quantizing noise in the majority of actual cases.

The spectral distribution of the error signal is more difficult to discern. The spectrum of the elementary error signal in Figure 1.13, $E_\tau(f)$, can be derived from its derivative. Using expressions (1.22) and then (1.12), the following is obtained:

$$E_\tau(f) = \frac{1}{\mathrm{j}2\pi f} q \left[\frac{\sin(\pi f\tau)}{\pi f\tau} - \cos(\pi f\tau)\right] \tag{1.65}$$

It appears that the majority of the energy is found around the frequency $1/\tau$. Under these conditions, the spectral distribution of the error signal depends both on the slope of the elementary signal – that is, on the statistical distribution of the derivative $s'(t)$ of the signal – and on the size of the step $q$ relative to the signal. Reference [7] gives the calculation of this spectrum for a noise signal and shows the spread as a function of frequency, when the quantizing step is sufficiently small, to be on a range several hundred times the width of the signal band. If the signal to be quantized is not random, the spectrum of the error signal will be concentrated on certain frequencies – for example, the harmonics of a sinusoidal signal.

When converting an analog signal into digital form, quantization occurs, along with the sampling, as the two operations are carried out in succession. While sampling is normally carried out first, it is equally valid to carry out the quantization first and the sampling second, at a frequency $f_s$ which is usually a little over twice the bandwidth of the signal. Under these conditions, the error signal often has a spectrum which extends beyond the sampling frequency, and since it is actually the sum of the signal and the error signal which is sampled, aliasing occurs, and the whole energy of the error signal is recovered in the frequency band ( $-f_s/2, f_s/2$). In the majority of cases, the spectral energy density of the quantizing noise is constant, and the following statement can be made:

> *The noise produced during uniform quantization with a step q has a power which is generally expressed by $B = q^2/12$ and shows a constant spectral distribution in the frequency band $(-f_s/2, f_s/2)$.*

It should be noted that the quantization of small signals (those with amplitude of the same order of magnitude as step $q$) depends critically on the centering of the characteristic. For example, with the centering in Figure 1.11, a sinusoidal signal with an amplitude of less than $q/2$ is totally suppressed. It is possible, nevertheless, to suitably code these small signals by superimposing on them a large-amplitude auxiliary signal which is removed later in the process. Thus, the coding introduces limits on the small amplitudes of the signal. Equally, however, other limits appear in the case of large amplitudes, as will be seen below.

## 1.10 The Coding Dynamic Range

The signal which is sampled and quantized in amplitude is represented by a set of numbers which are almost always in binary form. If each number has $N$ bits, the maximum number of quantized amplitudes that it is possible to represent is $2^N$. Thus, the range of amplitudes that can be coded is subject to a twofold limitation – at low values, it is limited by the quantum $q$, and at larger values by $2^N q$. Any amplitude that exceeds this value cannot be represented and the signal is clipped. This results in degradation, so that, for example, if the signal is sinusoidal, harmonic distortion is introduced.

If the range of amplitudes to be coded covers the domain [ $-A_m, A_m$], we get:

$$A_m = 2^N q/2 \tag{1.66}$$

and, with rounding, the error signal $e(t)$ is:

$$|e(t)| \leq A_m 2^{-N}$$

By definition, the peak power of a coder is the power of that sinusoidal signal which has the maximum possible amplitude, $A_m$, which the coder will pass without clipping. It is expressed by:

$$P_c = \frac{1}{2}\left[\frac{2^N q}{2}\right]^2 = 2^{2N-3} q^2$$

Figure 1.14 illustrates this signal together with the quantizing step and the decision amplitudes.

The coding dynamic range is defined as the ratio between this peak power and the power of the quantizing noise; this is, in fact, the maximum value of the signal-to-noise ratio (SNR) for a sinusoidal signal with uniform coding. The following formula expresses this dynamic range:

$$P_c/B = (S/B)_{max} = 2^{2N-3} \cdot 12 = \frac{3}{2} \cdot 2^{2N}$$

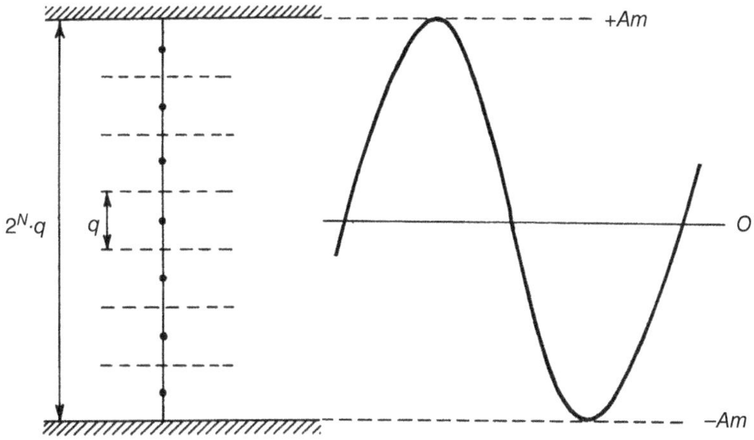


**Figure 1.14** Peak power of the coder.

This can be expressed more conveniently in dB as:

$$P_c/B = 6.02N + 1.76\text{dB} \tag{1.67}$$

This formula, which is of great practical use, relates the number of bits in the coding to the range of amplitudes which can be coded.

The signal to be coded is, in general, not sinusoidal. However, it is still possible to treat this case if an equivalent peak power can be defined for the signal, which can then be taken as the peak power of the coder. This occurs, for example, with multiplexed telephone signals. The case of random Gaussian signals is of particular importance because they conveniently represent a large number of signals which are encountered in practice. The maximum amplitude of the coder must then be correctly positioned relative to the signal amplitude so that the distortion introduced by peak limitation remains within the imposed limits.

It is evident from the table given in Appendix 2 that the probability is less than $10^{-3}$ that a signal exceeds $3.4\sigma$ when it has a zero mean and a power of $\sigma^2$. Figure 1.15 gives an example of coding where $\sigma = q$. It would appear that the probability of clipping is less than $5 \times 10^{-4}$ for the chosen parameters.

Finally, in order to obtain the maximum SNR, it is necessary to consider the peak SNR expressed by:

$$(\text{SNR})_c = \frac{A_\text{m}^2 q^2}{12} = 3.2^N$$

and subtract the peak factor $F_c$. Thus, the general expression for the maximum SNR is:

$$(\text{SNR})_{\text{max}} = 6.02\,N + 4.77 - F_c\,dB \tag{1.67bis}$$

This result is used not only in specifying analog signal coders but also in digital processing for determining memory sizes and scaling internal data.

The coding dynamic range for a given number of bits can be considerably increased if coding is carried out with a quantizing step which varies with the amplitude of the signal. This is called nonlinear coding. Numerous variations can be envisaged, but of particular importance is the 13-segment coding law which has been standardized by the International Telegraph and Telephone Consultative Committee (CCITT) for the coding of signals in telecommunications networks [8].

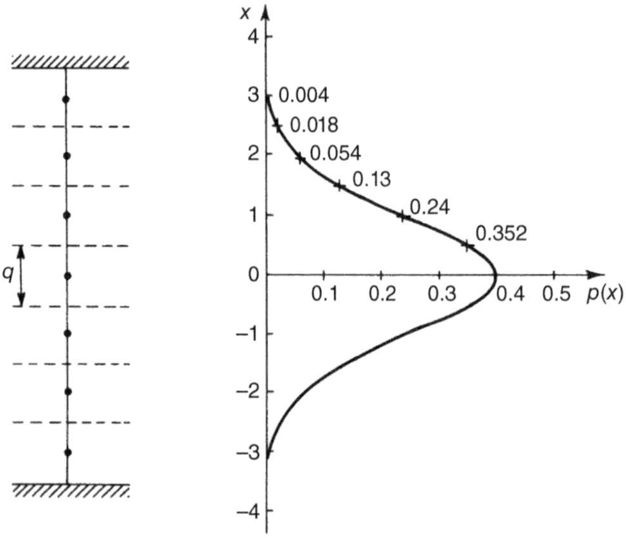


**Figure 1.15** Coding of a Gaussian signal.

## 1.11 Nonlinear Coding with the 13-segment A-law

In nonlinear coding using the 13-segment $A$-law, the positive and negative amplitudes which are to be coded are divided into 7 ranges, each with a quantizing step which is related to an elementary step $q$ by some power of 2. This operation can be regarded as resulting from linear coding which was preceded by a compression in which the signal $x$ is transformed into the signal $y$ according to the following relations:

$$y = \text{sign}(x)\frac{1+\log A\,|\,x\,|}{1+\log\ A} \quad \text{for} \quad \frac{1}{A} \leqslant |\,x\,| \leqslant 1$$

$$y = \text{sign}(x)\frac{A\,|\,x\,|}{1+\log A} \quad \text{for} \quad 0 \leqslant |\,x\,| \leqslant \frac{1}{A} \tag{1.68}$$

The parameter $A$ controls the dynamic range increase; the value used is $A = 87.6$.

The compression characteristic of the 13-segment $A$-law is shown in Figure 1.16.

The operation can be described as follows:

$$\text{if } 0 < x < \frac{1}{64}, \quad \text{then } y = 16x$$

$$\frac{1}{64} \leqslant |\,x\,| \leqslant \frac{1}{32} \qquad y = 8x + \frac{1}{8}$$

$$\frac{1}{32} \leqslant |\,x\,| \leqslant \frac{1}{16} \qquad y = 4x + \frac{1}{4}$$

$$\frac{1}{16} \leqslant |\,x\,| \leqslant \frac{1}{8} \qquad y = 2x + \frac{3}{8}$$

$$\frac{1}{8} \leqslant |\,x\,| \leqslant \frac{1}{4} \qquad y = x + \frac{1}{2}$$

$$\frac{1}{4} \leqslant |\,x\,| \leqslant \frac{1}{2} \qquad y = \frac{1}{2}x + \frac{5}{8}$$

$$\frac{1}{2} \leqslant |\,x\,| \leqslant 1 \qquad y = \frac{1}{4}x + \frac{3}{4}$$

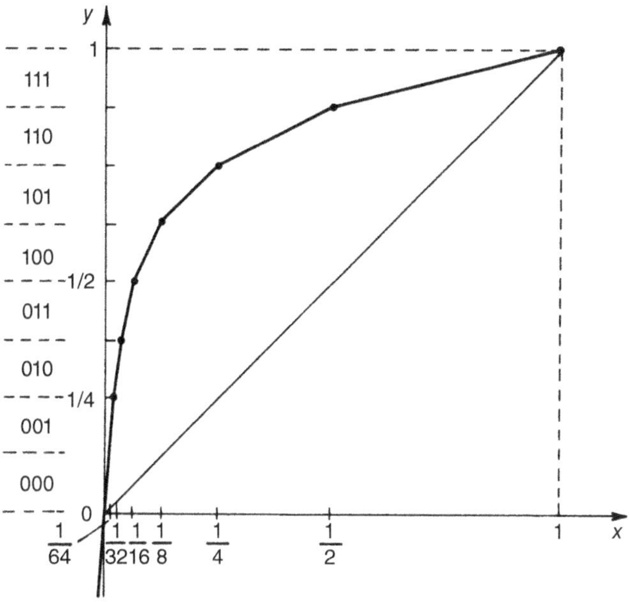


**Figure 1.16** The 13-segment compression A-law.

This characteristic causes seven straight-line segments to appear in both the positive and negative quadrants. As the two segments which encompass the origin are colinear, the characteristic has a total of 13 segments.

Since the quantization of the $y$ amplitudes is carried out with the quantum $q$, quantization of the $x$ amplitudes near the origin is based on a quantum $q/16$ – that is, the dynamic of the coder is increased by 24 dB. Amplitudes close to unity are less well quantized as the step is multiplied by 4. The power of the quantization noise is a function of the signal amplitude – it is necessary to calculate an average for each value, and this interferes with the statistics of the signal.

Figure 1.17 gives the SNR for a Gaussian signal as a function of the signal level after coding into 8 bits for linear and nonlinear coding. The reference level for the signal (0 dB) is the peak power of the coder, and it is clearly apparent that the dynamic is extended by nonlinear coding. For low amplitudes, quantization in fact corresponds to 12 bits. In practice, the signal can be coded by linear quantization into 12 bits, followed by a process which is very close to the conversion of an integer into a floating-point number:

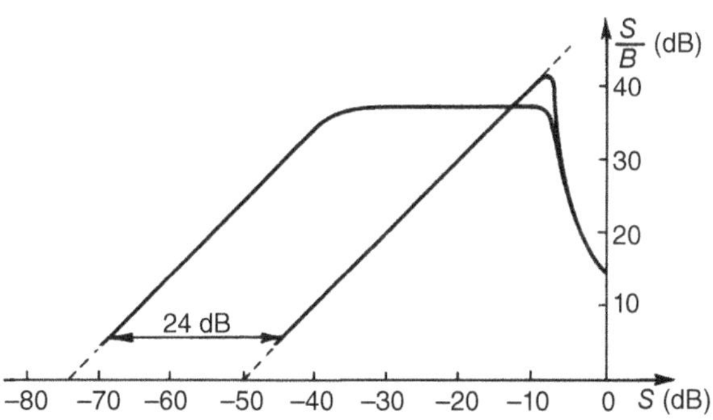


**Figure 1.17** Eight-bit linear and nonlinear coding of a Gaussian signal.

For example, the 12-bit number

+0 0 0 1 0 1 1 0 1 1 0

corresponds to the 8-bit one

+1 0 0 0 1 1 0

after application of the compression law.

The three bits which follow the sign give the code for the exponent; the four following bits give the position within the segment or the mantissa. The difference in the conversion from a whole number to a floating-point number appears in the vicinity of the origin.

The implementation can use an array of gates arranged in parallel, or a shift register combined with a 3-bit counter, for serial realization. Equally, one could use memories holding the conversion table.

Another nonlinear coding law is also widely used in telecommunications – the 15-segment $\mu$-law, which follows the relation:

$$y = \text{sign}(x)\frac{\log(1+\mu|x|)}{\log(1+\mu)} \quad \text{for } -1 \leqslant x \leqslant 1 \tag{1.69}$$

The compression parameter is $\mu = 255$.

## 1.12 Optimal Coding

The coding can be improved when the probability distribution $p(x)$ of the signal amplitude is known. For a given number of bits $N$, an optimal quantizing characteristic can be found, which minimizes the total quantizing distortion.

The signal amplitude range is divided into $M = 2^N$ subsets $(x_{i-1}, x_i)$ with $-(M/2)+1 \leqslant i \leqslant M/2$ and every subset is represented by a value $y_i$, as shown in Figure 1.18. The optimization consists of determining the set of values $x_i$ and $y_i$ which minimize the error signal power $E^2$ expressed by:

$$E^2 = \sum_{i=-(M/2)+1}^{M/2} \int_{x_i-1}^{x_i} (x-y_i)^2 p(x)dx$$

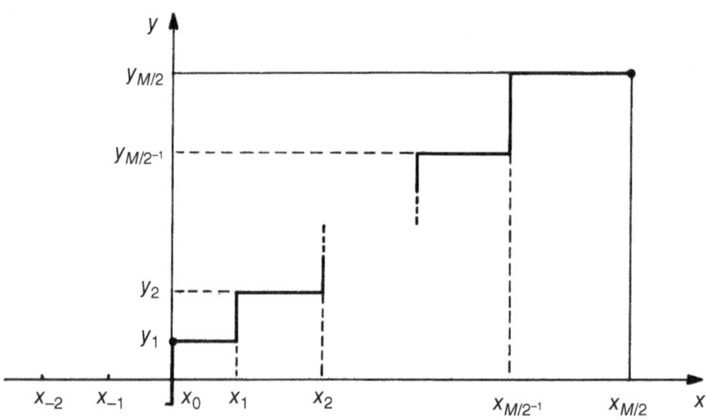


**Figure 1.18** Optimal quantization characteristic.

Taking the derivative with respect to $x_i$ and $y_i$, it can be shown that the following relations must hold:

$$x_i = \frac{1}{2}(y_i + y_{i+1}) \quad \text{for} -\frac{M}{2} + 1 \leqslant i \leqslant \frac{M}{2} - 1$$

$$\int_{x_i-0}^{x_i} (x - y_i)p(x)\mathrm{d}x = 0 \quad \text{for} \quad \in -\frac{M}{2} + 1 \leqslant i \leqslant \frac{M}{2}$$

$$p(x_{m/2}) = p(x_{-M/2}) = 0 \tag{1.70}$$

These relations lead to the determination of the quantizing characteristic. If $p(x)$ is an even function, $x_0 = 0$, and an iterative procedure is used, starting with an *a priori* choice for $y_1$. If relation (1.70) is not satisfied for $M/2$, another initial choice is made for $y_1$ and so on [9].

Table 1.1 gives the error signal power obtained with a Gaussian signal of unit power, for several numbers of bits $N$, for the optimal coding and for uniform coding with the best scaling of the quantizing characteristic [9]. Table 1.2, taken from Reference [10], gives, for the same conditions, the values which correspond to a signal probability density following expression (1.48).

The coding optimization can also be carried out with respect to the information content, by introducing the concept of entropy $H$ defined as follows [2, Vol. 3]:

$$H = -\sum_i p_i \log_2(p_i) \tag{1.71}$$

with

$$-\frac{M}{2} + 1 \leqslant i \leqslant \frac{M}{2},$$

and where $p_i$ designates the probability that the signal is in the amplitude subrange represented by $y_i$.

Considering that:

$$\sum_i p_i = 1$$

**Table 1.1** Coding of unitary Gaussian signal.

| N<br>E | 1 | 2 | 3 | 4 | 5 |
|---|---|---|---|---|---|
| Optimal coding | 0.3634 | 0.1175 | 0.03454 | 0.0095 | 0.0025 |
| Uniform coding | 0.3634 | 0.1188 | 0.03744 | 0.01154 | 0.00349 |
| Entropy $H$ | 1 | 1.911 | 2.825 | 3.765 | 4.730 |

**Table 1.2** Coding of unitary Laplacian signal.

| $N$<br>$E^2$ | 1 | 2 | 3 | 4 | 5 |
|---|---|---|---|---|---|
| Optimal coding | 0.5 | 0.1765 | 0.0548 | 0.0954 | 0.00414 |
| Uniform coding | 0.5 | 0.1963 | 0.0717 | 0.0254 | 0.0087 |

the entropy is zero when the amplitude is concentrated in a single subrange. It is maximal when the signal amplitude is uniformly distributed when it takes the value $H_{max}$ equal to the number of bits $N$ of the coder:

$$H_{max} = \log_2 M = N \tag{1.72}$$

In fact, the entropy measures the difference between a given distribution and the uniform distribution. The quantizing characteristic which maximizes the entropy is that which leads to amplitude subranges corresponding to a uniform probability distribution.

The last row in Table 1.1 shows that for a Gaussian signal, the quantizing law which minimizes the error signal power leads to entropy values close to the maximum $N$.

## 1.13 Quantity of Information and Channel Capacity

The results obtained for sampling and quantization can be used, inversely, to evaluate the quantity of information carried by a signal or to determine the capacity of a transmission channel.

A real channel of bandwidth $f_m$ can carry $2f_m$ independent samples per second, as shown in Figure 1.4, by replacing $\tau$ with $2f_m$. The quantity of information per sample depends on the relative powers of the useful signal, the noise, and their amplitude distributions.

An important particular case is the Gaussian channel [11].

Let us assume a set of $M$ symbols of $N$ bits each is to be transmitted by a channel in the presence of white Gaussian noise of power $B = \sigma_b^2$.

In a $M$-dimension hyperspace, the $M$ symbols occupy the volume of a hypersphere $V_M$ defined by:

$$V_M = \int_0^R r^{M-1}\, \mathrm{d}r \int_{i=1,..,M-1} \cdots \int f(\theta_i)\, \mathrm{d}\theta_i = \frac{R^M}{M} F_\theta \tag{1.73}$$

If a uniform distribution of the symbols in the hypersphere with radius R is assumed, the energy of the corresponding signal is:

$$E_S = \frac{1}{V_M} \int_0^R r^2 r^{M-1} \mathrm{d}r \int_{i=1} \cdots \int_{M-1} f(\theta_i) \mathrm{d}\theta_i = R^2 \frac{M}{M+2} \tag{1.74}$$

The quantity of transmitted information is $M{\cdot}N$ bits. In the hypersphere, it is possible to associate with each set of bits a volume $V_s$ expressed by:

$$V_s = \frac{V_M}{2^{MN}} = \frac{1}{M} F(\theta) \left(\frac{R}{2^N}\right)^M \tag{1.75}$$

Now, a $M$-component noise with energy $E_b = M\sigma_b^2$ is assigned to each set of bits. When $M$ tends toward infinity, the point representing the noise in the hypersphere comes close to a sphere with radius $\sigma_b^2 \sqrt{M}$ and centered on the point representing the set of bits. In fact, for $M$ Gaussian random variables $b(n)$, the variable $r = \sqrt{\sum_{n=1}^{M} b^2(n)}$ has the first-order moment $m_1 = \sigma_b \sqrt{\frac{M^2}{M+1}}$ and its variance $m_2 - m_1^2$ tends toward zero when M tends toward infinity. The volume of the sphere is:

$$V_b = \frac{(\sqrt{M}\, \sigma_b)^M}{M} F(\theta) \tag{1.76}$$

The condition for the absence of transmission errors is that the volume of the sphere be included in the volume assigned to each set of bits, which leads to:

$$\sqrt{M}\,\sigma_b < \frac{R}{2^N} \tag{1.77}$$

However, when $M$ tends toward infinity, according to (1.74), $R^2$ represents the energy of the sum of the signal with power $S$ and the noise:

$$R^2 = M\left(S + \sigma_b^2\right) \tag{1.78}$$

Then, there are no transmission errors if the following inequality is satisfied:

$$2^{2N} < \frac{S + \sigma_b^2}{\sigma_b^2} \tag{1.79}$$

Hence:

$$N < \frac{1}{2}\log_2\left(1 + \frac{S}{\sigma_b^2}\right) \tag{1.80}$$

Assuming a real channel of bandwidth $W$ and no distortion, the symbols may be emitted at the rate $2W$, and the asymptotic capacity of the channel, in bits per second, is:

$$C = W\log_2\left(1 + \frac{S}{B}\right) \tag{1.81}$$

The assumptions are worth emphasizing:

- distortion-free channel
- white Gaussian noise
- infinite transmission delay

In practice, channel equalization and error-correcting codes make it achievable to approach that limit with finite transmission delay.

## 1.14 Binary Representations

There are several ways of establishing the correspondence between the set of quantized amplitudes and the set of binary numbers which must represent them. As the signals to be coded generally have both positive and negative amplitudes, the preferred representations are those which preserve the sign information. The following are the most usual representations:

(1) Sign and absolute value
(2) Offset binary
(3) 1s-complement
(4) 2s-complement

Table 1.3 defines them for three bits.

The most useful representations for analog-to-digital conversion are sign and absolute value, and offset binary. The remaining two representations are mostly used in arithmetic circuits. As mentioned in Section 1.11, nonlinear coding brings a considerable increase of the dynamic range.

**Table 1.3** Binary representation for linear code.

| Number | Sign and value | Offset binary | 1s complement | 2s complement |
|---|---|---|---|---|
| +3 | 011 | 111 | 011 | 011 |
| +2 | 010 | 110 | 010 | 010 |
| +1 | 001 | 101 | 001 | 001 |
| +0 | 000 | 100 | 000 | 000 |
| −0 | 100 | – | 111 | – |
| −1 | 101 | 011 | 110 | 111 |
| −2 | 110 | 010 | 101 | 110 |
| −3 | 111 | 001 | 100 | 101 |
| | | (000) | | (100) |

Digital processing machines, and particularly general-purpose ones, often use floating-point representations in which each number has three parts: the sign bit, the mantissa, and the exponent. The mantissa represents the fractional part, and the exponent is a power of the base number, for example, with base 10: $+0.719 \times 10^5$.

The dynamic range extension comes from the multiplicative effect introduced by the exponent. For example, in base 2 for a 6-bit exponent and 16-bit mantissa, the dynamic range is: $2^{2^6} \times 2^{16} = 2^{80} \simeq 10^{24}$ – that is, 24 decimal numbers. Additional gain is achieved by choosing a base which is, itself, a power of two, such as 8 or 16, leading to octal or hexadecimal operations (Figure 1.19).

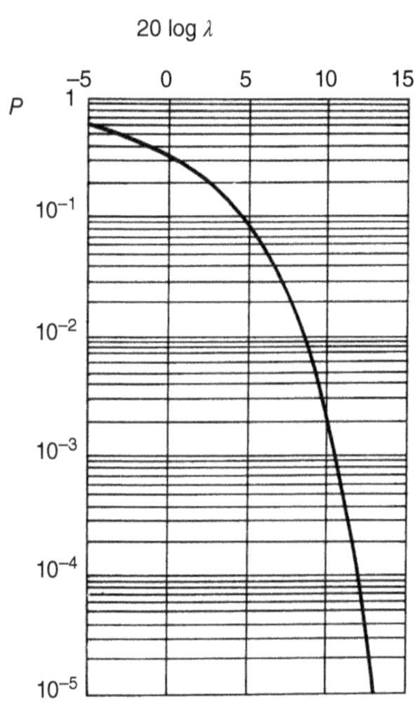


**Figure 1.19** The reduced normal distribution: P as a function of 20 log λ.

## 1.A Appendix 1: The Function $I(x)$

The set of values:

$$I(n) = \frac{\sin(\pi(n/20))}{\pi(n/20)}$$

for $0 \leqslant n \leqslant 159$ when $n = k + 20N$, is given in the table below.

| $k$ | $N = 0$ | $N = 1$ | $N = 2$ | $N = 3$ | $N = 4$ | $N = 5$ | $N = 6$ | $N = 7$ |
|---|---|---|---|---|---|---|---|---|
| 0 | 1 | 0 | 0 | 0 | 0 | 0 | 0 | 0 |
| 1 | 0.99589 | −0.04742 | 0.02429 | −0.01633 | 0.01229 | −0.00986 | 0.00823 | −0.00706 |
| 2 | 0.98363 | −0.08942 | 0.04684 | −0.03173 | 0.02399 | −0.01929 | 0.01613 | −0.01385 |
| 3 | 0.96340 | −0.12566 | 0.06721 | −0.04588 | 0.03482 | −0.02806 | 0.02350 | −0.02021 |
| 4 | 0.93549 | −0.15591 | 0.08504 | −0.05847 | 0.04455 | −0.03598 | 0.03018 | −0.02599 |
| 5 | 0.90032 | −0.18006 | 0.10004 | −0.06926 | 0.05296 | −0.04287 | 0.03601 | −0.03105 |
| 6 | 0.85839 | −0.19809 | 0.11196 | −0.07804 | 0.05989 | −0.04850 | 0.04088 | −0.03528 |
| 7 | 0.81033 | −0.21009 | 0.12069 | −0.08466 | 0.06520 | −0.05301 | 0.04466 | −0.03859 |
| 8 | 0.75683 | −0.21624 | 0.12614 | −0.08904 | 0.06880 | −0.05606 | 0.04730 | −0.04091 |
| 9 | 0.69865 | −0.21682 | 0.12832 | −0.09113 | 0.07065 | −0.05769 | 0.04874 | −0.04220 |
| 10 | 0.63662 | −0.21221 | 0.12732 | −0.09095 | 0.07074 | −0.05787 | 0.04897 | −0.04244 |
| 11 | 0.57162 | −0.20283 | 0.12329 | −0.08856 | 0.06910 | −0.05665 | 0.04800 | −0.04164 |
| 12 | 0.50455 | −0.18921 | 0.11643 | −0.08409 | 0.06581 | −0.05406 | 0.04587 | −0.03983 |
| 13 | 0.43633 | −0.17189 | 0.10702 | −0.07770 | 0.06099 | −0.05020 | 0.04265 | −0.03707 |
| 14 | 0.36788 | −0.15148 | 0.09538 | −0.06960 | 0.05479 | −0.04518 | 0.03844 | −0.03344 |
| 15 | 0.30011 | −0.12862 | 0.08185 | −0.06002 | 0.04739 | −0.03914 | 0.03335 | −0.02904 |
| 16 | 0.23387 | −0.10394 | 0.06682 | −0.04924 | 0.03989 | −0.03298 | 0.02751 | −0.02399 |
| 17 | 0.17001 | −0.07811 | 0.05071 | −0.03753 | 0.02980 | −0.02470 | 0.02110 | −0.01841 |
| 18 | 0.10929 | −0.05177 | 0.03392 | −0.02522 | 0.02007 | −0.01667 | 0.01426 | −0.01245 |
| 19 | 0.05242 | −0.02544 | 0.01688 | −0.01261 | 0.01006 | −0.00837 | 0.00716 | −0.00626 |

## 1.B Appendix 2: The Reduced Normal Distribution

| $f(x) = \frac{1}{\sqrt{2\pi}} e^{-x^2/2}$ | | $P = \frac{2}{\sqrt{2\pi}} \int_\lambda^\infty e^{-x^2/2} dx$ | | $P = \frac{2}{\sqrt{2\pi}} \int_\lambda^\infty e^{-x^2/2} dx$ | |
|---|---|---|---|---|---|
| $X$ | $10^5 \cdot f(x)$ | 100 $P$ | $\lambda$ | $\lambda$ | 100 $P$ |
| 0 | 39,894 | 100 | 0 | 0 | 100 |
| 0.2 | 39,104 | 95 | 0.0627 | 0.2 | 84.148 |
| 0.4 | 36,827 | 90 | 0.1257 | 0.4 | 68.916 |
| 0.6 | 33,322 | 85 | 0.1891 | 0.6 | 54.851 |
| 0.8 | 28,969 | 80 | 0.2533 | 0.8 | 42.371 |
| 1 | 24,197 | 75 | 0.3186 | 1 | 31.731 |

| $f(x)=\frac{1}{\sqrt{2\pi}}e^{-x^2/2}$ | | $P=\frac{2}{\sqrt{2\pi}}\int_\lambda^\infty e^{-x^2/2}dx$ | | $P=\frac{2}{\sqrt{2\pi}}\int_\lambda^\infty e^{-x^2/2}dx$ | |
|---|---|---|---|---|---|
| $X$ | $10^5 \cdot f(x)$ | 100 $P$ | $\lambda$ | $\lambda$ | 100 $P$ |
| 1.2 | 19,419 | 70 | 0.3853 | 1.2 | 23.014 |
| 1.4 | 14,973 | 65 | 0.4538 | 1.4 | 16.151 |
| 1.6 | 11,092 | 60 | 0.5244 | 1.6 | 10.960 |
| 1.8 | 7,895 | 55 | 0.5978 | 1.8 | 7.186 |
| 2 | 5,399 | 50 | 0.6745 | 2 | 4.550 |
| 2.2 | 3,547 | 45 | 0.7554 | 2.2 | 2.781 |
| 2.4 | 2,239 | 40 | 0.8416 | 2.4 | 1.640 |
| 2.6 | 1,358 | 35 | 0.9346 | 2.6 | 0.932 |
| 2.8 | 792 | 30 | 1.0364 | 2.8 | 0.511 |
| 3 | 443 | 25 | 1.1503 | 3 | 0.270 |
| 3.2 | 238 | 20 | 1.2816 | 3.2 | 0.137 |
| 3.4 | 123 | 15 | 1.4395 | 3.4 | 0.067 |
| 3.6 | 61 | 10 | 1.6449 | 3.6 | 0.032 |
| 3.8 | 29 | 5 | 1.9600 | 3.8 | 0.014 |
| 4 | 13 | 1 | 2.5758 | 4 | 0.006 |
| 4.2 | 5.9 | 0.1 | 3.2905 | 4.5 | 0.00068 |
| 4.4 | 2.5 | 0.01 | 3.8906 | 5 | 0.000057 |
| 4.6 | 1 | 0.001 | 4.4172 | 5.5 | 0.000004 |
| | | 0.0001 | 4.8916 | | |
| | | 0.00001 | 5.3267 | | |

Approximation for large values of $\lambda$:

$$P \approx \frac{3}{4}\frac{1}{\lambda}e^{-\lambda/2}$$

## Exercises

**1.1** Consider the Fourier series expansion of the periodic function $i(t)$ of period $T$, which is zero throughout the period except for the range $-\tau/2 \leqslant t \leqslant \tau/2$, where it has a value of 1.
Give the value of the coefficients for $\tau = T/2$ and $\tau = T/3$.
Verify that the expansion leads to $i(0) = 1$ and draw the function when the expansion is limited to 5 terms.

**1.2** Analyze the sampling at the frequency $f_s$ of the signal $s(t) = \sin(\pi f_s t + \varphi)$ when $\varphi$ varies from 0 to $\pi/2$.
Examine the reconstitution of this signal from the samples.

**1.3** Calculate the amplitude distortion introduced into a signal reconstituted by pulses with a width of half the sampling period.

**1.4** A signal occupies the frequency band $[f_1, f_2]$. What conditions should be imposed on the frequency $f_1$ so that this signal can be sampled directly at a frequency between $f_2$ and $2f_2$?

**1.5** Analyze the sampling of a signal given by:

$$s_i(t) = \sum_{n=1}^{3} \sin\left(2\pi \frac{n}{8}\frac{t}{T}\right)$$

and compare it with that of the signal:

$$s_r(t) = \sum_{n=1}^{3} \cos\left(2\pi \frac{n}{8}\frac{t}{T}\right)$$

Show by studying the spectra that the combination of the two sets of samplings forms the sampling of a complex signal.

**1.6** Let $s(t)$ be the signal defined by:

$$s(t) = 1 + 2\sum_{k=1}^{3} \cos\left(2\pi \frac{kt}{8} + \varphi_k\right) + \cos(\pi t + \varphi_4)$$

This signal is sampled with the period $T = 1$. What is the maximum value of $s(n)$, where $n$ is a whole number? Show that there exists a set of values $\varphi_k$ $(k = 1{,}2{,}3{,}4)$ which minimizes the maximum value of $s(n)$. Can this property be generalized?

**1.7** A digital frequency synthesizer is constructed from read-only memory of 16 kbits with an access time of 500 ns. Knowing that the numbers which represent the samplings of sinusoidal signals total 8 bits, what are the characteristics of the synthesizer, and the frequency range and increment step that can be obtained?

**1.8** What is the probability distribution of the amplitudes of the sinusoidal signal:

$$s(t) = A \cos\left(2\pi \frac{t}{T}\right)$$

Give its autocorrelation function. Give the autocorrelation function of a stationary random Gaussian function whose spectrum has a uniform distribution in the frequency band $(f_1, f_2)$.

**1.9** Calculate the spectrum of a set of impulses with width $T/2$, separated by $T$, the occurrence of each pulse having probability $p$. In particular, examine the case where $p = 1/2$.
What happens to the spectrum if these pulses form a pseudorandom sequence with a length $2^4-1 = 15$ produced by a 4-bit shift register, following the polynomial $g(x) = x^4 + x + 1$?

**1.10** A sinusoidal signal with frequency 1050 Hz is sampled at 8 kHz and coded into 10 bits. What is the maximum value of the SNR? What is the value of the signal-to-quantization-noise ratio measured in the frequency band 300–500 Hz? What are the values if the sampling frequency is increased to 16 kHz?

**1.11** The sinusoidal signal $\sin(2\pi t/8 + \varphi)$ with $0 \leqslant \varphi \leqslant \pi/2$ is sampled with period $T = 1$ and coded into 5 bits.
In the case where $\varphi = 0$, calculate the power and the spectrum of the quantization noise. How does this spectrum appear as a function of the phase $\varphi$?

**1.12** Consider a coding scale in which the quantizing step has a value $q$. What is the quantization of the signal $s_1(t) = \alpha q \sin(\omega_1 t)$ for $-1 \leqslant \alpha \leqslant 1$, as a function of the centering of the quantization characteristic? Show the envelope of the restored signal after decoding and narrow filtering around the frequency $\omega_1$.
The signal $s_2(t) = 10q \sin \omega_2 t$ is superimposed on $s_1(t)$. Show the envelope of the restored signal under these conditions.

**1.13** Assume a Gaussian signal is to be coded. How many bits would be required to have the signal-to-quantization-noise ratio greater than 50 dB? Can this number be reduced if signal clipping is allowed for 1% of this time?

**1.14** The signal $s(t) = A \sin(2\pi \cdot 810t)$ is coded into 8 bits. If the quantization step is $q$, trace the curve which shows the signal-to-quantization-noise ratio as a function of the amplitude $A$ when this amplitude varies from $q$ to $2^7 q$. Sketch the corresponding curve for nonlinear coding following the 13-segment $A$-law.

**1.15** Calculate the limits of the amplitude subranges for the optimal 2-bit coding of a unit Gaussian signal.

## References


1 A. Papoulis, *The Fourier Integral and its Applications*, McGraw-Hill, New York, 1962.
2 E. Roubine, *Introduction à la théorie de la communication, 3 vols*, Masson, Paris, 1970.
3 W. B. Davenport, *Probability and Random Processes*, McGraw-Hill, New York, 1970.
4 J. R. Rice, *The Approximation of Functions*, vol. 1, Addison-Wesley, Reading, Mass, 1964.
5 B. Picinbono, *Principles of Signals and Systems*, Artech House Inc., London, 1988.
6 W. Peterson, *Error Correcting Codes*, MIT Press, 1972.
7 W. B. Bennet, Spectra of quantized signals. *The Bell System Technical Journal*, 1948.
8 CCITT, Digital Networks — transmission systems and multiplexing equipment. *Yellow Book*, Vol. III, 3, Geneva, Switzerland, 1981.
9 J. Max, Quantizing for minimum distortion. *IRE Transactions on Information Theory* 6, 7–12, 1960.
10 M. D. Paez and T. H. Glisson, Minimum mean-squared error quantization in speech PCM and DPCM systems. *IEEE Transactions on Communications* 20, 225–30, 1972.
11 C. Shannon, *Communication in the presence of noise*, Proceedings of I.R.E., Vol. 37, pp. 10–21, Jan. 1949 (reprinted in: Proceedings of the IEEE, Sept. 1984 and Feb. 1998).

# 2

# The Discrete Fourier Transform

The discrete Fourier transform (DFT) is introduced when the Fourier transform of a function is to be calculated using a digital computer. This type of processor can handle only numbers and, in a quantity limited by the size of its memory. It follows that the Fourier transform:

$$S(f) = \int_{-\infty}^{\infty} s(t)e^{-j2\pi ft}dt$$

must be adapted, by replacing the signal $s(t)$ with the numbers $s(nT)$ which represent a sample of the signal, and by limiting to a finite value $N$ the set of numbers on which the calculations are carried out. The calculation then provides numbers $S^*(f)$ defined by

$$S * (f) = \sum_{n=0}^{N-1} s(nT)e^{-j2\pi fnT}$$

As the computer has limited processing power, it can only provide results for a limited number of values of the frequency $f$, and it is natural to choose multiples of a certain frequency step $\Delta f$. Thus,

$$S * (k\Delta f) = \sum_{n=0}^{N-1} s(nT)e^{-j2\pi nk\Delta fT}$$

The conditions under which the calculated values form a good approximation to the required values are examined below. An interesting simplifying choice is to take $\Delta f = 1/NT$. Then there are only $N$ different values of $S^*(k/NT)$, which is a periodic set of period $N$ since:

$$S * [(k + N)/NT] = S * (k/NT)$$

On the other hand, the transform thus calculated appears as discrete values and, as shown in Section 1.6, this property is characteristic of the spectrum of periodic functions. Thus, the set $S^*(k/NT)$ is obtained by the Fourier transform of the set $s(nT)$, which is periodic, with period $NT$. The DFT and the inverse transform establish the relations between these two periodic sets.

The definition, properties, methods of calculation, and applications of the DFT have been discussed in numerous publications. Overviews are given in References [1–4].

*Digital Signal Processing: Theory and Practice*, Tenth Edition. Maurice Bellanger.

## 2.1 Definition and Properties of the Discrete Fourier Transform

If two sets of complex numbers, $x(n)$ and $X(k)$, which are periodic with period $N$, are chosen, then the DFT and the inverse transform establish the following relationships between them:

$$X(k) = \frac{1}{N}\sum_{n=0}^{N-1} x(n)e^{-j2\pi nk/N} \tag{2.1}$$

$$x(n) = \sum_{k=0}^{N-1} X(k)e^{j2\pi kn/N} \tag{2.2}$$

The position of the scale factor $1/N$ is chosen so that the $X(k)$ are the coefficients of the Fourier series expansion of the set $x(n)$. This transformation has the following properties.

*Linearity:* If $x(n)$ and $y(n)$ are two sets with the same period and with transforms $X(k)$ and $Y(k)$, respectively, the set $v(n) = x(n) + \lambda y(n)$, where $\lambda$ is a scalar, has the transform

$$V(k) = X(k) + \lambda Y(k)$$

*A translation of the $x(n)$ implies a rotation of the phase of the $X(k)$.* If the transform $X_{n_0}(k)$ of the set $x(n - n_0)$ is calculated, then,

$$X_{n_0}(k) = \sum_{n=0}^{N-1} x(n - n_0)e^{-j2\pi nk/n} = X(k)e^{-j2\pi n_0 k/N}$$

A translation of the $x(n)$ from $n_0$ induces on $X(k)$ a rotation of the phase through an angle equal to $2\pi n_0 k/N$.

*Symmetry:* If the set $x(n)$ is real, the numbers $x(k)$ and $X(N - k)$ are complex conjugates:

$$\overline{X}(N - k) = \sum_{n=0}^{N} x(n)e^{j2\pi n(N-k)/N} = X(k)$$

If the set $x(n)$ is real and even, then so is the set $X(k)$. Indeed, if $x(N - n) = x(n)$ then, for example, for $N = 2P + 1$:

$$X(N - k) = x(0) + 2\sum_{n-1}^{p} x(n)\cos\left(2\pi\frac{nk}{N}\right) = X(k)$$

If the set $x(n)$ is real and odd, the set $X(k)$ is purely imaginary. In this case, $x(N - n) = -x(n)$ and $x(0) = x(N) = 0$. For example, for $N = 2P + 1$, this becomes:

$$X(k) = -2j\sum_{n=1}^{p} x(n)\sin\left(2\pi\frac{nk}{N}\right) = -X(N - k)$$

It should be noted that $X(0) = X(N) = 0$.

As any real signal can always be decomposed into odd and even parts, these last two symmetry properties are important.

*Circular convolution:* The transform of a convolution product is equal to the product of the transforms.

If $x(n)$ and $h(n)$ are two sets with period $N$, the circular convolution $y(n)$ can be defined by the equation:

$$y(n) = \sum_{l=0}^{N-1} x(l)h(n - l) \tag{2.3}$$

This is a set which has the same period $N$. Its transform is written as:

$$Y(k) = \sum_{n=0}^{N-1} \left[ \sum_{l=0}^{N-1} x(l)h(n-l) \right] e^{-j2\pi nk/N}$$

$$= \sum_{l=0}^{N-1} x(l) \left[ \sum_{n=0}^{N-1} h(n-l)e^{-j2\pi(n-l)k/N} \right] e^{-j2\pi lk/N}$$

$$Y(k) = \left( \sum_{n=0}^{N-1} h(n-l)e^{-j2\pi[(n-l)k/N]} \right) \left( \sum_{l=0}^{N-1} x(l)e^{-j2\pi lk/N} \right) = H(k)X(k) \tag{2.4}$$

This is an important property of the DFT. A direct application will be given later.

*Parseval's relation:* This relation states that the power of a signal is equal to the sum of the powers of its harmonics. Thus,

$$\frac{1}{N}\sum_{n=0}^{N-1} x(n)\bar{x}(n) = \frac{1}{N}\sum_{k=0}^{N-1} x(n) \sum_{k=0}^{N-1} \overline{X}(k)e^{-j2\pi kn/N}$$

$$\frac{1}{N}\sum_{n=0}^{N-1} |x(n)|^2 = \frac{1}{N}\sum_{k=0}^{N-1} \overline{X}(k) \frac{1}{N}\sum_{n=0}^{N-1} x(n)e^{-j2\pi kn/N}$$

$$\frac{1}{N}\sum_{n=0}^{N-1} |x(n)|^2 = \sum_{k=0}^{N-1} |X(k)|^2 \tag{2.5}$$

*Relations with Fourier series:* Due to the presence of the scale factor 1/N in the definition (2.1) of the DFT, the output terms $X(k)$ represent, except for spectrum aliasing effects, the coefficients of the Fourier series development of the periodic signal, when this signal exhibits no discontinuity. If it is not the case, noticeable discrepancies emerge. In fact, it can be shown that, if there is a discontinuity in function $x(t)$ at time $t_0$, the Fourier series development of $x(t)$ at time $t_0$ equals the average of the left and right limits of $x(t)$ when $t$ tends toward $t_0$. By contrast, the inverse DFT restores the exact original signal samples and, therefore, the terms $X(k)$ include the DFT of the distribution of the discontinuities with the inverse sign and scale factor 0.5.

*Illustration*: $x(t)$ is the following triangular signal:

$$x(t) = t;\, 0 \le t < 1$$

$$x(t+1) = x(t)$$

Coefficients of the Fourier series:

$$C_0 = \frac{1}{2}; \quad C_n = \frac{j}{2\pi n}; \quad n : \mathit{integer}; \quad n \neq 0$$

The magnitude of the discontinuity at the origin is 1 and the DFT of order $N$ gives the following values:

$$X(k) = -\frac{1}{N} + X'(k); \quad X'(k) \approx C_k$$

This specificity of the DFT is an undesirable effect when a development with the smallest number of non-negligible coefficients is sought, as in signal compression, for example. If the signal is real, a symmetric signal is appended, which cancels the discontinuity and leads to the discrete cosine transform described in Section 3.3.4.

However, the most important property of the DFT probably lies in the fact that it lends itself to efficient calculation techniques. This property has won it a prominent position in digital signal processing.

## 2.2 Fast Fourier Transform (FFT)

The equations defining the DFT provide a relationship between two sets of $N$ complex numbers. This is conveniently written in matrix form by setting

$$W = e^{-j2\pi/N}. \tag{2.6}$$

The coordinates of the numbers $W^n$, and the coefficients of the DFT, appear on the unit circle in the complex plane as shown in Figure 2.1, and are the roots of the equation $Z^N - 1 = 0$.

The matrix equation for the direct transform is as follows:

$$\begin{bmatrix} X_0 \\ X_1 \\ X_2 \\ \vdots \\ X_{N-1} \end{bmatrix} = \frac{1}{N} \begin{bmatrix} 1 & 1 & 1 & 1 & \cdots & 1 \\ 1 & W & W^2 & W^3 & \cdots & W^{N-1} \\ 1 & W^2 & W^4 & W^6 & \cdots & W^{2(N-1)} \\ \vdots & \vdots & \vdots & & \cdots & \vdots \\ 1 & W^{(N-1)} & W^{2(N-1)} & & \cdots & W^{(N-1)(N-1)} \end{bmatrix} \begin{bmatrix} x_0 \\ x_1 \\ x_2 \\ \vdots \\ x_{N-1} \end{bmatrix}$$

For the inverse transform, it is sufficient to remove the factor $1/N$ and change $W^n$ to $W^{-n}$.

The square matrix $T_N$ of order $N$ exhibits obvious features – rows and columns with the same index have the same elements and these elements are powers of a basic number such that $\mathbf{W}^N = \mathbf{1}$. Significant simplifications can be envisaged under these conditions, leading to algorithms for fast calculation. A FFT is said to have been carried out when the DFT is calculated using such algorithms.

An important case occurs when $N$ is a power of 2 because it leads to algorithms which are simple and particularly effective. These algorithms are based on a decomposition of the set to be transformed into a number of interleaved subsets. The case of interleaving in the time domain will be considered first, which leads to the so-called decimation-in-time algorithms.

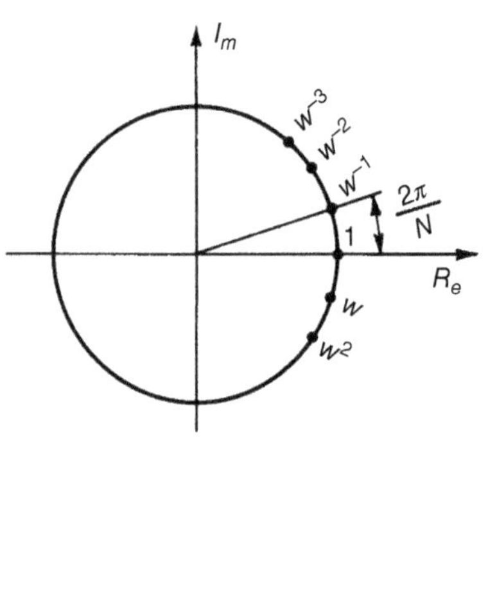


**Figure 2.1** Coordinates of the coefficients of a discrete Fourier transform.

### 2.2.1 Decimation-in-time Fast Fourier Transform

The set of elements $x(n)$ can be decomposed into two interleaved sets – those with even indices and those with odd ones. Using this decomposition, the first $N/2$ elements of the set $X(k)$ can be written as:

$$\begin{bmatrix} X_0 \\ X_1 \\ X_2 \\ \vdots \\ X_{N/2-1} \end{bmatrix} = \begin{bmatrix} 1 & 1 & \cdots & 1 \\ 1 & W^2 & \cdots & W^{2(N/2-1)} \\ 1 & W^4 & \cdots & W^{4(N/2-1)} \\ \vdots & \vdots & & \vdots \\ 1 & W^{2(N/2-1)} & \cdots & W^{2(N/2-1)(N/2-1)} \end{bmatrix} \begin{bmatrix} x_0 \\ x_2 \\ x_4 \\ \vdots \\ x_{2(N/2-1)} \end{bmatrix}$$

$$+ \begin{bmatrix} 1 & 1 & \cdots & 1 \\ W & W^3 & \cdots & W^{N-1} \\ W^2 & W^6 & \cdots & W^{2(N-1)} \\ \vdots & \vdots & & \vdots \\ W^{N/2-1} & W^{3(N/2-1)} & \cdots & W^{(N/2-1)(N-1)} \end{bmatrix} \begin{bmatrix} x_1 \\ x_3 \\ x_5 \\ \vdots \\ x_{N-1} \end{bmatrix}$$

If the matrix which multiplies the column vector of elements with even indices is denoted by $T_{N/2}$, then the matrix multiplying the vector of elements with odd indices can be factorized into the product $T_{N/2}$ and a diagonal matrix so that,

$$\begin{bmatrix} X_0 \\ X_1 \\ X_2 \\ \vdots \\ X_{N/2-1} \end{bmatrix} = T_{N/2} \begin{bmatrix} x_0 \\ x_2 \\ x_4 \\ \vdots \\ x_{2(N/2-1)} \end{bmatrix} + \begin{bmatrix} 1 & 0 & 0 & \cdots & 0 \\ 0 & W & 0 & \cdots & 0 \\ 0 & 0 & W^2 & \cdots & 0 \\ \vdots & \vdots & & & \vdots \\ 0 & 0 & \ldots & W^{N/2-1} & \end{bmatrix} T_{N/2} \begin{bmatrix} x_1 \\ x_3 \\ x_5 \\ \vdots \\ x_{N/2-1} \end{bmatrix}$$

Similarly, for the last $N/2$ elements of the set $X(k)$, remembering that $W^N = 1$, one can write:

$$\begin{bmatrix} X_{N/2} \\ X_{N/2+1} \\ X_{N/2+2} \\ \vdots \\ X_{N-1} \end{bmatrix} = T_{N/2} \begin{bmatrix} x_0 \\ x_2 \\ x_4 \\ \vdots \\ x_{2(N/2-1)} \end{bmatrix} - \begin{bmatrix} 1 & 0 & 0 & \cdots & 0 \\ 1 & W & 0 & \cdots & 0 \\ 0 & 0 & W^2 & \cdots & 0 \\ \vdots & \vdots & \vdots & & \vdots \\ 0 & 0 & 0 & \ldots & W^{N/2-1} \end{bmatrix} T_{N/2} \begin{bmatrix} x_1 \\ x_3 \\ x_5 \\ \vdots \\ x_{N-1} \end{bmatrix}$$

It is apparent that the calculation of $X(k)$ and $X(k + N/2)$ for $0 < k(N/2) - 1$ uses the same calculations with only a change in sign in the final sum. Hence the following diagram:

This shows that the calculation of a Fourier transform of order $N$ reduces to the calculation of two transforms of order $N/2$, to which $M/2$ complex multiplications are added. By iteration through a number of steps given by $\log_2 N - 1 = \log_2(N/2)$, transforms of order 2 are arrived at. These have the matrix:

$$T_2 = \begin{bmatrix} 1 & 1 \\ 1 & -1 \end{bmatrix}$$

and no multiplications are required.

As each stage involves $N/2$ complex multiplications, the complete transform requires $M_c$ complex multiplications, where:

$$M_c = (N/2)\log_2(N/2) \tag{2.7}$$

and $A_c$ complex additions, where:

$$A_c = N\log_2 N \tag{2.8}$$

In practice, the number of complex multiplications can again be reduced because some powers of $W$ have certain properties. For example, $W^0 = 1$ and $W^{N/4} = -j$ do not require complex multiplications, and

$$W^{N/8} = \frac{\sqrt{2}}{2}(1 - j)$$

only requires one complex half-multiplication each. Thus, three multiplications can be saved in the first stage, $3N/8$ can be eliminated in the penultimate one, and $2N/4$ in the last. The gain over all the stages is $5N/4 - 3$ and the minimum number of complex multiplications is given by:

$$m_c = N/2\left[\log_2(N/2) - \frac{5}{2}\right] + 3 \tag{2.9}$$

It should be noted that all of these calculation reductions cannot always be easily implemented either in software or in hardware.

The matrix for the fourth-order transform is:

$$T_4 = \begin{bmatrix} 1 & 1 & 1 & 1 \\ 1 & -j & -1 & +j \\ 1 & -1 & 1 & -1 \\ 1 & +j & -1 & -j \end{bmatrix} \tag{2.10}$$

The diagram for its reduction is shown in Figure 2.2. By convention, the arrows represent multiplications and the solid circles to the left of the elementary flow graphs, often called "butterflies," represent an addition (upper one) and a subtraction (lower one). The eighth-order transform is represented in Figure 2.3.

It is seen that in this treatment, the indices of the $X(k)$ appear in natural order, while those of the $x(n)$ are in a permuted one. This permutation is caused by the successive interleaving and results

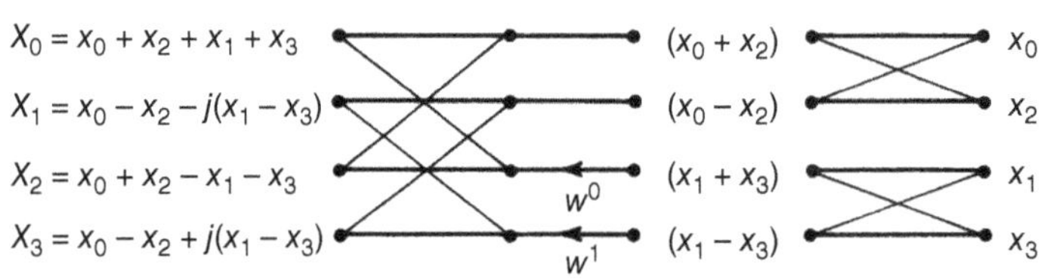


**Figure 2.2** Transform of order 4 with decimation in time.

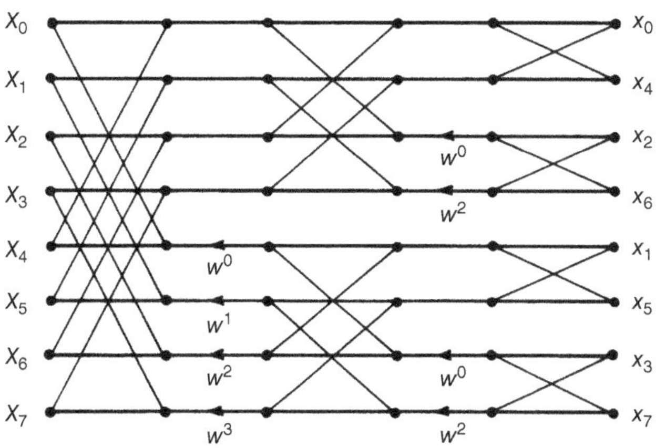


**Figure 2.3** Transform of order 8 with decimation in time.

in a reversal or inversion of the binary representation of the indices, which is often called "bit reversal." For example, for $N = 8$,

$x_0$ (000) corresponds to $x_0$ (000)
$x_4$ (100) corresponds to $x_1$ (001)
$x_2$ (010) corresponds to $x_2$ (010)
$x_6$ (110) corresponds to $x_3$ (011)
$x_1$ (001) corresponds to $x_4$ (100)
$x_5$ (101) corresponds to $x_5$ (101)
$x_3$ (011) corresponds to $x_6$ (110)
$x_7$ (111) corresponds to $x_7$ (111)

The amount of data memory required to calculate a transform of order $N$ is that needed to hold $N$ complex positions. Indeed, the calculations are performed on pairs of variables which undergo the operation represented by a butterfly and preserve their position in the set of variables at the end of the operation, as is clearly shown in the diagrams. This is called "in-place computation." The inverse transform is simply obtained by changing the sign of the exponent of $W$. The factor $1/N$ can be introduced, for example, by halving the results of the additions and subtractions made in the butterflies. This allows scaling of the numbers in the memories.

This type of interleaving can also be applied to $X(k)$ when a similar algorithm is obtained. It leads to the so-called decimation-in-frequency algorithms.

### 2.2.2 Decimation-in-frequency Fast Fourier Transform

The set of elements $X(k)$ can be decomposed into a pair of interleaved sets – one with even indices and the other with odd ones. For the even index elements, since $W^N = 1$, the following situation occurs after elementary factorization:

$$\begin{bmatrix} X_1 \\ X_3 \\ X_5 \\ \vdots \\ X_{N-1} \end{bmatrix} = \begin{bmatrix} 1 & W & W_2 & \cdots & W^{N/2-1} \\ 1 & W^3 & W_6 & \cdots & W^{3(N/2-1)} \\ 1 & W^5 & W_{10} & \cdots & W^{5(N/2-1)} \\ \vdots & \vdots & \vdots & \cdots & \vdots \\ 1 & W^{N-1} & W^{2(N-1)} & \cdots & W^{(N-1)(N/2-1)} \end{bmatrix} \begin{bmatrix} x_0 - x_{N/2} \\ x_1 - x_{N/2+1} \\ x_2 - x_{N/2+2} \\ \vdots \\ x_{N/2-1} - x_{N-1} \end{bmatrix}$$

For the elements with odd indices after a similar process, the corresponding equation is:

$$\begin{bmatrix} X_1 \\ X_3 \\ X_5 \\ \vdots \\ X_{N-1} \end{bmatrix} = \begin{bmatrix} 1 & W & W_2 & \cdots & W^{N/2-1} \\ 1 & W^3 & W_6 & \cdots & W^{3(N/2-1)} \\ 1 & W^5 & W_{10} & \cdots & W^{5(N/2-1)} \\ \vdots & \vdots & \vdots & \cdots & \vdots \\ 1 & W^{N-1} & W^{2(N-1)} & \ldots & W^{(N-1)(N/2-1)} \end{bmatrix} \begin{bmatrix} x_0 - x_{N/2} \\ x_1 - x_{N/2+1} \\ x_2 - x_{N/2+2} \\ \vdots \\ x_{N/2-1} - x_{N-1} \end{bmatrix}$$

In this case, the square matrix obtained is equal to the product of the matrix $T_{N/2}$ obtained for the elements with even indices, and the diagonal matrix whose elements are the powers $W^k$ with $0 \leqslant k \leqslant N/2-1$. Thus,

$$\begin{bmatrix} X_1 \\ X_3 \\ X_5 \\ \vdots \\ X_{N-1} \end{bmatrix} = \begin{bmatrix} 1 & 0 & 0 & \cdots & 0 \\ 0 & W & 0 & \cdots & 0 \\ 0 & 0 & W^2 & \cdots & 0 \\ \vdots & \vdots & \vdots & & \vdots \\ 0 & 0 & 0 & \ldots & W^{N/2-1} \end{bmatrix} \begin{bmatrix} x_0 - x_{N/2} \\ x_1 - x_{N/2+1} \\ x_2 - x_{N/2+2} \\ \vdots \\ x_{N/2-1} - x_{N-1} \end{bmatrix}$$

The elements $x(k)$ with even and odd indices are calculated using the square matrix $T_{N/2}$ for the transform of order $N/2$, and the following diagram is obtained:

By adopting the same notation for the butterflies as was used in the preceding section, similar diagrams are obtained. Figure 2.4 shows the diagram for $N = 8$.

In decimation-in-frequency algorithms, the number of calculations is the same as with time interleaving. The numbers $x(n)$ to be transformed appear in their natural order, while the transformed numbers $X(k)$ are permuted.

The algorithms which have been obtained so far are based on a decomposition of the transform of order $N$ into elementary second-order transforms which do not require multiplications. These algorithms are said to be radix-2 transforms. Other elementary transforms can also be used, the most important being the radix-4 one, which uses the elementary matrix $T_4$.

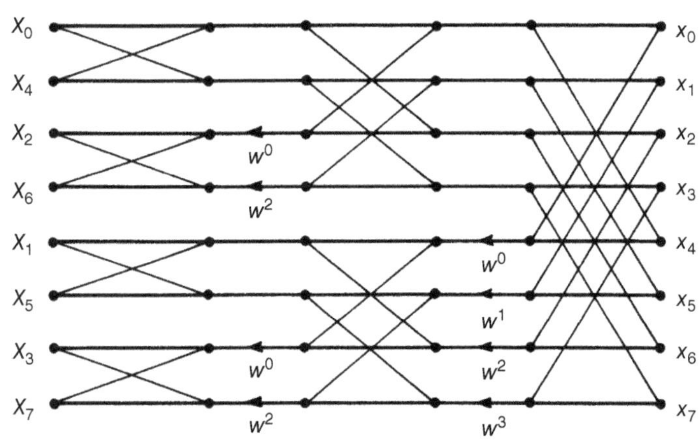


**Figure 2.4** Transform of order 8 with decimation in frequency.

### 2.2.3 Radix-4 FFT Algorithm

This algorithm can be used when $N$ is a power of 4. The set of numbers $x(k)$ is decomposed into four interleaved sets. Let us calculate the first $N/4$ values of $X(k)$ as an illustration of the decomposition. If $T_{N/4}$ denotes the square matrix of the transform of order $N/4$. and if $D_i$ $(i = 1, 2, 3)$ is the diagonal matrix whose elements are the powers $W^{ik}$ with $0 \leqslant k \leqslant N/4 - 1$, then,

$$\begin{bmatrix} X_0 \\ X_1 \\ X_2 \\ \vdots \\ X_{N/4-1} \end{bmatrix} = T_{N/4} \begin{bmatrix} x_0 \\ x_4 \\ x_8 \\ \vdots \\ x_{4(N/4-1)} \end{bmatrix} + D_1 T_{N/4} \begin{bmatrix} x_1 \\ x_5 \\ x_9 \\ \vdots \\ x_{N-3} \end{bmatrix} + D_2 T_{N/4} \begin{bmatrix} X_2 \\ X_6 \\ X_{10} \\ \vdots \\ X^{N-2} \end{bmatrix} + D_3 T_{N/4} \begin{bmatrix} X_3 \\ X_7 \\ X_{11} \\ \vdots \\ X_{N-1} \end{bmatrix}$$

The next $N/4$ terms of $X(k)$ are given by:

$$\begin{bmatrix} X_{N/4} \\ X_{N/4+1} \\ \vdots \\ X_{N/2-1} \end{bmatrix} = T_{N/4} \begin{bmatrix} x_0 \\ x_4 \\ \vdots \\ x_{4(N/4-1)} \end{bmatrix} - jD_1 T_{N/4} \begin{bmatrix} x_1 \\ x_5 \\ \vdots \\ x_{N-3} \end{bmatrix} - D_2 T_{N/4} \begin{bmatrix} x_2 \\ x_6 \\ \vdots \\ x_{N-2} \end{bmatrix} + jD_3 T_{N/4} \begin{bmatrix} x_3 \\ x_7 \\ \vdots \\ x_{N-1} \end{bmatrix}$$

This equation involves the same matrix calculations as the previous one, with the addition of the multiplications by the elements of the second row of the matrix $T_4$. It can hence be shown that the calculation of the transform results in the diagram in Figure 2.5.

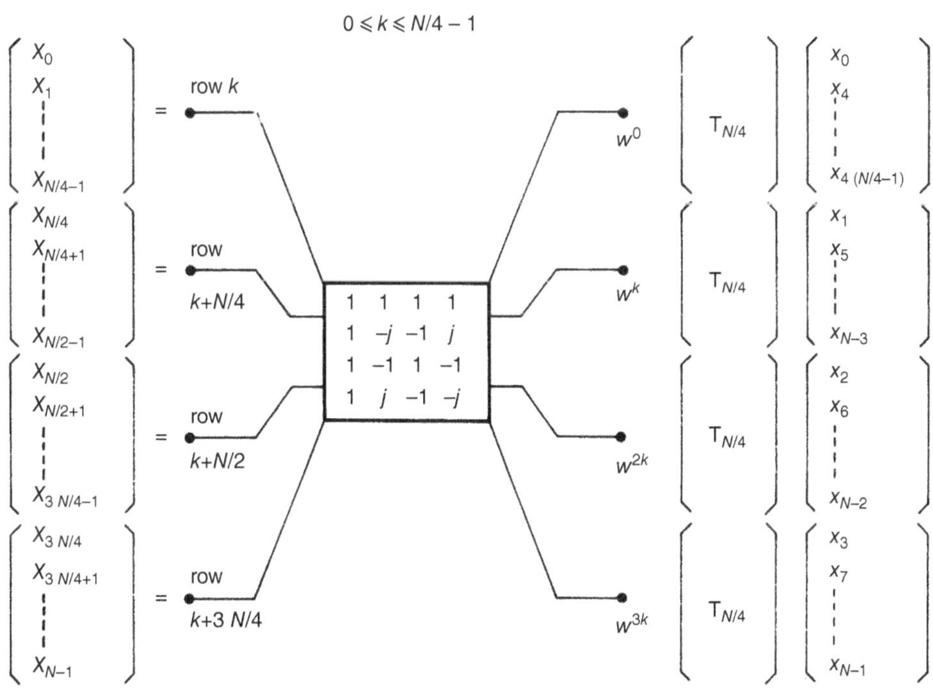


**Figure 2.5** Radix-4 transform.

This type of transform is carried out in $\log_4 N - 1 = \log_4 (N/4)$ stages. Each stage requires $3(N/4)$ complex multiplications, which results in a total of $M_{c4}$ multiplications, where:

$$M_{c4} = \frac{3}{4} N \log_4 \left(\frac{N}{4}\right) \tag{2.11}$$

The number of complex additions $A_{c4}$ is:

$$A_{c4} = 2N \log_4 N \tag{2.12}$$

It is apparent that the number of additions is the same in radix-2 and in radix-4 algorithms, in contrast to the complex multiplications, where calculation in radix-4 algorithms results in a saving of over 25%. Other radices can also be envisaged – for example, radix 8. In this case, there are multiplications in the elementary matrix, and the savings over radix 4 are negligible. Different radices can also be combined [5].

### 2.2.4 Split-radix FFT Algorithm

In a transform of order $N$, the set of odd-index transformed values can be decomposed into two subsets:

$$X(4k+1) = \frac{1}{N} \sum_{n=0}^{N-1} x(n) W^{n} W^{4kn}$$

and:

$$X(4k+3) = \frac{1}{N} \sum_{n=0}^{N-1} x(n) W^{3n} W^{4kn}$$

Using definition (2.6) for $W$, the summations can also be written as:

$$X(4k+1) = \frac{1}{N} \sum_{n=0}^{N/4-1} \left[\left[x(n) - x\left(n + \frac{N}{2}\right)\right] - j\left[x\left(n + \frac{N}{4}\right) - x\left(n + \frac{3N}{4}\right)\right]\right] W^{n} W^{4nk} \tag{2.13}$$

and:

$$X(4k+3) = \frac{1}{N} \sum_{n=0}^{N/4-1} \left[\left[x(n) - x\left(n + \frac{n}{2}\right)\right] + j\left[x\left(n + \frac{N}{4}\right) - x\left(n + \frac{3N}{4}\right)\right]\right] W^{3n} W^{4kn} \tag{2.14}$$

The set of even index transformed values is:

$$X(2k) = \frac{1}{N} \sum_{n=0}^{N/2-1} \left[x(n) + x\left(n + \frac{N}{2}\right)\right] W^{2nk} \tag{2.15}$$

The above equations show that the first step in the decimation-in-time approach can be replaced by the calculation of a transform of order $N/2$ and two transforms of order $N/4$. The split-radix algorithm is obtained through repeated applications of the above procedure:

In the decimation-in-frequency approach, the split-radix algorithm is based on the following decomposition:

$$X(k) = \sum_{n=0}^{N/2-1} x(2n) W^{2nk} + W^{k} \sum_{n=0}^{N/4-1} x(4n+1) W^{4nk} + W^{3k} \sum_{n=0}^{N/4-1} x(4n+3) W^{4nk} \tag{2.16}$$

For a transform of order $N$, the number of complex multiplications $M_{c2/4}(N)$ is given by the recurrence derived from the above equations:

$$M_{c2/4}(N) = M_{c2/4}\left(\frac{N}{2}\right) + 2M_{c2/4}\left(\frac{N}{4}\right) + \frac{N}{2} \tag{2.17}$$

with the initial values $M(2) = M(4) = 0$. The value obtained in this way is slightly smaller than for a radix-4 algorithm.

In practice, taking trivial operations into account and implementing complex multiplications with 3 real multiplications and 5 real additions, as explained later on, it can be shown that the $N$-order transform needs $N\log_2(N) - 3N + 4$ real multiplications and $3N\log_2(N) - 3N + 4$ additions [6].

The algorithms which have been presented for decimation-in-time and -frequency, and for radix-2 and 4, are elements of a large set of algorithms. A unified presentation of FFT algorithms is given in the next chapter, so that the most appropriate can be selected for each application. In actual calculations, however, operations are carried out with limited precision. This results in some degradation of the signal.

## 2.3 Degradation Arising from Wordlength Limitation Effects

The equipment used introduces limitations caused by the finite precision of arithmetic units and the limited memory capacity. Firstly, the coefficients are held in a memory, with a limited number of bits. Thus, the memory contents represent approximations of the actual coefficients, which are generally obtained by rounding. Secondly, as the calculation proceeds, rounding is performed so as to keep the data wordlength within the capacities of the memory locations, or of the arithmetic units. It is important to analyze the degradations introduced by these two types of wordlength limitation in order to be able to precisely determine the hardware needed to produce a transform with specific performance.

To begin with, the rounding of coefficients is considered.

The coefficients actually used by the machine represent an approximation of the theoretical coefficients, which have values of the real and imaginary parts within the range $[-1, +1]$.

For the coefficient $e^{-j2\pi n/N}$, digitization into $b_c$ bits involves a quantization error $\delta(n) = \delta_R(n) + j\delta_I(n)$ such that, if rounding is employed,

$$|\delta_R(n)| \leqslant 2^{-b_c} \quad \text{and} \quad |\delta_I(n)| \leqslant 2^{-b_c}$$

The calculation of each transformed number $X(k)$ from the data $x(n)$ is performed with an error $\Delta(k)$ such that:

$$X(k) + \Delta(k) = \frac{1}{N}\sum_{n=0}^{N-1} x(n)[e^{-j2\pi nk/N} + \delta(nk)]$$

or

$$\Delta(k) = \frac{1}{N}\sum_{n=0}^{N-1} x(n)\delta(nk)$$

As the $x(n)$ and $X(k)$ are related by equation (2.2):

$$x(n) = \sum_{k=0}^{N-1} X(k)e^{j2\pi nk/N}$$

this becomes,

$$\Delta(k) = \sum_{i=0}^{N-1} X(i)\varepsilon(i, k) \tag{2.18}$$

with,

$$\varepsilon(i,k) = \frac{1}{N}\sum_{n=0}^{N-1}\delta(nk)e^{j2\pi ni/N}$$

Consequently, for the transformed number $x(k)$, rounding the coefficients of the transform introduces a perturbation $\Delta(k)$ obtained by summing the elementary perturbations, each of which is equal to the product of a transformed number by a factor representing its contribution. The transformed numbers interact with each other and are no longer strictly independent.

It is possible to calculate the $\varepsilon(i,k)$ for each transformation. In general, it is important to know the maximum value $\varepsilon_m$ that the $|\varepsilon(i,k)|$ can have for a given order of transformation and for a given number of bits $b_c$.

The inequality $|\delta(n)| < 2^{-b_c}\sqrt{2}$ provides a maximum for $\varepsilon_m$ as

$$\varepsilon_m \leqslant 2^{b_c}\sqrt{2}$$

In practice, the values found for $\varepsilon_m$ are much lower than this maximum. For example, for $N = 64$, it is found that $\varepsilon_m \simeq 0.6 \times 2 - b_c$, and this value is also found for higher values of $N$ [7].

Another performance degradation arises from computation noise. At DFT computer input, data have a limited number of bits, and this number grows with every multiplication and addition. In general, the number of bits assigned to internal data in the computer is fixed and it is necessary to limit word lengths. Most of the time, limitations are obtained just by rounding, because, since overflow is generally not acceptable, the scaling is selected at the beginning of the transform so that the whole of the calculation can be performed without any risk of overflow. Two cases can be distinguished for the radix-2 transform:

- Direct transform: Due to the global scale factor 1/N, it is sufficient to halve the results of additions and subtractions in each butterfly to keep the correct scaling.
- Inverse transform: The scaling at the beginning of the transform is selected so that no overflow can occur in the calculations.

Regarding round-off noise power, detailed estimation requires identification of the sources, with corresponding quantization steps, and summations [8].

## 2.4 Calculation of a Spectrum Using the DFT

The calculation of a spectrum by the DFT requires that certain approximations and a suitable choice of parameters be made to attain the desired performance. Before considering any application, it is therefore useful to look carefully at the function fulfilled by the DFT.

### 2.4.1 The Filtering Function of the DFT

Let us examine the relationship between the outputs $X(k)$ and the inputs $x(n)$, considered as the result of sampling a signal $x(t)$ with period $T$, which establishes the DFT. For $k = 0$ this is:

$$X(0) = \frac{1}{N}\sum_{n=0}^{N-1}x(n)$$

The signal $X(0)$ thus defined results from the convolution of the signal $x(t)$ with the distribution $\varphi_0(t)$ such that:

$$\phi_0(t) = \frac{1}{N}\sum_{n=0}^{N-1}\delta(t - nT)$$

The Fourier transform of this distribution is given by:

$$\Phi_0(f) = \frac{1}{N}\sum_{n=0}^{N-1} e^{-j2\pi nf/T} = \frac{1}{N}\frac{1 - e^{-j2\pi f/NT}}{1 - e^{j2\pi fT}}$$

or

$$\Phi_0(f) = e^{-j\pi f(N-1)T}\phi(f)$$

with,

$$\Phi(f) = \frac{1}{N}\frac{\sin(\pi fNT)}{\sin(\pi fT)} \tag{2.19}$$

Now, a convolution operation in the time domain corresponds to a product in the frequency domain – that is, $X(0)$ is a signal obtained by filtering the input signal by the function $\Phi_0(f)$. Figure 2.6 shows the function $\Phi(f)$ and the function $\phi(t)$ of which it is the Fourier transform. The function $\Phi(f)$ is zero at points on the frequency axis which are whole multiples of $1/NT$, except for multiples of $1/T$. It is periodic and has a period $1/T$, conforming to the laws of sampling. It is simply the spectrum of a sampled impulse of width $NT$. Similarly, the output $X(k)$ has the corresponding function $\phi_k(t)$ such that:

$$\phi_k(t) = \frac{1}{N}\sum_{n=0}^{N-1} e^{j2\pi nk/N}\delta(t - nT)$$

$$\Phi_k(f) = \frac{1}{N}\sum_{n=0}^{N-1} e^{-j2\pi nk/N}e^{-j2\pi fnT}$$

In compact form, after simplification, this becomes:

$$\Phi_k(f) = (-1)^k e^{-j\pi f(N-1)T}e^{j\pi k/N}\phi\left(f\frac{k}{NT}\right) \tag{2.19b}$$

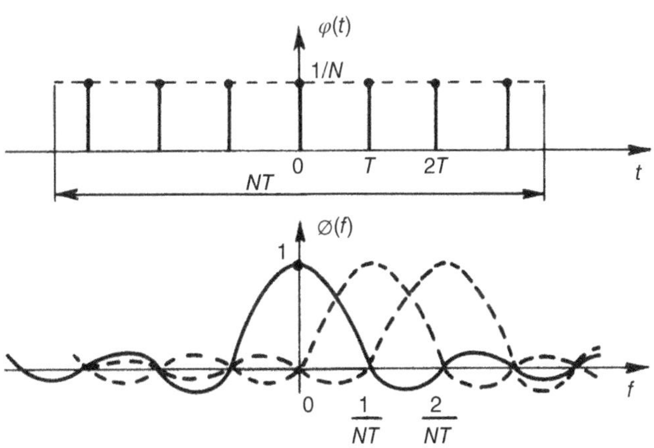


**Figure 2.6** The filtering function of a discrete Fourier transform.

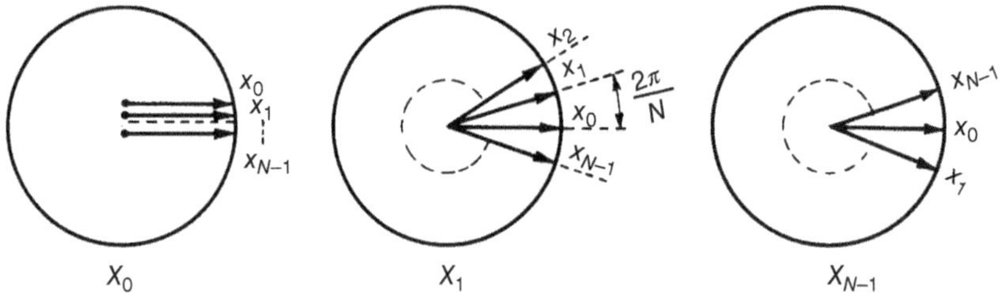


**Figure 2.7** Filtering by phase shift in a discrete Fourier transform.

The output $X(k)$ provides the signal filtered according to the function $\Phi_0(f)$, but translated by $k/NT$ along the frequency axis.

*Thus, the DFT forms a set of N identical filters, or a bank of filters, distributed uniformly over the frequency domain at intervals of 1/NT.*

If the input signal is periodic, then, from the definition of the DFT, this bank of filters is frequency sampled at intervals of $1/NT$, and it should be noted that there is no interference between the outputs $X(k)$. Strictly speaking, this property is lost if the coefficients are rounded, as has been shown earlier.

The DFT function described above also illustrates the problem of scaling numbers in the memory of an FFT calculator. Let us suppose that the numbers of transform $x(n)$ result from the sampling of a random signal for which the amplitude probability distribution has the variance $\sigma^2$. If the signal has a uniform distribution of its energy spectrum, its power is uniformly distributed between the $X(k)$, and each has a variance equal to $\sigma^2/N$. By contrast, if the signal has a spectral distribution which can be concentrated on one $X(k)$, this $X(k)$ has the same probability distribution as the $x(n)$ – in particular, the variance $\sigma^2$. Scaling the numbers by dividing by 2 at each stage of an FFT calculation is suitable for the handling of such signals.

A different view of the filtering process is provided by observing that the outputs $X(k)$ of the DFT are the sums of the inputs $x(n)$ after phase shifting. In effect, the output $X(0)$ is the sum of the $x(n)$ with zero phase shift, and the output $X(k)$ is the sum of $x(n)$ with phase shifts which are multiples of $2\pi(k/N)$, as shown in Figure 2.7. At each output, in-phase components of the resulting signals add while the others cancel. For example, if the $x(n)$ are complex numbers with the same phase and modulus, all the $X(k)$ become zero except for $X(0)$. The input signal is thus found to be decomposed according to the base represented by the $N$ vectors $e{-}j2\pi(k/N)$ with $0 \leqslant k \leqslant N-1$.

This result is useful in studying banks of filters which include a DFT processor.

### 2.4.2 Spectral Resolution

Spectrum analysis is used in many fields and is often performed on recorded data. By definition, the DFT establishes a relation between two periodic sets, the $x(n)$ and $X(k)$, each of which includes $N$ different elements. In order to use it, it is therefore necessary to examine this double periodicity.

The periodicity in the frequency is introduced by the process of signal sampling. The data to be processed are either in digital form, with sampling and coding being performed within the signal source, or in analog form, which then has to be converted into digital. The choice of sampling frequency $f_s = 1/T$ should be such that components of the signal with frequencies greater than or equal to $f_s/2$ are negligible, and, in any case, less than the tolerable error in the amplitude of the useful components. Fulfillment of this condition can be ensured by prefiltering the signal.

Periodicity in time is introduced artificially by assuming that the signal is repeated outside the time interval $\theta = NT$ which actually corresponds to the data being processed. Under these conditions, the DFT supplies a sample of the spectrum with a frequency period $\Delta f$, equal to the inverse of the duration of the data, which constitutes the frequency resolution of the analysis. The relation $\Delta f = NT$ expresses the Heisenberg uncertainty principle for spectrum analysis. A more accurate analysis can be obtained by increasing the duration of the data collection – for example, by making it $N'T$ (with $N' > N$) with zero additional samples. The additional frequency samples are obtained simply from an interpolation of the others. This procedure is commonly used to provide a number $N'$ of data points which is a power of 2, thus allowing fast algorithms to be used. On the other hand, the fact that the signal is not formed solely of lines at frequencies which are multiples of $1/NT$ introduces interference between the spectral components obtained. Indeed, the filter function $\Phi(f)$ of the DFT, which is given in Section 2.4.1, introduces ripples throughout the frequency band and, if the signal has a spectral component $S(f_0)$ at frequency $f_0$ such that $k/NT < f_0 < (k+1)/NT$, then:

$$X(k) = S(f_0)\Phi(k/NT - f_0), \quad 0 \leqslant k \leqslant N-1 \tag{2.20}$$

As a result, all the transform outputs can assume nonzero values as shown in Figure 2.8. Thus, limitations appear for the resolution of the analyzer. This effect can be reduced by modifying the filter function of the DFT by weighting the signal samples before transforming them.

This operation amounts to replacing the rectangular time window $\phi(t)$ with a function whose Fourier transform results in smaller ripples. Numerous functions are used, of which the simplest is the raised cosine window:

$$\phi(t) = \frac{1}{2}\left(1 + \cos 2\pi \frac{t}{NT}\right) \tag{2.21}$$

and the Hamming window:

$$\phi(t) = 0.54 + 0.46 \cos 2\left(\pi \frac{t}{NT}\right) \tag{2.22}$$

The latter function has 99.96% of its energy in the main lobe. The peak side lobe ripple is about 40 dB below the main lobe peak. Other time windows can be used, and several efficient functions are introduced in Reference. [9].

Let $\Phi(f)$ be the spectrum of the time window $\varphi(t)$ after sampling; the expression (2.20) can be extended to any signal with spectrum $S(f)$, using the definition of the convolution and taking account of the periodicity of $\Phi(f)$:

$$X(k) = \int_0^{1/T} \left[\frac{1}{T}\sum_{n=-\infty}^{\infty} S\left(u - \frac{n}{T}\right)\right] \Phi\left(\frac{k}{NT} - u\right) du$$

The signal spectrum aliasing due to sampling with period $T$ is apparent.

To cope better with interference between the calculated spectral components, it is necessary to employ a bank of more selective filters, like that presented in Chapter 10.

The DFT can also be used indirectly in the calculation of convolutions.

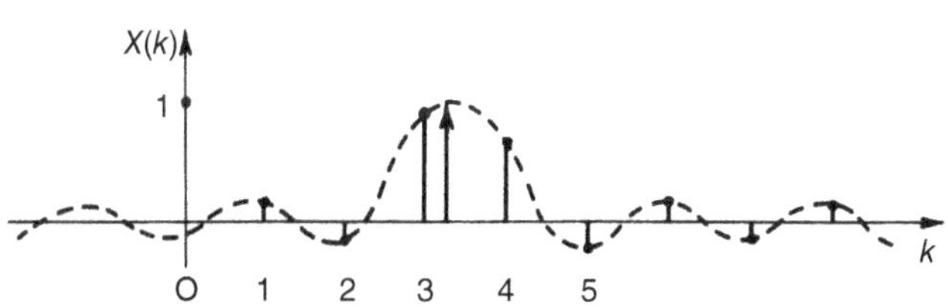


**Figure 2.8** Analysis of a signal with a frequency which is not a multiple of 1/NT.

## 2.5 Fast Convolution

The efficiency of FFT algorithms leads to the use of the DFT in cases other than spectrum analysis and, in particular, in convolutions.

Although, in general, this is not the most efficient of approaches, it can be useful in applications where an FFT processor is available.

One of the properties of the DFT is that the transform of a convolution product is equal to the product of its transforms. Given two sets $x(n)$ and $h(n)$ of period $N$, with transforms $X(k)$ and $H(k)$, the circular convolution

$$y(n) = \sum_{m=0}^{N-1} h(m)x(n-m)$$

is a set of the same period whose transform is written as:

$$Y(k) = H(k)X(k)$$

Fast convolution consists of calculating the set $y(n)$ by applying a DFT to the set $Y(k)$. As one of the convolution terms is normally constant, the operation requires one DFT, one product, and one inverse DFT. This technique is applied to sets of finite length. If $x(n)$ and $h(n)$ are two sets of $N_1$ and $N_2$ nonzero terms, the set $y(n)$ defined by

$$y(n) = \sum_{m=0}^{n} h(m)x(n-m)$$

is a set of finite length, having $N_1 + N_2 - 1$ terms. The fast convolution is applied by considering that the three sets $y(n)$, $x(n)$, and $h(n)$ have the period $N$ such that $N \geqslant N_1 + N_2 - 1$. It is then sufficient to complete each set with a suitable number of zero terms. It is of particular interest if a power of 2 is chosen for $N$.

Nevertheless, in practice, convolution is a filtering operation, where the $x(n)$ represents the signal and the $h(n)$ the coefficients. The set of the $x(n)$ is much longer than that of the $h(n)$, and it is necessary to subdivide the calculation. To do this, the set of the $x(n)$ is regarded as a superposition of elementary sets $x_k(n)$, each of $N_3$ terms. That is,

$$x(n) = \sum_{k} x_k(n)$$

with $x_k(n) = x(n)$ for $kN_3 \leqslant n \leqslant (k+1)N_3 - 1$ and $x_k(n) = 0$ elsewhere.

We can then write:

$$y(n) = \sum_{m=0}^{n} h(m) \sum_{k} x_k(n-m)$$

$$y(n) = \sum_{k} \sum_{m=0}^{n} h(m)x_k(n-m) = \sum_{k} y_k(n)$$

Each set $y_k(n)$ contains $N_3 + N_2 - 1$ nonzero terms. Thus, the convolutions involve $N_3 + N_2 - 1$ terms. Figure 2.9 shows the sequence of operations. The sets $y_k(n)$ and $y_{k+1}(n)$ have $N_2 - 1$ terms which are superposed. The same operations can be performed by decomposing the set $x(n)$ into sets $x_k(n)$ such that $N_2 - 1$ terms are superposed.

In this process, the number of calculations to be performed on each element of the set $y(n)$ increases as $\log_2 (N_2 + N_3 - 1)$ and $N_3$ must not be chosen as too large. Also, if $N_3 < N_2$, no terms in the set $y(n)$ can be obtained directly. Consequently, there is an optimal value for $N_3$. The number

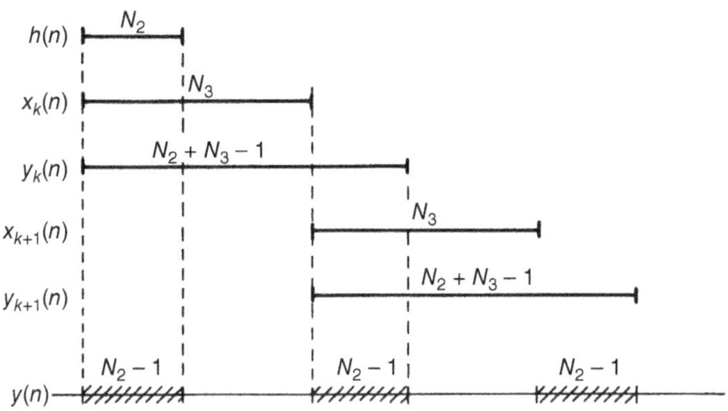


**Figure 2.9** Sequence of the operations in fast convolution.

of memory locations needed increases as $N_3 + N_2 + 1$, and a good compromise is reached by taking for $N_3$ the first value above $N_2$ such that $N_3 + N_2 - 1$ is a power of 2.

## 2.6 Calculations of a DFT Using Convolution

In certain applications, only those operators which can form convolutions are available to calculate a DFT. Such is the case for circuits using charge transfer devices which allow calculations to be performed on the sampled signals in analog form at speeds compatible with the frequencies encountered, for example, in radar applications.

The definition of a DFT can be written as:

$$X(k) = \frac{1}{N}\sum_{n=0}^{N-1} x(n)e^{-j2\pi(nk/N)}$$

By writing

$$nk = \frac{1}{2}[n^2 + k^2 - (n-k)^2] \quad \text{and} \quad W = e^{-j(2\pi/N)}$$

this becomes

$$X(k) = W^{k^2/2}\sum_{n=0}^{N-1} x(n)W^{n^2/2}W^{-(n-k)^2/2} \tag{2.24}$$

This equation expresses the circular convolution product of the sets $x(n)W^{n^2/2}$ and $W^{-n^2/2}$. It follows that the calculation of $X(k)$ can be performed in three stages comprising the following operations:

(1) Multiply the data $x(n)$ by the coefficients $W^{n^2/2}$
(2) Form the convolution product with the set of coefficients $W^{n^2/2}$
(3) Multiply the results by the coefficients $W^{k^2/2}$

This process is represented in Figure 2.10, and the method can be extended to the case where $W$ is a complex number with a non-unit modulus [10].

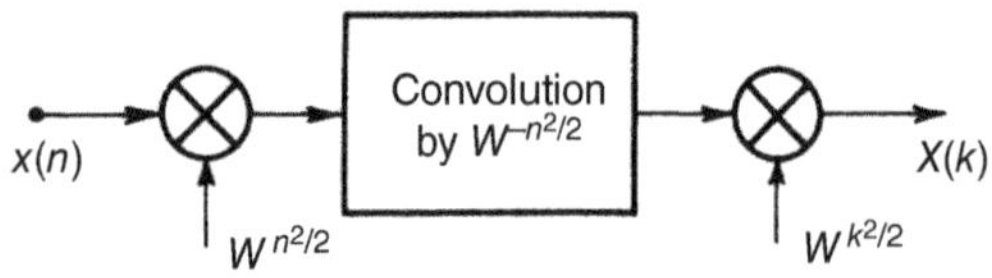


**Figure 2.10** Calculation of a discrete Fourier transform by convolution.

## 2.7 Implementation

In order to implement DFT algorithms, the following elements must be employed:

(1) A memory unit to store the input–output data and the intermediate results
(2) A memory unit to store the coefficients of the transform
(3) An arithmetic unit which can add and multiply complex numbers
(4) A control unit to link the various operations

These fundamental elements are found in every machine designed for digital signal processing, whether it uses hardwired or programmed logic. Implementation of the FFT has two main characteristics:

(1) A large number of arithmetic calculations have to be performed.
(2) Permutations are required on the data, resulting in complicated calculations for the indices.

The constraints increase with the order of the transform.

The problem of implementation is to find efficient procedures for the algorithms described in the above sections. Various circuits and methods of organizing the calculations can be devised. An important possibility is worth mentioning – a complex multiplication can be implemented with the help of 3 real multiplications and 5 additions according to the equation:

$$(a_R + ja_I)(b_R + jb_I) = [(a_R + a_I)b_R - a_I(b_R + b_I)] + j[(a_R + a_I)b_R + a_R(b_I - b_R)] \tag{2.25}$$

This option is frequently invoked in complexity estimations.

When a degree of optimization is being sought, particularly to adapt the circuits to the algorithms, it is preferable to look for algorithms which are compatible with the constraints imposed on the circuitry by technology. Moreover, in order to further reduce the number of arithmetic calculations, algorithms can be developed which are more sophisticated than the FFT. Also, significant simplifications can be made if the data display any particular feature (for example, if they are real numbers, or if symmetries appear).

These topics will be examined in the following chapter after a unified presentation of FFT algorithms.

## Exercises

**2.1** Calculate the DFT of the set comprising $N = 16$ terms such that:

$$x(0) = x(1) = x(2) = x(14) = x(15) = 1$$

$$x(n) = 0 \quad \text{for} \quad 3 \leqslant n \leqslant 13$$

and of the set

$$x(0) = x(1) = x(2) = x(3) = x(4) = 1$$

$$x(n) = 0 \quad \text{for} \quad 5 \leqslant n \leqslant 15$$

Compare the results obtained. Carry out the inverse transform of these results.

**2.2** Establish the diagram for the FFT algorithm of order 16 with time and frequency interleaving. What is the minimum number of multiplications and additions that are required?

**2.3** Calculate the DFT of the set comprising $N = 128$ terms such that:

$$x(0) = x(1) = x(2) = x(126) = x(127) = 1$$

$$x(n) = 0 \quad \text{for} \quad 3 \leqslant n \leqslant 125$$

Compare the results with those in Exercise 2.1.
The set $X(k)$ obtained forms an approximation to the Fourier series expansion of a set of impulses. Compare the results obtained with the figures in the table in Appendix 1, Chapter 1. Account for the differences.

**2.4** We wish to develop a DFT of order 64 with a minimum of arithmetic operations. Determine the number of multiplications and additions required with algorithms with radix 2, 4, and 8.

**2.5** Analyze the power of the rounding noise produced in a transform of order 32. Using the results in Section 2.3, show how the results vary at the different outputs. Calculate the distortion introduced by limiting the coefficients to 8 bits.

**2.6** Show that each output of a DFT, $X(k)$, can be obtained from the inputs $x(n)$ by a recurrence relation. Calculate the number of multiplications that would be required.

**2.7** Carry out a DFT of order 64 on data which are 16-bit numbers. Calculate the degradation of signal-to-noise ratio when a cascade of direct and inverse transforms is used on a 16-bit machine.

**2.8** Assume that the bandwidth occupied by a signal for analysis is from 0 to 10 kHz. The spectral resolution required is 1 Hz. What length of recording is required in order to carry out such an analysis? What memory capacity is required to store the data, assuming they are coded into 8 bits? Determine the characteristics of a computer capable of performing such a spectral analysis – memory capacity, memory cycle, addition, and multiplication times.

**2.9** Calculate the DFT of the set $x(n)$ which is defined by:

$$x(n) = \sin(2\pi n/3.5) + 0.2\sin(2\pi n/6.5) \quad \text{with} \quad 0 \leqslant n \leqslant 15$$

The following windows are used to improve the analysis:

$$g(n) = \frac{1}{2}[1 - \cos(2\pi n/16)]$$

$$g(n) = 0.54 - 0.46\cos\frac{2\pi n}{16} \quad \text{(Hamming)}$$

$$g(n) = 0.42 - 0.5\cos\frac{2\pi n}{16} + 0.08\cos\frac{4\pi n}{16} \quad \text{(Blackman)}$$

Compare the results.

## References


1 Special issue on FFT and applications. *IEEE Transactions*, 15(2), 1967, https://ieeexplore.ieee.org/xpl/tocresult.jsp?isnumber=26059&punumber=8337, 1–113.
2 A. Oppenheim and R. Schafer, *Digital Signal Processing*, Prentice Hall, Englewood Cliffs NJ, 1974, Chapters 3 and 6.
3 L. Rabiner and B. Gold, *Theory and Application of Digital Signal Processing*, Prentice Hall, Englewood Cliffs NJ, 1975, Chapters 6 and 10.
4 C. S. Burrus and T. W. Parks, *DFT/FFT and Convolution Algorithms*, Wiley, New York, 1985.
5 P. Duhamel and H. Hollmann, Split radix FFT algorithm. *Electronics Letters*, 1984, 20 (1), 14–16.
6 H. Sorensen, M. Heideman and S. Burrus. On computing the split-radix FFT, *IEEE Transactions*, 34(1), 1986, 152–156.
7 D. W. Tufts, H. S. Hersey and W. E. Mosier, Effects of FFT Coefficient quantization on bin frequency response. *Proceedings of IEEE*, 1972, 60, 1, pp. 146–147, 10.1109/PROC.1972.8582.
8 T. Thong and B. Liu, Fixed point FFT error analysis. *IEEE Transactions*, 24(6), 563–573, 1976.
9 A. Eberhard, An optimal discrete window for the calculation of power spectra. *IEEE Transactions*, 21(1), 37–43, 1973.
10 L. Rabiner, R. Schafer and C. Rader, The chirp z-transform algorithm. *IEEE Transactions on Audio and Electroacoustics*, 17, 1969. 17, 2, pp. 86–92, 10.1109/TAU.1969.1162034

# 3

# Other Fast Algorithms for the FFT

Algorithms for the fast calculation of a discrete Fourier transform (DFT) are based on factorization of the matrix of the transform. We have already seen such factorization in the sections on decimation-in-time and decimation-in-frequency algorithms, in the preceding chapter, which are particular examples of a large group of algorithms.

In order to use these fast algorithms and thus to exploit to the full both the characteristics of the signals to be processed and the various technological possibilities, one must use a suitable mathematical tool – the Kronecker product of matrices. By combining this product with the conventional product, it is possible to factorize the matrix of the DFT in a simple way.

## 3.1 Kronecker Product of Matrices

The Kronecker product is a tensor operation which is a generalization of the multiplication of a matrix by a scalar [1]. Knowing two matrices $A$ and $B$ with $m$ and $p$ rows and $n$ and $q$ columns respectively, the Kronecker product of $A$ by $B$ (written $A \times B$) is a new matrix with $mp$ rows and $nq$ columns, which is obtained by replacing each element $b_{ij}$ of the matrix $B$ by the following array $b_{ij}A$:

$$\begin{matrix} b_{ij}a_{11} & b_{ij}a_{12} & \cdots & b_{ij}a_{1n} \\ \vdots & & & \vdots \\ b_{ij}a_{m1} & b_{ij}a_{m2} & \cdots & b_{ij}a_{mn} \end{matrix}$$

This product is generally not commutative:

$$A \times B \neq B \times A$$

As an example of the product, if the matrix $B$ is

$$B = \begin{bmatrix} b_{11} & b_{12} \\ b_{21} & b_{22} \end{bmatrix}$$

the Kronecker product of the matrix $A$ by the matrix $B$ is

$$A \times B = \begin{bmatrix} b_{11}A & b_{12}A \\ b_{21}A & b_{22}A \end{bmatrix} \tag{3.1}$$

In particular, it should be noted that the Kronecker product of the unit matrix $I_N$ by the $M$-dimensional unit matrix $I_M$ is equal to the unit matrix of dimension $MN$:

$$I_N \times I_M = I_{NM} \tag{3.2}$$

*Digital Signal Processing: Theory and Practice*, Tenth Edition. Maurice Bellanger.

Similarly, the Kronecker product of a diagonal matrix by another diagonal matrix is once again a diagonal matrix.

The Kronecker product can be combined with conventional matrix products, and thus we have the following properties which will be used in the coming sections, provided that the dimensions are compatible:

(1) The Kronecker product of a product of matrices with the unit matrix is equal to the product of the Kronecker products of each matrix with the unit matrix:

$$(ABC) \times I = (A \times I)(B \times I)(C \times I) \tag{3.3}$$

(2) The product of Kronecker products is equal to the Kronecker product of the products:

$$(A \times B \times C)(D \times E \times F) = (AD) \times (BE) \times (CF) \tag{3.4}$$

(3) The inverse of a Kronecker product is equal to the Kronecker product of the inverses:

$$(A \times B \times C)^{-1} = A^{-1} \times B^{-1} \times C^{-1} \tag{3.5}$$

The property of transposition is similar to that of inversion:

$$(A \times B \times C)^t = A^t \times B^t \times C^t \tag{3.6}$$

The transpose of the matrix of a Kronecker product is the Kronecker product of the transposes of the matrices.

These properties can be easily demonstrated using some simple examples. They are used to factorize matrices with redundant elements and, in particular, for DFT matrices [2]. Decimation-in-frequency will be considered first.

It should be noted that the scale factor $1/N$ is ignored throughout the rest of this chapter.

## 3.2 Factorizing the Matrix of a Decimation-in-Frequency Algorithm

In the algorithms examined in the previous chapter, one of the sets – either the input or the output – was permuted. The matrix which represents this algorithm is derived from the matrix $T_N$ by permutation of the rows or the columns depending upon whether the decimation in frequency or decimation in time is considered [3].

Let $T'_N$ denote the matrix corresponding to decimation in frequency. This is obtained by permutation of the rows of $T_N$ as follows. The rows are numbered, and each number is expressed in binary notation; then the binary numbers are reversed, and the resulting number denotes the position of that row in the new matrix. For example, for $N = 8$ we obtain:

$$T_8 = \begin{bmatrix} 1 & 1 & 1 & 1 & 1 & 1 & 1 & 1 \\ 1 & W & W^2 & W^3 & -1 & -W & -W^2 & -W^3 \\ 1 & W^2 & -1 & -W^2 & 1 & W^2 & -1 & -W^2 \\ 1 & W^3 & -W^2 & W & -1 & -W^3 & W^2 & W \\ 1 & -1 & 1 & -1 & 1 & -1 & 1 & -1 \\ 1 & -W & W^2 & -W^3 & -1 & W & -W^2 & W^3 \\ 1 & -W^2 & -1 & W^2 & 1 & -W^2 & -1 & W^2 \\ 1 & -W^3 & -W^2 & -W & -1 & W^3 & W^2 & W \end{bmatrix} \begin{matrix} 0\,0\,0 = 0 \\ 0\,0\,1 = 1 \\ 0\,1\,0 = 2 \\ 0\,1\,1 = 3 \\ 1\,0\,0 = 4 \\ 1\,0\,1 = 5 \\ 1\,1\,0 = 6 \\ 1\,1\,1 = 7 \end{matrix}$$

$$T'_8 = \begin{bmatrix} 1 & 1 & 1 & 1 & 1 & 1 & 1 & 1 \\ 1 & -1 & 1 & -1 & 1 & -1 & 1 & -1 \\ 1 & W^2 & -1 & -W^2 & 1 & W^2 & -1 & -W^2 \\ 1 & -W^2 & -1 & W^2 & 1 & -W^2 & -1 & W^2 \\ 1 & W & W^2 & W^3 & -1 & -W & -W^2 & -W^3 \\ 1 & -W & W^2 & -W^3 & -1 & W & -W^2 & W^3 \\ 1 & W^3 & -W^2 & W & -1 & -W^3 & W^2 & -W \\ 1 & -W^3 & -W^2 & -W & -1 & W^3 & W^2 & W \end{bmatrix} \begin{matrix} 000 = 0 \\ 100 = 4 \\ 010 = 2 \\ 110 = 6 \\ 001 = 1 \\ 101 = 5 \\ 011 = 3 \\ 111 = 7 \end{matrix}$$

Note that for $N = 2$, the matrix $T'_2$ is equal to $T_2$.

The matrix $T'_2$ is factorized by finding the matrix $T'_{N/2}$ and the diagonal matrix $D_{N/2}$ whose elements are the numbers $W^k$ with $0 \leqslant k \leqslant N/2 - 1$. Thus,

$$T'_N = \begin{bmatrix} T'_{N/2} & T'_{N/2} \\ T'_{N/2}D_{N/2} & -T'_{N/2}D_{N/2} \end{bmatrix}$$

This decomposition appears clearly for $T'_8$. If $I_{N/2}$ denotes the unit matrix of order $N/2$, we can write:

$$T'_N = \begin{bmatrix} T'_{N/2} & 0 \\ 0 & T'_{N/2} \end{bmatrix} \begin{bmatrix} I_{N/2} & I_{N/2} \\ D_{N/2} & -D_{N/2} \end{bmatrix}$$

or

$$T'_N = \begin{bmatrix} T'_{N/2} & 0 \\ 0 & T'_{N/2} \end{bmatrix} \begin{bmatrix} I_{N/2} & 0 \\ 0 & D_{N/2} \end{bmatrix} \begin{bmatrix} I_{N/2} & I_{N/2} \\ I_{N/2} & -I_{N/2} \end{bmatrix}$$

By using the Kronecker products of the matrices, for $T'_N$ we obtain:

$$T'_N = \left(T'_{N/2} \times I_2\right) \Delta_N \left(I_{N/2} \times T'_2\right) \tag{3.7}$$

where $\Delta_N$ is a diagonal square matrix of order $N$, in which the first $N/2$ elements have the value 1 and the subsequent elements are powers of $W$, $W^k$ with $0 \leqslant k \leqslant N/2 - 1$.

The complete factorization is obtained by iteration:

$$T'_N = \left(T'_2 \times I_{N/2}\right) \left(\Delta_4 \times I_{N/4}\right) \left(I_2 \times T'_2 \times I_{N/4}\right)$$
$$\cdots\cdots\cdots\cdots\cdots\cdots\cdots\cdots\cdots\cdots\cdots\cdots\cdots\cdots\cdots\cdots$$
$$\left(\Delta_{N/2} \times I_2\right) \left(I_{N/4} \times T'_2 \times I_2\right)$$
$$\Delta_N \left(I_{N/2} \times T'_2\right)$$

or

$$T'_N = \prod_{i=1}^{\log_2 N} \left(\Delta_2 i \times I_{N/2^i}\right) \left(I_{2^{i-1}} \times T'_2 \times I_{N/2^i}\right) \tag{3.8}$$

This expression shows that the transform is calculated in $\log_2 (N)$ stages, each containing:

(1) One part involving the ordering of the data corresponding to the factor $\left(I_{2^{i-1}} \times I_{N/2^i}\right)$, which contains only additions and subtractions.

(2) One part which involves the multiplications by the coefficients represented in the matrix $\left(\Delta_{2^i} \times T'_2 \times I_{N/2^i}\right)$. The stage corresponding to $i = 1$ does not involve any multiplications. It can be verified that all the matrices indeed have the dimension $N$.

In order to see how factorization is generalized to radix 4, it is interesting to examine the matrix $T'_{16}$, which is obtained from $T_{16}$ by the following permutation of the rows. The rows are numbered to base 4 and the order of the digits in the row numbers are reversed. The value obtained shows the number of the row in the new matrix. Following this permutation, we obtain $T_4 = T'_4$.

If $D_4$ denotes the diagonal matrix

$$D_4 = \begin{bmatrix} 1 & 0 & 0 & 0 \\ 0 & W & 0 & 0 \\ 0 & 0 & W^2 & 0 \\ 0 & 0 & 0 & W^3 \end{bmatrix}$$

then the matrix of the transform of order 16 thus obtained is:

$$T'_{16} = \begin{bmatrix} T_4 & T_4 & T_4 & T_4 \\ T_4D_4 & T_4(-\mathrm{j})D_4 & T_4(-1)D_4 & T_4(+\mathrm{j})D_4 \\ T_4D_4^2 & T_4(-1)D_4^2 & T_4D_4^2 & T_4(-1)D_4^2 \\ T_4D_4^3 & T_4(+\mathrm{j})D_4^3 & T_4(-1)D_4^3 & T_4(-\mathrm{j})D_4^3 \end{bmatrix}$$

$$T'_{16} = \begin{bmatrix} T_4 & 0 & 0 & 0 & I_4 & 0 & 0 & 0 \\ 0 & T_4 & 0 & 0 & 0 & D_4 & 0 & 0 \\ 0 & 0 & T_4 & 0 & 0 & 0 & D_4^2 & 0 \\ 0 & 0 & 0 & T_4 & 0 & 0 & 0 & D_4^3 \end{bmatrix} \times \begin{bmatrix} I_4 & I_4 & I_4 & I_4 \\ I_4 & -\mathrm{j}I_4 & -I_4 & +\mathrm{j}I_4 \\ I_4 & -I_4 & I_4 & -I_4 \\ I_4 & +\mathrm{j}I_4 & -I_4 & -\mathrm{j}I_4 \end{bmatrix}$$

This expression is written in Kronecker product form as:

$$T'_{16} = (T_4 \times I_4)\Delta_{16}(I_4 \times T_4) \tag{3.9}$$

where $\Delta_{16}$ is a diagonal matrix in which the first four terms have the value 1, the next four terms $W^k$ with $0 \leqslant k \leqslant 3$, and the subsequent terms $(W^2)^k$ and $(W^3)^k$ with $0 \leqslant k \leqslant 3$.

Factorization as Kronecker products forms the basis of algorithms which have various properties – notably the order of presentation/extraction of data, and the linking of operations. It also applies to partial transforms, which are of great practical importance.

## 3.3 Partial Transforms

The transforms which have been studied in the above sections relate to sets of $N$ numbers which may be complex. In a fine spectrum analysis, it can happen that the order of the transform $N$ becomes very large, though we are interested in knowing only a reduced number of points in the spectrum. The limitation of the calculation to useful single points can then allow for significant savings.

Let us calculate the partial transform defined by the following equation, where $r$ is a factor of $N$:

$$\begin{bmatrix} X_p \\ X_{p+1} \\ X_{p+2} \\ \vdots \\ X_{p+r-1} \end{bmatrix} = \begin{bmatrix} 1 & W^P & W^{2P} & \dots & W^{(N-1)P} \\ 1 & W^{P+1} & W^{2(P+1)} & \dots & W^{(N-1)(P+1)} \\ 1 & W^{P+2} & W^{2(P+2)} & \dots & W^{(N-1)(P+2)} \\ \vdots & \vdots & & & \vdots \\ 1 & W^{(P+r-1)} & \dots & \dots & W^{(N-1)(P+r-1)} \end{bmatrix} \begin{bmatrix} x_0 \\ x_1 \\ x_2 \\ \vdots \\ x_{N-1} \end{bmatrix} \tag{3.10}$$

From the whole set of data, one can form $N/r$ subsets, each containing $r$ terms:

$$(x_0, x_{N/r}, x_{2N/r}, \quad , \ x_{(r-1)N/r})$$

$$(x_1, x_{(N/r)+1}, \quad , \ x_{(r-1)N/r+1})$$

$$\dots\dots\dots\dots\dots\dots\dots\dots\dots\dots\dots$$

$$(x_{(N/r)-1}, x_{(2N/r)-1}, \quad , \ x_{N-1})$$

Assume $D_r$ is the diagonal matrix of dimension $r$, whose elements are the powers of $W$, $W^k$ with $0 \leqslant k \leqslant r-1$.

The matrix of the partial transform can be separated into $N/r$ submatrices which can each be applied to one of the sets which were defined earlier, and the matrix equation of the transform is written as:

$$[X]_{p,r} = \sum_{i=0}^{(N/r)-1} D_r^i T_r (W^p)^i D_r^{(N/r)} [x]_{i,r} \tag{3.11}$$

where $[X]_{p,r}$ denotes the set of $r$ numbers $X_k$ with $p \leqslant k \leqslant p+r-1$ and $[x]_{i,r}$ the set of data $x_k$ with $k = nN/r + i$ and $n = 0, 1, \dots, r-1$. The transform $T_r$ is that of order $r$.

Consequently, if $r$ is a factor of $N$, a partial transform relating to $\boldsymbol{r}$ points is calculated using $N/r$ transforms of order $r$ with which the appropriate diagonal matrices are associated.

If $N$ and $r$ are powers of 2, the number $M_P$ of complex multiplications to be made is given by:

$$M_P = \frac{N}{r}\left(\frac{r}{2}\log_2\frac{r}{2} + 2r\right) = N\left[\frac{1}{2}\log_2\left(\frac{r}{2}\right) + 2\right] \tag{3.12}$$

This result is equally valid when it is the number of points to be transformed which is limited, as is often the case in spectrum analysis. A common example of a partial transform is that applied to real data.

### 3.3.1 Transform of Real Data and Odd DFT

If the data to be transformed are real, the properties listed in Chapter 2 show that the transformed numbers $X(k)$ and $X(N-k)$ are complex conjugates – that is, $X(k) = X(N-\text{k})$. Thus, it is only necessary to calculate the set of the $X_k$ with $0 \leqslant k \leqslant N/2-1$ and the above result can be applied:

$$[X]_{0,N/2} = T_{N/2}[x]_{0,N/2} + D_{N/2} T_{N/2} [x]_{1,N/2} \tag{3.13}$$

In this particular case, the transform $T_{N/2}$ has only to be calculated once, taking advantage of the following property of the DFT – if the set to be transformed $x_k$ is purely imaginary, and the transformed set is such that:

$$X(k) = -\overline{X}(N-k)$$

Under these conditions, the procedure for calculating the transform of a real set is as follows:

(1) Using the $x(k)$, form a complex set of $N/2$ terms $y(k) = x(2k) + jx(2k+1)$ with $0 \leqslant k \leqslant N/2-1$.
(2) Calculate the transform $Y(k)$ of the set $y(k)$ with $0 \leqslant k \leqslant N/2-1$.
(3) Calculate the required numbers using the expression

$$X(k) = \frac{1}{2}\left[Y(k) + \overline{Y}\left(\frac{N}{2}-k\right)\right] + \frac{1}{2}je^{-j2\pi(k/N)}\left[\overline{Y}\left(\frac{N}{2}-k\right) - Y(k)\right] \tag{3.14}$$

with $0 \leqslant k \leqslant N/2$ and $Y\left(\frac{N}{2}\right) = Y(0)$.

It is convenient to rewrite equation (3.14) as follows:

$$X(k) = A(k)Y(k) + B(k)\overline{Y}\left(\frac{N}{2} - k\right) \tag{3.15}$$

$$A(k) = \frac{1}{2}(1 - j\,W^k); \quad B(k) = \frac{1}{2}(1 + j\,W^k)$$

The inverse transform is obtained by, first, calculating:

$$Y(k) = \overline{A}(k)X(k) + \overline{B}(k)\overline{X}\left(\frac{N}{2} - k\right); \quad 0 \le k \le \frac{N}{2} - 1 \tag{3.16}$$

then, calculating the inverse DFT of size $N/2$ and, finally, taking the real parts as odd index data and the imaginary parts as even index data.

If $N$ is a power of 2, the number of complex multiplications $Mc$ to be performed is

$$Mc = \frac{N}{4}\log_2\left(\frac{N}{4}\right) + \frac{N}{2} = \frac{N}{4}\log_2 N \tag{3.17}$$

Memory locations are required for $N$ real numbers. An algorithm for real data is described in detail in Reference [4]. Another method of calculating the transforms of real numbers is to use odd transforms [5].

The odd DFT, by definition, establishes the following relations between two sets of $N$ complex numbers $x(n)$ and $X(k)$:

$$X(k) = \frac{1}{N}\sum_{n=0}^{N-1} x(n)e^{-j2\pi(2k+1)n/(2N)} \tag{3.18}$$

$$x(n) = \sum_{k=0}^{N-1} X(k)e^{j2\pi(2k+1)n/(2N)}$$

The coefficients of this transform have as their coordinates the points $M$ of a unit circle such that the vector $\overrightarrow{\boldsymbol{OM}}$ forms an angle with the abscissa which is an odd multiple of $2\pi/2N$, as shown in Figure 3.1.

By setting $W = e^{-j(\pi/N)}$, the matrix of this transform is written:

$$T_N^I = \begin{bmatrix} 1 & W & W^2 \ldots & W^{N-1} \\ 1 & W^3 & W^6 \ldots & W^{3(N-1)} \\ 1 & W^5 & W^{10} \ldots & W^{5(N-1)} \\ \vdots & \vdots & & \vdots \\ 1 & W^{(2N-1)} & \ldots & \ldots W^{(2N-1)(N-1)} \end{bmatrix}$$

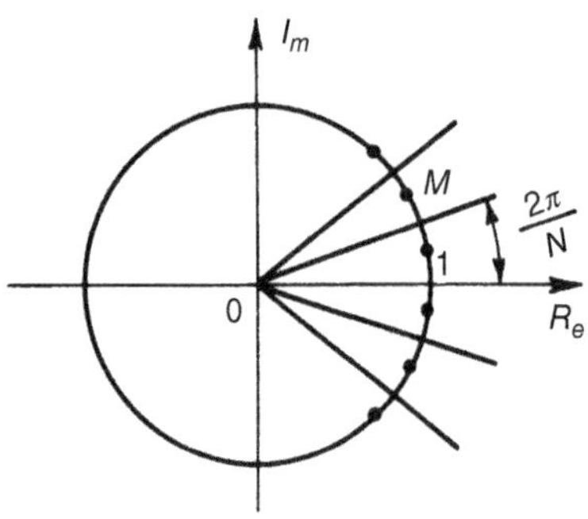


**Figure 3.1** Coefficients of the odd discrete Fourier transform.

If the $x(n)$ are real numbers, one can write:

$$X(N-1-k) = \frac{1}{N}\sum_{n=0}^{N-1} x(n)e^{-j2\pi[2(N-1-k)+1]/(2N)}$$
$$= \frac{1}{N}\sum_{n=0}^{N-1} x(n)e^{j2\pi(2k+1)n/(2N)} \tag{3.19}$$

Thus,

$$X(N-1-k) = \overline{X(k)} \tag{3.20}$$

Consequently, since the $X(k)$ with even and odd indices are complex conjugates, it is sufficient to calculate the $X(k)$ with an even index in order to perform a transform on real numbers. Such a transform is the matrix $T_R$ given by:

$$T_R = \begin{bmatrix} 1 & W & W^2 \dots & W^{N/2} & \dots & W^{N-1} \\ 1 & W^5 & W^{10} \dots & W^{5N/2} & \dots & W^{5(N-1)} \\ \vdots & \vdots & & \vdots & & \vdots \\ 1 W^{2N-3} & \dots & \dots & W^{(2N-3)N/2} & \dots & W^{(2N-3)(N-1)} \end{bmatrix}$$

Let $D_{N/2}$ be the diagonal matrix whose elements are $W^k$ with $0 \leqslant k \leqslant N/2-1$ and let $T_{N/2}$ be the matrix of the transform of order $N/2$. Allowing for the fact that $W^{2N} = 1$ and $W^{N/2} = -j$, this becomes:

$$T_R = [T_{N/2}D, -jT_{N/2}D] = (T_{N/2}D) \times [1, -j] \tag{3.21}$$

The odd transform of the real data is then calculated by carrying out a transform of order $N/2$ on the set of complex numbers:

$$y(n) = \left[x(n) - jx\left(\frac{N}{2} + n\right)\right] W^n \quad \text{with} \quad 0 \leqslant n \leqslant \frac{N}{2} - 1 \tag{3.22}$$

The number of calculations is the same as in the method illustrated at the beginning of this section, but the structure is simpler. It should be noted that the transformed numbers give a frequency sample of the signal spectrum represented by the $x(n)$, displaced by a half step on the frequency axis.

An important case where significant simplifications are introduced is that of real symmetrical sets. Reductions in the calculations are illustrated by using the doubly odd transform [6].

### 3.3.2 The Odd-time Odd-frequency DFT

The odd-time odd-frequency DFT, by definition, establishes the following relations between two sets of $N$ complex numbers $x(n)$ and $X(k)$:

$$X(k) = \frac{1}{N}\sum_{n=0}^{N-1} x(n)e^{-j2\pi(2k+1)(2n+1)/(4N)} \tag{3.23}$$

$$x(n) = \sum_{k=0}^{N-1} X(k)e^{j2\pi(2k+1)(2n+1)/(4N)} \tag{3.24}$$

The coefficients of this transform are based on the points $M$ of a unit circle such that the vector $\overrightarrow{OM}$ forms an angle with the abscissa which is an odd multiple of $2\pi/4N$ as shown in Figure 3.2.

If the $x(n)$ are real numbers, the following holds:

$$X(N-1-k) = -\overline{X}(k)$$

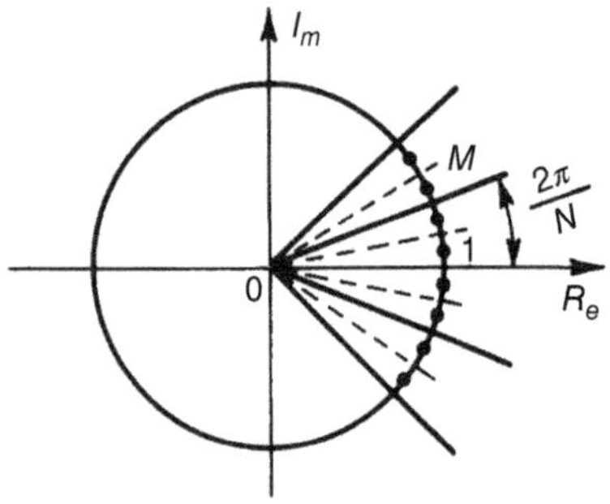


**Figure 3.2** Coefficients of the doubly odd discrete Fourier transform.

Similarly, if the $X(k)$ are real numbers, then:

$$x(N-1-n) = -\bar{x}(n)$$

By assuming, as before, that $W = e^{-j(\pi/N)}$, the matrix of the transform is written as:

$$T_N^{\mathrm{II}} = \begin{bmatrix} W^{1/2} & W^{3/2} & W^{5/2}\dots & W^{N-1/2} \\ W^{3/2} & W^{9/2} & W^{15/2}\dots & W^{3(N-1/2)} \\ W^{5/2} & W^{15/2} & W^{25/2}\dots & W^{5(N-1/2)} \\ \vdots & \vdots & \vdots & \vdots \\ W^{N-1/2} & W^{3(N-1/2)} & \dots & \dots W^{(2N-1)(N-1/2)} \end{bmatrix} \tag{3.25}$$

This transform is factorized as follows:

$$T_N^{\mathrm{II}} = W^{1/2} \begin{bmatrix} 1 & 0 & 0 & \dots & 0 \\ 0 & W & 0 & \dots & 0 \\ \vdots & \vdots & & & \vdots \\ \vdots & \vdots & & & \vdots \\ 0 & 0 & & \dots & W^{N-1} \end{bmatrix} \begin{bmatrix} 1 & 1 & 1 & \dots & 1 \\ 1 & W^2 & W^4 & \dots & W^{2(N-1)} \\ 1 & W^4 & W^8 & \dots & W^{4(N-1)} \\ \vdots & \vdots & & & \vdots \\ 1 & W^{2(N-1)} & & \dots & W^{2(N-1)(N-1)} \end{bmatrix} \times \begin{bmatrix} 1 & 0 & 0 & \dots & 0 \\ 1 & W & 0 & \dots & 0 \\ \vdots & \vdots & & & \vdots \\ 0 & 0 & & \dots & W^{N-1} \end{bmatrix}$$

That is,

$$T_N^{\mathrm{II}} = W^{1/2} D_N T_N D_N \tag{3.26}$$

Let us consider the case where the set of data $x(n)$ is real and antisymmetric – that is, $x(n) = -x(N-1-n)$. Then the same applies to the set $X(k)$. The set of the $x(n)$ for even $n$ is equal to the set of the $x(n)$ for odd $n$, except for the sign. The situation is the same for the set of $X(k)$.

In order to calculate the transform, it is sufficient in this case to carry out the calculations for the $x(2n)$ with $0 \leqslant n \leqslant N/2-1$, since the $X(k)$ are real numbers. Alternatively, it is sufficient to perform the calculations on the $X(2k)$ with $0 \leqslant k \leqslant N/2-1$.

The corresponding matrix $T_{\mathrm{RR}}$ is written as:

$$T_{\mathrm{RR}} = W^{1/2} \begin{bmatrix} 1 & 0 & 0 & \dots & & 0 \\ 0 & W^2 & 0 & \dots & & 0 \\ \vdots & \vdots & & & & \vdots \\ 0 & 0 & \dots & \dots & & W^{2(N/2-1)} \end{bmatrix} \begin{bmatrix} 1 & 1 & \dots & 1 & 1 \\ 1 & W^2 & \dots & W^{8[(N/2)-1]} & 0 \\ 1 & W^{16} & \dots & \vdots & \vdots \\ \vdots & \vdots & & \vdots & \vdots \\ 1 & W^{8[(N/2)-1]} & & W^{8[(N/2)-1][(N/2)-1]} & 0 \end{bmatrix} \begin{bmatrix} 0 & \dots & 0 \\ W^2 & \dots & 0 \\ \vdots & & \vdots \\ \vdots & & \vdots \\ 0 & & W^{2[(N/2)-1]} \end{bmatrix}$$

Allowing for $W^{2N} = 1$, this becomes:

$$T_{\mathrm{RR}} = W^{1/2} D_{N/2} \begin{bmatrix} T_{N/4} & T_{N/4} \\ T_{N/4} & T_{N/4} \end{bmatrix} D_{N/2} \tag{3.27}$$

**Table 3.1** Arithmetic complexities of the various fast fourier transforms.

| | Complex multiplications | Complex additions | Memory positions |
|---|---|---|---|
| Complex DFT | $\frac{N}{2}\log_2\left(\frac{N}{2}\right)$ | $N\log_2 N$ | $2N$ |
| Odd DFT - real data | $\frac{N}{4}\log_2(N)$ | $\frac{N}{2}\log_2\left(\frac{N}{2}\right)$ | $N$ |
| Doubly odd DFT - symmetrical real data | $\frac{N}{8}\log_2(2N)$ | $\frac{N}{4}\log_2\left(\frac{N}{4}\right)$ | $\frac{N}{2}$ |

and, as $W^{2N/4} = -j$, this calculation can be carried out with one performance of the operations represented by the matrix $T_{N/4}$, on the set of numbers $x(2n) - jx(2n + N/2)$ with $0 \leqslant n \leqslant N/4-1$. The $N/4$ numbers obtained are complex ones whose real parts form the set of the desired $X(2k)$ with $0 \leqslant k \leqslant N/4-1$. By carrying out the operation defined by $T_{RR}$ for the transformed numbers of rank $2k+N/2$ with $0 \leqslant k \leqslant N/4-1$, it can be verified that we have obtained the earlier numbers multiplied by $-$j. That is, the imaginary part of the numbers obtained previously furnishes the set of the $X(2k + N/2)$. It follows that, if the doubly odd transform is applied to a real and antisymmetric set of $N$ terms, or to a symmetric set which becomes antisymmetric through a suitable change of sign, it is reduced to the equation:

$$\left[X(2k) + \text{j}X\left(2k + \frac{N}{2}\right)\right] = W^{1/2} D_{N/4} T_{N/4} D_{N/4} \left[x(2n) - \text{j}x\left(2n + \frac{N}{2}\right)\right] \tag{3.28}$$

with $0 \leqslant k \leqslant N/4-1$, $0 \leqslant \text{n} \leqslant N/4-1$, and where $D_{N/4}$ is a diagonal matrix, whose elements are $W^{2i}$ with $0 \leqslant i \leqslant N/4-1$.

The number of complex multiplications $Mc$ which are necessary is:

$$Mc = \frac{N}{8}\log_2\left(\frac{N}{8}\right) + 2\frac{N}{4} = \frac{N}{8}\log_2(2n) \tag{3.29}$$

Comparisons using the different transforms are given in Table 3.1 to illustrate the amount of calculations for each type.

The importance of the odd transforms can be easily seen. It should, however, be noted that other algorithms allow greater reductions to be made for real data and for symmetric real data [7], but these are not as simple to use, especially for practical implementations.

One feature of the doubly odd transform, when applied to a real antisymmetric set, is that it is identical to the inverse transform. Apart from the scale factor $1/N$, there is no distinction, in this case, between the direct and inverse transforms.

The Fourier transform of a real symmetric set is introduced, for example, when deriving the power spectrum density of a signal from its autocorrelation function.

### 3.3.3 Sine and Cosine Transforms

The transforms considered so far have complex coefficients. Discrete transforms of the same family can be obtained using the real and imaginary parts of the complex coefficients.

The following transforms can be defined:

(1) The cosine DFT (cos-DFT):

$$X_{\text{CF}}(k) = \frac{1}{N}\sum_{n=0}^{N-1} x(n)\cos\frac{2\pi nk}{N} \tag{3.30}$$

(2) The sine DFT (sin-DFT):

$$X_{SF}(k) = \frac{1}{N}\sum_{n=0}^{N-1} x(n) \sin \frac{2\pi nk}{N} \tag{3.31}$$

(3) The discrete cosine transform (DCT):

$$X_{DC}(0) = \frac{\sqrt{2}}{N}\sum_{n=0}^{N-1} x(n)$$

$$X_{DC}(k) = \frac{2}{N}\sum_{n=0}^{N-1} x(n) \cos\left(\frac{2\pi(2n+1)k}{4N}\right) \tag{3.32}$$

The inverse transform is given by:

$$x(n) = \frac{1}{\sqrt{2}} X_{DC}(0) + \sum_{n=1}^{N-1} X_{DC}(k) \cos \frac{2\pi(2n+1)k}{4N}$$

(4) The discrete sine transform (DST):

$$X_{DS}(k) = \sqrt{\left(\frac{2}{N+1}\right)}\sum_{n=0}^{N-1} x(n) \sin\left[\frac{2\pi(n+1)(k+1)}{2N+2}\right] \tag{3.33}$$

Through algebraic manipulations, as in the previous sections, it is possible to establish relationships between the standard DFT and these transforms, and also among these transforms themselves.

For example, from the definitions, we have:

$$DFT(N) = \cos\text{-}DFT(N) - j \sin\text{-}DFT(N)$$

Now, considering the cosine DFT:

$$X_{CF}(k) = \sum_{n=0}^{N/2-1} x(2n) \cos \frac{2\pi nk}{N/2} + \sum_{n=0}^{N/4-1} [x(2n+1) + x(N-2n-1)] \cos\left[\frac{2\pi(2n+1)k}{4N/4}\right]$$

it is clear that the cosine DFT of order $N$ can be completed with the help of a cosine DFT of order $N/2$ and a DCT of order $N/4$. In concise form:

$$\cos DFT(N) = \cos DFT(N/2) + DCT(N/4)$$

Similarly, the DCT is expressed by:

$$X_{DC}(k) = \frac{2}{N}\sum_{n=0}^{N/2-1}\left[x(2n) \cos \frac{2\pi(4n+1)k}{4N} + x(2n+1) \cos \frac{2\pi[4(N-n-1)+1]k}{4N}\right]$$

and now, taking

$$y(n) = x(2n); \quad 0 \leqslant n \leqslant \frac{N}{2} - 1$$

$$y(N-n-1) = x(2n+1)$$

we have:

$$X_{DC}(k) = \frac{2}{N}\sum_{n=0}^{N-1} y(n) \cos \frac{2\pi(4n+1)k}{4N}$$

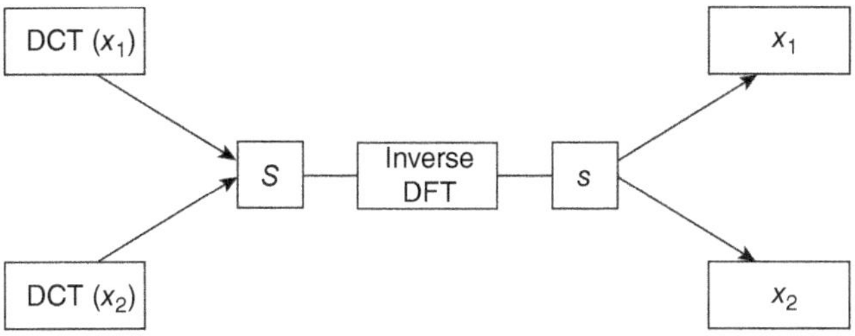


**Figure 3.3** Computing 2 inverse DCTs with the help of a single same-size inverse DFT.

Expanding the cosine function yields:

$$DCT(N) = \cos\frac{2\pi k}{4N}\cos DFT(N) - \sin\frac{2\pi k}{4N}\sin -DFT(N)$$

In concise form

$$DCT(x) = 2\,c(k)\,\mathrm{Re}\left\{DFT(y)\,e^{-j\frac{\pi k}{2N}}\right\} \tag{3.34}$$

$$c(0) = \frac{1}{\sqrt{2}};\quad c(k) = 1;\quad 1 \le k \le N-1$$

Finally, the DCT of order $N$ can be completed with the help of a DFT of the same order.

Taking into account the fact that only the real part is exploited in the above equation, it is even possible to compute two DCTs, using the imaginary part. Similarly, it is possible to compute two inverse DCTs with a single inverse DFT, as illustrated in Figure 3.3.

At the input of the inverse DFT, the relations between variables are the following [8]:

$$S_0 = \frac{C_0(x_1) + jC_0(x_2)}{\sqrt{2}};\qquad S_{N/2} = \frac{C_{N/2}(x_1) + jC_{N/2}(x_2)}{\sqrt{2}}$$

$$\begin{aligned}2S_k = &\left\{[C_k(x_1) + C_{N-k}(x_2)]\cos\left(\frac{\pi k}{2N}\right) + [C_{N-k}(x_1) - C_{N-k}(x_2)]\sin\left(\frac{\pi k}{2N}\right)]\right\}\\ &+j\left\{[C_k(x_1) + C_{N-k}(x_2)]\sin\left(\frac{\pi k}{2N}\right) + [C_k(x_2) - C_{N-k}(x_1)]\cos\left(\frac{\pi k}{2N}\right)]\right\}\end{aligned}$$

$$1 \le k \le N-1;\qquad k \ne N/2$$

Similarly, at the output of the inverse DFT:

$$x_1(2p) = \mathrm{Re}\{s(p)\};\quad x_1(2p+1) = \mathrm{Re}\{s(N-p-1)\}$$

$$x_2(2p) = \mathrm{Im}\{s(p)\};\quad x_2(2p+1) = \mathrm{Im}\{s(N-p-1)\}$$

The scheme yields computation gains in image compression, for example.

Among the transforms based on real coefficients, the discrete Hartley transform (DHT) is worth mentioning. It is defined by [9]:

$$X_{\mathrm{DH}}(k) = \frac{1}{N}\sum_{n=0}^{N-1} x(n)\left[\cos 2\pi\frac{nk}{N} + \sin 2\pi\frac{nk}{N}\right] \tag{3.35}$$

and the inverse transform:

$$x(n) = \sum_{k=0}^{N-1} X_{\mathrm{DH}}(k)\left[\cos 2\pi\frac{nk}{N} + \sin 2\pi\frac{nk}{N}\right]$$

The connection with the DFT is given by:

$$X(k) = \frac{1}{2}[X_{DH}(k) + X_{DH}(N-1-k) - j(X_{DH}(k) - X_{DH}(N-1-k))] \tag{3.36}$$

The discrete sine and cosine transforms have been introduced for information compression, particularly in image processing. It is worth pointing out that, for images, they provide reasonably accurate approximations of the Eigen transform, which yields a signal representation with the minimum number of parameters.

This compression power stems from the cancellation of discontinuities at the edge of the DFT, as pointed out in Section 2.1, because the DFT is performed on a symmetric sequence. However, this implies that the signal sample with index zero is zero, which is obtained by using the sequence $u(n)$, defined as follows:

$$u(2p) = 0; \quad 0 \le p \le 2N-1$$

$$u(2p+1) = u(4N-2p-1) = x(p); \quad 0 \le p \le N-1$$

The $4N$ size DFT of the sequence $u(n)$ leads, after simplifications, to expression (3.32).

### 3.3.4 The Two-dimensional DCT

For a set of $(N \times N)$ real data, the two-dimensional DCT (2D-DCT) is defined by:

$$X(k_1, k_2) = \frac{4e(k_1)e(k_2)}{N^2} \sum_{n_1=0}^{N-1} \sum_{n_2=0}^{N-1} x(n_1, n_2) \cos \frac{2\pi(2n_1+1)k_1}{4N} \cos \frac{2\pi(2n_2+1)k_2}{4N} \tag{3.37}$$

and:

$$x(n_1, n_2) = \sum_{k_1=0}^{N-1} \sum_{k_2=0}^{N-1} e(k_1)e(k_2)X(k_1, k_2) \cos \frac{2\pi(2n_1+1)k_1}{4N} \cos \frac{2\pi(2n_2+1)k_2}{4N}$$

with:

$$e(k) = \frac{1}{\sqrt{2}}; \quad k = 0$$

$$e(k) = 1; \quad k \ne 0$$

That transform is separable, and it can be computed as follows:

$$X(k_1, k_2) = \frac{2}{N} e(k_2) \sum_{n_2=0}^{N-1} \cos \frac{2\pi(2n_2+1)k_2}{4N} \times \left[ \frac{2}{N} e(k_1) \sum_{n_1=0}^{N-1} x(n_1, n_2) \cos \frac{2\pi(2n_1+1)k_1}{4N} \right]$$

Thus, the 2D-transform can be computed using $2N$ times the 1D-DCT, and the number of real multiplications is of the order of $N^2 \log_2 (N)$. In fact, that amount can be reached with the help of an algorithm based on the decomposition of a DCT of order $N$ into two DCTs of order $N/2$ [10]. It is even possible to reach the value $3/4\, N^2 \log_2 N$, through extension of the decimation technique and splitting the set of $(N \times N)$ data into subsets of $(N/2 \times N/2)$ data [11].

## 3.4 Lapped Transform

The filtering function of the DFT is improved when the transform is performed on overlapping blocks of data [12].

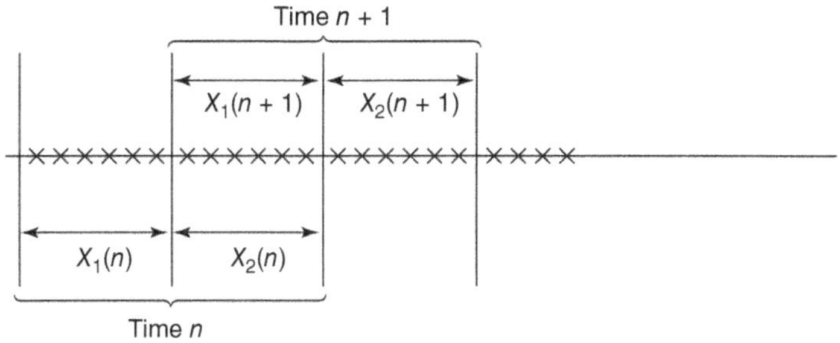


**Figure 3.4** Overlapping data blocks.

Let us consider a data sequence of length $2M$ and a transform whose matrix is $M \times 2M$. At time $n$, the block of processed data can be represented by two $M$-element vectors denoted $\boldsymbol{X}_1(\boldsymbol{n})$ and $\boldsymbol{X}_2(\boldsymbol{n})$, as shown in Figure 3.4.

At time $n+1$, a block of data includes half the data of the previous block – that is, $\boldsymbol{X}_1(\boldsymbol{n}+\mathbf{1})=\boldsymbol{X}_2(\boldsymbol{n})$. The lapped transform allows the two vectors to be recovered. The block of data is multiplied by the matrix $[A\ B]$, $A$ and $B$ being $M \times M$ matrices:

$$U_1 = [A, B]\begin{bmatrix} X_1(n) \\ X_2(n) \end{bmatrix}; \quad U_2 = [A, B]\begin{bmatrix} X_1(n+1) \\ X_2(n+1) \end{bmatrix} \tag{3.38}$$

Then, the results, $U_1$ at time $n$ and $U_2$ at time $n+1$, are multiplied by the transpose matrix.

$$\begin{bmatrix} \boldsymbol{Y}_1(\boldsymbol{n}) \\ \boldsymbol{Y}_2(\boldsymbol{n}) \end{bmatrix} = \begin{bmatrix} \boldsymbol{A}^t \\ \boldsymbol{B}^t \end{bmatrix} \boldsymbol{U}_1; \quad \begin{bmatrix} \boldsymbol{Y}_1(\boldsymbol{n}+\mathbf{1}) \\ \boldsymbol{Y}_2(\boldsymbol{n}+\mathbf{1}) \end{bmatrix} = \begin{bmatrix} \boldsymbol{A}^t \\ \boldsymbol{B}^t \end{bmatrix} \boldsymbol{U}_2 \tag{3.39}$$

which yields:

$$[\boldsymbol{Y}_2(\boldsymbol{n}) + \boldsymbol{Y}_1(\boldsymbol{n}+\mathbf{1})] = [\boldsymbol{B}^t\boldsymbol{A}\boldsymbol{X}_1(\boldsymbol{n}) + \boldsymbol{B}^t\boldsymbol{B}\boldsymbol{X}_2(\boldsymbol{n}) + \boldsymbol{A}^t\boldsymbol{A}\boldsymbol{X}_1(\boldsymbol{n}+\mathbf{1}) + \boldsymbol{A}^t\boldsymbol{B}\boldsymbol{X}_2(\boldsymbol{n}+\mathbf{1})] \tag{3.40}$$

At time $n+1$, the input data are restored by addition: $\boldsymbol{X}_2(\boldsymbol{n}) = \frac{1}{2}[\boldsymbol{Y}_2(\boldsymbol{n}) + \boldsymbol{Y}_1(\boldsymbol{n}+\mathbf{1})]$, provided the following conditions are satisfied:

$$\boldsymbol{B}^t\boldsymbol{A} = \boldsymbol{A}^t\boldsymbol{B} = \mathbf{0}; \quad \frac{1}{2}[\boldsymbol{B}^t\boldsymbol{B} + \boldsymbol{A}^t\boldsymbol{A}] = \boldsymbol{I}_M \tag{3.41}$$

For example, these conditions are met if the elements of the matrix $[A\ B]$ are:

$$\boldsymbol{a}_{nk} = \boldsymbol{h}(\boldsymbol{n})\sqrt{\frac{2}{M}}\cos\left[\left(\boldsymbol{n} + \frac{\boldsymbol{M}+\mathbf{1}}{2}\right)\left(\boldsymbol{k} + \frac{1}{2}\right)\frac{\pi}{\boldsymbol{M}}\right]$$

$$0 \le \boldsymbol{n} \le 2\boldsymbol{M} - 1; \quad 0 \le \boldsymbol{k} \le \boldsymbol{M} - 1; \quad \boldsymbol{h}(\boldsymbol{n}) = -\sin\left(\boldsymbol{n} + \frac{1}{2}\right)\frac{\pi}{2\boldsymbol{M}}$$

In fact, a bank of $M$ orthogonal filters has been obtained and the terms $h(n)$ are the coefficients of the prototype filter, which, in the present case, is a low-pass filter whose frequency response, given in Section 5.8, is more selective than that of the DFT filter.

In image compression, lapped transforms introduce smoothing and attenuate the so-called blocking effects.

## 3.5 Other Fast Algorithms

Fast Fourier transform algorithms form a technique for calculating a DFT of order $N$ using a number of multiplications of the order of $N \log_2 N$. It has been shown in the previous sections that these algorithms have a relatively simple structure and offer sufficient flexibility for good adaptation to the operating constraints and to technological characteristics to be achieved. Hence, they are of great interest for practical applications.

Nevertheless, they are not the only method of fast calculation of a DFT, and algorithms can be elaborated which involve, at least in certain cases, a shorter calculation time or a lower number of multiplications, or which are applicable for values of the order $N$ which are not necessarily powers of two.

A first approach consists of replacing the complex multiplications, which are costly in terms of circuitry or time, with a set of operations which are simpler to put into operation. Reference [13] describes a technique which uses one property of the DFT mentioned in Section 2.4.1 – namely the fact that multiplications by the coefficients $W^k$ correspond to phase rotations.

The technique known as CORDIC (coordinate rotation digital computer) enables these rotations to be realized by linking simple operations – to rotate a vector $(x,y)$ through an angle $\theta$ with an accuracy of $0/2^n$, a sequence of $n$ elementary rotations of angle $d\theta_i$ is performed such that $\tan d\theta_i = 2^{-i}$ with $0 \le i \le n-1$ and $-\pi/2 \le 0 \le \pi/2$. The coordinates $x_i$ and $y_i$ of the vector at iteration $i$ yield the coordinates at iteration $i+1$ by using the expressions:

$$\begin{aligned} x_{i+1} &= x_i + \text{sign}[\theta_i] \cdot y_i 2^{-i} \\ y_i + 1 &= y_i - \text{sign}[\theta_i] \cdot x_i 2^{-i} \\ \theta_i + 1 &= \theta_i - \text{sign}[\theta_i] \cdot d\theta_i \end{aligned} \tag{3.42}$$

The function sign $[\theta_i]$ is the sign of $\theta$, and $\theta_0 = -\theta$. These operations consist only of additions and shifts and can therefore be more advantageous than complex multiplication of the same precision.

One method which is particularly interesting is used to obtain a multiplication volume of order $N$, instead of $N \log_2 N$, for a DFT of order $N$. This method depends on factorizing the matrix $T_N$ in a particular way. It is decomposed into a product of three factors:

$$T_N = B_N C_N A_N$$

where $A_N$ is a $J \times N$ matrix, $J$ is a whole number, $C_N$ is a diagonal matrix of dimension $J$, and $B_N$ is an $N \times J$ matrix. The special feature of this factorization is that the elements of the matrices $A_N$ and $B_N$ are 0, 1, or – 1. Under these conditions, the calculation requires only $J$ multiplications.

For example, the complex multiplication takes the form:

$$\begin{bmatrix} a & -b \\ b & a \end{bmatrix} = \begin{bmatrix} 1 & -1 & 0 \\ 1 & 0 & 1 \end{bmatrix} \begin{bmatrix} a & 0 & 0 \\ 0 & a+b & 0 \\ 0 & 0 & a-b \end{bmatrix} \begin{bmatrix} 1 & 1 \\ 0 & 1 \\ 1 & 0 \end{bmatrix}$$

which shows that 3 real multiplications are sufficient, as shown in Section 2.7.

This decomposition is obvious for $J = N^2$; for example for $N = 3$, we obtain:

$$T_3 = \begin{bmatrix} 1&1&1&0&0&0&0&0&0 \\ 0&0&0&1&1&1&0&0&0 \\ 0&0&0&0&0&0&1&1&1 \end{bmatrix} \times \begin{bmatrix} 1 &&&&&&&& \\ & 1 &&&&& \text{O} && \\ && 1 &&&&&& \\ &&& 1 &&&&& \\ &&&& W &&&& \\ &&&&& W^2 &&& \\ &&&&&& 1 && \\ && \text{O} &&&&& W^2 & \\ &&&&&&&& W \end{bmatrix} \begin{bmatrix} 1&0&0 \\ 0&1&0 \\ 0&0&1 \\ 1&0&0 \\ 0&1&0 \\ 0&0&1 \\ 1&0&0 \\ 0&1&0 \\ 0&0&1 \end{bmatrix}$$

With some low values of $N$, certain factorizations are available in which J is of the order of $N$, in which case there are the same number of multiplications. In order to generalize this property and to illustrate a suitable factorization of $T_N$, it is necessary to perform a permutation of the data before and after transformation. For example, for $N = 12$, by assuming:

$$X' = \begin{bmatrix} X_0 \\ X_3 \\ X_6 \\ X_9 \\ X_4 \\ X_7 \\ X_{10} \\ X_1 \\ X_8 \\ X_{11} \\ X_2 \\ X_5 \end{bmatrix} \quad \text{and} \quad x' = \begin{bmatrix} X_0 \\ X_9 \\ X_6 \\ X_3 \\ X_4 \\ X_1 \\ X_{10} \\ X_7 \\ X_8 \\ X_5 \\ X_2 \\ X_{11} \end{bmatrix}$$

and by using the Kronecker products of the matrices, it can be shown that:

$$X' = (T_3 \times T_4)x'$$

Similarly, if $N$ has $L$ factors such that:

$$N = N_L N_{L-1} \cdots N_1$$

it can be shown that:

$$X' = (T_{N_L} \times T_{N_{L-1}} \times \cdots T_{N_1})x' \tag{3.43}$$

By using the factorization defined earlier for the matrices $T_{N_L}$ and the algebraic properties of the Kronecker products, this becomes:

$$X' = (B_{N_L} \times B_{N_{L-1}} \times \cdots \times B_{N_1}(C_{N_L} \times C_{N_{L-1}} \times \cdots \times C_{N_1}) \times (A_{N_L} \times A_{N_{L-1}} \times \cdots \times A_{N_1})x'$$

This result defines a type of algorithm called the Winograd algorithm.

It can be clearly seen that the algorithm of order $N$ is deduced from algorithms of order $N_i$ with $1 \leqslant i \leqslant L$. Herein lies the importance of algorithms with small numbers of multiplications for small values of $N$. Reference [14] gives algorithms for $N = 2, 3, 4, 5, 7, 8, 9$, and 16, where the number of multiplications is of the order of $N$, as shown in Table 3.2. In the multiplication column, the figures in parentheses give the number of multiplications by coefficients different from 1. Further, these are complex multiplications which correspond to two real multiplications. The number of additions is comparable to that of FFT algorithms.

The algorithms for low values of $N$ are obtained by calculating the Fourier transform as a set of correlations:

$$X_k = \sum_{n=1}^{N-1} (x_n - x_0) W^{nk}; \quad k = 1, \ldots, N-1$$

and by using the algebraic properties of the set of exponents of $W$ which are defined modulo $N$.

For example, for $N = 4$, the sequence of operations is as follows:

$$t_1 = x_0 + x_2, \qquad t_2 = x_1 + x_3$$
$$m_0 = 1(t_1 + t_2), \quad m_1 = 1(t_1 - t_2)$$
$$m_2 = 1(x_0 - x_2), \quad m_3 = j(x_1 - x_3)$$

**Table 3.2** Arithmetic complexities of low-order Winograd algorithms.

| Order of the DFT | Multiplications | Additions |
|---|---|---|
| 2 | 2 (0) | 2 |
| 3 | 3 (2) | 6 |
| 4 | 4 (0) | 8 |
| 5 | 6 (5) | 17 |
| 7 | 9 (8) | 36 |
| 8 | 8 (2) | 26 |
| 9 | 11 (10) | 44 |
| 16 | 18 (10) | 74 |

$$X_0 = m_0$$
$$X_1 = m_2 + m_3$$
$$X_2 = m_1$$
$$X_3 = m_2 - m_3$$

For $N = 8$:

$$t_1 = x_0 + x_4, \quad t_2 = x_2 + x_6, \quad t_3 = x_1 + x_5,$$
$$t_4 = x_1 - x_5, \quad t_5 = x_3 + x_7, \quad t_6 = x_3 - x_7,$$
$$t_7 = t_1 + t_2, \quad t_8 = t_3 + t_5,$$

$$m_0 = 1(t_7 + t_8), \quad m_1 = 1(t_7 - t_8),$$
$$m_2 = 1(t_1 - t_2), \quad m_3 = 1(x_0 - x_4),$$
$$m_4 = \cos\left(\frac{\pi}{4}\right)(t_4 - t_6), \quad m_5 = \mathrm{j}(t_3 - t_5),$$
$$m_6 = \mathrm{j}(x_2 - x_6), \quad m_7 = \mathrm{j}\sin\left(\frac{\pi}{4}\right)(t_4 + t_6),$$

$$s_1 = m_3 + m_4, \quad s_2 = m_3 + m_4, \quad s_3 = m_6 + m_7, \quad s_4 = m_6 + m_7$$
$$X_0 = m_0, \quad X_1 = s_1 + s_3, \quad X_2 = m_2 + m_5, \quad X_3 = s_2 - s_4,$$
$$X_4 = m_1, \quad X_5 = s_5 + s_4, \quad X_6 = m_2 - m_5, \quad X_7 = s_1 - s_3.$$

Finally, Winograd algorithms generally introduce a reduction in the amount of computation. This reduction can be quite significant when compared to FFT algorithms. The situation is similar for other algorithms, such as those using the polynomial transforms [15].

These techniques have a broad range of potential applications and are of considerable importance in certain cases. However, it should be noted that they may require a larger memory capacity and a more complicated sequence of operations, which results in an increase in the size of the system's control unit, or in the volume of the program memory.

Another attractive path toward the optimization of processing and machines consists of avoiding multiplications, as with the Hadamard transform and number theoretic transforms.

## 3.6 Binary Fourier Transform - Hadamard

The DFT and its variants require a computation load that can be seen as unaffordable in certain applications, such as fast real-time processing of large images. This being the case, transforms with similar properties but without multiplications can be used [16].

The order $N = 2^M$ Hadamard transform is defined by the matrix $H_N$ derived from the size-2 DFT, $T_2$, through Kronecker products:

$$H_2 = T_2 = \begin{bmatrix} 1 & 1 \\ 1 & -1 \end{bmatrix};\; H_4 = T_2 \times T_2 = \begin{bmatrix} 1 & 1 & 1 & 1 \\ 1 & -1 & 1 & -1 \\ 1 & 1 & -1 & -1 \\ 1 & -1 & -1 & 1 \end{bmatrix};\; H_N = T_2 \times H_{N/2} \tag{3.44}$$

It is real, symmetric, and orthogonal:

$$\frac{1}{N} H_N H_N = I_N$$

The algorithm structure is the same as that of the FFT, but there are no multiplications in the butterflies. Regarding filtering, the filter bank is much less selective because of the presence of harmonics of basic frequencies. As an illustration, Figure 3.5 shows the Fourier transform of row 31 in matrix $H_{64}$.

Hadamard matrices have applications in certain types of wireless communication systems that resort to spread spectrum to cope with adverse environments.

## 3.7 Number-Theoretic Transforms

Fourier transformation involves performing arithmetic operations on the field of complex numbers. The machines which carry out these operations generally use binary representations which are approximations of the data and the coefficients. The precision of the calculations is a function of the number of bits available in the machine.

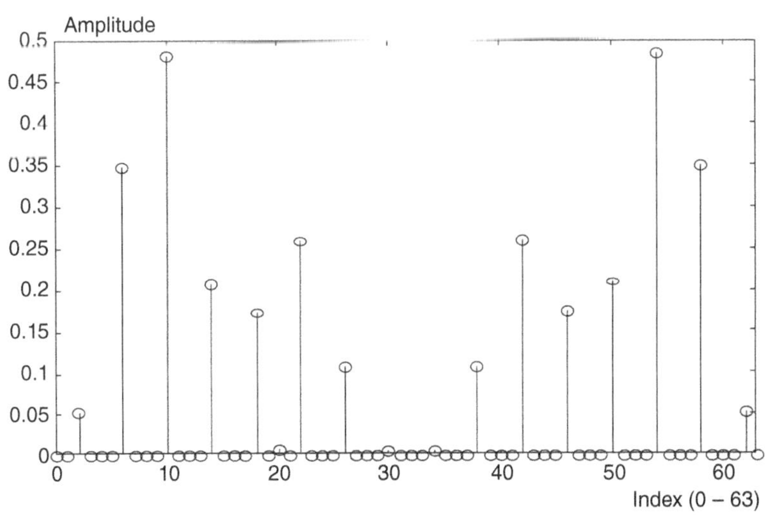


**Figure 3.5** Spectrum of the Hadamard code 31/64.

In practice, a machine with $B$ bits carries out its operations on the set with $2^B$ integers: 0, 1, …, $2^B - 1$. In this set, the usual laws of addition and multiplication cannot be applied, as shifting and truncation must be introduced, which lead to approximations in the calculations, as was shown in Section 2.3.

The first condition to be fulfilled to ensure exact calculations in a set $E$ is that the product or the sum of two elements of the set $E$ must also belong to this set. This condition is satisfied in the set of integers 0, 1, …, $M - 1$ if the calculations are made modulo $M$. By appropriate selection of the modulus $M$, it is possible to define transformations with properties which are comparable to those of the DFT and which allow error-free calculation of convolutions with fast calculation algorithms.

The definition of such transformation rests on the algebraic properties of integers modulo $M$, for certain values of $M$. They are called number-theoretic transforms.

The choice of the modulus $M$ is governed by the following considerations:

(1) Simplicity of the calculations in the modular arithmetic. In principle, modular arithmetic implies a division by the modulus $M$. This division is trivial for $M = 2^m$. It is very simple for $M = 2^m \pm 1$, because the result is obtained by adding a carry-bit (1's complement arithmetic) or subtracting it.
(2) The modulus must be sufficiently large. The result of the convolution must be capable of representation without ambiguity in modulo $M$ arithmetic. For example, a convolution with 32 terms with 12-bit data and 8-bit coefficients requires $M > 2^{25}$.
(3) Suitable algebraic properties. The set of modulo $M$ integers should have algebraic properties allowing the definition of transformations comparable to the DFT.

First, there should be periodic elements in order that the fast algorithms can be elaborated; the set must have an element $\alpha$ such that:

$$\alpha^N = 1$$

A transformation can then be defined by the expression:

$$X(k) = \sum_{n=0}^{N-1} x(n)\alpha^{nk} \tag{3.45}$$

For the existence of the inverse transformation which is defined by the expression:

$$x(n) = N^{-1}\sum_{k=0}^{N-1} X(k)\alpha^{-nk} \tag{3.46}$$

it is first necessary for $N$ and the powers of $\alpha$ to have inverses.

It can be shown that $N$ has an inverse modulo $M$ if $N$ and $M$ are prime relative to each other. The element $\alpha$ should be prime relative to the $M$ and of order $N$ – that is, $\alpha^N = 1$.

The existence of the inverse transformation implies a further condition for the $\alpha^i$, that:

$$\sum_{k=0}^{N-1} \alpha^{ik} = N\delta(i) \quad \text{with} \quad \begin{array}{l} \delta(i) = 1 \quad \text{if} \quad i = 0 \text{ modulo } N \\ \delta(i) = 0 \quad \text{if} \quad i \neq 0 \text{ modulo } N \end{array}$$

This condition reflects the fact that each element $(1 - \alpha^i)$ must have an inverse. It can be shown that all of the conditions for the existence of a transformation and its inverse reduce to the following one. For each prime factor $P$ of $M$, $N$ must be a factor of $P - 1$. Thus, if $M$ is prime, $N$ must divide into $M - 1$.

Fast algorithms can be elaborated if $N$ is a composite number – in particular, if $N$ is a power of 2. These algorithms are similar to those of the FFT:

$$N = 2^m$$

The calculations to be performed in the transformation are considerably simplified in the particular case when $\alpha = 2$.

Finally, an interesting choice for the modulus $M$ is:

$$M = 2^{2m} + 1$$

where $M$ is prime. These are the Fermat numbers.

A transform based on the Fermat numbers is defined as follows:

(1) Modulus $M = 2^{2^m} + 1$
(2) Order of the transform: $N = 2^{m+1}$
(3) Direct transform:

$$X(k) = \sum_{n=0}^{N-1} x(n)2^{nk}$$

(4) Inverse transform:

$$x(n) = (2^t)\sum_{k=0}^{N-1} X(k)2^{-nk} \quad \text{with} \quad t = 2^{m+1} - m - 1$$

*Example:* $m = 3$, $2^m = 8$, $M = 257$, $N = 16$, $t = 12$.

This transform allows the calculation of convolutions of real numbers, as with the DFT, but with the following advantages:

(a) The result is obtained without approximation
(b) The operations relate to real numbers
(c) The calculation of the transform and its inverse does not require any multiplication. The only multiplications which remain are in the transformed space.

This technique does, nevertheless, have some significant limitations. As the calculations are exact, the modulus $M$ needs to be sufficiently large, resulting in numbers which are long.

The relations between the parameters $M$ and $N$ given above require that the calculations be carried out with a number of bits $B$ of the order of $N/2$. That is, the number of terms in the convolution is approximately twice the number of bits of data in the calculation. The application is consequently restricted to convolutions which contain a small number of terms.

The field of application of number theoretic transforms can be widened by employing numbers other than the Fermat numbers, or by treating the convolutions with many terms as two-dimensional convolutions [17]. Reference [18] describes a practical example.

Number-theoretic transforms are used in error-correction coding – particularly in Reed–Solomon codes.

## Exercises

**3.1** Find the Kronecker product $A \times I_3$ for the matrix $A$ such that

$$A = \begin{bmatrix} a_{11} & a_{12} \\ a_{21} & a_{22} \end{bmatrix}$$

and the unit matrix $I_3$ of dimension 3.

Find the product $I_3 \times A$ and compare it with the above.

**3.2** By taking square matrices of dimension 2, verify the properties of the Kronecker products given in Section 3.1.

**3.3** Give the factorizations of the matrix of a DFT of order 64 on base 2, base 4, and base 8, following the procedure given in Section 3.2.
Calculate the required number of multiplications for each of these three cases and compare the results.

**3.4** Using the decimation-in-time approach, factorize the matrix of the DFT of order 12. What is the minimum number of multiplications and additions required? Write a computer program for the calculation.

**3.5** Factorize the matrix of a DFT of order 16 on base 2 for the following two cases:
(a) When the data appear at both input and output in natural order.
(b) When the stages in the calculation are identical.
For the latter case, devise an implementation scheme involving the use of shift registers as memories.

**3.6** Calculate a DFT of order 16 relating to real data
Give the algorithm which uses a DFT of order 8 for this calculation.
Give the algorithm based on the odd DFT.
Compare these algorithms and the number of operations.

**3.7** Calculate a DFT of order 12 by using a factorization of the type given in Section 3.4 and with the given permutations for the data
Evaluate the number of operations and compare it with the values found in Exercise 3.4.

**3.8** To perform a circular convolution of the two sets $x = (2, -2, 1, 0)$ and $h = (1, 2, 0, 0)$ a number-theoretic transform of modulus $M = 17$ and coefficient $\alpha = 4$ is used
As $N = 4$, verify that $\alpha^N = 1$. Give the matrices of the transformation and the inverse transformation. Prove that the desired result is the set $y = (2, 2, -3, 2)$.

## References


1 M. C. Pease, *Methods of Matrix Algebra*. Academic Press, New York, 1965.
2 C. S. Burrus and T. W. Parks, *DFT/FFT and Convolution Algorithms*, Wiley, New York, 1985.
3 H. Sloate, Matrix representations for sorting and the fast Fourier transform, *IEEE Transactions*, 21(1), 109–1161974.
4 G. Bergland, A fast Fourier transform algorithm for real valued series, *Communications of the ACM*, 11(10), 1968 703–710.
5 J. L. Vernet, Real signals FFT by means of an odd discrete Fourier transform. *Proceedings of IEEE*, 1971 5 (10), 1531–1532.
6 G. Bonnerot and M. Bellanger, Odd-time odd-frequency DFT for symmetric real-valued series. *Proceedings of IEEE* 64 (3), 392–393, 1976.

7 H. Ziegler, A fast transform algorithm for symmetric real valued series. *IEEE Transactions,* 20(5), 1972.

8 C. Diab, M. Oueidat and R. Prost, A new IDCT-DFT relationship reducing the IDCT computational cost. *IEEE Transactions on Signal Processing*, 50(7), 1681–84, 2002.

9 R. Bracewell, The fast Hartley transform. *Proceedings of IEEE* 72(8), 1010–1018, 1984.

10 M. Vetterli, H. J. Nussbaumer, Simple FFT and DCT algorithms with reduced number of operations. *Signal Processing*, 6(4), 267–278, 1984.

11 M. A. Haque, A two-dimensional fast cosine transform. *IEEE Transactions,* 33(6), 1532–1539, 1985.

12 H. S. Malvar, *Signal Processing with Lapped Transforms,* Artech House, Norwood MA, 1992.

13 A. Despain, Very fast Fourier transform algorithms hardware for implementation. *IEEE Transactions on Computers*, C-28(5), 333–341, 1979.

14 H. Silverman, An introduction to programming the Winograd Fourier transform algorithm. *IEEE Transactions*, 25(2), 152–165, 1977.

15 H. Nussbaumer, *Fast Fourier Transform and Convolution Algorithms*, Springer, Berlin, 1981.

16 B. Fino and R. Algazi, Unified matrix treatment of fast Walsh-Hadamard transform. *IEEE Transactions on Computers*, C-25 (11), 1142–1146, 1976.

17 R. C. Agrawal and C. S. Burrus, Number theoretic transforms to implement fast digital convolution. *Proceeding of IEEE*, 63(4), 550–560, 1975.

18 J. H. McClellan, Hardware realization of a Fermat number transform. *IEEE Transactions*, 24(3), 216–225, 1976.

# 4

# Time-Invariant Discrete Linear Systems

Time-invariant discrete linear systems represent a very important area for digital signal processing – digital filters with fixed characteristics. These systems are characterized by the fact that their behavior is governed by a convolution equation. Their properties are analyzed using the $Z$-transform, which plays the same role in discrete systems as the Laplace or Fourier transforms do in continuous systems. In this chapter, the elements which are most useful for studying such systems will be briefly introduced. To supplement this discussion, reference should be made to References [1–5].

## 4.1 Definition and Properties

A discrete system is one which converts a set of input data $x(n)$ into a set of output data $y(n)$. It is linear if the set $x_1(n) + ax_2(n)$ is converted to the set $y_1(n) + ay_2(n)$, and it is time-invariant if the set $x(n - n_0)$ is converted to the set $y(n - n_0)$ for any integer $n_0$.

Assume $u_0(n)$ is a unit set, as shown in Figure 4.1, and defined by:

$$u_0(n) = 1 \quad \text{for} \quad n = 0$$
$$u_0(n) = 0 \quad \text{for} \quad n \neq 0 \tag{4.1}$$

The set $x(n)$ can be decomposed into a sum of suitably shifted unit sets:

$$x(n) = \sum_{m=-\infty}^{\infty} x(m)u_0(n - m) \tag{4.2}$$

Further, if $h(n)$ is the set forming the system's response to the unit set $u_0(n)$, then $h(n - m)$ corresponds to $u_0(n - m)$ because of the time-invariance. Linearity then implies the following relation:

$$y(n) = \sum_m x(m)h(n - m) = \sum_m h(m)x(n - m) = h(n) * x(n) \tag{4.3}$$

This is the convolution equation which represents the linear time-invariant (LTI) system. Such a system is completely defined by the values of the set $h(n)$, which is called the system's impulse response.

This system has the property of causality if the output with index $n = n_0$ depends only on inputs with indices $n < n_0$. This property implies that $h(n) = 0$ for $n < 0$, and the output is given by:

$$y(n) = \sum_{m=0}^{\infty} h(m)x(n - m) \tag{4.4}$$

*Digital Signal Processing: Theory and Practice*, Tenth Edition. Maurice Bellanger.

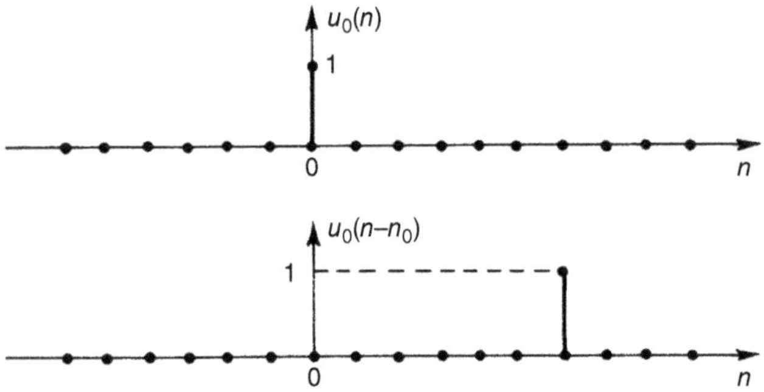


**Figure 4.1** Unit series.

An LTI system is stable if each input with a bounded amplitude has a corresponding bounded output. A necessary and sufficient condition for stability is given by the inequality:

$$\sum_n |h(n)| < \infty \tag{4.5}$$

To show that the condition is necessary, it is sufficient to apply to the system the input set $x(n)$ such that:

$$x(n) = +1 \quad \text{if} \quad h(n) \geqslant 0$$
$$-1 \quad \text{if} \quad h(n) < 0$$

Then, for $n = 0$, we obtain:

$$y(0) = \sum_m |h(m)|$$

If the inequality (4.5) is not satisfied, then $y(0)$ is not bounded, and the system is not stable. If the input set is bounded – that is,

$$|x(n)| \leqslant M \quad \text{for all} \quad n$$

then we have

$$\Big| y(n) \leqslant \sum_m |h(m)||x(n-m)| \leqslant M \sum_m |h(m)|$$

If the inequality (4.5) is satisfied, then $y(n)$ is bounded, and the condition is sufficient.

In particular, the LTI system defined by the response:

$$h(m) = a^m \quad \text{with} \quad m \geqslant 0$$

is stable for $|a| < 1$. The properties of LTI systems will be studied using the Z-transform.

## 4.2 The Z-Transform

The Z-transform, $X(Z)$, of the set $x(n)$ is defined by the following relation:

$$X(Z) = \sum_{n=-\infty}^{\infty} x(n) Z^{-n} \tag{4.6}$$

$Z$ is a complex variable and the function $X(Z)$ has a convergence region which, in general, is an annular ring centered on the origin, with radii $R_1$ and $R_2$. That is, $X(Z)$ is defined for $R_1 < |Z| < R_2$. The values $R_1$ and $R_2$ depend on the set $x(n)$. If the set $x(n)$ represents the samples of a signal taken with period $T$, its Fourier transform is written as:

$$S(f) = \sum_{n=-\infty}^{\infty} x(n)e^{-j2\pi fnT}$$

Consequently, for $Z = e^{j2\pi fT}$, the $Z$-transform of the set $x(n)$ coincides with its Fourier transform. That is, the analysis of a discrete system can be performed with the $Z$-transform, and, in order to find a frequency response, it is sufficient to replace $Z$ with $e^{j2\pi fT}$.

This transform has an inverse. Assuming $\Gamma$ is a closed contour containing the origin and all singular points, or poles, of $X(Z)$, we can write:

$$Z^{m-1}X(Z) = \sum_{n=-\infty}^{\infty} x(n)Z^{m-1-n} = x(m)Z^{-1} + \sum_{n\neq m} Z^{m-1-n}x(n)$$

and from the theory of residues:

$$x(m) = \frac{1}{2\pi j}\int_{\Gamma} Z^{m-1}X(Z)dZ \tag{4.7}$$

For example, if $X(Z) = 1/(1 - pZ^{-1})$, direct application of the above equation yields

$$x(n) = p^n \quad \text{for} \quad n \geqslant 0$$

$$x(n) = 0 \quad \text{for} \quad n < 0$$

Similarly, for $X(Z)$ defined by:

$$X(Z) = \sum_{i=1}^{N} \frac{a_i}{1 - p_i Z^{-1}}$$

there is a corresponding set $x(n)$ such that:

$$x(n) = \sum_{i=1}^{N} a_i p_i^n \quad \text{for } n \geqslant 0$$

$$x(n) = 0 \qquad \text{for } n < 0$$

A stability condition appears quite simply from observing that the set $x(n)$ is bounded if and only if $|p_i| < 1$ for $1 \leqslant i \leqslant N$ – that is, the poles of $X(Z)$ are inside the unit circle.

In these examples, the terms of the set $x(n)$ can be obtained directly by series expansion. When $X(Z)$ is a rational fraction, a very simple method of obtaining the first values of the set $x(n)$ is direct division. For example, if

$$X(Z) = \frac{1 + 2Z^{-1} + Z^{-2} + Z^{-3}}{1 - Z^{-1} - 8Z^{-2} + 12Z^{-3}}$$

direct division gives:

$$X(Z) = 1 + 3Z^{-1} + 12Z^{-2} + 25Z^{-3} + \cdots$$

and thus:

$$x(0) = 1; \quad x(1) = 3; \quad x(2) = 12; \quad x(3) = 25$$

The $Z$-transform has the property of linearity. Also, the $Z$-transform of the delayed set $x(n - n_0)$ is written:

$$X_{n0}(Z) = Z^{-n0}X(Z) \tag{4.8}$$

These two properties are used to calculate the $Z$-transform, $Y(Z)$, of the set $y(n)$ obtained at the output of a discrete linear system by convolution of the sets $x(n)$ and $h(n)$ which have transforms $X(Z)$ and $H(Z)$, respectively.

By calculating the $Z$-transform for the two components of the convolution equation (4.3):

$$y(n) = \sum_{m} h(m)x(n - m)$$

we find,

$$Y(Z) = \sum_{m} h(m)Z^{-m}X(Z) = H(Z)X(Z) \tag{4.9}$$

Consequently, the $Z$-transform of a convolution product is the product of the transforms. The function $H(Z)$ is called the $Z$-transfer function of the LTI system being considered.

The $Z$-transform of the product of two sets $x_3(n) = x_1(n)x_2(n)$ is the function $X_3(Z)$ defined by:

$$X_3(Z) = \frac{1}{2\pi j}\int_{\Gamma} X_1(v)X_2\left(\frac{Z}{v}\right) v^{-1}dv \tag{4.10}$$

The integration contour is inside the region of convergence of the functions $X_1(v)$ and $X_2(Z/v)$.

When applied to causal sets, the one-sided $Z$-transform is introduced. The one-sided $Z$-transform of the set $x(n)$ is written:

$$X(Z) = \sum_{n=0}^{\infty} x(n)Z^{-n} \tag{4.11}$$

The properties are the same as for the transform defined by equation (4.6), except for delayed sets where the transform of the set $x(n - n_0)$ is written:

$$X_{n_0}(z) = \sum_{n=0}^{\infty} x(n - n_0)z^{-n} = z^{-n_0}X(z) + \sum_{n=1}^{n_0} x(-n)z^{-(n_0-n)} \tag{4.12}$$

The value of this transform in the study of system response is that account can be taken of the initial conditions and that the transient response can be brought to light. It also allows for determination of the extreme values of the set $x(n)$ from the $x(z)$. The initial value $x(0)$ is written as:

$$x(0) = \lim_{z\to\infty} X(Z) \tag{4.13}$$

and the final value is:

$$x(\infty) = \lim_{z\to 1}(Z - 1)X(Z) \tag{4.14}$$

More on the $Z$-transform and its applications can be found in Reference [6]. The above results can be applied to the calculation of the power of discrete signals.

## 4.3 Energy and Power of Discrete Signals

Let us calculate the energy $E$ of a signal represented by the set $x(n)$, whose $Z$-transform is written $X(Z)$. By definition:

$$E = \sum_{n=-\infty}^{\infty} |x(n)|^2$$

The set $x_3(n)$ defined by:

$$x_3(n) = |x(n)|^2$$

can be regarded as the product of two sets $x_1(n)$ and $x_2(n)$ such that:

$$x_1(n) = x(n); \quad x_2(n) = \overline{x}(n)$$

The transform $X_3(Z)$ is calculated from the functions $X_1(Z)$ and $X_2(Z)$ using formula (4.10) given in the preceding section for the $Z$-transform of the product of two sets:

$$X_3(Z) = \frac{1}{2\pi j}\int_\Gamma X_1(v)X_2\left(\frac{Z}{v}\right)\frac{dv}{v}$$

The evaluation of $X_3(Z)$ at the point $Z = 1$ leads to the equation:

$$X_3(1) = \sum_{n=-\infty}^{\infty} |x(n)|^2 = \frac{1}{2\pi j}\int_\Gamma X_1(v)X_2\left(\frac{1}{v}\right)\frac{dv}{v}$$

If $\Gamma$ is the unit circle, $1/v = \overline{v}$ and consequently:

$$X_2\left(\frac{1}{v}\right) = X_2(\overline{v}) = \sum_{n=-\infty}^{\infty} \overline{x}(n)(\overline{v})^{-n} = \overline{X}(v)$$

Since $v = e^{j2\pi f}$, we have:

$$E = \sum_{n=-\infty}^{\infty} |x(n)|^2 = \int_{-1/2}^{1/2} |X(e^{j2\pi f})|^2 df \tag{4.15}$$

This is the Bessel–Parseval relation given in Section 1.1.1, which expresses the conservation of energy for discrete signals. The energy of the signal is equal to the energy of its spectrum.

The calculations above provide an expression which is useful for the norm $\|X\|_2$ of the function $X(f)$. Indeed, by definition,

$$\|X\|_2^2 = \int_{-1/2}^{1/2} |X(f)|^2 df$$

This becomes:

$$\|X\|_2^2 = \frac{1}{2\pi j}\int_{|Z|=1} X(Z)X(Z^{-1})\frac{dZ}{Z} \tag{4.16}$$

If $X(Z)$ is a holomorphic function of the complex variable in a domain which contains the unit circle, the integral is calculated by the method of residues and directly yields the value of $\|X\|_2$, which is also the $L_2$ norm of the discrete signal $x(n)$.

Let us now assume we want to calculate the energy $E_y$ of the signal $y(n)$ at the output of the LTI system with impulse response $h(n)$, to which the signal $x(n)$ is applied.

The signal $x(n)$ is assumed to be deterministic. Using equation (4.15), by setting $\omega = 2\pi f$, we can write:

$$E_y = \frac{1}{2\pi}\int_{-\pi}^{\pi} |Y(e^{j\omega})|^2 d\omega$$

Equation (4.9) directly provides the following result:

$$E_y = \frac{1}{2\pi}\int_{-\pi}^{\pi} |H(e^{j\omega})|^2|X(e^{j\omega})|^2 d\omega \tag{4.17}$$

These results also apply to random signals.

## 4.4 Filtering of Random Signals

If the signal $x(n)$ is random and has a moment of order 1, which is $E[x(n)]$, then the expected value of the output $y(n)$ for the LTI system can be calculated as:

$$E[y(n)] = \sum_m h(m)E[x(n-m)] \tag{4.18}$$

If the expected value of $x(n)$ is stationary, then so is that of $y(n)$, provided that the system is stable – that is, it satisfies equation (4.5).

If the input signal $x(n)$ is stationary and of the second order, with an autocorrelation function $r_{xx}(n)$, then the autocorrelation function $r_{yy}(n)$ of the output of the LTI system can be calculated. Using the equation of definition (1.58), we have:

$$r_{yy}(n) = E[y(i)y(i-n)] = \sum_m h(m)E[x(i-m)y(i-n)]$$

If the correlation function $r_{xy}(n)$ between $x(n)$ and $y(n)$ is:

$$r_{xy}(n) = E[x(i)y(i-n)] \tag{4.19}$$

we can write:

$$r_{yy}(n) = h(n) * r_{xy}(n) \tag{4.20}$$

Then,

$$r_{xy}(n) = \sum_m h(m)E[x(i-n-m)] = h(-n) * r_{xx}(n)$$

Finally, one obtains:

$$r_{yy}(n) = h(n) * h(-n) * r_{xx}(n) \tag{4.21}$$

Then, the $Z$-transforms $\Phi_{xx}(Z)$ and $\Phi_{yy}(Z)$ are related by:

$$\Phi_{yy}(Z) = H(Z)H(Z^{-1})\Phi_{xx}(Z) \tag{4.22}$$

This expression can provide a more useful approach than equation (4.21) for calculating the autocorrelation function of the output signal by inverse transformation. With the Fourier transform, one has:

$$r_{yy}(n) = \frac{1}{2\pi}\int_{-\pi}^{\pi} |H(e^{j\omega})|^2 \Phi_{xx}(e^{j\omega})e^{jn\omega}\,d\omega \tag{4.23}$$

In particular, the output power of the signal can be written as:

$$P_y = r_{yy}(0) = \frac{1}{2\pi}\int_{-\pi}^{\pi} |H(e^{j\omega})|^2 \Phi_{xx}(e^{j\omega})e^{j\omega}\,d\omega \tag{4.24}$$

This is equivalent to equation (4.17) for random signals. If the signal $x(n)$ is assumed to be a white noise with variance $\sigma_x^2$, the variance $\sigma_y^2$ of the output signal $y(n)$ is given by:

$$\sigma_y^2 = \frac{\sigma_x^2}{2\pi}\int_{-\pi}^{\pi} |H(e^{j\omega})|^2\,d\omega \tag{4.25}$$

or using the equality (4.15):

$$\sigma_y^2 = \sigma_x^2\left[\sum_n h^2(n)\right] \tag{4.26}$$

These results are of great practical importance and will be frequently used in later sections (for example, when evaluating the powers of the round-off noise in the filters).

## 4.5 Systems Defined by Difference Equations

The most interesting LTI systems are those where the input and output sets are related by a linear difference equation with constant coefficients. On the one hand, they represent simple examples, and on the other, they offer an excellent representation of many natural systems.

A system of this type of order $N$ is defined by the equation:

$$y(n) = \sum_{i=0}^{N} a_i x(n-i) - \sum_{i=1}^{N} b_i y(n-i) \tag{4.27}$$

By applying the $Z$-transform to both sides of this equation, and by denoting the transforms of the sets $y(n)$ and $x(n)$ by $Y(Z)$ and $X(Z)$, we obtain:

$$Y(Z) = \sum_{i=0}^{N} a_i Z^{-i} X(Z) - \sum_{i=1}^{N} b_i Z^{-1} Y(Z) \tag{4.28}$$

Hence:

$$Y(Z) = H(Z)X(Z)$$

with:

$$H(Z) = \frac{a_0 + a_1 Z^{-1} + \cdots + a_N Z^{-N}}{1 + b_1 Z^{-1} + \cdots b_N Z^{-N}} \tag{4.29}$$

The transfer function $H(Z)$ is a rational fraction. The $a_i$ and $b_i$ are the coefficients of the system. Some coefficients can be zero, as is the case, for example, when the two summations of expression (4.27) have different numbers of terms. To find the frequency response, it is sufficient to replace the variable $Z$ by $e^{j2\pi f}$ in $H(Z)$.

The function $H(Z)$ is written in the form of a quotient of two polynomials $N(Z)$ and $D(Z)$ of degree $N$ which have $N$ roots $Z_i$ and $P_i$, respectively, with $1 \leqslant i \leqslant N$.

By using these roots, another expression for $H(Z)$ appears:

$$H(Z) = \frac{N(Z)}{D(Z)} = a_0 \frac{\prod_{i=1}^{N}(1 - Z_i Z^{-1})}{\prod_{i=1}^{N}(1 - P_i Z^{-1})} \tag{4.30}$$

where $a_0$ is a scale factor. Thus, we can write:

$$H(Z) = a_0 \frac{\prod_{i=1}^{N}(Z - Z_i)}{\prod_{i=1}^{N}(Z - P_i)} \tag{4.31}$$

In the complex plane, $Z$ is the affix of a moving point $M$, $P_i$ and $Z_i$ $(1 \leqslant i \leqslant N)$ are affixes of the poles, and zeros of the function $H(Z)$. Then:

$$Z - Z_i = MZ_i e^{j\theta i} \quad \text{and} \quad Z - P_i = MP_i e^{j\theta i}$$

Consequently, the transfer function is expressed by:

$$H(Z) = a_0 \prod_{i=1}^{N} \frac{MZ_i}{MP_i} e^{j} \sum_{i=1}^{N} (\theta_i - \phi_i) \tag{4.32}$$

From this, a graphic interpretation in the complex plane can be developed. The frequency response of the system is obtained when the moving point $M$ lies on the unit circle. Figure 4.2 shows the example of a system of order $N = 2$.

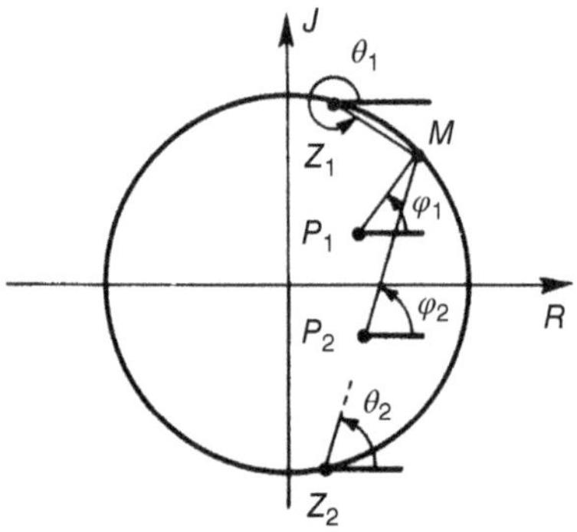


**Figure 4.2** Graphical interpretation of a transfer function.

The modulus of the transfer function is thus equal to the ratio of the product of the distances between the moving point $M$ and the roots $Z_i$ to the product of the distances between $M$ and the poles $P_i$. The phase is equal to the difference between the sum of the angles between the vectors $\overrightarrow{P_iM}$ and the real axis, and the sum of the angles between the vectors $\overrightarrow{Z_iM}$ and the real axis, following the convention introduced in Chapter 1. This graphic interpretation is frequently used in practice because it offers a very simple visualization of a system's frequency response.

Analysis of a system using its frequency response corresponds to steady-state operation. It is adequate only insofar as transient phenomena can be neglected. If this is not the case, initial conditions have to be introduced, to represent, for example, the status of the equipment and the contents of its memory at switch on.

Consider the behavior for values with index $n \geqslant 0$ of the system defined by equation (4.27), to which the set $x(n)$ ($x(n) = 0$ for $n < 0$) is applied. The $y(n)$ are completely determined if the values $y(-i)$ with $1 \leqslant i \leqslant N$ are known. These values correspond to the initial conditions, and, in order to introduce them, the one-sided $Z$-transform has to be used.

The one-sided $Z$-transform is applied to both sides of equation (4.27), assuming that the input $x(n)$ is causal – that is, that $x(n) = 0$ for $n < 0$. Allowing for equation (4.12), which gives the transform $Y_i(Z)$ of the delayed set $y(n-i)$:

$$Y_i(Z) = Z^{-i}Y(Z) + \sum_{n=1}^{i} y(-n)Z^{-(i-n)}$$

we obtain:

$$Y(Z) = \sum_{i=0}^{N} a_i Z^{-i}X(Z) - \sum_{i=1}^{N} b_i Z^{-i}Y(Z) - \sum_{i=1}^{N} b_i \sum_{n=1}^{i} y(-n)Z^{-(i-n)}$$

or

$$Y(Z) = H(Z)X(Z) - \frac{\sum_{i=1}^{N} b_i \sum_{n=1}^{i} y(-n)Z^{-(i-n)}}{1 + \sum_{i=1}^{N} b_i Z^{-i}} \tag{4.33}$$

The system response with index $n$, $y(n)$, is obtained by series expansion or by inverse transformation.

It should be noted that the values $y(-i)$ represent the state of the system at switch on, provided that only the set of output numbers is contained in the system memory. However, the memory often contains other internal variables which can be introduced into the analysis for generalization and to provide other features relating, in particular, to the implementation.

## 4.6 State Variable Analysis

The state of a system of order $N$ at time $n$ is defined by a set of at least $N$ internal variables represented by a vector $U(n)$ called the state vector. Its operation is governed by the relations between this state vector and the input and output signals. The behavior of a linear system to which the input set $x(n)$ is applied, and which provides the output set $y(n)$, is described by the following pair of equations, called state equations [7]:

$$\begin{aligned} U(n+1) &= AU(n) + Bx(n) \\ y(n) &= C^t U(n) + \mathrm{d}x(n) \end{aligned} \tag{4.34}$$

$A$ is called the matrix of the system, $B$ is the control vector, $C$ the observation vector, and $d$ is the transition coefficient. The set $x(n)$ is the innovation and $y(n)$ is the observation. The reasons why they are so called will be outlined later, particularly in Chapters 6 and 15. The matrix $A$ is a square matrix of order $N$. The vectors $B$ and $C$ are $N$-dimensional.

The state of the system at time $n$ is obtained from the initial state at time zero by the equation:

$$U(n) = A^n U(0) + \sum_{i=1}^{n} A^{n-i} Bx(i-1) \tag{4.35}$$

Consequently, the behavior of such a system depends on successive powers of the matrix $A$.

The $Z$-transfer function of the system is obtained by taking the $Z$-transform of the state equations (4.34). Thus:

$$\begin{aligned} (ZI - A)U(Z) &= BX(Z) \\ Y(Z) &= C^t U(Z) + \mathrm{d}X(Z) \end{aligned}$$

and consequently:

$$H(Z) = C^t (ZI - A)^{-1} B + d \tag{4.36}$$

The poles of the transfer function thus obtained are the values of $Z$ for which the determinant of the matrix $(ZI - A)$ is zero. That is, they are the roots of the characteristic polynomial of $A$. Consequently, the poles of the transfer function are the eigenvalues of the matrix $A$ and have absolute values less than unity to ensure stability. This result agrees with the equation for the operation of the system (4.35). Indeed, by diagonalizing the matrix $A$, it can be seen that it is the condition whereby the vector $U(n) = A^n U(0)$ tends toward zero as $n$ tends toward infinity – a situation which corresponds to the free evolution of the system from the initial state $U(0)$.

Examination of the transfer function of the system (4.36) shows by another route that when a system is specified by the input–output equation, there is some latitude in the choice of the state parameters. Indeed, only the eigenvalues of the matrix $A$ are imposed, and the matrix of the system can be replaced by another matrix $A' = M^{-1}AM$, which has the same eigenvalues. Then, in order to preserve the same output set, using equation (4.35), the following criteria are necessary:

$$A' = M^{-1}AM; \quad C'^t M; \quad B' = M^{-1}B$$

The matrix $A$ can also be replaced by its transpose $A'$. The system is then described by a system of equations, parallel to that of equation (4.34), corresponding to the state vector $V(n)$, such that:

$$\begin{aligned} V(n+1) &= A^t V(n) + Cx(n) \\ y(n) &= B^t V(n) + \mathrm{d}x(n) \end{aligned} \tag{4.37}$$

This state representation produces another method of realization.

The results obtained in this section are used subsequently for studying the properties of, and finding the structures for, the realization of LTI systems.

## Exercises

**4.1** Consider an LTI system with an impulse response $h(n)$ such that:

$$h(n) = 1 \quad 0 \leqslant n \leqslant 3$$
$$h(n) = 0 \quad \text{elsewhere}$$

Calculate the response $y(n)$ to the set $x(n)$ given by

$$x(n) = a^n \quad \text{with } a = 0.7 \quad \text{for } 0 \leqslant n \leqslant 5$$
$$x(n) = 0 \quad \text{elsewhere}$$

Calculate the response to the set:

$$x(n) = \cos(2\pi n/8) \quad \text{for } 0 \leqslant n \leqslant 7$$
$$x(n) = 0 \qquad \text{elsewhere}$$

**4.2** Show that the $Z$-transform of the causal sequence $x(n)$ defined by:

$$x(n) = nTe^{-anT} \quad \text{for } n \geqslant 0$$
$$x(n) = 0 \quad \text{for } n < 0$$

is

$$X(Z) = \frac{Te^{-aT}Z^{-1}}{(1 - e^{aT}Z^{-1})^2}$$

Calculate the inverse transform of $\ln(Z - a)$, $Z/\{(Z - a)(Z - b)\}$ and establish the conditions on $a$ and $b$ such that the resulting sequence converges.

**4.3** Calculate the $Z$-transform for the impulse response:

$$h(n) = r^n \frac{\sin[(n+1)\theta]}{\sin(\theta)} \quad n \geqslant 0$$
$$h(n) = 0 \quad n < 0$$

What is the domain of convergence of the function obtained? Plot its poles and zeros in the $Z$-plane.

**4.4** Consider an LTI system whose transfer function $H(Z)$ is written:

$$H(Z) = \frac{1}{1 - 1.6Z^{-1} + 0.92Z^{-2}}$$

and to which a signal with a uniform spectrum and unit power is applied. Calculate the power of the signal at the output of the system and give the spectral distribution.

**4.5** Use the one-sided $Z$-transform to calculate the response of the system defined by the difference equation:

$$y(n) = x(n) + y(n-1) - 0.8y(n-2)$$

with the initial conditions $y(-1) = a$ and $y(-2) = b$, to the set $x(n)$ defined by:

$$x(n) = e^{jn\omega} \quad \text{for } n \geqslant 0$$

$$x(n) = 0 \quad \text{for } n < 0$$

Show the response due to the initial conditions and the steady state response.

## References


1 L. R. Rabiner and B. Gold, *Theory and Application of Digital Signal Processing*. Chapter II, Prentice Hall, 1975.

2 A. V. Oppenheim and R. W. Schafer, *Digital Signal Processing*. Chapter II, Prentice Hall, 1974.

3 M. Kunt, *Digital Signal Processing*, Artech House, Norwood, MA, 1986.

4 F.J. Taylor, *Digital Filter Design Handbook*, Dekker, New York, 1983.

5 A. Peled and B. Liu, *Digital Signal Processing: Theory, Design and Implementation*, Wiley, New York, 1976.

6 E. I. Jury, *Theory and Application of the Z-Transform Method*, Wiley, 1964.

7 J. A. Cadzow, *Discrete Time Systems*, Prentice-Hall, 1973.

# 5

# Finite Impulse Response (FIR) Filters

Digital FIR filters are discrete linear time-invariant systems in which an output number, representing a sample of the filtered signal, is obtained by weighted summation of a finite set of input numbers, representing samples of the signal to be filtered. The coefficients of the weighted summation constitute the filter's impulse response and only a finite number of them take nonzero values. This is a "finite memory" filter – that is, it determines its output as a function of input data of limited age. It is frequently called a non-recursive filter because, unlike the infinite impulse response filter, it does not require a feedback loop in its implementation.

The properties of FIR filters will be illustrated by two simple examples.

## 5.1 FIR Filters

Consider a signal $x(t)$ represented by its samples $x(nT)$, taken at frequency $f_s = 1/T$, and examine the effect on its spectrum of replacing the set $x(nT)$ with the set $y(nT)$ defined by the equation:

$$y(nT) = \frac{1}{2}[x(nT) + x(n-1)T] \tag{5.1}$$

This set is also obtained by sampling the signal $y(t)$ such that:

$$y(t) = \frac{1}{2}[x(t) + x(t-T)]$$

If $Y(f)$ and $X(f)$ denote the Fourier transforms of the signals $y(t)$ and $x(t)$, then:

$$Y(f) = \frac{1}{2}X(f)(1 + e^{\ j2\pi fT})$$

Such an operation corresponds to the transfer function:

$$H(f) = Y(f)/X(f)$$

where

$$H(f) = e^{-j\pi fT}\cos(\pi fT) \tag{5.2}$$

This is called cosine filtering, and it conserves the zero frequency component and eliminates that at $f_s/2$, as can be readily verified.

In the expression for $H(f)$, the complex term $e^{-j\pi fT}$ represents a delay $\tau = T/2$, which is the propagation time of the signal through the filter.

*Digital Signal Processing: Theory and Practice*, Tenth Edition. Maurice Bellanger.

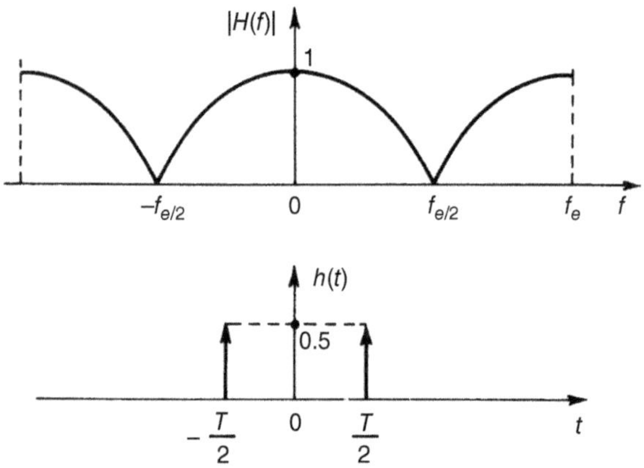


**Figure 5.1** Cosine filtering.

The impulse response $h(t)$ which corresponds to the filter transfer function $|H(f)|$ is written as:

$$h(t) = \frac{1}{2}\left[\delta\left(t + \frac{T}{2}\right) + \delta\left(t - \frac{T}{2}\right)\right]$$

Figure 5.1 shows the characteristics of the filter.

Another simple operation associates the set $x(nT)$ with the set $y(nT)$ by:

$$y(nT) = \frac{1}{4}[x(n\mathrm{T}) + 2x[(n-1)T] + [(n-2)T]] \tag{5.3}$$

As in the previous case, this equation conserves the component with zero frequency and eliminates the one with frequency $f_s/2$. This corresponds to the transfer function:

$$H(f) = \frac{1}{4}(1 + e^{-j2\pi f2T} + e^{-j2\pi f2T}) = e^{-j2\pi f2T}\frac{1}{2}(1 + \cos 2\pi fT) \tag{5.4}$$

This is a raised-cosine filter. Its propagation delay is $\tau = T$, and $|H(f)|$ corresponds to the impulse response $h(t)$ such that:

$$h(t) = \frac{1}{4}\delta(t+T) + \frac{1}{2}\delta(t) + \frac{1}{4}\delta(t-T)$$

This is a more selective low-pass filter than the preceding one, and it is evident that an even more selective filtering function can be obtained merely by increasing the number of terms in the set $x(nT)$ over which the weighted summation is carried out (Figure 5.2).

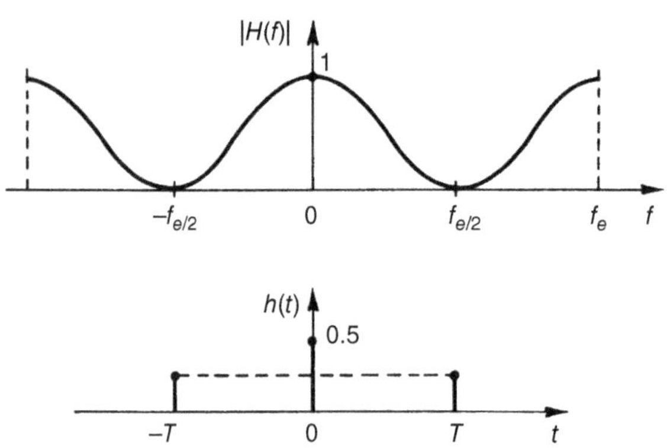


**Figure 5.2** Raised-cosine filtering.

These two examples have served to illustrate the following properties of FIR filters:

(1) The input set $x(n)$ and the output set $y(n)$ are related by an equation of the following type (the defining relation):

$$y(n) = \sum_{i=0}^{N-1} a_i x(n-i) \tag{5.5}$$

The filter defined in this way comprises a finite number $N$ of coefficients $a_i$. If it is regarded as a discrete system, its response $h(i)$ to the unit set is:

$$h(i) = a_i \quad \text{if} \quad 0 \leqslant i \leqslant N-1$$
$$0 \qquad \text{elsewhere}$$

That is, the impulse response is simply the set of the coefficients.

(2) The transfer function of the filter is:

$$H(f) = \sum_{i=0}^{N-1} a_i e^{-j2\pi f i T} \tag{5.6}$$

or, expressed in terms of $Z$:

$$H(Z) = \sum_{i=0}^{N-1} a_i Z^{-i} \tag{5.7}$$

(3) The function $H(f)$, the filter's frequency response, is periodic with period $f_s = 1/T$. The coefficients $a_i$ $(0 \leqslant i \leqslant N-1)$ form the Fourier series expansion of this function.
In view of the Bessel–Parseval relation given in Section 1.1.1, the following can be written:

$$\sum_{i=0}^{N-1} |a_i|^2 = \frac{1}{f_s}\int_0^{f_s} |H(f)|^2 df \tag{5.8}$$

(4) If the coefficients are symmetrical, the transfer function can be written as the product of two terms, of which one is a real function and the other a complex number with modulus 1, representing a constant propagation delay $\tau$ which is a whole multiple of half the sampling period. Such a filter is said to have a linear phase.

## 5.2 Practical Transfer Functions and Linear Phase Filters

A digital filter which processes numbers representing samples of a signal taken with period $T$ has a periodic frequency response of period $f_s = 1/T$. As a consequence, this function $H(f)$ can be expanded into a Fourier series:

$$H(f) = \sum_{n=-\infty}^{\infty} \alpha_n e^{+j2\pi f n T} \tag{5.9}$$

with

$$\alpha_n = \frac{1}{f_s}\int_0^{f_s} H(f) e^{-j2\pi f n T} df \tag{5.10}$$

The coefficients $\alpha_n$ of the expansion are, except for a constant factor, the samples taken with period $T$ of the Fourier transform of the function $H(f)$ over a frequency interval of width $f_s$. As they form the impulse response, the condition (4.5) for the stability of the filter implies that the $\alpha_n$ tends toward zero when $n$ tends toward infinity. Consequently, the function $H(f)$ can be approximated by an expansion reduced to a limited number of terms:

$$H(f) \simeq \sum_{n=-p}^{Q} \alpha_n e^{j2\pi fnT} = H_L(f)$$

where $P$ and $Q$ are finite integers. The approximation is improved as the numbers increase.

The property of causality, which expresses the fact that in a real filter, the output cannot precede the input in time, implies that the impulse response $h(n)$ is zero for $n < 0$. Using relations (5.5) and (5.6), if the filter is causal, then $Q = 0$ and we obtain:

$$H_L(f) = \sum_{n=0}^{P} a_n e^{-j2\pi fnT}$$

As a result, any causal digital filtering function can be approximated by the transfer function of an FIR filter.

The fact that transfer functions with linear phase can be realized is an important property which is exploited in spectral analysis or in data-transmission applications. For FIR filters, this capability leads to a reduction in complexity, because it implies symmetry in the coefficients which is used to halve the number of multiplications needed to produce each output number.

By definition, a linearphase filter has the following frequency response:

$$H(f) = R(f)e^{-j\phi(f)} \tag{5.11}$$

where $R(f)$ is a real function and the phase $\phi(f)$ is a linear one: $\phi(f) = \phi_0 + 2\pi f\tau$, when $\tau$ is a constant giving the propagation delay through the filter.

In fact, the phase is not rigorously linear, due to the sign changes of $R(f)$ which introduce discontinuities of $\pi$ in the phase. However, the filter is said to be linear phase.

The impulse response of this filter is written as:

$$h(t) = e^{-j\phi_0} \int_{-\infty}^{\infty} R(f)e^{j2\pi f(t-\tau)} df \tag{5.12}$$

By assuming $\phi_0$ to be zero and by decomposing the real function $R(f)$ into an even part $P(f)$ and an odd part $I(f)$, this becomes:

$$h(t+\tau) = 2\int_0^{\infty} P(f)\cos(2\pi ft)df + 2j\int_0^{\infty} I(f)\sin(2\pi ft)df$$

If the condition is imposed that the function $h(t)$ must be real, this becomes:

$$h(t+\tau) = 2\int_0^{\infty} P(f)\cos(2\pi ft)df$$

This relation shows that the impulse is symmetric about the point $t = \tau$ on the time axis. Such a condition is satisfied in a filter with real symmetric coefficients. Two configurations are available, depending upon whether the number of coefficients $N$ is even or odd:

(1) $N = 2P + 1$: the filter has a propagation time $\tau = PT$. The transfer function is:

$$H(f) = e^{-j2\pi fPT}\left[h_0 + 2\sum_{i=1}^{P} h_i \cos(2\pi fiT)\right] \tag{5.13}$$

(2) $N = 2P$: the filter has a propagation delay $\tau = \left(P - \frac{1}{2}\right) T$. The transfer function is:

$$H(f) = e^{-j2\pi f(P-1/2)T} 2 \left( \sum_{i=1}^{P} h_i \cos \left[ 2\pi f \left( i - \frac{1}{2} \right) T \right] \right) \tag{5.14}$$

The filter coefficients $h_i$ form the digital filter's response to the unit set. They can also be regarded as samples, taken with period $T$, of the continuous impulse response $h(t)$ of a filter which has the same frequency response as the digital filter in the range $(-T/2, T/2)$ but has no periodicity on the frequency axis. Figures 5.3 and 5.4 illustrate this for odd and even $N$.

These filters have the even function $P(f)$ in their frequency response. With real coefficients, a frequency response can also be obtained which corresponds to $I(f)$, the odd part of $R(f)$.

As the function $h(t)$ must be real, this category of filter has the transfer function:

$$H(f) = -je^{-j2\pi f\tau} I(f) = e^{-j(\pi/2)} e^{-j2\pi f\tau} I(f)$$

A fixed phase shift $\phi_0 = \pi/2$, which corresponds to a quadrature, is added to the frequency-proportional phase shift. This possibility is important in certain types of modulation and will be examined later. The impulse response is zero at the point $t = \tau$ on the time axis and is antisymmetric about it. Figures 5.5 and 5.6 show the configurations when the number of coefficients $N$ is even or odd.

**Figure 5.3** Symmetrical filter with odd order.

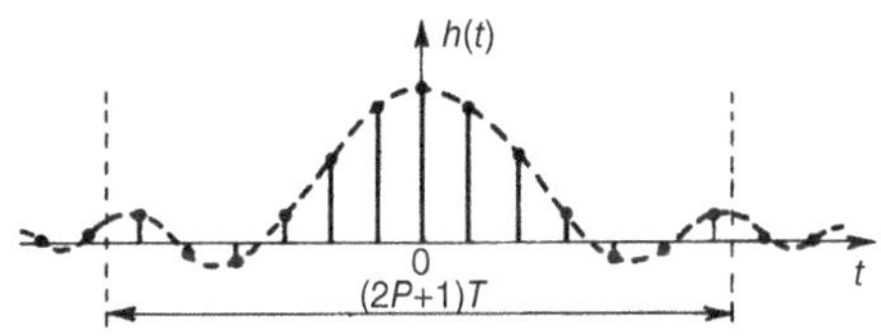


**Figure 5.4** Symmetrical filter with even order.

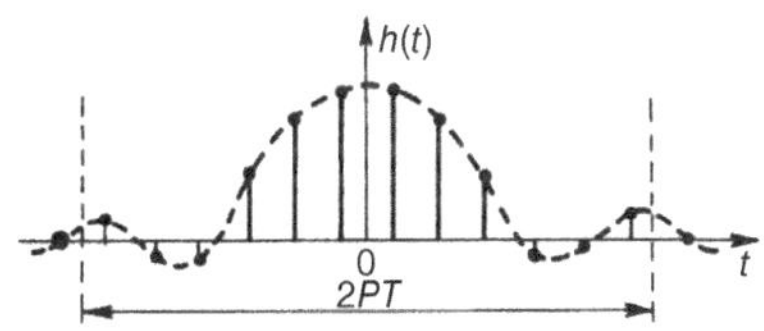


**Figure 5.5** Antisymmetrical filter with odd order.

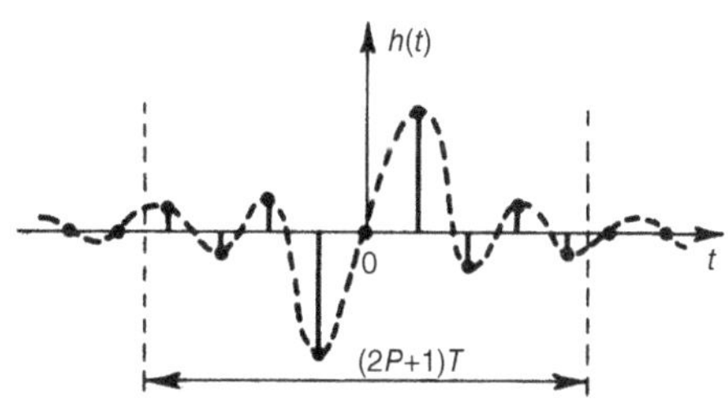


**Figure 5.6** Antisymmetrical filter with even order.

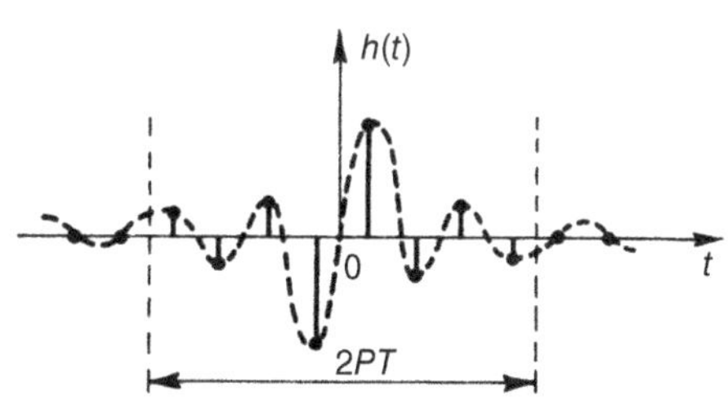

(1) $N = 2P + 1$: the filter has a propagation delay $\tau = PT$

$$H(f) = -je^{-j2\pi f\tau} 2\sum_{i=1}^{P} h_i \sin(2\pi + iT) \tag{5.15}$$

(2) $N = 2P$: this becomes $\tau = \left(P - \frac{1}{2}\right) T$

$$H(f) = -je^{j2\pi f\tau} 2\sum_{i=1}^{P} h_i \sin\left[2\pi f\left(i - \frac{1}{2}\right) T\right] \tag{5.16}$$

As $h_0 = 0$, the transfer function has the same form in both cases. It is not difficult to envisage that fixed phase shifts other than $\phi_0 = 0$ and $\phi_0 = \pi/2$ can be obtained for filters with complex coefficients.

We now turn our attention to the calculation of the coefficients of FIR filters, assuming the phase to be linear, which applies to a majority of applications, and is the case when specifications are given for the frequency response.

## 5.3 Calculation of Coefficients by Fourier Series Expansion for Frequency Specifications

The frequency specifications are generally provided through a mask.

For a low-pass filter, for example, the absolute value of the transfer function is required to approximate the value 1 with precision $\delta_1$, in the frequency band $(0, f_1)$, called the pass band, and the value 0 with precision $\delta_2$ in the band $(f_2, f_s/2)$, which is called the stop (or rejection) band. The corresponding limits are represented in Figure 5.7. The range $\Delta f = f_2 - f_1$ is called the transition band, and the steepness of the cutoff is described by the parameter $R_c$ such that:

$$R_{c=}\frac{f_1 + f_2}{2(f_2 - f_1)} \tag{5.17}$$

A very simple method of obtaining the coefficients $h_i$ is to expand the periodic function $H(f)$ into a Fourier series to produce an approximation:

$$h_i = \frac{1}{f_s}\int_0^{f_s} H(f)e^{-j2\pi if/f_s} df$$

In the case of a low-pass filter corresponding to the mask given in Figure 5.7, expression (1.5) leads to:

$$h_i = \frac{f_1 + f_2}{f_s} \frac{\sin \pi i[(f_1 + f_2)/f_s]}{\pi i[(f_1 + f_2)]/f_s} \tag{5.18}$$

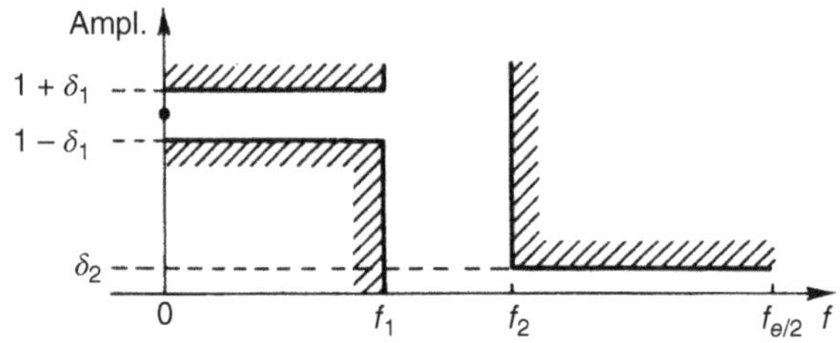


**Figure 5.7** Mask for a low-pass filter.

The table given in Appendix 1 of Chapter 1 can then be used to provide an initial estimation of the coefficients of an FIR filter. The optimized values are, in fact, not very different.

For the filter to be realizable, it is necessary to limit the number of coefficients to $N$. This operation boils down to multiplying the impulse response $h(t)$ by a time window $g(t)$ so that:

$$g(t) = 1 \text{ for } -\frac{NT}{2} \leqslant t \leqslant \frac{NT}{2}$$
$$0 \quad \text{elsewhere}$$

Using equation (1.10), the Fourier transform of this function is written as:

$$G(f) = NT\frac{\sin(\pi fNT)}{\pi fNT} \tag{5.19}$$

Figure 5.8 shows these functions.

The real filter, with a limited number of coefficients $N$, has the following convolution product $H_R(f)$ as its transfer function:

$$H_R(f) = \int_{-\infty}^{\infty} H(f')G(f - f')\,df'$$

By limiting the number of coefficients, ripples are introduced and the steepness of the cutoff of the filter is limited, as shown in Figure 5.9, which corresponds to an ideal low-pass filter with a cutoff frequency $f_c$.

The ripples depend on those of the function $G(f)$. In order to reduce their amplitude, it is sufficient to choose as the time window functions whose spectra introduce smaller ripples than that of

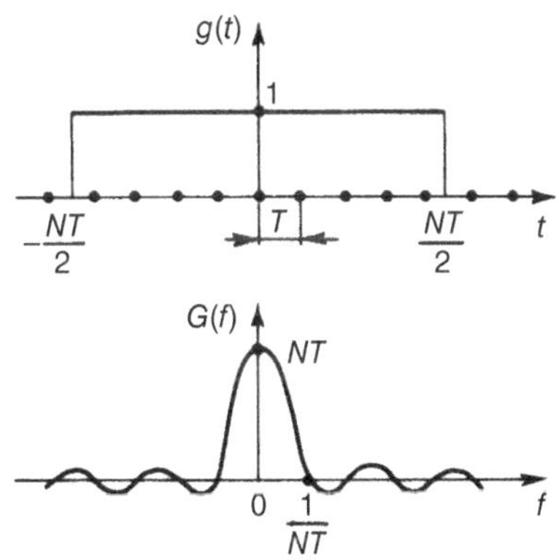


**Figure 5.8** Rectangular window.

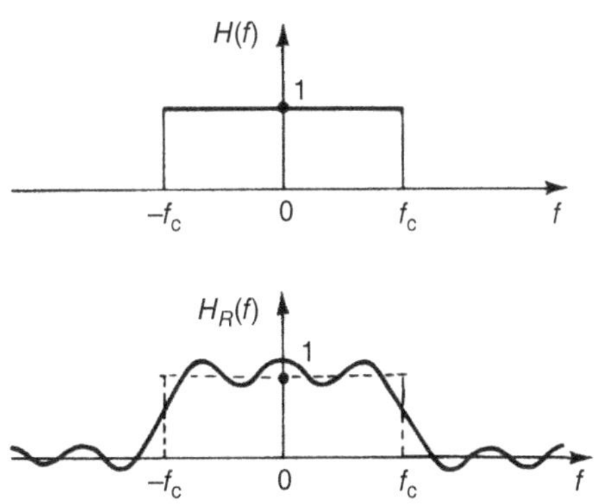


**Figure 5.9** Effect of limiting the number of coefficients.

the rectangular window given above. This situation has been encountered in Section 2.4.2 for spectrum analysis and the same functions can be employed. One example is the Hamming window, which is defined as:

$$g(t) = 0.54 + 0.46\cos(2\pi i/NT) \quad \text{for } |t| \leqslant NT/2$$
$$0 \quad \text{for } |t| > NT/2$$

The consequence of reducing the ripples in the pass and stop bands is an increase in the width of the transition band.

The function which presents the smallest ripples for a given width of the principal lobe is the Dolph–Chebyshev function:

$$G(x) = \frac{\cos[K\cos^{-1}(Z_0\cos\pi x)]}{ch[K\ ch^{-1}(Z_0)]} \quad \text{for } x_0 \leqslant x \leqslant 1 - x_0$$
$$\frac{ch[K\,ch^{-1}(Z_0\cos\pi x)]}{ch[K\,ch^{-1}(Z_0)]} \quad \textit{for } 0 \leqslant x \leqslant x_0 \text{ and } 1 - x_0 \leqslant x \leqslant 1 \tag{5.20}$$

with $x_0 = (1/\pi)\cos^{-1}(1/Z_0)$, $K$ an integer, and $Z_0$ a parameter. This function, as shown in Figure 5.10, presents a main lobe of width $B$, given by:

$$B = 2x_0 = \frac{2}{\pi}\cos^{-1}\left(\frac{1}{Z_0}\right)$$

and secondary lobes of constant amplitude given by:

$$A = \frac{1}{ch[K\,ch^{-1}(Z_0)]}$$

It is periodic and its inverse Fourier transform is formed of a set of $K + 1$ discrete nonzero values, used to weight the coefficients of the Fourier series expansion of the filter function to be approximated.

It is worth noting that if the ripples in the function $G(x)$ are of constant amplitude, those in the resulting filter decrease in amplitude, with distance from the pass band and stop band edges. The ripples in the pass and stop bands, however, are the same.

The technique of Fourier series expansion of the function to be approximated leads to a simple method of determining the coefficients of the filter, but it involves two important restrictions:

(1) The ripples of the filter in the pass and stop bands are equal.
(2) The amplitude of the ripple is not constant.

The first limitation can be eliminated by using a method which preserves the simplicity of direct calculation: the least-squares method. Further, this corresponds exactly to the objectives to be achieved in a number of applications.

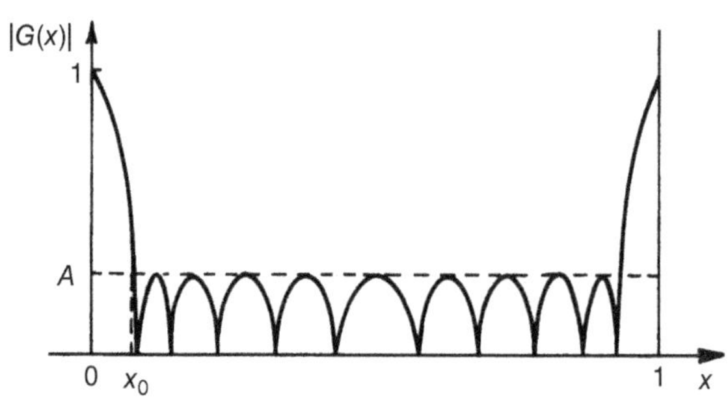


**Figure 5.10** The Dolph–Chebyshev function.

## 5.4 Calculation of Coefficients by the Least-Squares Method

Let us calculate the $N$ coefficients $h_i$ of an FIR filter according to the leastsquares criterion so that the transfer function approximates a given function.

The calculation can be carried out directly from the relationship between coefficients and frequency response, as shown in Section 5.16 for 2-dimensional filters. However, it is interesting to provide a more general iterative approach, operating in the frequency domain from an approximate solution, which can cope with non-quadratic cost functions and can be used for other types of filters – in particular, IIR filters.

The discrete Fourier transform applied to the set $h_i$, with $(0 \leqslant i \leqslant N-1)$, produces a set $H_k$ such that:

$$H_k = \frac{1}{N}\sum_{i=0}^{N-1} h_i e^{-j2\pi(ik/N)} \tag{5.21}$$

The set of $H_k$, $0 \leqslant k \leqslant N-1$ forms a sample of the filter's frequency response with period $f_s/N$. Conversely, the coefficients $h_i$ are related to the set $H_k$ by the equation:

$$H_k = \sum_{k=0}^{N-1} h_k e^{-j2\pi(ik/N)} \tag{5.22}$$

Consequently, the problem of calculating the $N$ coefficients is equivalent to that of determining the filter's frequency response at $N$ points in the range $(0, f_s)$. The function $H(f)$ is then obtained by the interpolation formula which expresses the convolution product of the set of the samples $H_k\delta(f - [k/N]f_s)$ by the Fourier transform of the rectangular sampling window, calculated in Section 2.4:

$$H(f) = \sum_{k=0}^{N-1} H_k \frac{\sin\{\pi N[(f/f_s) - (k/N)]\}}{N\sin\{\pi[(f/f_s) - (k/N)]\}} \tag{5.23}$$

It should be noted that this expression simply forms a different type of series expansion of the function $H(f)$ with a limited number of terms.

Given the function to be approximated, $D(f)$, a first possibility is to choose the $H_k$ such that:

$$H_k = D\left(\frac{k}{N}f_s\right) \quad \text{for} \quad 0 \leqslant k \leqslant N-1$$

The transfer function of the filter $H(f)$, obtained by interpolation, shows ripples in the pass and stop bands, as shown in Figure 5.11.

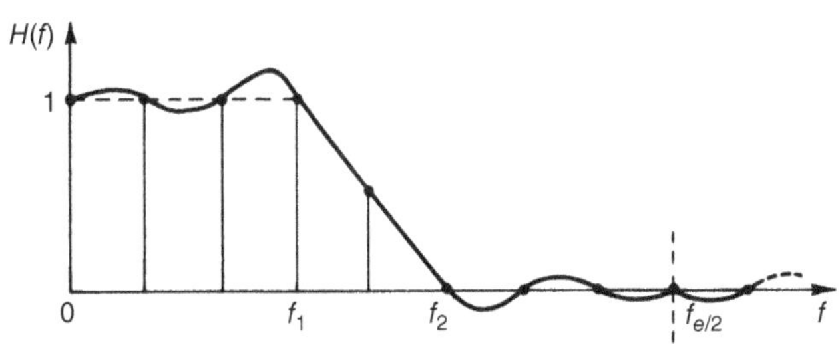


**Figure 5.11** Interpolated transfer function.

The divergence between this function and that given represents an error $e(f) = H(f) - D(f)$ which can be minimized by the least-squares criterion. The procedure begins by evaluating the squared error $E$ which is the $L_2$ norm of the divergence function. To do this, the response $H(f)$ is sampled with a frequency step $\Delta$ less than $f_s/N$, in a way that produces interpolated values, for example:

$$\Delta = \frac{f_s}{NL} \quad \text{with } L \text{ as an integer greater than 1}$$

The function $e(f)$ is calculated at frequencies which are multiples of $\Delta$. Generally, when evaluating the squared error $E$, only one part of the band $(0, f_s/2)$ is taken into account. For a low-pass filter this can be the pass band, the stop band or the set of both bands. In order to demonstrate the principle of the calculation, it is assumed that the minimization relates to the pass band $(0, f_1)$ of a low-pass filter, whence:

$$E = \sum_{n=0}^{N_0-1} e^2\left(n\frac{f}{NL}\right) \quad \text{with} \frac{f_1}{f_s}NL < N_0 \leqslant \frac{f_1}{f_s}NL + 1$$

Further, it is often useful to find a weighting factor $P_0(n)$ for the error element of index $n$, so that the frequency response can be modeled. Thus:

$$E = \sum_{n=0}^{N_0-1} P_0^2(n)e^2\left(n\frac{f_s}{NL}\right) = \sum_{n=0}^{N_0-1} P_0^2(n)e^2(n) \tag{5.24}$$

With the error function obtained from the interpolation formula (5.23), the squared error $E$ is a function of the set of $H_k$ with $0 \leqslant k \leqslant N-1$ and is expressed by $E(H)$. If these samples of the frequency response are given increases of $\Delta H_k$, a new squared error value is obtained:

$$E(H + \Delta H) = E(H) + \sum_{K=0}^{N-1} \frac{\partial E}{\partial H}\Delta H_k + \frac{1}{2}\sum_{k=0}^{N-1}\sum_{l=0}^{N-1} \frac{\partial^2 E}{\partial H_K \partial H_1}\Delta H_k \Delta H_1 \tag{5.25}$$

From the defining equation for $E$ and the interpolation equation (5.23), this becomes:

$$\frac{\partial E}{\partial H_k} = 2\sum_{n=0}^{N_0-1} P_0^2(n)e(n)\frac{\partial e(n)}{\partial H_k}$$

$$\frac{\partial^2 E}{\partial H_k \partial H_1} = 2\sum_{n=0}^{N_0-1} P_0^2(n)\frac{\partial e(n)}{\partial H_1}\frac{\partial e(n)}{\partial H_k}$$

These equations can be written in matrix form where $A$ is the matrix with $N$ rows and $N_0$ columns such that:

$$A = \begin{bmatrix} a_{00} & a_{01} & \cdots & a_{0(N_0-1)} \\ a_{10} & a_{11} & \cdots & a_{1(N_0-1)} \\ \vdots & \vdots & & \vdots \\ a_{0(N-1)} & a_{1(N-1)} & \cdots & a_{(N-1)(N_0-1)} \end{bmatrix} \quad \text{with} \quad a_{ij} = \frac{\partial e(j)}{\partial H_i}$$

Let $P_0$ be the diagonal matrix of order $N_0$ whose elements are the weighting factor $P_0(n)$. Then:

$$\left[\frac{\partial E}{\partial H_k}\right] = 2AP_0^2[e(n)] \tag{5.26}$$

The set of terms $\partial^2 E/\partial H_k \partial H_e$ form a square matrix of order $N$ such that:

$$\frac{\partial^2 E}{\partial H_k \partial H_1} = 2AP_0^2 A^t \tag{5.27}$$

The condition for $E(H+\Delta H)$ to be a minimum is that all its derivatives with respect to $H_k$ $(0 \leqslant k \leqslant N-1)$ be zero. Now,

$$\frac{\partial}{\partial H_k} E(H+\Delta H) = \frac{\partial E}{\partial H_k} + \sum_{l=0}^{N-1} \frac{\partial E}{\partial H_k} \frac{\partial E}{\partial H_l} \Delta H_l$$

The least-squares condition can be written as:

$$AP_0^2[e(n)] + AP_0^2 A^t[\Delta H] = 0 \tag{5.28}$$

Under these conditions, the increments $\Delta H_k$ $(0 \leqslant k \leqslant N-1)$ which transfer the initial values of the samples to the optimal values of the frequency response form a column vector which is written as:

$$[\Delta H] = -\left[AP_0^2 A^t\right]^{-1} AP_0^2[e(n)] \tag{5.29}$$

To summarize, the following operations are required for calculating the coefficients of the filter by the least-squares method:

(1) Sample the function to be approximated at $N$ points in order to obtain $N$ numbers $H_k$ $(0 \leqslant k \leqslant N-1)$.
(2) In the frequency band where the error should be minimized, interpolate the response between the $H_k$ in order to obtain $N_0$ numbers $e(n)$ $(0 \leqslant n \leqslant N_0-1)$ which represent the divergence between the response of the filter and the function to be approximated.
(3) Determine $N_0$ weighting coefficients $P_0(n)$ as a function of the constraints of the approximation.
(4) Using the interpolation equation, calculate the elements of the matrix $A$.
(5) Solve the matrix equation which gives the $\Delta H_k$.
(6) Perform an inverse Fourier transformation on the set of numbers $(H_k + \Delta H_k)$ with $0 \leqslant k \leqslant N-1$ to obtain the coefficients for the filter.

The weighting coefficients $P_0(n)$ allow certain constraints to be introduced – for example, to obtain ripples in the pass and stop bands which are in a given ratio, or to force the frequency response to a particular value. This latter condition can also be taken into account by reducing the number of degrees of freedom by 1; this is more elegant but more complicated to program.

Implementation of this calculation does not present any particular difficulty and permits the calculation of a filter in a direct way. However, the filter obtained does not have constant amplitude. This is a commonly required feature, and to achieve it, an iterative technique is employed.

If the matrix inversion in equation (5.29) is difficult or impossible, the optimum can still be reached by replacing the matrix with a small constant and iterating the process. This is the gradient algorithm, which is dealt with in Chapter 14.

## 5.5 Calculation of Coefficient by Discrete Fourier Transform

A first iterative approach consists of using the discrete Fourier transform, which is calculated efficiently by a fast algorithm.

Consider calculation of a linear-phase filter with $N$ coefficients meeting the specification of Figure 5.7. A discrete Fourier transform of order $N_0$ with $N_0 \simeq 10\,N$ will be used.

The procedure consists of taking initial values for the coefficients – for example, the terms $h$ given by (5.18) for $-P \leq i \leq P$, if $N = 2P + 1$. This set of $N$ values is completed symmetrically with zeros to obtain a set of $N_0$ real values, symmetrical with respect to the origin.

A DFT calculation then gives the response $H(f)$ at $N_0$ points on the frequency axis:

$$H(f) = H_{\mathrm{id}}(f) + E(f)$$

where $H_{\mathrm{id}}(f)$ is the ideal response and $E(f)$ is the deviation from this response.

A reduction of the deviation $E(f)$ is then performed, by replacing $H(f)$ with the function $G(f)$ such that:

$$G(f) = H_{\mathrm{id}}(f) + E_{\mathrm{L}}(f) \quad \text{if} \quad H(f) > H_{\mathrm{id}}(f) + E_{\mathrm{L}}(f)$$
$$G(f) = H_{\mathrm{id}}(f) - E_{\mathrm{L}}(f) \quad \text{if} \quad H(f) < H_{\mathrm{id}}(f) - E_{\mathrm{L}}(f)$$

where $E_{\mathrm{L}}(f)$ represents the limit of the deviation given by the characteristic – for example, $\delta_1$ and $\delta_2$ for the low-pass filter in Figure 5.7.

Calculation of the inverse DFT gives $N_0$ terms; the $N$ values which encircle the origin are retained and the others are discarded. The procedure is repeated by taking the DFT of the $N_0$ values obtained in this way.

Denoting the sum of the squares of the $N_0 - N$ terms discarded in the time domain at iteration $k$ by $J(k)$, a decreasing function is obtained if the specification of the filter is compatible with the number of coefficients $N$. The procedure is terminated when $J(k)$ falls below a fixed threshold.

By applying the method for different numbers of coefficients $N$, an optimum solution can be approached and even achieved in special cases. All types of linear phase filters can be designed in this way.

To obtain the optimum filter, a method based on the Chebyshev approximation is used.

## 5.6 Calculation of Coefficients by Chebyshev Approximation

The objective is to produce a filter whose frequency response has constant amplitude ripple in such a way as to best approximate the desired characteristic. An example was given in Figure 5.8 for a low-pass filter whose ripples must not exceed amplitude $\delta_1$ in the pass band and $\delta_2$ in the stop band. This problem depends upon the approximation of a function by a polynomial in the Chebyshev sense, and $L_\infty$ is the norm to be considered for the error function.

Using the expression for the transfer function of a linear-phase FIR filter, the calculation of the coefficients corresponds to determining the function $H_{\mathrm{R}}(f)$ which is written as:

$$H_{\mathrm{R}}(f) = \sum_{i=0}^{r-1} h_i \cos(2\pi f i T) \tag{5.30}$$

when the number of coefficients is $N = 2r - 1$. The technique which will be presented is valid in all cases, whether $N$ is even or odd, or whether the coefficients are symmetric or antisymmetric. It is based on the following theorem [1].

A necessary and sufficient condition for $H_{\mathrm{R}}(f)$ to be the unique and, in the Chebyshev sense, the best approximation to a given function $D(f)$ over a compact subset $A$ for the range $\left[0, \frac{1}{2}\right]$ is that the error function $e(f) = H_{\mathrm{R}}(f) - D(f)$ present at least $(r + 1)$ extremal frequencies on $A$. That is, there exist $(r + 1)$ extremal frequencies $(f_0, f_1, \ldots, f_r)$ such that $e(f_i) = -e(f_{i-1})$, with $1 \leqslant i \leqslant r$ and

$$|e(f_i)| = \max_{f \in A} |e(f)|$$

This result is still valid if a weighting function $P_0(f)$ for the error is introduced.

Thus, the problem is equivalent to the solution of the system of $(r+1)$ equations:

$$P_0(f_i)[D(f_i) - H_R(f_i)] = (-1)^i \delta$$

The unknowns are the coefficients $h_i$ $(0 \leqslant i \leqslant r-1)$ and the maximum of the error function: $\delta$. In matrix form, by writing the unknowns as a column vector and by normalizing the frequencies so that $f_s = 1/T$, this becomes:

$$\begin{bmatrix} D(f_0) \\ D(f_1) \\ \vdots \\ D(f_r) \end{bmatrix} = \begin{bmatrix} 1 & \cos(2\pi f_0) \ldots & \cos[2\pi f_0(r-1)] & \dfrac{1}{P_0(f_0)} \\ 1 & \cos(2\pi f_1) \ldots & \cos[2\pi f_1(r-1)] & \dfrac{1}{P_0(f_1)} \\ \vdots & \vdots & \vdots & \vdots \\ 1 & \cos(2\pi f_r) \ldots & \cos[2\pi f_r(r-1)] & \dfrac{1}{P_0(f_r)} \end{bmatrix} \begin{bmatrix} h_0 \\ h_1 \\ \vdots \\ h_{r-1} \\ \delta \end{bmatrix}$$

This matrix equation results in the determination of the coefficients of the filter, under the condition that the $(r+1)$ extremal frequencies $f_i$ are known.

An iterative procedure based on an algorithm called the Remez algorithm is used to find the extremal frequencies. In this algorithm, each stage has the following phases:

(1) Initial values are assigned or are available for the parameters $f_i$ $(0 \leqslant i \leqslant r)$.
(2) The corresponding value $\delta$ is calculated by solving the system of equations, which leads to the following formula:

$$\delta = \frac{a_0 D(f_0) + a_1 D(f_1) + \cdots + a_r D(f_r)}{a_0/P_0(f_0) - a_1/P_0(f_1) + \cdots + (-1)^r a_r/P_0(f_r)}$$

with:

$$a_k = \prod_{\substack{i=0 \\ i \neq k}}^{r} \frac{1}{\cos(2\pi f_k) - \cos(2\pi f_i)}$$

(1) The values of the function $H_R(f)$ are interpolated between the $f_i (0 \leqslant i \leqslant r)$ in order to calculate $e(f)$.
(2) The extremal frequencies obtained in the previous phase are taken as the initial values for the next stage.

Figure 5.12 shows the evolution of the error function through one stage of the calculation. The procedure is halted when the difference between the value $\delta$ is calculated on the basis of the new extremal frequencies, iteratively. The new $\delta$ value is compared with the preceding value. When the difference between two consecutive values of $\delta$ falls below a certain threshold, the procedure is halted. Usually, this occurs after a few iterations.

The convergence of this procedure is related to the choice of initial values for the frequencies $f_i$. For the first iteration, the extremal frequencies obtained using a different method of calculation can be used as the filter coefficients. Even more simply, a uniform distribution of the extremal frequencies over the frequency range can be assumed.

As in the least squares method presented in the previous section, the values $H_R(f)$ have to be interpolated. Because of the non-uniform distribution of the extremal frequencies, it is more convenient

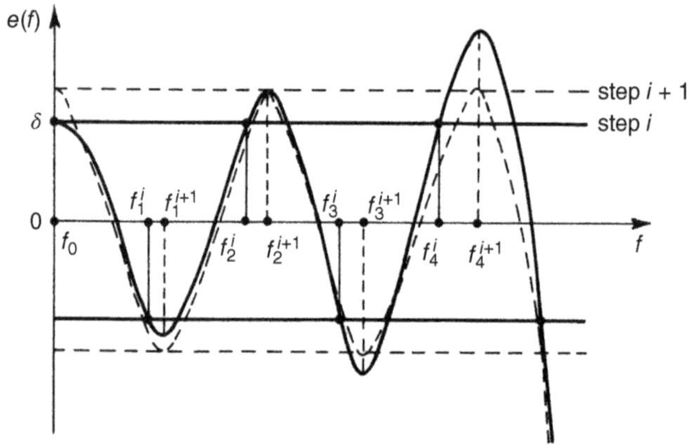


**Figure 5.12** Evolution of the error function in a stage of the Remez algorithm.

to use the Lagrange interpolation formulae:

$$H_R(f) = \frac{\sum_{k=0}^{r-1} \{\beta_k/(x - x_k)\}[D(f_k) - (-1)^k \delta / P_0(f_k)]}{\sum_{k=0}^{r-1} \beta_k (x - x_k)} \tag{5.31}$$

with:

$$\beta_k = \prod_{\substack{i=0 \\ i \neq k}}^{r-1} \frac{1}{x_k - x_i} \quad \text{and} \quad x = \cos(2\pi f)$$

At the end of the iteration, the extremal frequencies obtained are used to produce samples with a constant frequency increment which, by an inverse discrete Fourier transformation, produces the coefficients of the filter.

Filters having several hundred coefficients can be designed using this technique, and it applies to low-pass, high-pass, and band-pass filters with or without fixed phase shift [2].

## 5.7 Relationships Between the Number of Coefficients and the Filter Characteristic

In the calculation techniques which have been considered previously, the number of coefficients $N$ of the filter was assumed to have been given *a priori*. In practice, then, $N$ is an important parameter (for example, in projects where the computational complexity needed to implement a digital filter satisfying given specifications must be determined).

For a low-pass filter, as shown in Figure 5.7, the specifications concern the pass-band ripple $\delta_1$, the stop-band ripple $\delta_2$, the pass-band edge $f_1$, and the transition band $\Delta f = f_2 - f_1$. An analysis of the results obtained in the design of a large number of filters with a wide range of specifications shows that the number of coefficients is proportional to the logarithms of $1/\delta_1$ and $1/\delta_2$, and also to the ratio of the sampling frequency $f_s$ to the transition band $\Delta f$. The adjustment of parameters leads to the following estimate $N_e$ for the filter order:

$$N_e = \frac{2}{3} \log\left(\frac{1}{10\delta_1\delta_2}\right) f_s/\Delta f \tag{5.32}$$

This simple estimate is sufficient for most practical cases. It clearly indicates the relative importance of the various parameters. The transition band $\Delta f$ is the most sensitive, with the pass-band and stop-band ripples having less significant impact. For example, if $\delta_1 = \delta_2 = 0.01$, dividing one of these figures by 2 leads to an increase of only 10% in the filter order. Moreover, it is worth emphasizing that, according to the estimation, the filter complexity is independent of the pass-band width.

*Examples:*

($f_s$ is taken to be 1.0 unless otherwise stated.)

(1) Calculation for a filter with 39 coefficients ($N = 39$) leads to the following values:

$$\delta_1 = 0.017; \quad \delta_2 = 0.034; \quad f_1 = 0.14375;$$
$$f_2 = 0.14375; \quad \Delta f = 0.04$$

With these values for the parameters, the estimate gives $N_e = 40$.

(2) A filter with 160 coefficients ($N = 160$) has the following parameters:

$$\delta_1 = 2.24 \times 10^{-2}; \quad \delta_2 = 1.12 \times 10^{-4}; \quad f_1 = 0.053125;$$
$$f_2 = 0.071875; \quad \Delta f = 0.01875$$

The estimate gives $N_e = 164$.

It should be noted that non-negligible differences can occur between the value $N$ which is necessary in practice and the estimated value $N_e$, when the transition-band limits approach 0 and 0.5, or when $N$ is a few units. A set of more complex formulae is given in Reference [3]. As indicated in Chapter 7, a high-pass filter can be derived from a low-pass one by inverting the sign of every second coefficient. Consequently, estimate (5.32) can also apply to high-pass filters. When the mask specifies different ripples for the pass and stop bands, an overestimation of the number of coefficients can be obtained by assuming the most rigid constraints on $\delta_1$ and $\delta_2$ in the pass and stop bands, respectively.

For band-pass filters, it is necessary to introduce several transition bands.

Figure 5.13 gives the characteristic of such a filter, which has two transition bands $\Delta f_1$ and $\Delta f_2$. Experience shows that the number of coefficients $N$ depends essentially on the smallest bandwidth, $\Delta f_m = \min(\Delta f_1, \Delta f_2)$. Then estimate (5.32) can be applied, with $\Delta f = \Delta f_m$. An upper

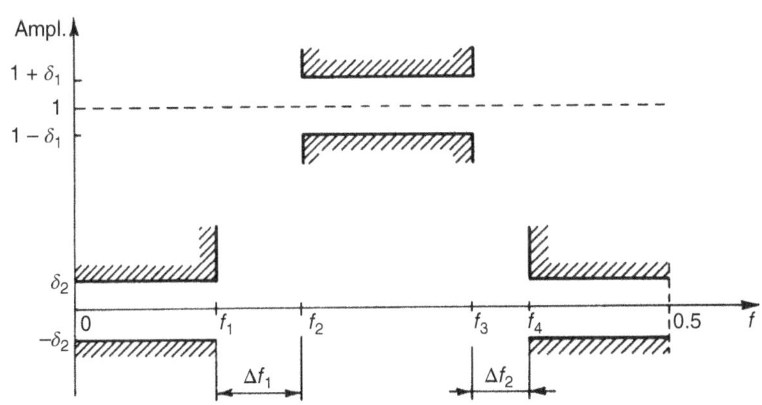


**Figure 5.13** Mask of a band-pass filter.

bound is obtained by considering the band-pass filter as a cascade of low-pass and high-pass filters and summing the estimates.

*Example*: A band-pass filter with 32 coefficients ($N = 32$) has the following characteristics:

$$\delta_1 = 0.015; \quad \delta_2 = 0.0015; \quad f_1 = 0.1; \quad f_1 = 0.2;$$

$$f_3 = 0.35; \quad f_4 = 0.425; \quad \Delta f_m = 0.075$$

The estimate using equation (5.32) with $\Delta f = \Delta f_{\mathrm{m}}$ gives $N_{\mathrm{e}} = 32$. A set of elaborate formulae to estimate band pass filter orders is given in Reference [3].

The estimation formulae can be used to complete a computer program for the filter coefficient calculations by determining the number $N$ at the beginning of the program. Expression (5.32) is very useful in practice for complexity assessment.

When the frequency responses of filters are examined in the transition band, it is evident that they approximate to raised-cosine form; the approximation becomes closer as the pass-band and attenuation-band ripples become closer. In fact, this type of response corresponds to the specifications imposed on data transmission using Nyquist filters, and it represents another approach to linear-phase FIR filters.

## 5.8 Raised-Cosine Transition Filter

The frequency response $H(f)$ of a filter whose transition band has raised-cosine form is represented in Figure 5.14(a).

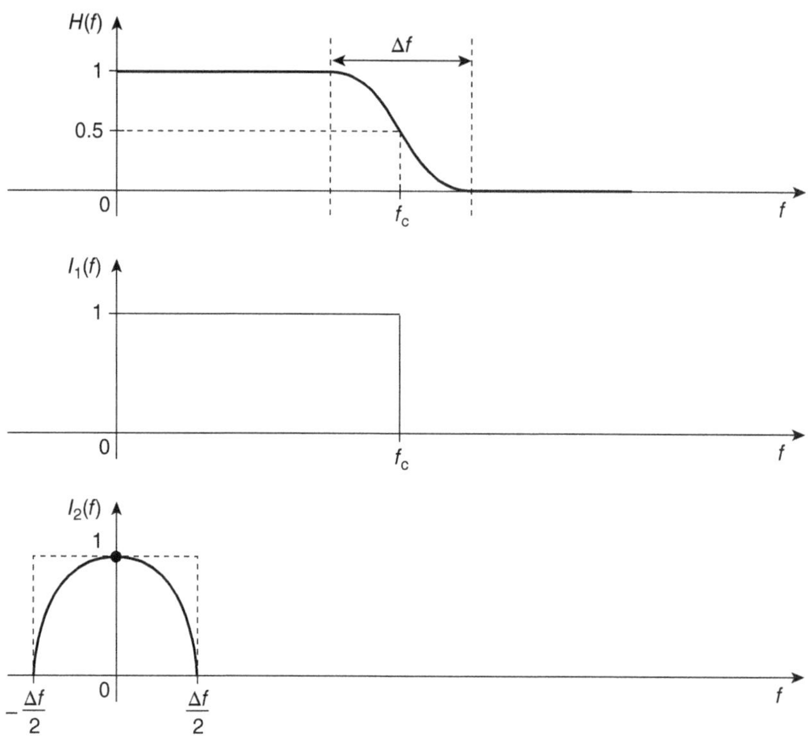


**Figure 5.14** (a) Response with raised-cosine transition; (b) frequency impulse of width $2f_c$; (c) frequency impulse for the transition.

It is expressed as the following convolution product:

$$H(f) = I_1(f) * \left[ I_2(f)\frac{\pi}{2\Delta f}\cos\left(\frac{\pi f}{\Delta f}\right)\right] \tag{5.33}$$

where $I_1$ is an impulse of width $2f_c$ and $I_2(f)$ is an impulse of width $\Delta f$.

Under these conditions, the impulse response $h(t)$ can be written as the product of two impulse responses $i_1(t)$ and $i_2(t)$ given by:

$$i_1(t) = 2f_c\frac{\sin \pi 2f_c t}{2\pi f_c t}$$

and

$$i_2(t) = \frac{1}{2}\frac{1\sin \pi\Delta f(t + 1/2\Delta f)}{2\Delta f(t + 1/2/\Delta f)} + \frac{1}{2}\frac{1\sin \pi\Delta f(t - 1/2\Delta f)}{2\Delta f(t - 1/2/\Delta f)}$$

After simplification, this gives:

$$h(t) = 2f_c = \frac{\sin 2\pi f_c t}{2\pi f_c t}\frac{\cos \pi f t}{1 - 4\Delta f^2 t^2} \tag{5.34}$$

The total number of coefficients of the filter is determined principally by the function $i_2(t)$ and the width of its main lobe, equal to $3/\Delta f$.

Hence, in a raised-cosine transition band filter, the number of coefficients can be estimated using:

$$N \approx \frac{3f_s}{\Delta f} \tag{5.35}$$

This estimate can be considered as a first approach, when it is compared with equation (5.32).

The digital filter coefficients are obtained by sampling $h(t)$ with $|i| \leq P$ and $N = 2P + 1$:

$$h_i = \frac{2f_c}{f_s}\frac{\sin 2\pi i\frac{f_c}{f_s}}{2\pi i\frac{f_c}{f_s}}\frac{\cos \pi i\frac{\Delta f}{f_s}}{1 - 4\left(\frac{\Delta f}{f_s}\right)^2 i^2} \tag{5.36}$$

Note that this expression can be applied to any filter and provides a more accurate estimation of the coefficients than (5.18).

The above results can be generalized to any symmetrical transition band:

$$H(f_c + f) = 1 - H(f_c - f);\quad |f| \leq \frac{\Delta f}{2}$$

These filters, called Nyquist filters, play a crucial role in digital transmission. Thc impulse response is zero at multiples of $1/2f_c$, which determines the interference free transmission rate.

An important particular case is the half-band filter such that $f_c = f_s/4$. Then, even coefficients are null and, for $N = 4M + 1$ coefficients, the input–output relationship is:

$$y(n) = \frac{1}{2}\left[x(n - 2M) + \sum_{i=1}^{M} h_{2i-1}[x(n - 2M + 2i - 1) + x(n - 2M - 2i + 1)]\right]$$

The frequency response is:

$$H(f) = e^{-j2\pi 2Mf}\frac{1}{2}\left[1 + 2\sum_{i=1}^{M} h_{2i-1}\cos(2\pi(2i - 1)f)\right]$$

With its reduced number of computations, this filter is a basic building block in multirate filtering.

In communications, the filtering function is generally shared between transmitter and receiver, which leads to the half-Nyquist filter – for example, with cosine transition band:

$$H^{\frac{1}{2}}(f) = 1; \qquad |f| \leq f_c - \frac{\Delta f}{2}$$

$$H^{\frac{1}{2}}(f) = \frac{\cos\left(\pi\left[f - \left(f_c - \frac{\Delta f}{2}\right)\right]\right)}{2\Delta f}; \quad f_c - \frac{\Delta f}{2} \leq f \leq f_c + \frac{\Delta f}{2}$$

$$H^{\frac{1}{2}}(f) = 0; \qquad |f| \leq f_c + \frac{\Delta f}{2}$$

The impulse response is:

$$h^{\frac{1}{2}}(t) = \frac{4\frac{\Delta f}{\pi}\cos 2\pi t\left(f_c + \frac{\Delta f}{2}\right) + \frac{1}{\pi t}\sin 2\pi t\left(f_c - \frac{\Delta f}{2}\right)}{1 - (4\Delta f t)^2} \tag{5.37}$$

As mentioned above, digital filter coefficients are obtained by sampling – that is, by substituting $i/f_s$ for $t$ in (5.37).

## 5.9 Structures for Implementing FIR Filters

FIR filters are composed of circuits which carry out the three fundamental operations of storage, multiplication, and addition. They are arranged to produce from the set of data $x(n)$ an output set $y(n)$ according to the equation of definition of the filter. As no real operation can be instantaneous, the equation which is implemented in place of equation (5.5) in Section 5.1 is:

$$y(n) = \sum_{i=0}^{N-1} a_i x(n - i - 1) \tag{5.38}$$

$N$ data memories are required, and for each output number, $N$ multiplications and $N-1$ additions have to be performed. Various arrangements of the circuits can be devised to put these operations into effect [4–6].

Figure 5.15(a) outlines the filter in what is called the "direct" structure. The transposition of the diagram of this scheme produces the "transposed" structure. This is represented in Figure 5.15(b), where the same operators are arranged differently. This structure allows multiplication of each piece of data $x(n)$ by all of the coefficients in succession. The memories store partial sums, and, in effect, at time $n$, the first memory stores the number $a_{N-1}\, x(n)$, the next memory stores $a_{N-1}\, x(n-1) + a_{N-2}\, x(n)$, and the last memory stores the sum $y(n)$. The difference between the two structures lies in the position of the memories. An intermediate structure having two memories per coefficient can also be envisaged, the internal data being stored for the duration $T/2$ in each. The cascade structure so obtained has local connections only.

In linear-phase filters, the symmetry of the coefficients can be exploited to halve the number of multiplications to be performed for each output number. This is very important for the complexity of the filter and justifies the almost universal use of linear-phase filters.

The complexity of the circuits depends on the number of operations to be performed and on their precision. Thus, the multiplication terms should have the minimum possible number of bits, which will reduce the amount of memory necessary both for the coefficients and for the data. These limitations modify the characteristics of the filter.

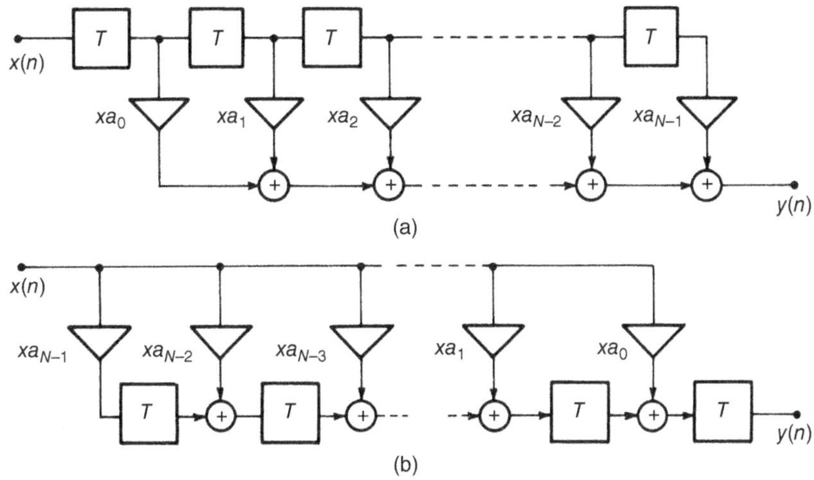


**Figure 5.15** Realization of FIR filters, (a) Direct structure; (b) transposed structure.

## 5.10 Limitation of the Number of Bits for Coefficients

Limitation of the number of bits in the coefficients of a filter produces a change in the frequency response, which appears as the superposition of an error function. The consequences will be analyzed for linear-phase filters, and it is a straightforward matter to extend the results to any FIR filter.

Consider a linear-phase filter with $N = 2P + 1$ coefficients, for which the transfer function is written according to Section 5.2, equation (5.13), as:

$$H(f) = e^{-j2\pi fPT}\left[h_0 + 2\sum_{i=1}^{P} h_i \cos(2\pi fiT)\right]$$

The limitation of the number of bits in the coefficients introduces an error $\delta h_i$ $(0 \leqslant i \leqslant P)$ in the coefficient $h_i$ which, under the assumption of rounding with a quantization step $q$, is such that:

$$|\delta h_i| \leq q/2$$

This results in the superposition on the function $H(f)$ of an error function $e(f)$ such that:

$$e(f) = e^{-j2\pi fPT}\left[\delta h_0 + 2\sum_{i=1}^{P} \delta h_i \cos(2\pi fiT)\right] \tag{5.39}$$

The amplitude of this function must be limited so that the response of the real filter is within the required specification. An upper limit is obtained from:

$$|e(f)| \leqslant |\delta h_0| + 2\sum_{i=1}^{P} |\delta h_i|\,|\cos(2\pi fiT)|$$

$$|e(f)| \leqslant \frac{q}{2}N \tag{5.40}$$

In general, this limit is much too large, and a more realistic estimate is obtained by statistical methods [7].

When a large number of filters with a wide range of specifications are considered and when general results are required, the variables $\delta h_i$ $(0 \leqslant i \leqslant P)$ can be viewed as random independent variables, uniformly distributed over the range $[-q/2, q/2]$. In these conditions, they have variance $q^2/12$. The function $e(f)$ can equally be regarded as a random variable. Assume that $e_0$ is the effective value of the function $e(f)$ over the frequency range $[0, f_s]$ – that is,

$$e_0^2 = \frac{1}{2}\int_0^{f_s} |e(f)|^2 df \tag{5.41}$$

In fact, the function $e(f)$ is periodic and defined by its Fourier series expansion. The Bessel–Parseval equation allows us to write, according to equation (5.8):

$$\frac{1}{f_s}\int_0^{f_s} |e(f)|^2 df = \sum_{i=0}^{N-1} (\delta h_i)^2$$

As a consequence, the variance $\sigma^2$ of the random variable $e_0$ can be written as:

$$\sigma^2 = E\left[e_0^2\right] = N(q^2/12)$$

By assuming the second-order moment to be independent of frequency and by using equation (5.41), the variable $e(f)$ can be regarded as a random variable with variance such that:

$$\sigma^2 = q/2\sqrt{(N/3)} \tag{5.42}$$

This relation provides an estimate of $|e(f)|$ which is much less than the bound equation (5.40). In fact, $e(f)$, resulting from equation (5.39) as a weighted sum of variables which are assumed to be independent, can be regarded as a Gaussian variable with a zero mean if quantization is performed by rounding and with variance $\sigma^2$. In order to determine the quantization step $q$, one can also base the reasoning on the confidence levels. For example, the probability that $|e(f)|$ will exceed the value $2\sigma$ is less than 5% using the table given in Appendix 2 of Chapter 1.

The above results will now be used to estimate the number of bits $b_c$ needed to represent the coefficients of a filter specified by a mask.

Given a filter specification, let $\delta_m$ be the required value for the amplitude of the ripple and $\delta_0$ the amplitude of the ripple before the number of bits in the coefficients is limited. The parasitic function $e(f)$ must be such that:

$$|e(f)| < \delta_m - \delta_0$$

The level of confidence in the estimate is deemed sufficient if $q$ is chosen as:

$$\frac{q}{2}\sqrt{\frac{N}{3}} < \frac{\delta_m - \delta_0}{2}$$

or

$$q < (\delta_m - \delta_0)\sqrt{(3/N)} \tag{5.43}$$

The number of bits $b_c$ needed to represent the coefficients depends on the largest of the values $h_i$ $(0 \leqslant i \leqslant P)$, and, allowing for the sign, the quantization step is given by:

$$q = 2^{1-bc}[\max_{0\leqslant i\leqslant p}|h_i|] \tag{5.44}$$

If the filter is a low-pass one with a frequency response approaching unity in the pass band and corresponding to the mask in Figure 5.7, initial approximations of the values of the coefficients can be calculated by equation (5.18). Under these conditions, the maximum for $h_0$ is obtained with:

$$h_0 = (f_1 + f_2)f_s \tag{5.45}$$

and equations (5.41) and (5.42) lead to the following estimate:

$$b_c \approx 1 + \log_2 \left[ \frac{f_1 + f_2}{f_s} \cdot \sqrt{\frac{N}{3}} \cdot \frac{1}{\delta_m - \delta_0} \right] \tag{5.46}$$

where:

$b_c$ is the number of bits of the coefficients (including the sign),
$N$ is the number of coefficients of the filter,
$f_1$ is the pass-band edge,
$f_2$ is the stop-band edge,
$f_s$ is the sampling frequency,
$\delta_m$ is the limit imposed on the amplitude of the ripple,
$\delta_0$ is the amplitude of the ripple of the filter before limitation of the number of bits in the coefficients.

In practice, equation (5.46) can be simplified. First, the tolerance provided by the filter mask is generally distributed equally between the ripples before limitation of the number of bits in the coefficients and before the supplementary error caused by this limitation (that is, $\delta_0 = \delta_m/2$). Further, the filters to be realized are generally such that:

$$0.5 f_s/\Delta f \leq N/3 \leq 1.5 f_s/\Delta f \tag{5.47}$$

which, using equation (5.32), corresponds to a large range of values for the parameters $\delta_1$ and $\delta_2$. Under these conditions, using estimation (5.35) and substituting $\frac{f_s}{\Delta f}$ for N/3, a suitable estimate of the number of bits of the coefficients is given by:

$$b_c \approx 1 + \log_2 \left[ \frac{f_1 + f_2}{f_s} \cdot \sqrt{\frac{f_s}{\Delta f}} \cdot \frac{2}{\delta_m} \right]$$

By introducing the steepness of the cutoff of the filter (equation (5.17)) and the normalized transition band, one finally obtains:

$$b_c \simeq 3 + \log_2[(f_1 + f_2)/2\Delta f] - \frac{1}{2}\log(f_s/\Delta f) + \log_2(1/\min\{\delta_1, \delta_2\}) \tag{5.48}$$

Thus, the number of bits for the coefficients is directly related to the filter specifications. It should be stated that filters with a narrow pass band require fewer bits than filters with a wide band. Equations (5.46) and (5.48) also apply to high-pass and band-pass filters.

The above analysis was performed under the assumption that the two operations of the calculation of the coefficients and limitation of the number of bits are separate. It is equally possible to take an overall view of these two operations, and techniques exist for the appropriate calculations, although they are somewhat complicated [8].

## 5.11 *Z*-Transfer Function of an FIR Filter

The *Z*-transfer function for an FIR filter with $N$ coefficients is a polynomial of degree $N-1$, which is written (equation (5.7)) as:

$$H(Z) = \sum_{i=0}^{N-1} a_i Z^{-i}$$

This polynomial has $N-1$ roots $Z_i \cdot (1 \leqslant i \leqslant N-1)$ in the complex plane and can be written as the product:

$$H(Z) = a_0 \prod_{i=1}^{N-1} (1 - Z_i Z^{-1}) \tag{5.49}$$

These roots have certain characteristics because of the properties of FIR filters.

Firstly, if the coefficients are real, each complex root $Z_i$ has a corresponding complex conjugate root $\overline{Z}_i$. Hence, $H(Z)$ can be written as a product of first- and second-degree terms with real coefficients. Each second-degree term is thus written as:

$$H(Z) = 1 - 2\text{Re}(Z_i)Z^{-1} + |Z_i|^2 Z^{-2} \tag{5.50}$$

Secondly, the symmetry of the coefficients of a linear-phase filter must appear in the decomposition into products of factors. For a second-degree term with real coefficients, it is necessary that $|Z_i| = 1$ if the roots are complex – that is, the root must lie on the unit circle. For a fourth-degree term with real coefficients, the four complex roots have to be $Z_i, Z_i, \overline{Z}_i, 1/Z_i, 1/\overline{Z}_i$. That is,

$$\begin{aligned} H_4(Z) = 1 - 2\text{Re}\left(Z_i + \frac{1}{Z_i}\right) Z^{-1} + \left[|Z_i|^2 + \frac{1}{Z_i^2}\right] + 4\text{Re}\left(\frac{1}{Z_i}\right)\Bigg] Z^{-2} \\ - 2\text{Re}\left(Z_i + \frac{1}{Z_i}\right) Z^{-3} + Z^{-4} \end{aligned} \tag{5.51}$$

Under these conditions, an FIR linear-phase filter can be decomposed into a set of elementary filters of second or fourth degree, having the symmetry properties of the coefficients.

The roots have been calculated for a low-pass filter with 15 coefficients. The coordinates of the 14 roots are:

$$\begin{aligned} Z_1 &= -0.976 \pm j0.217 & Z_5 &= 0.492 \pm j0.266 \\ Z_2 &= -0.797 \pm j0.603 & Z_6 &= 1.573 \pm j0.851 \\ Z_3 &= -0.512 \pm j0.859 & Z_7 &= 0.165 \\ Z_4 &= -0.271 \pm j0.962 & Z_8 &= 6.052 \end{aligned}$$

Their positions in the complex plane are shown in Figure 5.16. This illustrates the characteristics of the frequency response of the filter. The pairs of roots characteristic of the linear phase can be clearly seen. The configuration of the roots is modified if this constraint is no longer imposed.

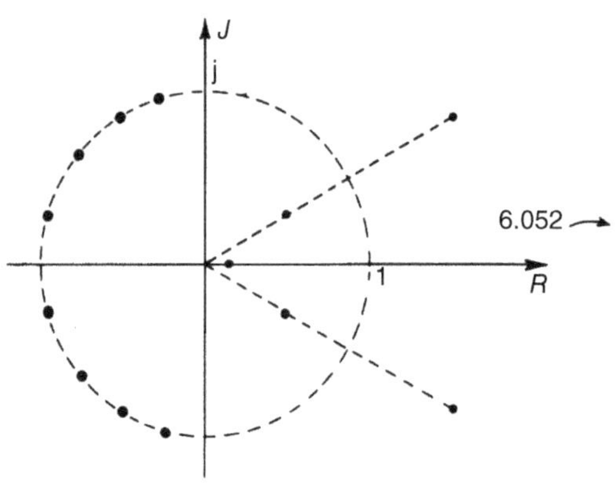


**Figure 5.16** Configuration of the zeros in an FIR filter.

It is interesting to observe that the frequency response of a linear phase filter of even length is zero at half sampling frequency, as shown by expression (5.14). Thus, such a filter has a zero at $f_s/2$. Similarly, if an odd-length filter has a zero at $f_s/2$, this is a double zero, which ensures the symmetry of the frequency response about $f_s/2$.

## 5.12 Minimum-Phase Filters

The propagation delay through a linear-phase filter can be excessive for some applications. Also, it is not always possible or useful to use symmetry of the coefficients of a linear-phase filter to simplify the calculations [9]. Therefore, if phase linearity is not a strictly imposed constraint, the complexity of the filter may be reduced by abandoning it. A linear-phase transfer function can be considered as the product of a minimum-phase function and a pure phase shift. The condition for a $Z$-transfer function to be minimum phase is that its roots must be within or on the unit circle. This point is developed in Chapter 9.

The calculation methods developed for linear-phase filters can be adapted. However, the coefficients for a minimum-phase filter can be obtained in a simple way using the coefficients of an optimal linear-phase filter.

Let the frequency response of a linear-phase filter with $N = 2P + 1$ coefficients be written as:

$$H(f) = e^{-j2\pi fPT}\left[h_0 + 2\sum_{i=1}^{P} h_i \cos(2\pi f iT)\right]$$

The ripples in the pass and stop bands are $\delta_1$ and $\delta_2$, respectively. Let us consider the filter obtained by adding $\delta_2$ to the above response and by rescaling it to approach unity in the pass band. Its response $H_2(f)$ will be:

$$H_2(f) = e^{-j2\pi fPT}\frac{1}{1+\delta_2}\left[h_0 + 2\sum_{i=1}^{P} h_i \cos(2\pi f iT)\right] \tag{5.52}$$

In the pass band, there are ripples with amplitude $\delta_1'$ such that:

$$\delta_1' = \frac{\delta_1}{(1+\delta_2)}$$

Its response in the stop band is represented in Figure 5.17; the ripples are limited to $\delta_2' = 2\delta_2/(1+\delta_2)$.

This is a linear-phase filter because the symmetry of the coefficients is conserved. In contrast, it can be noticed that the roots of the $Z$-transfer function which are on the unit circle are double because $H_2(f)$ never becomes negative. Under these conditions, the configuration of the roots is as shown in Figure 5.18.

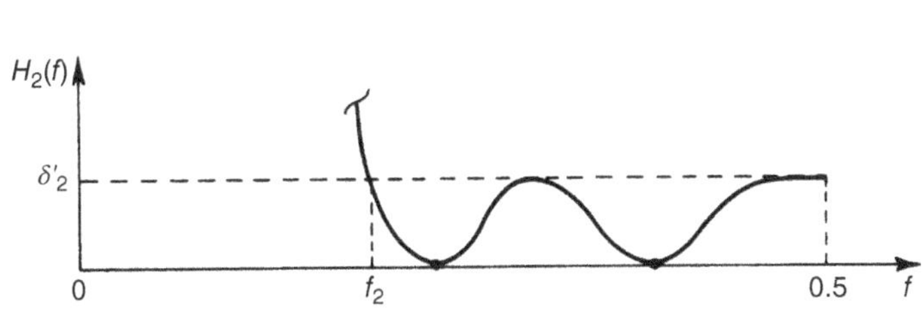


**Figure 5.17** Ripple in the stop band of the raised filter.

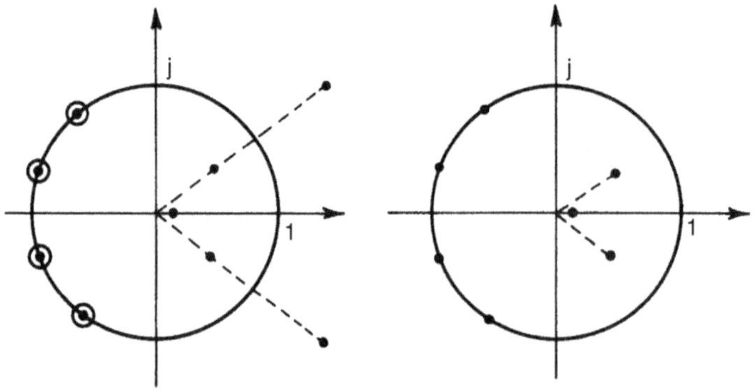


**Figure 5.18** Configuration of the zeros of $H_2(f)$ and of a minimum-phase filter.

The roots which are not on the unit circle are not double. However, the absolute value of the function $H_2(f)$ is not modified, except for a constant factor, if the roots $Z_i$ outside the unit circle are replaced by the roots $1/Z_i$, which are inside the unit circle and thus also become double. This operation amounts simply to multiplication by $G(Z)$ such that:

$$G(Z) = \frac{[1 - (Z^{-I}/Z_i)][1 - (Z^{-1}/\overline{Z}_i)]}{(1 - Z_i Z^{-1})(1 - \overline{Z}_i Z^{-1})}$$

Since $Z^{-1} = \overline{Z}$ on the unit circle, the symmetry with respect to the real axis yields:

$$\left| \frac{(Z^{-1} - Z_i)(Z^{-1} - \overline{Z}_i)}{(Z - Z_i)(Z - \overline{Z}_i)} \right|_{Z=e^{j\infty}} = 1 \tag{5.53}$$

and thus:

$$|G(e^{j2\pi f})| = \frac{1}{|Z_i|^2}$$

Under these conditions, we can write:

$$H_2(f) = H_m^2(f)K \text{ ($K$ is a constant)}$$

where $H_m(f)$ is the response of a filter which has a $Z$-transfer function with $P$ roots which are single and inside or on the unit circle. This filter satisfies the minimum-phase condition, it has $P + 1$ coefficients, and the amplitudes of the ripples in the pass and stop band are $\delta_{m_1}$ and $\delta_{m_2}$ such that:

$$\delta_{m_1} = \sqrt{1 + \frac{\delta_1}{1 + \delta_2}} - 1 \approx \frac{1}{2}\frac{\delta_1}{1 + \delta_2} \approx \frac{\delta_1}{2} \tag{5.54}$$

$$\delta_{m_2} = \sqrt{\frac{2\delta_2}{1 + \delta_2}} \approx \sqrt{2\delta_2} \tag{5.55}$$

To design this filter, it is sufficient to start from the linear-phase filter whose parameters $\delta_1$ and $\delta_2$ are determined from $\delta_{m_1}$ and $\delta_{m_2}$ and to follow the procedure described above. One drawback is that the extraction of the roots of a polynomial of degree $N - 1$ is required, which limits the values that can be envisaged for $N$. Other procedures can also be used [10, 11].

An estimate $N_e'$ of the order of the FIR filter with minimum phase shift can be deduced from equation (5.32). According to the procedure described earlier for the specifications of $\delta_1$ and $\delta_2$, this becomes:

$$N_e' \approx \frac{1}{2}\cdot\frac{2}{3}\log\left(\frac{1}{10\delta_1\delta_2^2}\right)\frac{f_3}{\Delta f}$$

or

$$N_e' \approx N_e - \frac{1}{3}\log\left(\frac{1}{10\delta_1}\right)\frac{f_s}{\Delta f} \tag{5.56}$$

The validity of this formula is naturally limited to the case where $\delta_1 \ll 0.1$. The improvement in the order of the filter with minimum phase is a function of the ripple in the pass band. It generally remains relatively low.

*Example*: Assume the following specifications for a low-pass filter:

$$\delta_{m_1} = 0.0411;\quad \delta_{m_2} = 0.0137;\quad \Delta f = 0.0115$$

whence

$$\delta_1 = 0.0822;\quad \delta_2 = 0.0000938$$

The number of coefficients needed for the corresponding linear-phase filter is estimated at $N_e = 24$, which leads to $N_e' = 12$. Actually, it can be shown that the minimum-phase filter satisfying the characteristic requires 11 coefficients instead of the 15 for the linear-phase filter.

In conclusion, when the symmetries provided by the linear phase cannot be used, it may be advantageous to resort to minimum-phase filters.

## 5.13 Design of Filters with a Large Number of Coefficients

Optimization techniques become difficult to use, or they no longer converge, when the number of filter coefficients is very large – perhaps one thousand or more – corresponding to extremely narrow transition bands of the order of thousandths. One can then use suboptimal techniques which require only the calculation of filters with a reduced number of coefficients. This is the case in the method described as frequency masking [12].

Consider a filter $H(Z)$ whose transition band $\Delta f$ is centered on the cutoff frequency $f_c$. One starts by designing a low-pass filter $H_0(Z^M)$ with a reduced sampling frequency of $f_s/M$, where $M < f_s/4\Delta f$, such that the transition band of one of the alias frequencies of this filter coincides with the transition band of the required filter, as shown in Figure 5.19(b).

Two complementary filters are then constructed from $H_0(Z^M)$ as shown in Figure 5.19(c); this requires an odd number of coefficients, $2P + 1$, for $H_0(Z^M)$.

A diagram with two branches is obtained, to which are applied the filters $G_1(Z)$ and $G_2(Z)$; $G_1$ and $G_2$ are described as interpolators, and they have the responses given in Figure 5.19(c). It is then sufficient to sum the outputs to obtain the desired filter of Figure 5.19(a). The overall arrangement is shown in Figure 5.20.

The procedure thus requires three filters having transition bands of $M\ \Delta f$, $f_c - kf_s/M$ and $(k+1)f_s/M - f_c$, where $k$ is the integer which permits the cutoff frequency $f_c$ to be included.

The transfer function $H(Z)$ of the required filter takes the form:

$$H(Z) = H_0(Z^M)G_1(Z) + [Z^{-PM} - H_0(Z^M)G_2(Z)] \tag{5.57}$$

which provides the coefficient values.

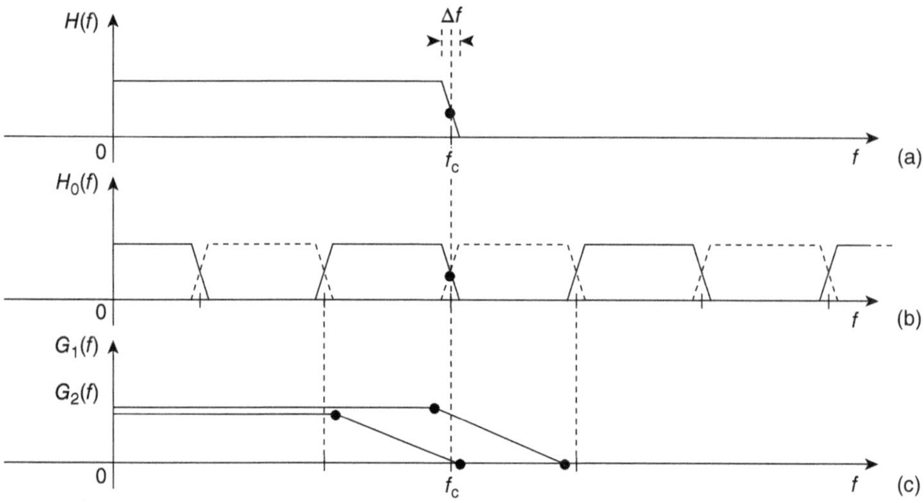


**Figure 5.19** The principle of frequency masking: (a) desired filter, (b) downsampled filter, (c) interpolating filters.

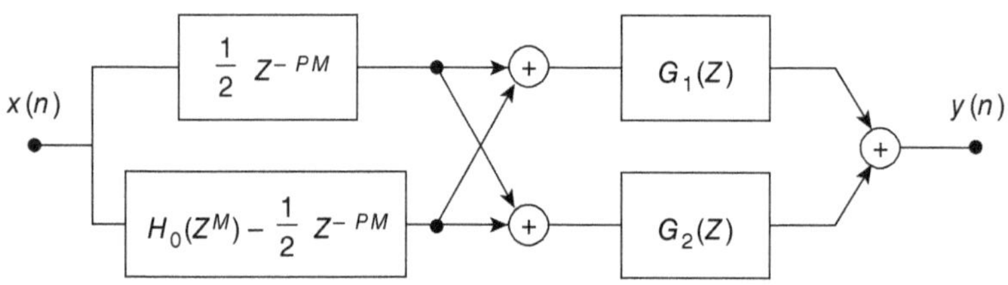


**Figure 5.20** Diagram of filter using the frequency-masking technique.

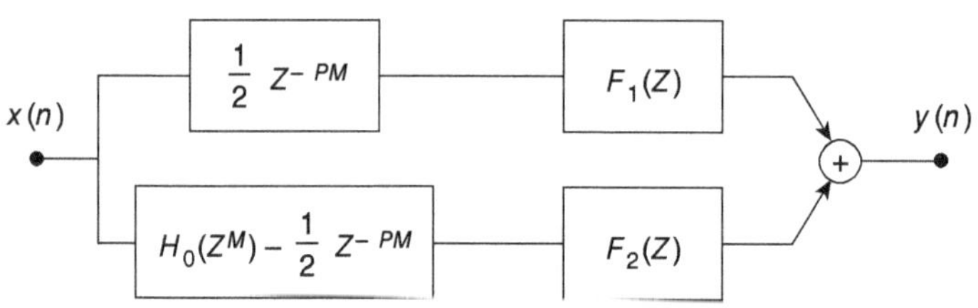


**Figure 5.21** Simplified diagram of the frequency-masking filter.

Note that the arrangement shown in Figure 5.20 provides an efficient realization of the overall filter since the filter $H_0(Z^M)$ has $M-1$ zero coefficients among two nonzero coefficients. This arrangement can be simplified as shown in Figure 5.21. The interpolating filters can be taken as $F_1(Z) = G_1(Z) + G_2(Z)$ and $F_2(Z) = G_1(Z) - G_2(Z)$ but they can also be derived directly from their specifications, as deduced from Figure 5.19.

## 5.14 Two-Dimensional FIR Filters

A two-dimensional FIR filter is defined by a relationship between the input $x(n, m)$ and the output $y(n, m)$ expressed by:

$$y(n,m) = \sum_{i=0}^{N_1-1} \sum_{j=0}^{N_2-1} a_{ij} x(n-i, m-j) \tag{5.58}$$

The set of coefficients $a_{ij}$ takes the form of a $N_1 \times N_2$ matrix denoted $A_{N_1N_2}$. The corresponding two-variable transfer function:

$$H(Z_1, Z_2) = \sum_{i=0}^{N_1-1} \sum_{j=0}^{N_2-1} a_{ij} Z_1^{-i} Z_2^{-j} \tag{5.59}$$

is given in vector notation by:

$$\boldsymbol{H(Z_1, Z_2)} = \left[1, \boldsymbol{Z_1^{-1}}, \dots, \boldsymbol{Z_1^{-(N_1-1)}}\right] [\boldsymbol{A_{N_1N_2}}] \begin{bmatrix} \boldsymbol{1} \\ \boldsymbol{Z_2^{-1}} \\ \vdots \\ \boldsymbol{Z_2^{-(N_2-1)}} \end{bmatrix} \tag{5.60}$$

The coefficient matrix $\boldsymbol{A_{N_1N_2}}$ is also called the mask. As an illustration, the following high-pass filters are often used in image processing:

$$A' = \begin{bmatrix} -1 & 0 & 1 \\ -1 & 0 & 2 \\ -1 & 0 & 1 \end{bmatrix}; \quad A'' = \begin{bmatrix} 1 & 1 & 1 \\ 1 & -2 & 1 \\ -1 & -1 & -1 \end{bmatrix}$$

In some edge-extraction processes, the SOBEL filter $A'$ and the PREVITT filter $A''$ are employed twice – once as above and once after a 90° rotation.

The coefficients of the two-dimensional filters can be computed directly from specifications in the two-dimensional frequency domain. When the impulse response is an even function, with respect to the two variables, the frequency response and the coefficients can be derived from those of a linear-phase one-dimensional filter. Let $H(\omega)$ be the frequency response of such a filter.

Equation (5.13), when the phase term is neglected, yields:

$$H(\omega) = h_0 + 2\sum_{i=1}^{P} h_i \cos i\omega$$

A polynomial relation exists between $\cos i\omega$ and $\cos \omega$:

$$\cos i\omega = T_i(\cos \omega) \tag{5.61}$$

where $T_i(x)$ is the Chebyshev polynomial of degree $i$. Under these conditions, $H(\omega)$ can also be written as:

$$H(\omega) = \sum_{i=0}^{P} g_i (\cos \omega)^i \tag{5.62}$$

Changing variables as follows:

$$\cos \omega = H_1(\omega_1, \omega_2) = \sum_{k=0}^{k-1} \sum_{l=0}^{L-1} t(k, l) \;\; \cos k\omega_1 \cos l\omega_2 \tag{5.63}$$

yields a two-variable function:

$$H\left(e^{\omega_1}, e^{\omega_2}\right) = \sum_{i=0}^{P} g_i \left( \sum_{k=0}^{k-1} \sum_{l-0}^{L-1} t(k, l) \cos k\omega_1 \cos l\omega_2 \right)^i \tag{5.64}$$

which can be rewritten as:

$$H(\omega_1, \omega_2) = \sum_{i=0}^{N_1-1} \sum_{j=0}^{N_2-1} h_{ij} \cos i\omega_1 \cos j\omega_2 \tag{5.65}$$

where:

$$N_1 = 2KP + 1; \quad N_2 = 2LP + 1$$

The function $t(k, l)$ can be chosen so as to map the points in the frequency response of the one-dimensional filter into contours in the $(\omega_1, \omega_2)$ plane. For example, the circular symmetry is approximately achieved by:

$$\cos\omega = \frac{1}{2}[\cos\omega_1 + \cos\omega_2 + \cos\omega_1 \cos\omega_2 - 1] \tag{5.66}$$

as can be seen from a series expansion of $\cos\omega_1$, with $\omega_1$ being small. The frequency response of a filter designed that way is shown in Figure 5.22.

The implementation of a two-dimensional filter can be obtained through straight application of equation (5.58). For filters derived from a one-dimensional function, expression (5.64) suggests an important simplification: in the one-dimensional FIR filter with $P+1$ coefficients $g_i$, the delays are replaced by two-dimensional sections corresponding to the function $H_1(\omega_1, \omega_2)$ [13].

Separable filters are particularly simple to produce; in this case, the coefficient matrix is dyadic – that is,

$$A_{N_1N_2} = V_1 V_2^t$$

where $V_1$ and $V_2$ are the vectors. Then, in accordance with equation (5.60), the transfer function factorizes:

$$H(z_1, z_2) = H_1(z_1)H_2(z_2) \tag{5.67}$$

The specifications of such filters are subject to limitations. First, they must have quadrantal symmetry along the coordinate axes. As shown in Figure 5.23, the useful frequency domain is divided into four parts: low pass/low pass (LL), low pass/high pass (LH), high pass/low pass (HL), and high pass/high pass (HH).

Consequently, the ripple specifications must be defined. For a two-dimensional low-pass filter, the HH domain is subjected to the attenuation of two filters, horizontal and vertical. An illustration is given in Figure 5.24 which shows the frequency response of a two-dimensional separable filter based on the half-band filter with coefficients: $a = [0.5\ 0.314\ 0\ -0.094\ 0\ 0.045\ 0\ -0.022]$.

Such filters can be realized by following the definition exactly – that is, a data table representing an image can be processed row by row with the horizontal filter and column by column with the vertical filter.

When the image is subjected to a horizontal scan as in television, the signal appears as one-dimensional and can be processed as such. If each row contains $N$ points, the transfer

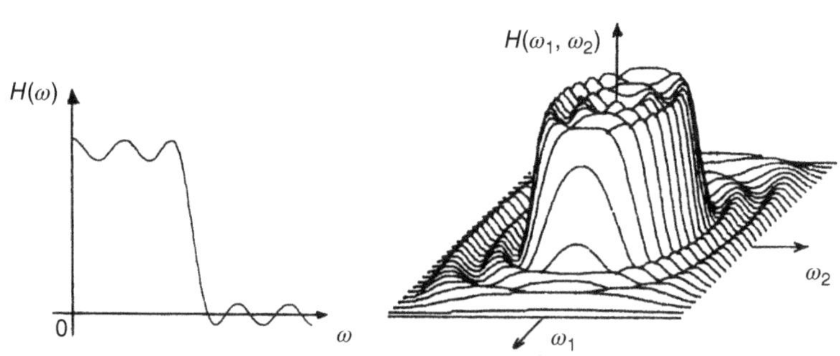


**Figure 5.22** Two-dimensional FIR filter designed from a one-dimensional linear-phase filter.

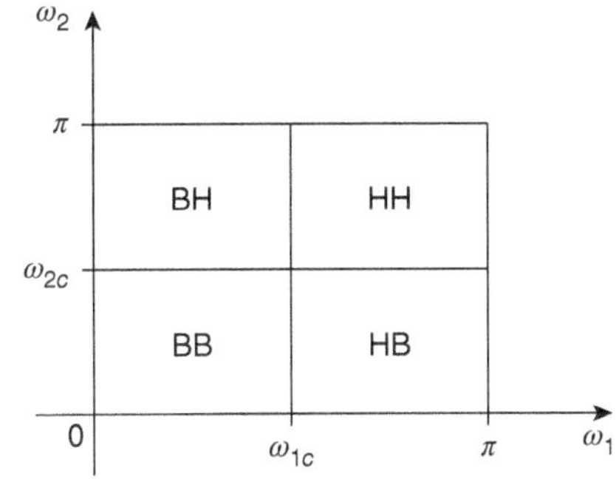


**Figure 5.23** Frequency domains for a two-dimensional separable filter.

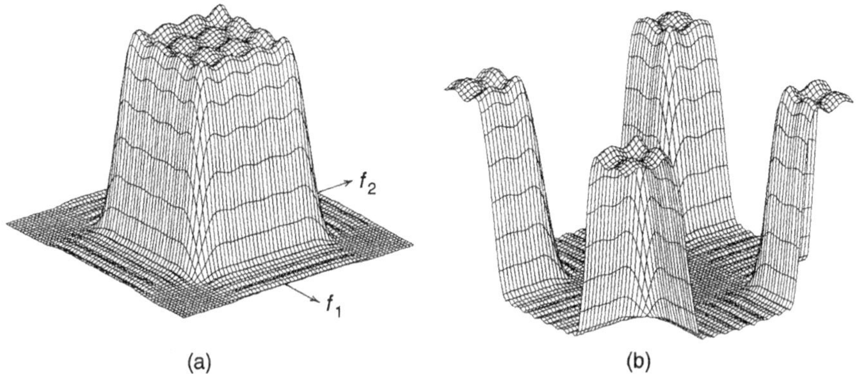


**Figure 5.24** Two-dimensional half-band separable filter: (a) low pass/low pass, (b) high pass/high pass.

function can be written as:

$$H(z_1, z_2) = H_1(z)H_2(z^N) \tag{5.68}$$

For example, for the Sobel filter $A'$, one has:

$$A' = \begin{bmatrix} 1 \\ 2 \\ 3 \end{bmatrix} \begin{bmatrix} -1 & 0 & 1 \end{bmatrix}$$

and the corresponding circuit is given in Figure 5.25. Realization is particularly simple as the circuits do not contain multipliers.

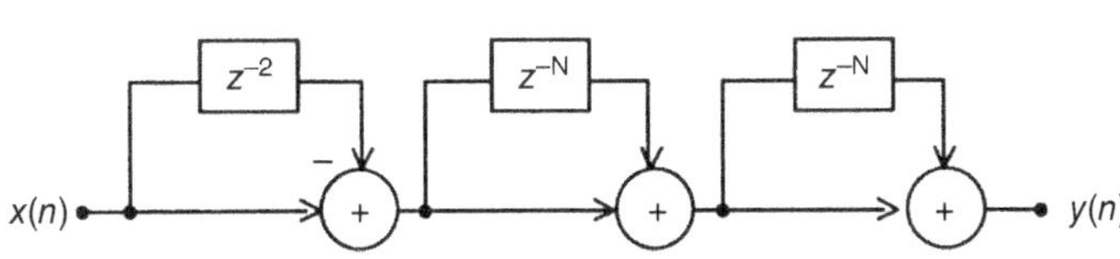


**Figure 5.25** Realization of a filter by contour extraction.

## 5.15 Coefficients of Two-Dimensional FIR Filters by the Least-Squares Method

The method will be developed for an important special case – filters with quadrantal symmetry. Two types of filters correspond to this category: rectangular and lozenge filters with the frequency domains shown in Figure 5.26.

The frequency response of a zero-phase filter having $(2M+1)\times(2N+1)$ coefficients with quadrantal symmetry is expressed by:

$$H(\omega_1,\omega_2) = h_{00} + 2\sum_{i=1}^{M} h_{i0}\cos i\omega_1 + 2\sum_{j=1}^{N} h_{0j}\cos j\omega_2 + 4\sum_{i=1}^{M}\sum_{j=1}^{N} h_{ij}\cos i\omega_1 \cos j\omega_2 \tag{5.69}$$

In total, the filter has $(1+M+N+MN)$ coefficients $h_{ij}$ with different values.

The least-squares method with weighting will be applied directly, to approach the desired response, $D(\omega_1,\omega_2)$. With an oversampling factor of $k$, the quadratic deviation function, or cost function, to be minimized is:

$$J = \sum_{m=0}^{K_M}\sum_{n=0}^{K_N}\left| H\left(\frac{m\pi}{K_M},\frac{n\pi}{K_N}\right) - D\left(\frac{m\pi}{K_M},\frac{n\pi}{K_N}\right)\right|^2 W\left(\frac{m\pi}{K_M},\frac{n\pi}{K_N}\right) \tag{5.70}$$

with $K_M = k(M+0.5)$ and $K_N = k(N+0.5)$ in order to cover all the useful frequency domains.

The weighting function $W(\omega_1,\omega_2)$ enables the approximation to be adjusted in accordance with the ripple specifications, for example.

With simplified notation, this gives:

$$J = \sum_{m=0}^{K_M}\sum_{n=0}^{K_N} e^2(m,n)W(m,n) \tag{5.71}$$

The minimum of the cost function is obtained for:

$$\sum_{m=0}^{K_M}\sum_{n=0}^{K_N} e^2(m,n)W(m,n)\frac{\partial e(m,n)}{\partial h_{ij}} = 0 \tag{5.72}$$

which yields a system with $(1+M+N+MN)$ equations.

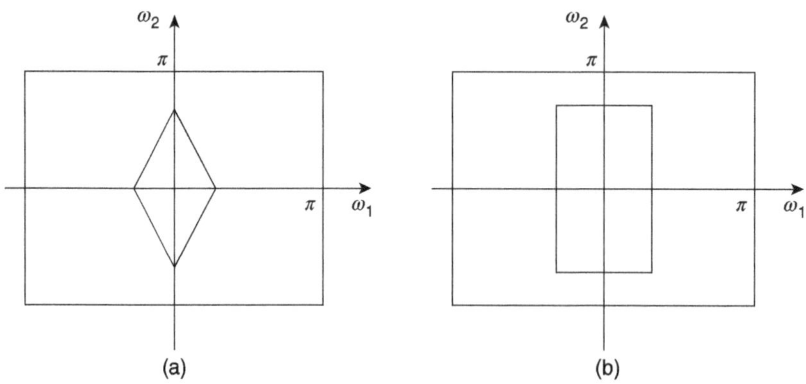


**Figure 5.26** Filters: (a) two-dimensional lozenge and (b) rectangle.

Designating the coefficient vector by $[h_{ij}]$ and the frequency vector by $V(m, n)$:

$$V^t(m,n) = \left[1, \ldots, 2\cos\left(i\frac{m\pi}{K_M}\right), \ldots, 2\cos\left(j\frac{n\pi}{K_N}\right), \ldots, 4\cos\left(i\frac{n\pi}{K_M}\right)\cos\left(j\frac{n\pi}{K_N}\right), \ldots\right]$$

the solution can be written as:

$$[h_{ij}] = \left[\sum_{m=0}^{K_M}\sum_{n=0}^{K_N} W(m,n)V(m,n)V^t(m,n)\right]^{-1}\left[\sum_{m=0}^{K_M}\sum_{n=0}^{K_N} W(m,n)V(m,n)D(m,n)\right] \quad (5.73)$$

If the number of coefficients is even, it is necessary to modify the parameters. For a filter with $(2M) \times (2N+1)$ coefficients, it is necessary to take:

$$V^t(m,n) = \left[\ldots, 2\cos\left((i-0.5)\frac{m\pi}{K_M}\right), \ldots, 4\cos\left((i-0.5)\frac{n\pi}{K_M}\right)\cos\left(j\frac{n\pi}{K_N}\right)\right] \quad (5.74)$$

with $K_M = kM$ and $K_N = k(N+0.5)$. The coefficient vector obtained in this case has $(M + MN)$ elements.

An important characteristic of filters used in image processing is the response to a unit step. Ringing at the transition can produce repetitions of contours and thus degrade the image. It is possible to reduce ringing by modifying the desired response $D(\omega_1, \omega_2)$ using a slope at the end of the pass band and the start of the attenuation band.

The method is illustrated by the design of a rectangular filter with $(2M+1) \times (2N+1) = 9 \times 9$ coefficients, with 0.125 and 0.25 as the end of the pass band and the start of the attenuation band on

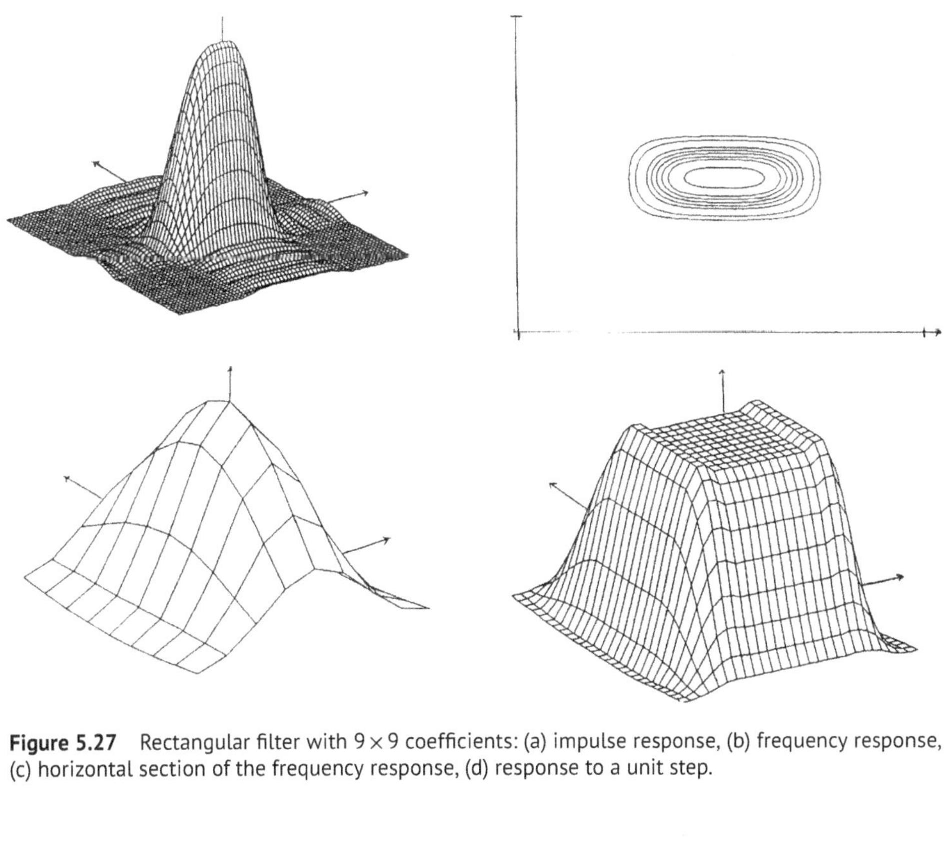

**Figure 5.27** Rectangular filter with $9 \times 9$ coefficients: (a) impulse response, (b) frequency response, (c) horizontal section of the frequency response, (d) response to a unit step.

the horizontal frequency axis and 0.0625 and 0.125 on the vertical axis. The 25 different coefficients obtained are:

$$h_{ij} = \begin{bmatrix} 0.052427 & 0.0419028 & 0.0184534 & -0.0002861 & -0.006258 \\ 0.0491981 & 0.0393451 & 0.0173566 & -0.0002629 & -0.0059292 \\ 0.041534 & 0.0332908 & 0.0147612 & -0.000261 & -0.005282 \\ 0.0299102 & 0.0240605 & 0.107414 & -0.0002704 & -0.0041828 \\ 0.0180912 & 0.0146366 & 0.0065523 & -0.0003836 & -0.0031209 \end{bmatrix}$$

and the corresponding frequency response is given in Figure 5.27. Evidently, this response is very close to that of a separable filter. Considering now a lozenge filter with $(2M+1)\times(2N) = 9\times 8$ coefficients, a pass band ending at 0.125, and an attenuation band starting at 0.25 on the horizontal and vertical axes, the coefficients for one quadrant are:

$$h_{ij} = \begin{bmatrix} 0.0763835 & 0.0680674 & 0.0403862 & 0.0130039 & 0.000071 \\ 0.0642979 & 0.03951 & 0.0217936 & 0.0008111 & -0.002745 \\ 0.0276109 & 0.0195655 & 0.0068997 & -0.0068997 & -0.0110481 \\ 0.0065124 & 0.0011002 & 0.0085984 & 0.0085984 & -0.0073724 \end{bmatrix}$$

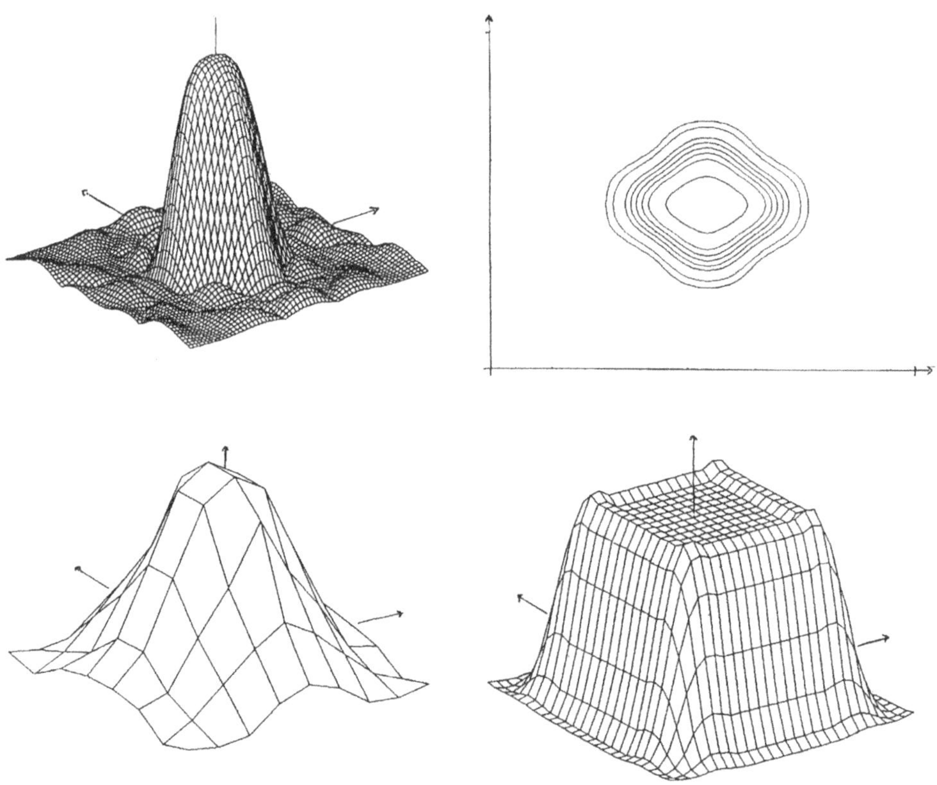

**Figure 5.28** Lozenge filter with 9 × 8 coefficients: (a) impulse response, (b) frequency response, (c) horizontal section of the frequency response, (d) response to a unit step.

The frequency response is given in Figure 5.28. Calculation has been guided by attempting to reduce the unit step response $g(i,j)$ defined by:

$$g(i,j) = \sum_{i_1=-M}^{i} \sum_{j_1=-N}^{j} h(i_1,j_1) \tag{5.75}$$

This response is also shown in the figure, where it has been repeated in the four quadrants to provide a complete picture.

See References [14, 15] for complementary developments in the design techniques for two-dimensional FIR filters, including those with limited precision coefficients and constraints on the unit step response.

## Exercises

**5.1** Consider the 17 coefficients of a low-pass filter with cutoff frequency $f_c = 0.25 f_s$ given in Figure 5.12. How many of these coefficients assume different values? Give the expression for the frequency response $H(f)$. Determine the frequencies for which it is zero and give the maximum ripple. Determine the zeros of the filter $Z$-transfer function.

**5.2** Consider a filter for which the sampling frequency is taken as the reference ($f_s = 1$) and whose frequency response $H(f)$ is such that:

$$H(k \times 0.0625) = 1 \quad \text{for} \quad k = 0, 1, 2, 3$$
$$H(0.25) = 0.5$$
$$H(k \times 0.0625) = 0 \quad \text{for} \quad k = 5, 6, 7, 8$$

Using the discrete Fourier transform, calculate the 17 coefficients of this filter. Draw the frequency response and determine the zeros of the $Z$-transfer function.

**5.3** Using the equations given in Section 5.7, determine the ripple of a low-pass filter with 17 coefficients for which the upper frequency of the pass band is $f_1 = 0.2$ and the lower frequency of the stop band is $f_2 = 0.3$. Compare the results obtained with those in the preceding exercises.

**5.4** Consider a filter with a transfer function $H(f)$ which, except for the phase shift, is given by the equation:

$$H(f) = h_0 + 2\sum_{i=0}^{4} h_{2i-1} \cos[2\pi f(2i-1)T]$$

Give the direct and transposed structures which allow this filter to be achieved with a minimum number of elements. What simplifications are involved if the sampling frequency of the output is divided by two?

**5.5** A narrow-band low-pass filter is defined by the equation

$$y(n) = \sum_{i=0}^{N-1} a_i x(n-i)$$

How is the frequency response modified if the coefficients $a_i$ are replaced by $a_i(-1)^i$ and by $a_i\cos(i\pi/2)$? How are the $Z$-transfer function zeros affected?

**5.6** Consider a low-pass filter which satisfies Figure 5.7 with the following values for the parameters

$$f_1 = 0.05; \quad f_2 = 0.15; \quad \delta_1 = 0.01; \quad \text{and} \quad \delta_2 = 0.001$$

How many coefficients are needed and how many bits are required to represent them? If the input data have 12 bits and if the signal-to-noise ratio degradation is limited to $\Delta SN = 0.1$ dB, what is the internal data wordlength?

## References


1 T. W. Parks and J. H. McClellan, Chebyshev approximation for non recursive digital filters with linear phase. *IEEE Transactions on Circuits and Systems*, 19(2), 189–194, 1972.

2 J. McClellan, T. Parks and L. Rabiner, "A computer program for designing optimum FIR linear phase digital filters," *IEEE Transactions on Audio and Electroacoustics*, 21, 6, pp. 506–526, 1973, 10.1109/TAU.1973.1162525.

3 J. Shen and G. Strang, The asymptotics of optimal (equiripple) filters. *IEEE Transactions*, 47(4), 1087–1098, 1999.

4 R. E. Crochiere and A. V. Oppenheim, "Analysis of linear digital networks," *Proceedings of the IEEE*, vol. 63, no. 4, pp. 581–595, 1975, 10.1109/PROC.1975.9793.

5 W. Schussler, *On Structures for Non Recursive Digital Filters*, Archiv der elektrischen Übertragung, 1972.

6 M. Bellanger and G. Bonnerot, Premultiplication scheme for digital FIR filters. *IEEE Transactions*, 26(1), 50–55, 1978.

7 D. Chan and L. Rabiner, "Analysis of quantization errors in the direct form for finite impulse response digital filters," *IEEE Transactions on Audio and Electroacoustics*, vol. 21, no. 4, pp. 354–366, 1973, 10.1109/TAU.1973.1162497.

8 F. Grenez, Synthèse des filtres numériques non récursifs à coefficients quantifiés. *Annales des Télécommunications*, 34, 1–2, 1979.

9 M. Feldmann, J. Henaff, B. Lacroix and J. C. Rebourg, Design of minimum phase charge-transfer transversal filters, *Electronic Letters*, 15(8), 1979.

10 R. Boite and H. Leich, A new procedure for the design of high order minimum phase FIR filters. *Signal Processing*, 3(2), 101–108, 1981.

11 Y. Kamp and C. J. Wellekens, Optimal design of minimum-phase FIR filters. *IEEE Transactions*, 31(4), 922–926, 1983.

12 Y. C. Lim and Y. Lian, The optimum design of one and two-dimensional FIR filters using the frequency response masking technique. *IEEE Transactions on Circuits and Systems II: Express Briefs*, 40(2), 88–95, 1993.

13 D. Dudgeon and R. Mersereau, *Multidimensional Digital Signal Processing*, Prentice-Hall, Englewood Cliffs, NJ, 1984.

14 D. E. Pearson, *Image Processing*, McGraw-Hill, UK, 1991.

15 V. Ouvrard and P. Siohan, Design of 2D video filters with spatial constraints, *Proceedings of EUSIPCO-92*, North Holland, Brussels, August 1992, 1001–1004.

# 6

# Infinite Impulse Response (IIR) Filter Sections

Digital filters with an infinite impulse response (IIR) are discrete linear systems which are governed by a convolution equation based on an infinite number of terms. In principle, they have infinite memory. This memory is achieved by feeding the output back to the input, so they are known as recursive filters. Each element of the set of output numbers is calculated by weighted summation of a certain number of elements of the input set and of the previous output set.

In general, this IIR allows for much more selective filtering functions to be obtained than with finite impulse response (FIR) filters of similar complexity. However, the feedback loop complicates the study of the properties and design of these filters and leads to parasitic phenomena.

When examining IIR filters, it is simpler initially to consider them in terms of first- and second-order sections. Not only are these simple structures useful in introducing the properties of IIR filters. They also represent the most frequently used type of implementation. Indeed, even the most complex IIR filters appearing in practice are generally formed from a set of such sections.

## 6.1 First-Order Section

Consider a system which, for the set of data $x(n)$, produces the set $y(n)$ such that:

$$y(n) = x(n) + by(n-1) \tag{6.1}$$

where $b$ is a constant. This is a first-order filter section.

This system's response to the unit set $u_0(n)$ such that:

$$\begin{aligned} u_0(n) &= 1 \ \text{ for } \ n = 0 \\ &\phantom{=}\ 0 \ \text{ for } \ n \neq 0 \end{aligned}$$

is the set $y_0(n)$ such that:

$$\begin{aligned} y_0(n) &= 0 \ \text{ for } \ n < 0 \\ &\phantom{=}\ b^n \ \text{ for } \ n \geq 0 \end{aligned}$$

This set constitutes the impulse response of the filter. It is infinite and the stability condition is written as:

$$\sum_{n=0}^{\infty} |b|^n < \infty$$

Hence $|b| < 1$.

*Digital Signal Processing: Theory and Practice*, Tenth Edition. Maurice Bellanger.

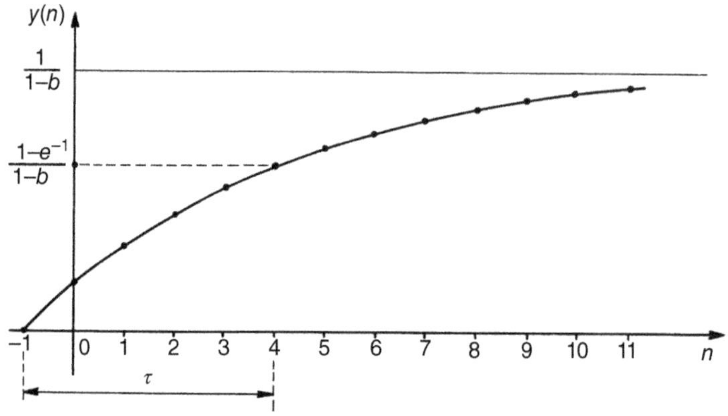


**Figure 6.1** Response of the first-order section to a unit step

The response of the system for the set $x(n)$ such that:

$$x(n) = 0 \text{ for } n < 0$$
$$1 \text{ for } n \geq 0$$

is the set $y(n)$ such that:

$$y(n) = 0 \quad \text{for } n < 0$$
$$\frac{(1-b^{n+1})}{(1-b)} \text{ for } n \geq 0 \tag{6.2}$$

which tends toward $1/(1 - b)$ as $n$ tends toward infinity if the system is stable. This response is shown in Figure 6.1.

By analogy with a continuous system with time constant $\tau$, sampled with period $T$, and whose response $y_c(n)$ is written as:

$$y_c(n) = [1 - e^{-(T/\tau)(n+1)}]$$

the time constant of the first-order digital filter is defined by:

$$e^{-T/\tau} = b$$

for $b > 0$. Then:

$$\tau = \frac{T}{\ln\left(\frac{1}{b}\right)} \tag{6.3}$$

For $b$ close to unity:

$$b = 1 - \delta \text{ with } 0 < \delta \ll 1$$

that is, for systems defined by:

$$y(n) = x(n) + (1 - \delta)y(n - 1) \tag{6.4}$$

we obtain:

$$\tau \approx \frac{T}{\delta} \tag{6.5}$$

This situation is encountered in adaptive systems, which are presented in Chapter 14.

If the input set $x(n)$ results for $n \geq 0$ from the sampling of the signal $x(t) = e^{j2\pi ft}$ or $x(t) = e^{j\omega t}$, with period $T = 1$, then:

$$y(n) = \frac{e^{jn\omega}}{1 - be^{-j\omega}} - \frac{b^{n+1}e^{-j\omega}}{1 - be^{-j\omega}} \tag{6.6}$$

This expression exhibits a transient and a steady-state term which corresponds to the frequency response $H(\omega)$ of the filter:

$$H(\omega) = \frac{1}{(1 - be^{-j\omega})} \tag{6.7}$$

The modulus and the phase of this function are:

$$|H(\omega)|^2 = \frac{1}{(1 - 2b\cos\omega + b^2)};\ \phi(\omega) = \tan^{-1}\frac{b\sin\omega}{1 - b\cos\omega} \tag{6.8}$$

The phase can also be written as:

$$\phi(\omega) = \tan^{-1}\left(\frac{\sin\omega}{(\cos\omega - b)}\right) - \omega$$

and the group delay:

$$\tau_g(\omega) = \frac{d\phi}{d\omega} = \frac{b\cos\omega - b^2}{1 - 2b\cos\omega + b^2} \tag{6.9}$$

It should be noted that for very small values of $\omega$:

$$\frac{|H(\omega)|^2 \approx 1}{\left\{(1 - b^2)\left[1 + \frac{b}{(1-b)^2}\omega^2\right]\right\}} \tag{6.10a}$$

This expression approximates the response $H_{RC}(\omega)$ of an $RC$ circuit, which is written as:

$$|H_{Rc}(\omega)|^2 = \frac{1}{1 + R^2C^2\omega^2} \tag{6.10b}$$

It appears that for frequencies which are very small in comparison to the sampling frequency, the digital circuit has a frequency response similar to that of an $RC$ network. Figure 6.2(a) shows the form of the frequency response for a digital first-order circuit. Figure 6.2(b) gives the phase response, and Figure 6.2(c) the group delay.

The phase can be written as:

$$\begin{aligned} \varphi(\omega) &= \tan^{-1}\left(\frac{\sin\omega}{\cos\omega - b}\right) - \omega & \cos\omega > b \\ \varphi_{\max} &\frac{\pi}{2}\cos^{-1}b & \cos\omega = b \\ \varphi(\omega) &= \pi + \tan^{-1}\left(\frac{\sin\omega}{\cos\omega - b}\right) - \omega & \cos\omega > b \end{aligned} \tag{6.11}$$

It passes through a maximum for $\omega$ such that $\cos\omega = b$, which corresponds to cancellation of the group delay. Thus, the coefficient $b$ directly controls the maximum phase of the section.

The transfer function of the first-order section can also be obtained using the $Z$-transform. Assume $Y(Z)$ and $X(Z)$ are the transforms of the output and input sets, respectively. Then:

$$Y(Z) = X(Z) + bZ^{-1}Y(Z)$$

and thus the $Z$-transfer function, $H(Z)$, is:

$$H(Z) = \frac{1}{(1 - bZ^{-1}) = \frac{Z}{(Z-b)}}$$

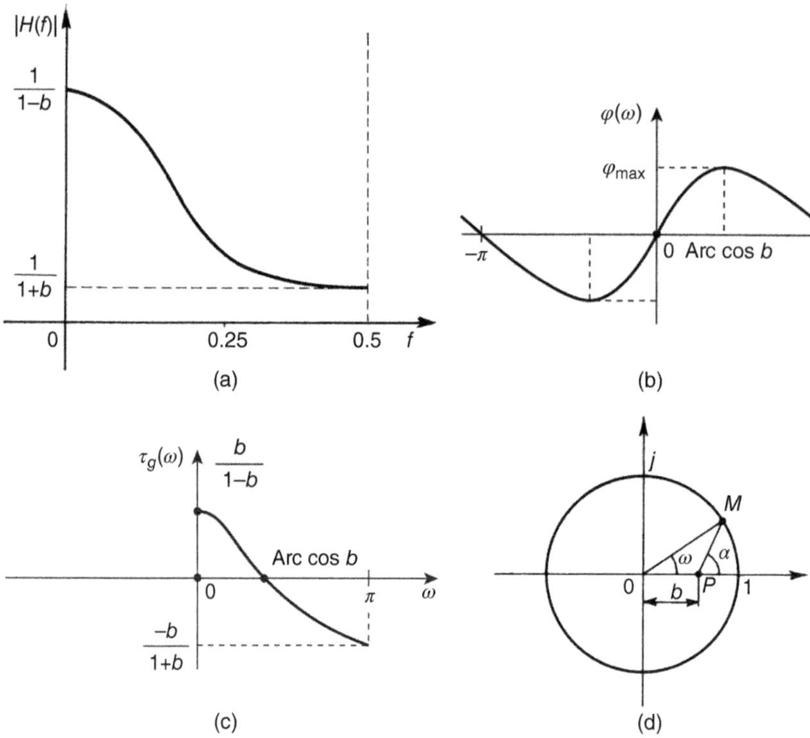


**Figure 6.2** First-order section: (a) frequency response, (b) phase response, (c) group delay, and (d) pole

The frequency response is obtained simply by replacing $Z$ with $e^{j\omega}$, in the expression for $H(Z)$, with $\omega = 2\pi f$.

A graphical interpretation is given in Figure 6.2(d), which represents the pole $P$ of this function in the complex plane. This is a point on the real axis with coordinate $b$.

Following this figure:

$$|H| = \frac{1}{MP} \text{ and } \phi = \alpha - \omega$$

The stability condition implies that the pole $P$ is inside the unit circle.

An interesting special case is the narrowband integrator, defined by the following transfer function:

$$H(\tau)\frac{\varepsilon}{1-(1-\varepsilon)Z^{-1}}_{int} \tag{6.12}$$

with $\varepsilon$ being small such that $0 < \varepsilon \ll 1$.

It can also be shown that the 3 dB bandwidth is approximately equal to $\varepsilon$ and the time constant equal to $1/\varepsilon$. The norm of the frequency response is: $||H||_2^2 \approx \frac{\varepsilon}{2}$.

The one-sided $Z$-transform generates the transient responses and introduces the initial conditions. Indeed:

$$\sum_{n=0}^{\infty} y(n)Z^{-n} = \sum_{n-0}^{\infty} x(n)Z^{-n} + b\sum_{n=0}^{\infty} y(n-1)Z^{-n}$$

$$Y(Z) = X(Z) + by(-1) + bZ^{-1}Y(Z)$$

Hence:

$$Y(Z) = \frac{X(Z)}{1 - bZ^{-1}} + \frac{by(-1)}{1 - bZ^{-1}}$$

If $x(n) = e^{jn\omega}$, $X(Z)$ is written:

$$Y(Z) = \sum_{n=0}^{\infty} e^{jn\omega} Z^{-n} = \frac{1}{1 - e^{j\omega} Z^{-1}} \tag{6.13}$$

The value $y(n)$ is obtained from the equation for the inverse $Z$-transform:

$$y(n) = \frac{1}{j2\pi} \int \Gamma Z^{n-1} \left[ \frac{1}{1 - e^{j\omega} Z^{-1}} \cdot \frac{1}{1 - bZ^{-1}} + \frac{by(-1)}{1 - bZ^{-1}} \right] dZ$$

By taking a circle with radius greater than unity as the contour of integration, $\Gamma$, the theory of residues gives:

$$y(n) = \frac{e^{jn\omega}}{1 - be^{-j\omega}} - \frac{b^{n+1} e^{-j\omega}}{1 - be^{-j\omega}} + y(-1) b^{n+1} \tag{6.14}$$

which can also be obtained directly from a series expansion of $Y(Z)$. This expression shows the steady-state and transient responses, and also the response due to the initial conditions. The last two items disappear when $n$ increases if $|b| < 1$ – that is, if the system is stable.

This analysis shows that the first-order filter offers restricted possibilities because it has only one pole, which must be real if the filter has real coefficients. Further, its frequency response is a monotonic function. The second-order filter has a wider variety of possibilities. It is the structure most commonly used in digital filtering because of the modularity it allows, even with the most complex filters, and because of its properties relating to limitations in the coefficient wordlengths and the round-off noise. We will first examine the purely recursive filter section.

## 6.2 Purely Recursive Second-Order Section

Consider a system which, for the dataset $x(n)$, produces the corresponding set $y(n)$ such that:

$$y(n) = x(n) - b_1 y(n-1) - b_2 y(n-2) \tag{6.15}$$

In this expression, the signs of the coefficients $b_1$ and $b_2$ have been changed compared to the previous section in order to facilitate the writing of the $Z$-transfer function of the system, $H(Z)$:

$$H(Z) = \frac{1}{1 + b_1 Z^{-1} + b_2 Z^{-2}} = \frac{Z^2}{Z^2 + b_1 Z + b_2}$$

This function has a double root at the origin and two poles $P_1$ and $P_2$ such that:

$$P_{1,2} = -\frac{b_1}{2} \pm \frac{1}{2} \sqrt{(b_1^2 - 4b_2)} \tag{6.16}$$

Two cases occur, depending on the sign of $b_1^2 - 4b_2$:

(1) $\boldsymbol{b_1^2 \geq 4b_2}$: both poles lie on the real axis of the complex plane. The transfer function is simply the product of two first-order functions with real coefficients. The corresponding filter is composed of two first-order sections in cascade and its properties are deduced accordingly. The amplitudes are multiplied, and the phases are added. The step response at the output of the second section is:

$$y_2(n) = \frac{1}{(1 - b_1)(1 - b_2)} \left[ 1 - b_2^{n+1} - (1 - b_2) \frac{b_1^{n+1} - b_2^{n+1}}{b_1 - b_2} \right] \tag{6.17}$$

where $b_1$ and $b_2$ are the coefficients. The corresponding time constant $\tau_{12}$ is given by:

$$\tau_{12} \approx \sqrt{2}\sqrt{\tau_1 \tau_2} \tag{6.18}$$

for coefficients sufficiently close to unity. For $N$ identical sections, the time constant $\tau_N$ can be approximated by:

$$\tau_N \approx \sqrt{(N)}\tau_1 \tag{6.19}$$

(2) $\boldsymbol{b_1^2 < 4b_2}$ : the two poles are complex conjugates written as $P$ and $\overline{\boldsymbol{P}}$, with

$$P = -\frac{b_1}{2} + j\frac{1}{2}\sqrt{\left(4b_2 - b_1^2\right)} \tag{6.20}$$

Figure 6.3 illustrates this, the most interesting case, and the remainder of this section will concentrate on this.

The relation between the position of the poles and the filter coefficients is very simple:

$$b_1 = -2\mathrm{Re}(P) \tag{6.21}$$

That is, the coefficient of the $Z^{-1}$ term in the expression for $H(Z)$ is equal in modulus to twice the real part of the pole and has the opposite sign. Then:

$$b_2 = |\mathrm{OP}|^2 \tag{6.22}$$

The coefficient of the $Z^{-2}$ term is equal to the square of the modulus of the pole or to the square of the distance from the pole to the origin. As will be seen later, both relations are very useful in determining filter coefficients.

If $M$ denotes the coordinate $e^{j\omega}$ in the complex plane, the modulus of the transfer function is:

$$|H(\omega)| = \frac{1}{MP \cdot M\overline{P}}$$

and the phase is:

$$\phi(\omega) = \alpha_1 + \alpha_2 - 2\omega$$

where $\alpha_1$ and $\alpha_2$ denote the angles between the vectors $\overrightarrow{PM}$ and $\overrightarrow{\overline{P}M}$ and the real axis.

The analytical expressions are deduced from $H(Z)$ by letting $Z = e^{j\omega}$. By using

$$H(Z) = \frac{1}{1 + b_1 Z^{-1} + b_2 Z^{-2}}$$

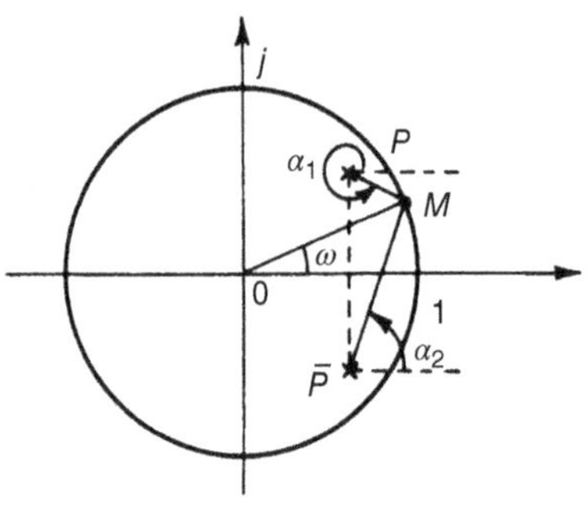


**Figure 6.3** Second-order section with complex poles

we have:

$$|H(\omega)|^2 = \frac{1}{1 + b_1^2 + b_2^2 + 2b_1(1 + b_2)\cos\omega + 2b_2\cos 2\omega} \tag{6.23a}$$

$$\phi(\omega) = -\arctan\left[\frac{b_1\sin\omega + b_2\sin 2\omega}{1 + b_1\cos\omega + b_2\cos 2\omega}\right] \tag{6.24a}$$

A very elegant form for the frequency response and the phase is obtained by representing the poles in polar coordinates, $P = re^{j\theta}$, and expressing $H(Z)$ as a factor product:

$$H(Z) = \frac{1}{(1 - PZ^{-1})(1 - \overline{P}Z^{-1})}$$

The coefficients $b_1$ and $b_2$ then become:

$$b_1 = -2r\cos\theta;\, b_2 = r^2$$

For $H(\omega)$, we obtain:

$$H(\omega) = \frac{1}{[1 - re^{j(\theta-\omega)}][1 - re^{-j(\theta+\omega)}]} \tag{6.25}$$

Hence:

$$|H(\omega)|^2 = \frac{1}{[1 + r^2 - 2r\cos(\theta - \omega)][1 + r^2 - 2r\cos(\theta + \omega)]} \tag{6.23b}$$

$$\phi(\omega) = \arctan\left[\frac{r\sin(\theta + \omega)}{1 - r\cos(\theta + \omega)}\right] - \arctan\left[\frac{r\sin(\theta - \omega)}{1 - r\cos(\theta - \omega)}\right] \tag{6.24b}$$

These expressions permit the curves for $|H(\omega)|$ and $\phi(\omega)$ to be plotted as a function of the frequency $\omega = 2\pi f$. It can be shown that $|H(\omega)|$ is an even function and that $\phi(\omega)$ is an odd function of $\omega$.

The values corresponding to the extrema of $|H(\omega)|$ are the roots of the following equation, which is obtained by taking the derivative of equation (6.23) with respect to $\omega$:

$$\sin\omega[b_1(1 + b_2) + 4b_2\cos\omega] = 0$$

The extremum frequencies are 0 and 0.5 and another extremal frequency $f_0$ exists if:

$$\left|\frac{b_1(1 + b_2)}{4b_2}\right| < 1 \tag{6.26a}$$

or, in polar coordinates:

$$\cos\theta < \frac{2r}{1 + r^2} \tag{6.26b}$$

In this case,

$$\cos(2\pi f_0) = \cos\omega_0 = -\frac{b_1(1 + b_2)}{4b_2} \tag{6.27}$$

The frequency $f_0$ is the resonance frequency of the filter section. The amplitude at the resonance is written as:

$$H_m = \frac{1}{1 - b_2}\sqrt{\left(\frac{4b_2}{4b_2 - b_1^2}\right)} \tag{6.28}$$

or, in polar coordinates:

$$H_m = \frac{1}{1 - r}\frac{1}{(1 + r)\sin\theta} \tag{6.29}$$

Thus, it appears that the frequency response at resonance is inversely proportional to the distance from the pole to the unit circle. This is a fundamental expression which will be used frequently in the following chapters.

It is also important for the second-order section to determine the 3-dB bandwidth, $B_3$, such that:

$$B_3 = f_2 - f_1 = (\omega_2 - \omega_1)/2\pi$$

with:

$$|H(\omega_1)|^2 = |H(\omega_2)|^2 = H_m^2/2$$

For a strongly resonant filter section ($r \approx 1$), using equations (6.22) and (6.23), the following approximation holds in the vicinity of the resonance frequency:

$$\left|H(\omega_1)^2\right| \approx \frac{1}{4\sin^2\theta}\frac{1}{1+r^2-2r\cos(\theta-\omega_1)}$$
$$= \frac{1}{2}\left[\frac{1}{(1-r^2)^2\sin^2\theta}\right]$$

whence:

$$\cos(\theta-\omega_1) = ((1+r^2)/2r) - (1-r^2)^2/4r$$

By expansion and limiting the number of terms, we derive:

$$|\theta-\omega_1| \approx 1-r$$

Hence, the approximation for a strongly resonant filter section is:

$$B_3 = (1-r)/\pi \tag{6.30a}$$

This result is used below for calculating the arithmetic complexity of filters.

Another characteristic is sometimes used for a purely recursive second-order section: the equivalent noise bandwidth $B_2$. This is the bandwidth of a noise source whose spectral density is assumed to be constant within this band and equal to $H_m^2$ and whose total noise power is equal to the power obtained at the output of the section when white noise of unit power is applied. By definition:

$$B_b.H_m^2 = \|H\|_2^2$$

Taking account of the expression for $B_b.H_m^2 = \|H\|_2^2$ given below (6.36), and expression (6.29) above, yields:

$$B_b = \frac{(1-r^2)\sin^2\theta}{1+r^4-2r^2\cos 2\theta} \tag{6.30b}$$

This expression is useful in spectral analysis, for example.

The main characteristics of a purely recursive second-order filter section can be illustrated by an example.

*Example*: Consider a second-order filter section having poles with coordinates:

$$P = 0.6073 + \text{j}0.5355$$
$$\overline{P} = 0.6073 - \text{j}0.5355$$

The various parameters are:

$$b_1 = -2\,\text{Re}(P) = -1.2146$$
$$b_2 = |OP|^2 = 0.6556$$

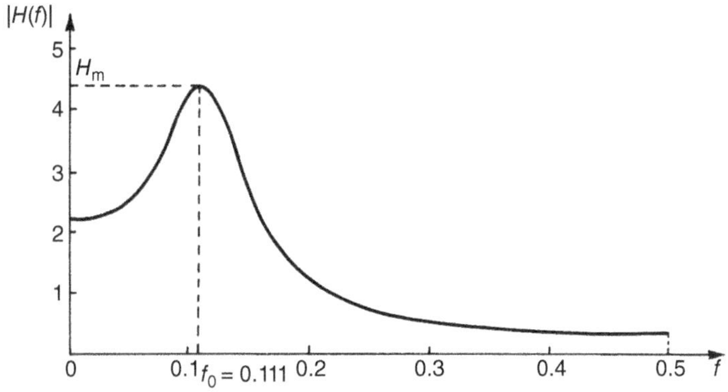


**Figure 6.4** Frequency response of a purely recursive second-order section

$$H(Z) = \frac{1}{1 - 1.2146Z^{-1} + 0.6556Z^{-2}}$$

$$|H(\omega)|^2 = \frac{1}{2.905 - 4.02 \cos \omega + 1.31 \cos(2\omega)}$$

$$\theta = 2\pi \times 0.1156; r = 0.81; f_0 = 0.111; H_m = 4.39; B_3 = 0.06$$

The modulus of the response is shown in Figure 6.4 as a function of the frequency.

The phase response of the second-order section can be considered using equation (6.24) for the function $\phi(\omega)$. In order to describe the variations in this function, it is useful first to calculate its derivative using equation (6.24b). Thence,

$$\frac{d\phi}{d\omega} = \frac{r \cos(\theta + \omega) - r^2}{1 - 2r \cos(\theta + \omega) + r^2} + \frac{r \cos(\theta - \omega) - r^2}{1 - 2r \cos(\theta - \omega) + r^2} \tag{6.31}$$

This derivative is interesting as it is the group delay of the filter. By definition (1.29):

$$\tau(\omega) = \frac{d\phi}{d\omega}$$

Thus, the group delay can be written as:

$$\tau(\omega) = \frac{r[\cos(\theta + \omega) - r]}{1 - 2r \cos(\theta + \omega) + r^2} + \frac{r[\cos(\theta - \omega) - r]}{1 - 2r \cos(\theta - \omega) + r^2} \tag{6.32}$$

The function $\tau(\omega)$ has a maximum in the vicinity of the resonance frequency. At the frequency of $f = \theta/2\pi$, this becomes:

$$\tau(\theta) = \frac{r}{1 - r}\left[1 + \frac{(1 - r)[\cos 2\theta - r]}{1 - 2r \cos 2\theta + r^2}\right] \approx \frac{r}{1 - r} \tag{6.33}$$

In physical systems, this function is positive and $\phi(\omega)$ is an increasing function which has a value of 0 at the origin and a multiple of $\pi$ at frequency 0.5.

*Example*: $r = 0.81$; $\theta = 2\pi \times 0.1156$

Figure 6.5 shows the curve $\tau(f)$ as a function of frequency. This curve has a maximum of 3.8 in the vicinity of the resonance. The unit of time is the sampling period $T$. The values obtained should be multiplied by $T$ if this period is different from unity.

The function $\tau(f)$ can be seen to have negative values. In fact, it is the theoretical group delay of the filter. The system, as it has been presented, however, cannot actually be realized. Each output element $y(n)$ is calculated by an addition which involves an input number $x(n)$. This operation cannot be instantaneous. To enable the system to be realized, $y(n)$ has to be delayed, for example,

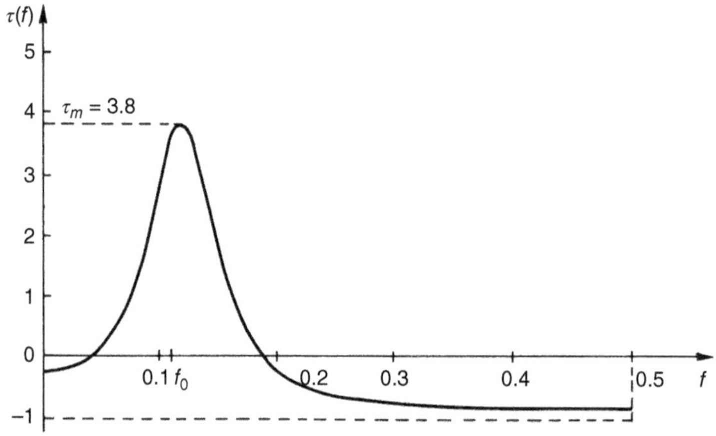


**Figure 6.5** Theoretical group delay of a purely recursive section

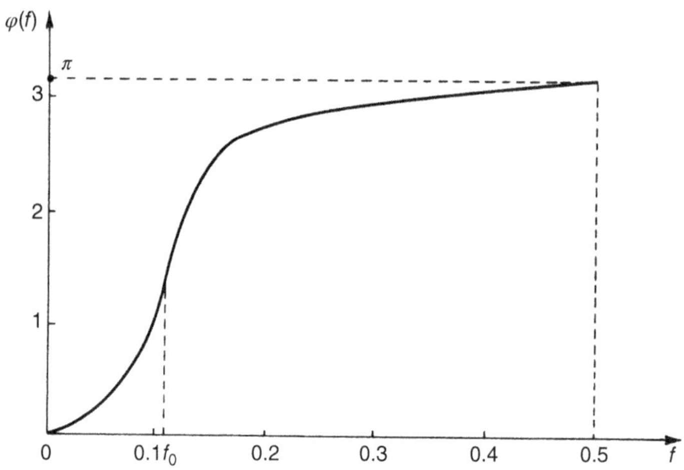


**Figure 6.6** Phase characteristic of a purely recursive section

by one period. The group delay will then be increased correspondingly, and it is necessary to add the value $\omega$ to the phase $\phi(\omega)$. The function $\phi(\omega)$ obtained under these conditions is represented in Figure 6.6, and the curve has maximum slope in the vicinity of the resonance.

The equations which were given for the functions $|H(\omega)|$, $\phi(\omega)$, and $\tau(\omega)$ are important because they allow the corresponding functions to be determined for filters realized by cascading second-order sections, either by multiplication for the modulus of the frequency response, or by addition for the phase and the group delay.

To introduce the initial conditions and find the transient responses, the one-sided $Z$-transform is used. From the equation of definition of the filter section, the following relation can be found between the one-sided transforms $Y(Z)$ and $X(Z)$:

$$Y(Z) = \frac{X(Z)}{1 + b_1Z^{-1} + b_2Z^{-2}} - \frac{b_1y(-1) + b_2[y(-2) + y(-1)Z^{-1}]}{1 + b_1Z^{-1} + b_2Z^{-2}} \tag{6.34a}$$

For $x(n) = e^{jn\omega}$, $y(n)$ is given by the equation:

$$y(n) = \frac{1}{j2\pi}\int_{\Gamma} Z^{n-1}\, Y(Z)dZ$$

with:

$$X(Z) = \frac{1}{1 - e^{j\omega}Z^{-1}}$$

taking a circle with radius greater than unity as the integration contour $\Gamma$.

The study of the purely recursive section has been carried out in the frequency domain. In the time domain, this section has an impulse response which is a set $h(n)$ determined by examining the response to the unit set, or by series expansion of the function $H(Z)$. For complex poles, one has:

$$H(Z) = \frac{1}{1 - PZ^{-1}}\frac{P}{P - \overline{P}} + \frac{1}{1 - \overline{P}Z^{-1}}\frac{\overline{P}}{\overline{P} - P} = \sum_{n=0}^{\infty} h(n)Z^{-n}$$

Hence,

$$h(n) = r^n \frac{\sin(n + 1)\theta}{\sin\,\theta} \tag{6.35}$$

Figure 6.7 shows the impulse response of the filter described in the above example. The response to a unit step can be derived from the definition, after several manipulations:

$$g(n) = \frac{1}{1 + b_1 + b_2}[1 + b_2 h(n) - h(n + 1)]$$

This then gives:

$$g(n) = \frac{1}{1 + r^2 - 2r\,\cos\theta}\left[1 + \frac{r^{i+1}}{\sin\,\theta}[r\,\sin(i + 1)\theta - \sin(i + 2)\theta]\right] \tag{6.34b}$$

The norm $\|H\|_2$ of the function $H(\omega)$ is used, and can be calculated by two methods, as was shown in Section 4.3. By summation of the series:

$$\|H\|_2^2 = \sum_{n=0}^{\infty} |h(n)|^2 = \frac{1}{\sin^2\theta}\sum_{n=0}^{\infty} r^{2n}\frac{1 - \cos[2(n + 1)\theta]}{2}$$

Using integration following the residues theorem:

$$\|H\|_2^2 = \frac{1}{j2\pi}\int_{|Z|=1} \frac{ZdZ}{(Z - P)(Z - \overline{P})(1 - PZ)(1 - \overline{P}Z)}$$

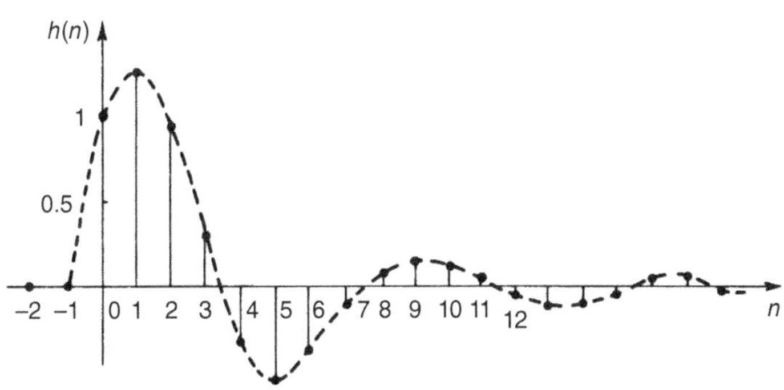


**Figure 6.7** Impulse response of a second-order section

Finally,

$$\|H\|_2^2 = \frac{1+r^2}{1-r^2}\frac{1}{1+r^4-2r^2\cos 2\theta} \tag{6.36}$$

The value $\|H\|_1$ is also used:

$$\|H\|_1 = \sum_{n=0}^{\infty} |h(n)|$$

This value is bounded by the inequality:

$$\|H\|_1 = \frac{1}{\sin\theta}\sum_{n=0}^{\infty} r^n |\sin[(n+1)\theta]| \leq \frac{1}{(1-r)\sin\theta} \qquad (0<\theta<\pi) \tag{6.37}$$

*Example*: When the poles are located on the imaginary axis in the $Z$-plane, $\theta = \pi/2$, and the impulse response is given by:

$$h(2p) = r^{2p}(-1)^p$$

then:

$$\|H\|_2^2 = \sum_{p=0}^{\infty} |h(2p)|^2 = \frac{1}{1-r^4}$$

$$\|H\|_1 = \sum_{p=0}^{\infty} h(2p) = \frac{1}{1-r^2}$$

The results obtained for the purely recursive second-order section can be extended to the general second-order section.

## 6.3 General Second-Order Section

The most general second-order filter introduces the input data $x(n-1)$ and $x(n-2)$ to the calculation of an element $y(n)$ of the output set at time $n$. Its equation of definition is written as:

$$y(n) = a_0x(n) + a_1x(n-1) + a_2x(n-2) - b_1y(n-1) - b_2y(n-2) \tag{6.38}$$

and it leads to the $Z$-transfer function:

$$H_T(Z) = \frac{a_0 + a_1Z^{-1} + a_2Z^{-2}}{1 + b_1Z^{-1} + b_2Z^{-2}}$$

which has two zeros which, since the numerator coefficients are real, are either real or complex conjugates. The position of these zeros, written as $Z_0$ and $\overline{Z}_0$, is often special, and two cases are encountered in practice. The first corresponds to a filtering function. The zeros are then almost always on the unit circle, both to optimize the attenuation of the filter by introducing an infinite attenuation frequency. and because, under these conditions, symmetry between the coefficients appears and the calculations can be simplified. The second case corresponds to a pure phase shifter, when the zeros are harmonic conjugates of the poles.

The filter case will be considered first. The transfer function is now written as:

$$H_T(Z) = a_0\frac{1 + a_1Z^{-1} + Z^{-2}}{1 + b_1Z^{-1} + b_2Z^{-2}} = a_0\frac{(Z-Z_0)(Z-\overline{Z}_0)}{(Z-P)(Z-\overline{P})} \tag{6.39}$$

or:

$$H_T(Z) = a_0 \frac{1 - 2\,\mathrm{Re}(Z_0)Z^{-1} + Z^{-2}}{1 - 2\,\mathrm{Re}(P)Z^{-1} + |P|^2 Z^{-2}}$$

The modulus of the frequency response of the general second-order section, when the zeros are placed on the unit circle, is expressed by:

$$|H_T(\omega)|^2 = \frac{(a_1 + 2a_0 \cos\omega)^2}{1 + b_1^2 + b_2^2 + 2b_1(1 + b_2)\cos\omega + 2b_2 \cos 2\omega} \tag{6.40}$$

Such a filter can be regarded as the cascade of a purely recursive IIR filter section and a linear-phase FIR filter section. Consequently, the phase characteristics and the group delay of the complete filter section are sums of the characteristics of the elementary components. That is,

$$\tau_T(\omega) = 1 + \frac{r\cos(\theta + \omega) - r^2}{1 - 2r\cos(\theta + \omega) + r^2} + \frac{r\cos(\theta - \omega) - r^2}{1 - 2r\cos(\theta - \omega) + r^2} \tag{6.41}$$

$$\phi_T(\omega) = \omega + \arctan\left[\frac{r\sin(\theta + \omega)}{1 - r\cos(\theta + \omega)}\right] - \arctan\left[\frac{r\sin(\theta - \omega)}{1 - r\sin(\theta - \omega)}\right] \tag{6.42}$$

These two expressions give the phase and the group delay of a second-order section when both zeros are on the unit circle. This filter section is usually called the second-order elliptic section, from the technique used for calculating the coefficients.

*Example*: To illustrate the properties of a general second-order filter section, let us use the example given earlier by completing the filter with two zeros:

$$Z_0 = 0.3325 + j0.943 \text{ and } Z_0 = 0.3325 - j0.943$$

The positions of the singularities in the complex plane are given by Figure 6.8(a). The transfer function $H_T(Z)$ is the quotient of the two second-order polynomials $N(Z)$ and $D(Z)$:

$$H_T(Z) = a_0 \frac{N(D)}{D(Z)}$$

with:

$$N(Z) = (1 - 0.665Z^{-1} + Z^{-2})$$

$$D(Z) = 1 - 1.2146Z^{-1} + 0.6556Z^{-2}$$

Figure 6.8(b) shows the frequency response for the filter. The contributions of the numerator and the denominator of the transfer are also indicated. The factor $a_0$ corresponds to a scaling factor which is calculated so that the response of the filter has a specified value at a given frequency. For example:

$$H_T(0) = 1 \text{ results in } a_0 = 0.33$$

The group delay and the phase are given by Figures 6.5 and 6.6, respectively. The norm $||H_T||_2$ of the function $H_T(\omega)$ is calculated as given earlier, and is:

$$||H_T||_2^2 = a_0^2 \frac{2 + a_1^2 + a_1^2 b_2 - 4a_1 b_1 + 2b_1^2 - 2b_2^2}{(1 - b_2)\left[(1 + b_2)^2 - b_1^2\right]} \tag{6.43}$$

An important particular case is the notch filter, which is used to remove a line in a spectrum without disturbing the other components. The transfer function is:

$$H_N(Z) = \frac{1 + a_1 Z^{-1} + Z^{-2}}{1 + a_1(1 - \varepsilon)Z^{-1} + (1 - \varepsilon)^2 Z^{-2}} \tag{6.44}$$

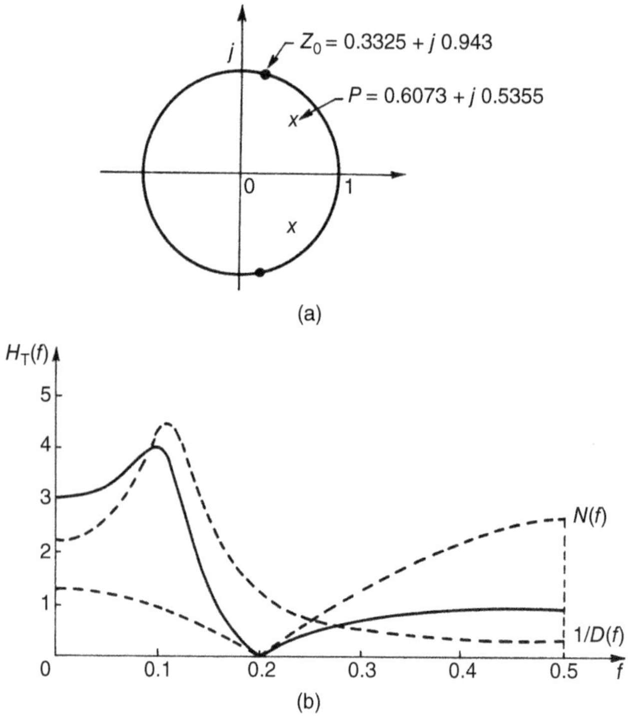


**Figure 6.8** (a) Poles and zeros of a general second-order section. (b) Frequency response of a general second-order section

where $\varepsilon$ is a small positive real value. As shown in Figure 6.9(a), $\varepsilon$ is the distance of the poles to the unit circle. For very small value of $\varepsilon$, the 3-dB attenuation bandwidth can be approximated by:

$$B_{3N} \approx \frac{\varepsilon}{\pi}$$

Outside the notch, the poles compensate for the zeros and the frequency response is almost flat and close to unity. Moreover, such a filter achieves a very small amplification of the input white noise, since equation (6.43) yields:

$$\|H_N\|_2^2 \approx \frac{2 - 3\varepsilon}{2 - 5\varepsilon} \approx 1 + \varepsilon$$

If the frequency of the signal to be removed is not precisely known, then the zeros have to be moved inside the unit circle in order to increase the notch width.

Another class of general second-order sections is that of phase-shifter circuits. The phase-shifter circuit is characterized by the fact that the numerator and the denominator of the transfer function have the same coefficients but in the reverse order:

$$H_D(Z) = \frac{b_2 + b_1 Z^{-1} + Z^{-2}}{1 + b_1 Z^{-1} + b_2 Z^{-2}} = \frac{N(Z)}{D(Z)} \tag{6.45}$$

The polynomials $N(Z)$ and $D(Z)$ are image polynomials. As a result, $|H_D(e^{j\omega})| = 1$ – that is, the circuit is a pure phase shifter.

The transfer function $H_D(Z)$ is written as a function of the poles and zeros as:

$$H_D(Z) = \frac{(P - Z^{-1})(\overline{P} - Z^{-1})}{(1 - PZ^{-1})(1 - PZ^{-1})}$$

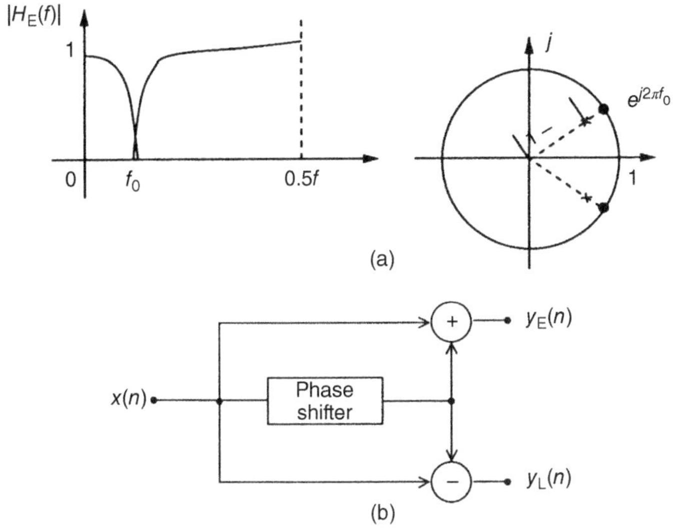


**Figure 6.9** (a) Second-order notch filter. (b) Implementation of the notch filter and its complement

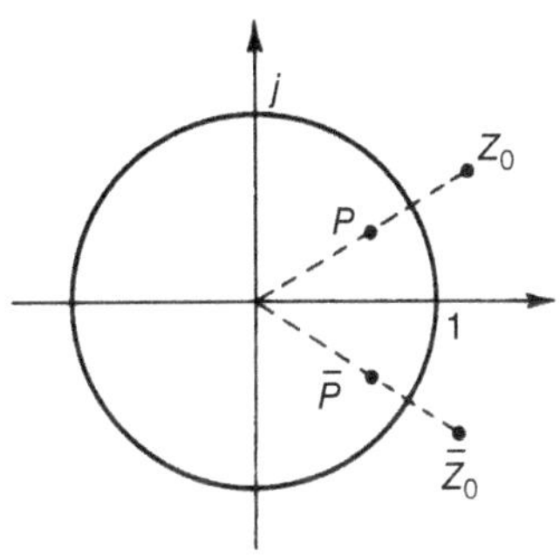


**Figure 6.10** Poles and zeros of the second-order phase shifter

It is clear that the poles and zeros are harmonic conjugates and Figure 6.10 shows their position in the $Z$-plane.

The calculation of the phase and group delay for this circuit can be very simply deduced from equations (6.24) and (6.32) for the purely recursive element as:

$$H_D(Z) = \frac{N(Z)}{D(Z)} = \frac{Z^{-2}(Z^{-1})}{D(Z)}$$

However,

$$|D(\omega)| = |D(-\omega)|; \ \ \phi(\omega) = -\phi(-\omega)$$

Hence,

$$\phi_D(\omega) = 2\phi(\omega) - 2\omega$$

After a few arithmetic manipulations, the group delay $\tau_g(\omega)$ of the phase shifter becomes:

$$\tau_g(\omega) = \frac{1 - r^2}{1 + 2r\cos(\theta - \omega) + r^2} + \frac{1 - r^2}{1 - 2r\cos(\theta + \omega) + r^2} \tag{6.46}$$

It is not difficult to show that, as $\omega$ varies from 0 to $\pi$, the phase $\phi_D(\omega)$ varies by $2\pi$:

$$\phi_D(\pi) = \int_0^\pi \tau(\omega) d\omega = 2 \int_0^\pi \frac{1 - r^2}{1 + r^2 - 2r \cos \alpha} d\alpha = 2\pi$$

An interesting application of this result is that the abovementioned notch filter can be implemented with the help of a phase shifter. In fact, two complementary filters can be obtained with a single all-pass section, as shown in Figure 6.9(b). The filter zero that sits on the unit circle corresponds to the phase shift $\pi$. In fact, it is even a set of two complementary filters that are obtained with a single second-order phase shifter [1, 2].

## 6.4 Structures for Implementation

The elements can be implemented by circuits which directly produce the operations represented in the expression for the transfer functions. The term $Z^{-1}$ corresponds to a delay of one elementary period and is achieved by one memory element. The coefficients used in the circuits are those of the transfer function, with the same sign for the numerator and the opposite sign for the denominator.

The circuit which corresponds directly to the equation for the definition of the purely recursive second-order section is given in Figure 6.11. The output numbers $y(n)$ are delayed twice, and multiplied by the coefficients $-b_1$ and $-b_2$ before being added to the input numbers $x(n)$. The circuit includes two memory locations for data and two for the coefficients. For each output number, two multiplications and two additions are required.

The general second-order filter section can be realized to conform with the equation of definition. However, two data memory locations are required for the input numbers and two for the output numbers. The structure obtained is not canonical, as it contains more than the minimum number of components. Indeed, only two data memories are necessary if the transfer function is factorized as:

$$H_T(Z) = \frac{N(Z)}{N(Z)} = \frac{1}{D(Z)} N(Z)$$

That is, the calculations involved in the denominator are performed first, followed by those for the numerator. This structure, called *D–N*, is shown in Figure 6.12. It corresponds to the introduction

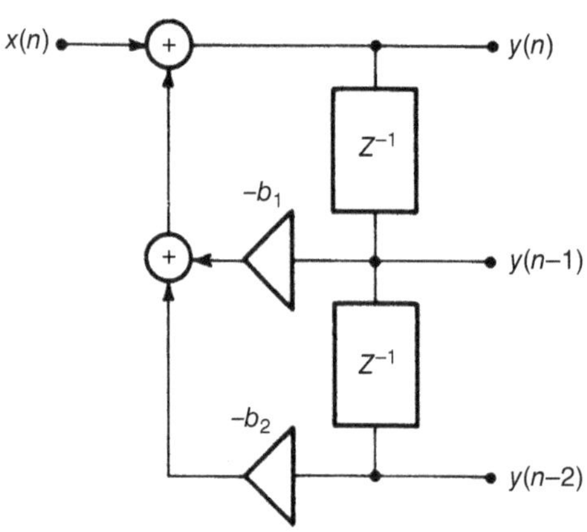


**Figure 6.11** Circuit of a purely recursive section

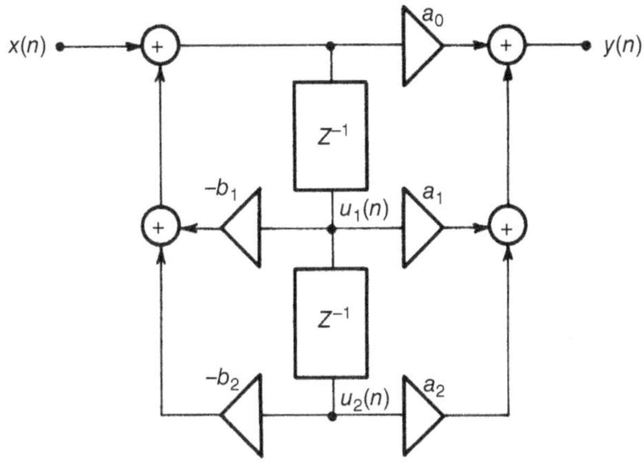


**Figure 6.12** Second-order section in the D–N structure

of two internal variables $u_1(n)$ and $u_2(n)$ forming a state vector $U(n)$ with $N = 2$ dimensions. The system is described by the following equations:

$$\begin{aligned} u_1(n+1) &= x(n) - b_1u_1(n) - b_2u_2(n) \\ u_2(n+1) &= u_1(n) \\ y(n) &= a_0x(n) - a_0b_1u_1(n) - a_0b_2u_2(n) + a_1u_1(n) + a_2u_2(n) \end{aligned}$$

or, in matrix form, conforming to equation (4.34):

$$\begin{aligned} u_1(n+1) &= x(n) - b_1u_1(n) - b_2u_2(n) \\ u_2(n+1) &= u_1(n) \\ y(n) &= a_0x(n) - a_0b_1u_1(n) - a_0b_2u_2(n) + a_1u_1(n) + a_2u_2(n) \end{aligned} \quad (6.47)$$

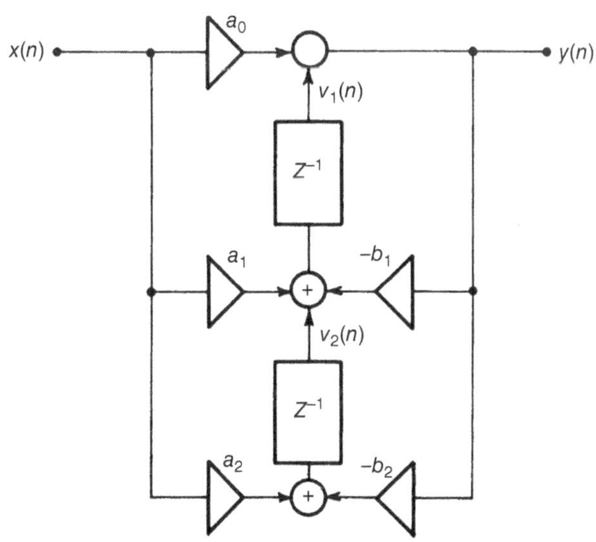


**Figure 6.13** Second-order section in the N–D structure

This representation thus results in a canonical realization, which has the minimum number of internal variables and consequently the minimum number of memory spaces.

Using the results given in Section 4.6, we see that there is a dual structure which corresponds to the internal variables $v_1(n)$and $v_2(n)$ such that:

$$\begin{bmatrix} v_1(n+1) \\ v_2(n+1) \end{bmatrix} = \begin{bmatrix} -b_1 & 1 \\ -b_2 & 0 \end{bmatrix} \begin{bmatrix} v_1(n) \\ v_2(n) \end{bmatrix} \begin{bmatrix} -a_0b_1 + a_1 \\ -a_0b_2 + a_2 \end{bmatrix} x(n)$$
$$y(n) = v_1(n) + a_0x(n) \tag{6.48}$$

This alternative canonical structure is represented in Figure 6.13. It corresponds to performing the operations on the numerator of the $Z$-transfer function first and is called the $N$–$D$ structure.

In order to optimize the circuit and to reduce the size of the multiplier, it is important to minimize the number of bits in each of the factors of the multiplication. The coefficients will be considered first.

## 6.5 Coefficient Wordlength Limitation

The limitation of the coefficient wordlength means that the coefficients can have only a limited number of values. It follows, therefore, that the poles have a limited number of possible positions inside the unit circle. The same effect occurs for zeros on the unit circle when the filter is elliptic. Thus, quantization of the absolute value of the coefficients to $b$ bits limits the number of positions that the poles can take in a quadrant of the unit circle to $2^{2b}$, and the number of frequencies of infinite attenuation to $2^b$. If the transfer function is calculated first, and the number of bits in the coefficients is then limited, for example, by rounding, the transfer function is modified by perturbations $e_N(Z)$ and $e_D(Z)$ in the numerator and the denominator [3]. The function $H_R(Z)$ is obtained:

$$H_R(Z) = \frac{N(Z) + e_N(Z)}{D(Z) + e_D(Z)} \tag{6.49}$$

If the round-off errors in the coefficients are denoted by $\delta a_i$ and $\delta b_i$ $(0 \le i \le 2)$, the perturbation transfer functions are written as:

$$e_N(f) = \sum_{i=0}^{2} \delta a_i e^{-j2\pi fi}; \quad e_D(f) = \sum_{i=1}^{2} \delta b_i e^{-j2\pi fi}$$

Let us consider the case of the elliptic filter element whose coefficients are quantized by rounding to $b_c$ bits, including the sign. Allowing for the inequalities:

$$|a_i| \le 2; \quad |b_i| < 2$$

the quantization step $q$ is written:

$$q = 2 \times 2^{1-b_c} = 2^{2-b_c}$$

Thus,

$$|e_D(f)| \le 2(q/2)2^{2-b_c} \tag{6.50}$$

The modifications of the transfer function caused by quantization of the coefficients of the denominator are a maximum for frequencies near the poles because the function $D(f)$ is then a minimum.

Neglecting the effect of rounding the numerator coefficients, we get:

$$H_R(f) = N(f)/[D(f) + e_D(f)] \approx N(f)[1 - \{e_D(f)/D(f)\}]/D(f)$$

The relative error, $e(f) = (H_R(f) - H(f))/H(f)$, is bounded by:

$$|e_D(f)| \leq q(1/|D(f)|) \tag{6.51}$$

This expression allows us to determine the number of bits needed to represent the coefficients of the denominator as a function of the tolerance of the frequency response and the values of the coefficients. It is used in the next chapter.

In the numerator, quantization of the coefficient $a_1$ of the elliptic filter section leads to a displacement of the zeros which lie on the unit circle. The displacement $df_i$ of the infinite attenuation point $f_i$ is such that:

$$|df_i| \leq \frac{1}{2\pi}\frac{2^{-bc}}{|\sin 2\pi fi|} \tag{6.52}$$

Quantization of the coefficient $a_0$ for the elliptic section results simply in a change in the gain of the filter.

## 6.6 Internal Data Wordlength Limitation

In the *D–N* structure, which is the one most frequently used, the second multiplication factor is the number held in the data memory. By necessity, this memory has a limited capacity; the feedback structure (Figure 6.12) implies that even if the input numbers $x(n)$ have a limited number of bits, and the memories are empty when the operation begins, the number of bits of the data to be stored in the memory increases indefinitely. Limitation, usually by rounding, is required. On the other hand, the filter sections can introduce large gains and major instabilities, so that logic saturation devices must be introduced to limit the amplitude of the data being stored in the memory. The elliptic second-order section with a logic saturation unit and quantization device is shown in Figure 6.14 for the *D–N* structure. For simplification, this figure assumes a single wordlength limitation device sited immediately before the memory.

The quantization device involves a degradation in the signal, which passes through the filter. This is the round-off noise. Following the circuit in Figure 6.14, it can be seen that quantization has the effect of superimposing, on the input signal $x(n)$, an error signal $e(n)$, which also passes through the

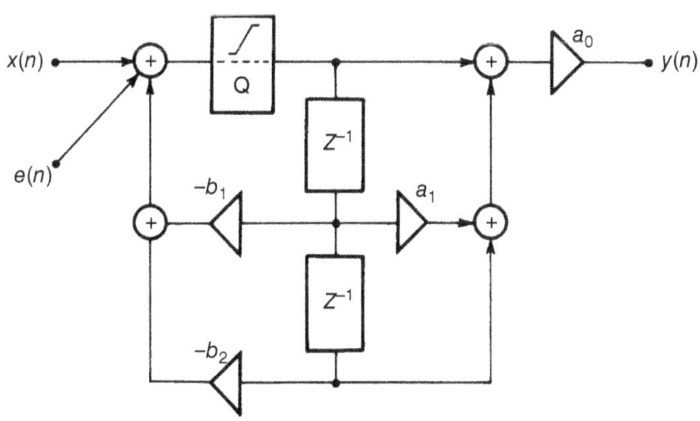


**Figure 6.14** Elliptic section with wordlength-limitation device

filter. If the quantization step has the value $q$, this error signal can be regarded as having a spectrum with uniform distribution and a power $q^2/12$. Under these conditions, the round off noise $N_c$ at the output can be determined if $f_s = 1$ by using equation (4.25) in Section 4.4, as:

$$N_c = \frac{q^2}{12} \int_0^1 \left| \frac{N(f)}{D(f)} \right|^2 df$$

or, as a function of the set $h(n)$, the impulse response of the filter:

$$N_c = \frac{q^2}{12} \sum_{n=0}^{\infty} |h(n)|^2$$

By using the results in the earlier sections, for a purely recursive element with complex poles and polar coordinates $(r, \theta)$, this becomes:

$$N_c = \frac{q^2}{12} \frac{1+r^2}{1-r^2} \frac{1}{1+r^4-2r^2\cos 2\theta} \tag{6.53}$$

and for the elliptic section:

$$N_c = \frac{q^2}{12} \frac{a_0^2\left(2+a_1^2+a_1^2 b_2 - 4a_1 b_1 + 2b_1^2 - 2b_2^2\right)}{(1-b_2)\left[(1+b_2)^2 - b_1^2\right]} \tag{6.54}$$

The quantization step $q$ is related to the number of bits of the internal data memories. This relation involves the amplitude of the frequency response of the purely recursive part. It is studied in detail in the following chapter, for a cascade of second-order sections.

In this section, only the $D$–$N$ structure has been considered. The calculations can be readily adapted to the $N$–$D$ structure [4]. The introduction of the quantization device also has consequences in the absence of a signal.

## 6.7 Stability and Limit Cycles

Even if there is no signal at the input of an IIR filter, there can still be a signal at the output. This is particularly likely if the coefficients are such that the filter is unstable.

The condition for stability of the filter is that the poles must lie inside the unit circle. This condition defines a stability domain in the plane $(b_1, b_2)$. From the results of Section 6.2, the domain of the complex poles is limited by the parabola:

$$b_2 = \frac{b_1^2}{4}$$

and the stability condition imposes:

$$0 \le b_2 < 1$$

If the poles are real, then:

$$-\frac{b_1}{2} + \frac{1}{2}\sqrt{\left(b_1^2 - 4b_2\right)} < 1; \quad -1 < -\frac{b_1}{2} - \frac{1}{2}\sqrt{\left(b_1^2 - 4b_2\right)}$$

Now, besides $b_2 < 1$, the stability condition is:

$$b_2 > -b_1 - 1; \quad b_2 > -1 + b_1; \quad \text{or} \quad |b_1| < 1 + b_2 \tag{6.55}$$

The domain of stability is, therefore, a triangle defined by the three straight lines:

$$b_2 = 1; \quad b_2 = -b_1 - 1; \quad b_2 = b_1 - 1$$

as shown in Figure 6.15.

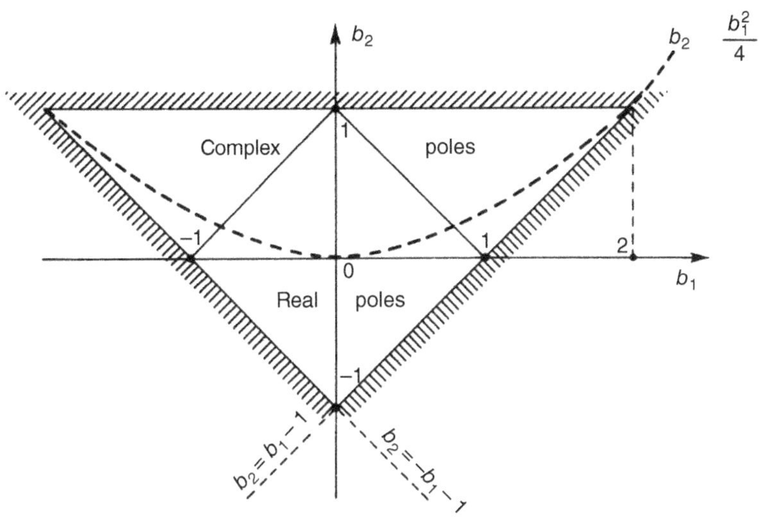


**Figure 6.15** Stability domain of the second-order section

Nevertheless, even if the stability condition is fulfilled, there may still be a signal at the output in the absence of an input. This is usually a constant or periodic signal which corresponds to an auto-oscillation of the filter, and which is often called a limit cycle. Such auto-oscillations can be produced with large amplitudes if overflow occurs when the capacity of the memories is exceeded in the absence of a logic saturation device. The equation for the system when there is no input signal is:

$$y(n) + b_1 y(n-1) + b_2 y(n-2) = 0$$

The natural condition for the absence of oscillations is given by the inequality:

$$|b_1 y(n-1) + b_2 y(n-2)| < 1$$

and thus the condition which is necessary and sufficient for the absence of large-amplitude auto-oscillations is:

$$|b_1| + |b_2| < 1 \tag{6.56}$$

This inequality determines a square in the plane $(b_1, b_2)$, within the triangle of stability of the filter element.

To eliminate all possibility of large-amplitude oscillations caused by overflow of the memory, one can show that it is sufficient to employ a logic saturation device as shown in Section 6.6 [5].

Limit cycles are also produced through quantization before storage in the memory. However, these have small amplitudes in well-designed systems. They arise through the fact that, in practice, the input signal is never zero, because, even in the absence of data $x(n)$, the error signal $e(n)$ caused by quantization of the internal data before storage in the memory is still applied to the filter. An estimation of the amplitude $A_a$ of the limit cycle is given by the expression:

$$A_a = \frac{q}{2} \max |H(\omega)|$$

where $H(\omega)$ is the transfer function for the filter section.

Application to a purely recursive second-order section with complex poles produces, according to equations (6.37) and (6.29):

$$|y(n)| \leq \frac{q}{2} \frac{1}{(1-r)\sin\theta} \tag{6.57}$$

$$A_a = \frac{q}{2} \frac{1}{(1-r^2)\sin\theta} \tag{6.58}$$

These signals often have a spectrum formed of lines with frequencies close to those at which $H(\omega)$ is maximum, which are either factors of, or in simple ratios to, the sampling frequency.

When designing filters, the number of bits in the internal data memories must be chosen to be sufficiently large, and the quantization step $q$ has to be chosen to be sufficiently small to prevent the limit cycles from being troublesome. It should be noted also that they can be eliminated by using a type of quantization other than rounding (for example, truncation of the absolute value) [2]. However, this is only achieved with an increase in the power of the round-off noise in the presence of a signal [6].

The results obtained in this chapter will be used in the next chapter, where second-order sections in cascade will be discussed.

## Exercises

**6.1** Study the first-order section

$$y(n) = x(n) + by(n-1)$$

in the following conditions:

$$x(n) = 0; \quad n < 0$$

$$x(n) = \cos n\omega; \quad n \geq 0; \quad b = -0.8; \quad \omega = \frac{\pi}{2}; \quad y(-1) = 0$$

Give the expression of $y(n)$ with transient and steady-state terms. From $y(n)$, compute the amplitude and phase responses and check with the results given in Section 6.1. Considering the steady-state term in $y(n)$, give the filter delay. Compare with the value obtained for the group delay. Justify the difference between the two values.

**6.2** Calculate the response of the system which is defined by the following equation:

$$y(n) = x(n) + x(n-1) - 0.8y(n-1)$$

to the unit set $u_0(n)$ and the set $x(n)$ such that:

$$x(n) = 0 \ \text{for}\ n < 0$$

$$1 \qquad \text{for}\ n \geq 0$$

Give the steady-state frequency response and the transient response.

**6.3** Assume a purely recursive second-order section which has the following coefficients:

$$b_1 = -1.56; \quad b_2 = -0.8$$

State the position of the poles. Calculate the frequency response, the phase response, and the group delay. How are the functions modified if two zeros are added at $j$ and $-j$? For this case, show the circuit diagram using the $D$–$N$ form and count the number of multiplications required for each output number.

**6.4** Give the expression for the impulse response of a purely recursive second-order filter section which has the following coefficients:

$$b_1 = -1.60; \quad b_2 = -0.98$$

Calculate the resonance frequency and amplitude of the response at resonance. Give the response $H(\omega)$ and calculate the norm $\|H\|_2$.
The zeros are added on the unit circle to produce an infinite attenuation at frequency $3f_s/8$. What are the coefficients of the filter? Calculate the new expression for $H(\omega)$ and the new value of $\|H\|_2$.
For this filter, find the amplitude of the limit cycles, using the $D$–$N$ form and then the $N$–$D$ form. Give an example of a limit cycle.
Does this filter produce large-amplitude oscillations when there is no logic saturation device? Give an example.

**6.5** How many bits are needed to represent the coefficients of the filter with the following $Z$-transfer function:

$$H(Z) = \frac{1 - 0.952Z^{-1} + Z^{-2}}{1 - 1.406Z^{-1} + 0.9172^{-2}}$$

in order that the frequency response is not modified by more than 1% in the vicinity of the poles? Calculate the displacement of the point of infinite attenuation.

**6.6** Assume the realization of a second-order phase shifter having poles $P_{1,2}$ such that:

$$P_{1,2} = 0.71 \pm j0.54$$

Calculate the coefficients and give the expression for the function $\tau_g(\omega)$. Show that an implementation scheme exists which produces a reduced number of multiplications. When there is no logic saturation device, can this element exhibit large-amplitude oscillations? Can it produce low-amplitude limit cycles?

## References


**1** S. K. Mitra and J. F. Kaiser, *Handbook for Digital Signal Processing*, John Wiley, New York, 1993.

**2** P. A. Regalia, S. K. Mitra and P. P. Vaidyanathan, All-pass filter: a versatile signal processing building block. *Proceedings of the IEEE*, 76(1), 19–37, 1988.

**3** J. B. Knowles and E. M. Olcayto, Coefficient accuracy and digital filter response. *IEEE Transactions on Circuit Theory*, 1968.

**4** L. B. Jackson, On the interaction of round-off noise and dynamic range in digital filters. *Bell Systems Technical Journal*, 1970.

**5** P. Ebert, J. Mazo and M. Taylor, Overflow oscillations in digital filters. *Bell Systems Technical Journal*, 1969.

**6** T. Claasen, W. Mecklenbrauker and J. Peek, Effects of quantizations and overflow in recursive digital filters. *IEEE Transactions*, ASSP24(6), 1976.

# 7

# Infinite Impulse Response Filters

Digital filters with an infinite impulse response (IIR), or recursive filters, have properties similar to those of analog filters, and consequently, their coefficients can be determined by similar techniques [1–3].

Before discussing the method for calculating the coefficients, it is useful to give some general expressions for the properties of these filters.

## 7.1 General Expressions for the Properties of IIR Filters

The general IIR filter is a system which, from the set of data $x(n)$, produces the set $y(n)$ such that:

$$y(n) = \sum_{l=0}^{L} a_1 x(n-1) - \sum_{k=1}^{K} b_k y(n-k) \tag{7.1}$$

The $Z$-transfer function for this system is written:

$$H(Z) = \frac{\sum_{l=0}^{L} a_1 Z^{-1}}{1 + \sum_{k=1}^{K} b_k Z^{-k}} \tag{7.2}$$

This is the quotient of two polynomials in $Z$, which are often of the same degree.

As the coefficients $a_l$ and $b_k$ are real numbers, $H(Z)$ is a complex number such that:

$$\overline{H(Z)} = H(\overline{Z})$$

and the frequency response of the filter can be written with the same conventions as in the earlier chapters:

$$H(\omega) = |H(\omega)|e^{-j\phi(\omega)}$$

The modulus and the phase are expressed in terms of $H(Z)$ by the following equations:

$$|H(\omega)|\,2 = [H(Z)H(Z^{-1})]_{Z=e^{j\omega}} \tag{7.3}$$

By squaring $H(\omega)$ and using equation (7.3),

$$\phi(\omega) = -\frac{1}{2}j \log\left[\frac{H(Z)}{H(Z^{-1})}\right]_{Z=e^{j\omega}} \tag{7.4}$$

and, by taking the derivative of $\phi(Z)$ with respect to the complex variable $Z$, we obtain:

$$\frac{d\phi}{dZ} = -\frac{1}{2}j\left[\frac{H'(Z)}{H(Z)} + \frac{1}{Z^2}\frac{H'(Z^{-1})}{H(Z^{-1})}\right]$$

*Digital Signal Processing: Theory and Practice*, Tenth Edition. Maurice Bellanger.

For $Z = e^{j\omega}$ this becomes:

$$\frac{d\phi}{dZ} = -\frac{1}{jZ} Re\left[Z\frac{d}{dZ}\log(H(Z))\right]$$

Thus, the equation for the group delay becomes:

$$\tau(\omega) = \frac{d\phi}{d\omega} = \frac{d\phi}{dZ} jZ = -Re\left[Z\frac{d}{dZ}\log(H(Z))\right]_{Z=e^{j\omega}} \tag{7.5}$$

Using equations (7.3)–(7.5), it is possible to analyze IIR filters of any order.

*Example*: Assume,

$$H(Z) = \frac{1}{D(Z)} = \frac{1}{1 + b_1 Z^{-1} + b_2 Z^{-2}}$$

Since

$$\tau(\omega) = Re\left[Z\frac{D'(Z)}{D(Z)}\right]_{Z=e^{j\omega}} = -Re\left[\frac{b_1 Z^{-1} + 2b_2 Z^{-2}}{1 + b_1 Z^{-1} + b_2 Z^{-2}}\right]_{Z=e^{j\omega}}$$

one has

$$\tau(\omega) = 1 - \frac{1 - b_2^2 + b_1(1 - b_2)\cos\omega}{1 + b_1^2 + b_2^2 + 2b_1(1 + b_2)\cos\omega + 2b_2\cos(2\omega)} \tag{7.6}$$

which is equivalent to the expression given in the previous chapter when the poles are complex with $b_1 = -2r\cos\vartheta$ and $b_2 = r^2$.

Other expressions for IIR filters can be obtained in terms of the poles and zeros of $H(Z)$. If the numerator and the denominator have the same degree $N$, and if $N$ is even, then:

$$H(Z) = a_0 \prod_{i=1}^{N/2} \frac{1 + a_1^i Z^{-1} + a_2^i Z^{-2}}{1 + b_1^i Z^{-1} + b_2^i Z^{-2}}$$

The square of the modulus of the transfer function is equal to the product of the squares of the moduli of the elementary functions. The phase and the group delay are the sums of the contributions of the sections:

$$|H(\omega)|\,2 = a_0 \prod_{i=1}^{N/2} |H_i(\omega)|\,2$$

$$\tau(\omega) = \sum_{i=1}^{N/2} \tau_i(\omega)$$

The general equations for IIR filters given above are used to calculate the coefficients.

## 7.2 Direct Calculations of the Coefficients Using Model Functions

A direct method for calculating the coefficients of an IIR filter is to use a model function, which is a real function defined on the frequency axis. The model functions which will be considered are those of Butterworth, Bessel, Chebyshev, and elliptic functions, all of which have known selectivity properties. They are also used to calculate analog filters and form a model for the square of the transfer function to be derived. However, one drawback to their use for calculating digital filters is that they are not periodic when the desired function has period $f_s$. It is therefore necessary to establish a map linking the real axis and the range $[0, f_s]$. Such mapping is supplied by a conformal transformation in the complex plane with the following properties:

(1) It transforms the imaginary axis onto the unit circle.
(2) It transforms a rational fraction of the complex variable $s$ into a rational fraction of the complex variable $Z$.
(3) It conserves stability.

### 7.2.1 Impulse Invariance

Consider the analog filter defined by the equation:

$$y'_a(t) = by_a(t) + x(t) \tag{7.7}$$

It has a transfer function:

$$H(s) = \frac{1}{s-b} \tag{7.8}$$

and an impulse response:

$$h(t) = e^{bt} \tag{7.9}$$

Sampling this response with period $T$ provides the sequence:

$$h(nT) = e^{bTn} \tag{7.10}$$

which has a $Z$-transform:

$$H(Z) = \frac{1}{1 - e^{bT}Z^{-1}} \tag{7.11}$$

The pole $b$ of the analog filter has become $e^{bT}$ for the digital filter. The method can be generalized to any number of poles.

This simple method is used in the simulation of analog systems by digital computers. It has a serious disadvantage for the design of filters due to aliasing of the frequency response. During the sampling operation, the frequency response of the analog filter is replicated in the useful band of the digital filter and the amplitude specification cannot be conserved.

Another approach consists of establishing a direct correspondence between the operation of the analog filter and the digital filter.

Reconsidering the differential equation of the analog filter, at time $nT$, one can write:

$$y_a(nT) = y_a(nT - T) + \int_0^T y'_a(nT - T + \tau)d\tau \tag{7.12}$$

Evaluating the integral using the trapezium rule leads to:

$$y_a(nT) - y_a(nT - T) = \frac{T}{2}\left[y'_a(nT) + y'_a(nT - T)\right] \tag{7.13}$$

Hence:

$$y_a(nT) - y_a(nT - T) = \frac{T}{2}[by_a(nT) + x(nT) + by_a(nT - T) + x(nT - T)]$$

Taking the $Z$-transform of both sides leads to the following $Z$-transfer function:

$$H(Z) = \frac{1}{\frac{\left(\frac{2}{T}\right)(1-Z^{-1})}{(1+Z^{-1})} - b} \tag{7.14}$$

which yields a relation between $s$ and $Z$, called the bilinear transform.

### 7.2.2 Bilinear Transform

Assume a transformation which transforms a point $Z$ on the complex plane into the point $s$ where:

$$s = \frac{2}{T}\frac{1 - Z^{-1}}{1 + Z^{-1}} \tag{7.15}$$

To each point on the unit circle $Z = e^{j\omega T}$, there is a corresponding point $s$ such that:

$$s = \frac{2}{T}\frac{1 - e^{-j\omega T}}{1 + e^{-j\omega T}} = j\frac{2}{T}\tan\frac{\omega T}{2} \tag{7.16}$$

Consequently, the imaginary axis corresponds to the unit circle. The equation which gives $Z$ as a function of $s$ is:

$$Z = \frac{(2/T) + s}{(2/T) - s} \tag{7.17}$$

A rational fraction involving $s$ is transformed into a rational fraction involving $Z$ in which the numerator and the denominator have the same degree.

If the real part of $s$ is negative, the modulus of $Z$ is less than 1 – that is, the part of the complex plane of the variable $s$ to the left of the imaginary axis is transformed inside the unit circle. This property ensures that the stability characteristics of the system are conserved.

In the definition of the transformation, the factor $2/T$ is a scale factor. $T = 1/f_s$ is the sampling period of the digital system. This factor controls the warping of the frequency axis which is introduced when the bilinear transform is used to obtain the $Z$-transfer function of a digital system using a complex function of $s$. Indeed, the digital system obtained has a frequency response which is a function of the variable $f_N$, which, in turn, is related by equation (7.16) to the values $jf_A$ of the initial function on the imaginary axis. Thus,

$$\pi f_A T = \tan(\pi f_N T) \tag{7.18}$$

Figure 7.1 illustrates this equation. Note that the warping is negligible for very low frequencies, which justifies the choice of the scale factor $2/T$.

Although other transforms can also be applied to the calculation of digital filter coefficients, the bilinear transform is the most commonly used. It allows for the calculation of digital filters

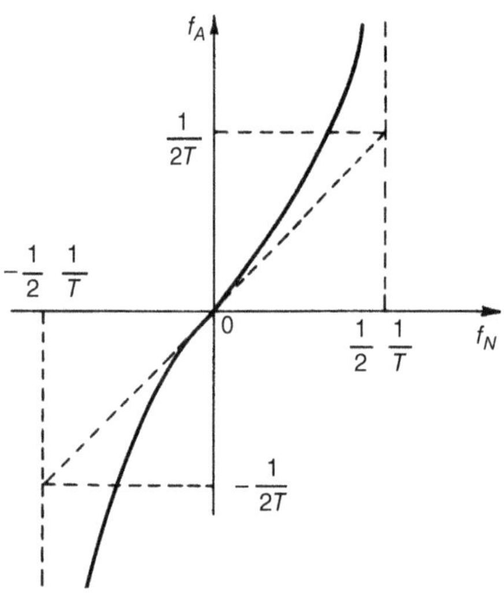


**Figure 7.1** Frequency warping introduced by the bilinear transform.

from the transfer functions for analog filters, or by using computer programs developed for analog filters.

It should be remembered, however, that the frequency response is warped, as shown above, since the analog and digital frequencies $\omega_{\text{A}}$ and $\omega_{\text{D}}$ are related by the equation:

$$\omega_{\text{A}} = \frac{2}{T}\tan\left[\frac{\omega_N T}{2}\right] \tag{7.19}$$

The group delay is also modified:

$$\tau_N = \tau_A\left[1+\left(\frac{\omega_A T}{2}\right)^2\right] \tag{7.20}$$

That is, an analog filter with a constant group delay is transformed into a digital filter which does not have this property.

In order to design a digital filter from a mask using this method, the mask must first be modified to take account of the subsequent frequency warping. The analog filter satisfying the new mask can then be calculated, and finally, the bilinear transform can be applied.

#### 7.2.2.1 Butterworth Filters

Two examples will be used to illustrate the calculation of the coefficients by a model function. The first is Butterworth filter functions, because of their simplicity, and the second is elliptic functions because they are the ones most frequently used.

A Butterworth function of order $n$ is defined by:

$$\left|F(\omega)\right|^2 = \frac{1}{1+\left(\frac{\omega}{\omega_c}\right)^{2n}} \tag{7.21}$$

The parameter $\omega_{\text{c}}$ gives the value of the variable for which the function has the value $\frac{1}{2}$. Figure 7.2 represents this function for various values of $n$.

By analytic extension, taking $\omega_{\text{c}} = 1$, this can be written as:

$$\left|F(\omega)\right|^2 = |H(j\omega)|^2 = |H(s)H(-s)|_{s=j\omega} = \frac{1}{1+\omega^{2n}}$$

$$H(s)H(-s) = \frac{1}{1+\left(\frac{s}{j}\right)^{2n}} = \frac{1}{1+(-s^2)^n}$$

The poles of this function lie on the unit circlc. For cxample, when $n$ is odd, one can write:

$$H(s)H(-s) = \frac{1}{\prod_{k=1}^{2n}(s-e^{j\pi(k/n)})}$$

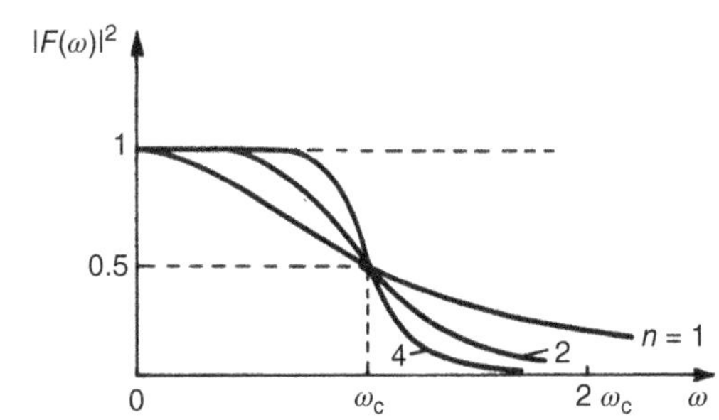


**Figure 7.2** Butterworth functions.

By setting to $H(s)$ the poles which are to the left of the imaginary axis, to obtain a stable filter, and after proper factorization to obtain first- and second-order sections with real coefficients, one has:

$$H(s) = \frac{1}{1+s} \prod_{k=1}^{(n-1)/2} \frac{1}{s^2 + 2\cos\left(\pi\left(\frac{k}{n}\right)\right) s + 1}$$

Similarly, for even values of $n$, one obtains:

$$H(s) = \prod_{k=1}^{\frac{n}{2}} \frac{1}{s^2 + 2\cos\left[\frac{\pi(2k-1)}{2n}\right] s + 1}$$

The corresponding digital filter is produced by change of variable in accordance with equation (7.15):

$$s = \frac{2}{\tan(\omega_c T/2)} \frac{1 - Z^{-1}}{1 + Z^{-1}}$$

The point on the frequency axis where the response of the digital filter has a value of $2^{-1/2}$ is $f_c$ with:

$$\omega_c = 2\pi f_c$$

By setting $u = 1/\tan(\pi f_0 T)$ and $\alpha_k = 2\cos(\pi k/n)$, the $Z$-transfer function for the digital filter of odd-order $n$ is obtained:

$$H(Z) = \frac{1 + Z^{-1}}{(1+u) + (1-u)Z^{-1}} \prod_{k=1}^{\frac{(n-1)}{2}} a_0^k \frac{(1+Z^{-1})^2}{1 + b_1^k Z^{-1} + b_2^k Z^{-2}}$$

with:

$$a_0^k = \frac{1}{1 + u\alpha_k + u^2}; \quad b_1^k = 2a_0^k(1 - u^2); \quad b_2^k = a_0^k(1 - u\alpha_k + u^2)$$

For even values of $n$, with $\alpha_k = 2\cos(\pi(2k-1)/2n)$, this becomes:

$$H(Z) = \prod_{k=1}^{n/2} a_0^k \frac{(1+Z^{-1})^2}{1 + b_1^k Z^{-1} + b_2^k Z^{-2}} \tag{7.22}$$

It would thus appear that the zeros of the $Z$-transfer function are all found at the point $Z = -1$, which can simplify the realization of the filter. On the other hand, the function is completely determined by the data for the parameters $n$ and $u$.

The order $n$ is calculated from the specification of the filter. Assume that a filter is to be produced with a frequency response greater than or equal to $1 - \delta_1$ in the band $[0, f_1]$ and less than or equal to $\delta_2$ in the band $[f_2, f_s/2]$. In terms of the model function $F(\omega)$, these constraints imply the following inequalities:

$$\frac{1}{1 + \left(\frac{\omega_1}{\omega_c}\right)^{2n}} \geq (1 - \delta_1)^2 \quad \text{and} \quad \frac{1}{1 + \left(\frac{\omega_2}{\omega_c}\right)^{2n}} \leq \delta_2^2$$

For small $\delta_1$ and $\delta_2$, the following equation for $n$ results:

$$n \geqslant \frac{\frac{1}{2}\log(2\delta_1) + \log(\delta_2)}{\log(\omega_1) - \log(\omega_2)}$$

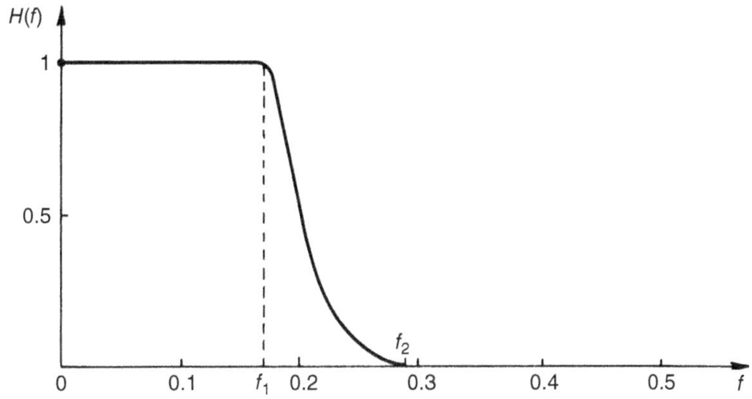


**Figure 7.3** Frequency response of a Butterworth filter of order 8.

The order $N$ of the digital filter is obtained using equation (7.18):

$$N \geq \frac{\log\left(\frac{1}{\delta_2}\sqrt{(2\delta_1)}\right)}{\log\tan(\pi f_2 T) - \log\tan(\pi f_1 T)} \tag{7.23}$$

Once $n$ is chosen, the parameter $u$ must lie in the interval:

$$\frac{1}{\tan(\pi f_2 T)}\left(\frac{1}{\delta_2}\right)^{1/n} \leq u \leq \frac{(2\delta_1)^{(1/2)n}}{\tan(\pi f_1 T)}$$

These parameters allow $H(Z)$ to be calculated.

*Example*: Assume the following characteristic, which was studied earlier for FIR filters:

$$\delta_1 = 0.045; \quad \delta_2 = 0.015; \quad f_s = 1; \quad f_1 = 0.1725; \quad f_2 = 0.2875$$

It is found that $N \approx 7.3$, so the value adopted is $N = 8$:

$$H(Z) = 0.00185\frac{(1+Z^{-1})^2}{1-0.36Z^{-1}+0.04Z^{-2}} \times \frac{(1+Z^{-1})^2}{1-0.39Z^{-1}+0.12Z^{-2}}$$
$$\times \frac{(1+Z^{-1})^2}{1-0.45Z^{-1}+0.31Z^{-2}} \times \frac{(1+Z^{-1})^2}{1-0.58Z^{-1}+0.69Z^{-2}}$$

The frequency response obtained is shown by Figure 7.3. When the filter transition band, $\Delta f = f_2 - f_1$, is sufficiently small, equation (7.23) can be simplified:

$$N \approx \log\frac{1}{\delta_2\sqrt{(2\delta_1)}}\frac{2.3}{2\pi}\frac{f_s}{\Delta f}\sin\left(2\pi\frac{f_1}{f_s}\right) \tag{7.24}$$

Thus, the filter order is proportional to the inverse of the transition band, as is the case for FIR filters. It follows that the selectivity of this kind of filter is rather limited in practice.

In conclusion, Butterworth filters are straightforward to calculate. Significant simplifications can be made in their realization because of the arrangement of the roots and, in certain cases, of the poles, but they are much less selective than elliptic filters.

#### 7.2.2.2 Elliptic Filters

The elliptic filter displays ripples in the pass band and in the stop band. It is optimal in that for a given order $n$ and for fixed ripple amplitudes, it has the smallest transition band. The model

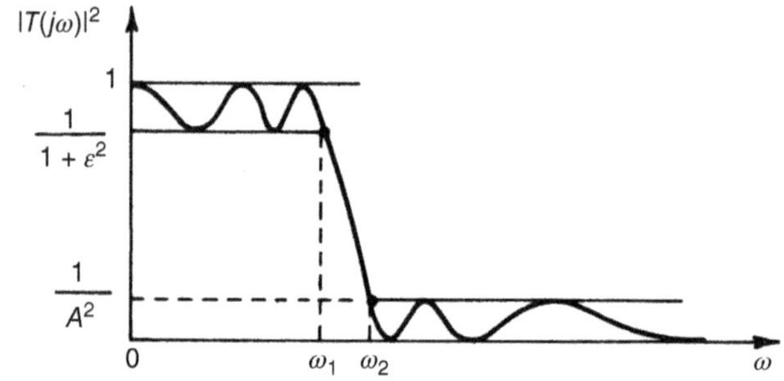


**Figure 7.4** Elliptic filtering function.

function involves elliptic functions and is written as:

$$T^2(u) = \frac{1}{1 + \varepsilon^2 sn^2(u, k_1)} \tag{7.25}$$

where $y = sn(u, k)$ is defined implicitly by the incomplete elliptic function of the first type:

$$u = \int_0^{\arcsin y} \frac{d\theta}{(1 - k^2 \sin^2 \theta)^{\frac{1}{2}}} \tag{7.26}$$

Figure 7.4 represents the function $T^2(u)$ for $u = j\omega$, and shows the parameters corresponding to $k_1$ such that:

$$k_1 = \frac{\varepsilon}{\sqrt{(A^2 - 1)}}$$

The function $sn^2(\omega, k)$ oscillates between 0 and 1 for $\omega < \omega_1$ and between $\sqrt{(A^2 - 1)}/\varepsilon$ and infinity for $\omega \geq \omega_2$.

One can show that the order $n$ of the filter is determined from the parameters $k_1$ and $k$, the selectivity factor:

$$k = \frac{\omega_1}{\omega_2}$$

by the expression:

$$n = \frac{K(k)K\left(\sqrt{(1 - k_1^2)}\right)}{K(k_1)K(\sqrt{(1 - k^2)})} \tag{7.27a}$$

where $K(k)$ is the complete elliptic integral of the first type:

$$K(k) = \int_0^{\frac{\pi}{2}} \frac{d\theta}{(1 - k^2 \sin^2 \theta)^{\frac{1}{2}}} \tag{7.27b}$$

This integral is calculated by the Chebyshev polynomial approximation method which results in an error of the order of $10^{-8}$ with a polynomial of degree 4. The inverse function to the incomplete integral of the first type is calculated as the quotient of two rapidly converging series.

A simplified equation for the order $n$ of the filter can be obtained from the general specification given in Figure 5.7. With the assumption of a ripple in the pass band of between 1 and $1 - 2\delta_1$, and with the following parameters (Figure 7.5):

$$\delta_2 = \frac{1}{A};\ 2\delta_1 = 1 - \frac{1}{\sqrt{(1+\varepsilon^2)}};\ f_s = 1$$

**Figure 7.5** Poles and zeros of an elliptic filter of order 4.

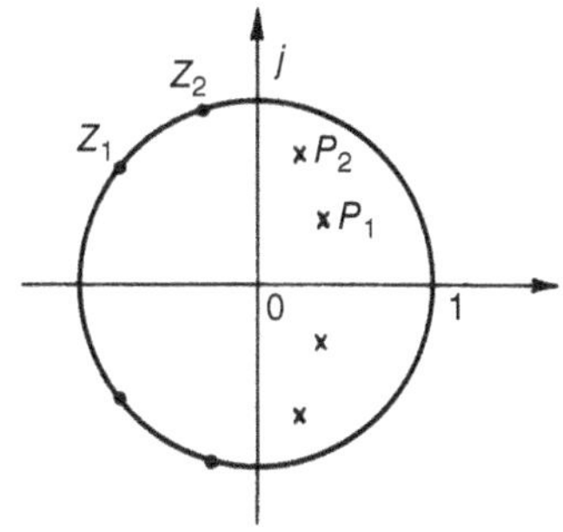


one has:

$$n \approx \left(\frac{2}{\pi^2}\right) \text{In}\left(\frac{2}{\delta_2\sqrt{\delta_1}}\right) \text{In}\left[\frac{8\omega_1}{(\omega_2-\omega_1)}\right] \tag{7.28}$$

The order $N$ of the digital filter satisfying the mask in Figure 5.7 is then given by:

$$N \approx \left(\frac{2}{\pi^2}\right) \text{In}\left(\frac{2}{\delta_2\sqrt{\delta_1}}\right) \times \text{In}\left[\frac{8\tan\left(\frac{\pi f_1}{f_s}\right)}{\left(\tan\left(\frac{\pi f_2}{f_s}\right)-\tan\left(\frac{\pi f_1}{f_s}\right)\right)}\right]$$

The transition band $\Delta f = f_2 - f_1$ is generally narrow, and thus (using logarithms to base 10),

$$N \approx 1.076 \log\left(\frac{2}{\delta_2\sqrt{\delta_1}}\right) \log\left[\left(\frac{f_s}{\Delta f}\right)\left(\frac{4}{\pi}\right)\sin\left(\frac{2\pi f_1}{f_s}\right)\right] \tag{7.29}$$

This relation should be compared with equation (5.32) for finite impulse response filters. It shows that, for elliptic IIR filters, the order is proportional to the logarithm of the inverse of the normalized transition band. This leads to much lower values than for FIR filters. Further, equation (7.29) shows that the width of the band is also involved. The maximum value of $N$ is found for $f_1$ close to $f_s/4$ – that is, for a pass band approximating half of the useful band. Further simplification can be obtained for filters with a narrow pass band. In this case, the order $N'$ of the filter is given by:

$$N' \approx 1.076 \log\left[\frac{2}{(\delta_2\sqrt{\delta_1})}\right] \log\left(\frac{8f_1}{\Delta f}\right) \tag{7.30}$$

and, as in analog filters, it is the steepness of the cutoff which is important here.

Once the filter order has been determined, the calculation procedure involves determining the poles and zeros of $T^2(u)$, which show double periodicity in the complex plane. By changing the variable and then applying the bilinear transform, the configuration of the poles and zeros of the digital filter in the $Z$-plane is obtained [4].

With this technique, the filter is specified by:

(1) The peak-to-peak amplitude of the pass band ripples, expressed in dB:

$$\text{BP} = -20\log(1-2\delta_1)$$

(2) The amplitude of the ripples in the attenuation band, expressed in dB:

$$\text{AT} = 20\log\left(\frac{1}{\delta_2}\right)$$

(3) The frequency of the end of the pass band, FB.

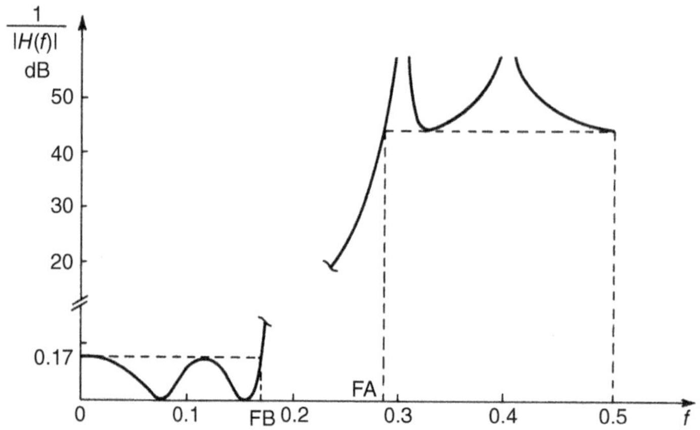


**Figure 7.6** Frequency response of an elliptic filter of order 4.

(4) The frequency of the start of the attenuation band, FA.
(5) The sampling frequency, FS.

*Example*: Consider the specification given in the previous section: BP = 0.4, AT = 36.5, FS = 1, FB = 0.1725, FA = 0.2875.

It is found that $N = 3.37$, and the adopted values in $N = 4$. The zeros and poles have co-ordinates (Figure 7.5):

$$Z_1 = -0.816 + j0.578 \quad Z_2 = -0.2987 + j0.954$$
$$P_1 = 0.407 + j0.313 \quad P_2 = 0.335 + j0.776$$

To demonstrate the point of infinite attenuation, the curve $1/|H(f)|$ giving the attenuation of the filter as a function of frequency is shown in Figure 7.6.

Figure 7.7 shows the group delay of the filter obtained. The curves can be compared with the results obtained for the same characteristic using the Butterworth filter. They demonstrate the advantage of the elliptic filter, which requires an order lower by a factor of 2 and produces a corresponding reduction in the complexity of the circuits.

The methods described allow for calculation of low-pass filters, from which suitable frequency transformations make it possible to obtain high-pass and band-pass filters.

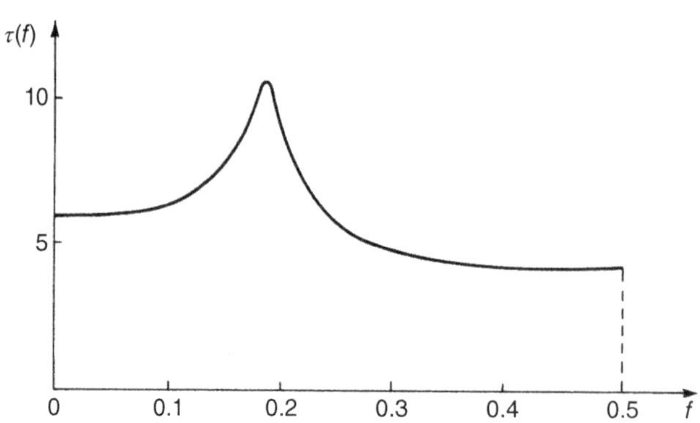


**Figure 7.7** Group delay of an elliptic filter of order 4.

#### 7.2.2.3 Calculating any Filter by Transformation of a Low-pass Filter

The calculation procedures presented in the previous sections result in a function $H(s)$ which provides, by bilinear transforms, the $Z$-transfer function for the digital filter. For $s = j\omega$, the function $H(\omega)$ is a low-pass filter function in the frequency domain which extends from zero to infinity. It is possible to apply transformations to it which result in other types of filters [5]. For example, in order to obtain a low-pass filter whose pass band extends from 0 to $\omega_1'$, using a function whose pass band covers the domain $[0, \omega_1]$, the following transformation can be made:

$$s \to s\frac{\omega_1}{\omega_1'}$$

By starting from a low-pass filter with a pass band limited to 1, the following filters can be obtained, by denoting the lower and upper limits of the pass band by $\omega B$ and $\omega H$:

(1) Another low-pass: $s \to \frac{s}{\omega_H}$
(2) High-pass: $s \to \frac{\omega_B}{s}$
(3) Band-pass: $s \to \frac{s^2+\omega_H\omega_B}{s(\omega_H-\omega_B)}$
(4) Band stop: $s \to \frac{s(\omega_H-\omega_B)}{s^2+\omega_H\omega_B}$

These transforms conserve the ripple in the response of the filter but result in frequency warping.

A more direct method consists of using transforms other than the bilinear transform to reach the function $H(Z)$. For example, the transform:

$$s = \frac{1}{T}\frac{1 - 2\cos(\omega_0 T)Z^{-1} + Z^{-2}}{1 - Z^{-2}} \tag{7.31}$$

allows a bandpass digital filter to be obtained from a low-pass filter function.

For $Z = ej\omega T$ this becomes:

$$s = j\frac{1}{T}\frac{\cos(\omega_0 T) - \cos(\omega T)}{\sin(\omega T)} \tag{7.32}$$

If the pass band of the digital filter extends from $\omega B$ to $\omega H$, $\omega 0$ must be chosen so that the abscissae of the transformed points are equal in absolute value but are of opposite sign:

$$\frac{\cos(\omega_o T) - \cos(\omega_B T)}{\sin(\omega_B T)} = -\frac{\cos(\omega_o T) - \cos(\omega_H T)}{\sin(\omega_H T)}$$

and thus:

$$\cos(\omega_o T) = \frac{\cos\left[(\omega_B + \omega_H)\frac{T}{2}\right]}{\cos\left[(\omega_B - \omega_H)\frac{T}{2}\right]}$$

This approach avoids adding a stage to the calculation procedure for a band-pass filter.

It is also possible to use transforms in the $Z$-plane which conserve the unit circle. The simplest is to transform from $Z$ to $-Z$, which changes a low-pass filter into a high-pass one.

The transform:

$$Z^{-1} \to \frac{(Z^{-1} - \alpha)}{(1 - \alpha Z^{-1})} \tag{7.33}$$

where $\alpha$ is a real number, changes a low-pass filter into another low-pass one. In fact, it can be shown that the most general transform is expressed by [5]:

$$Z^{-1} \to \pm\prod_{k=1}^{K}\frac{Z^{-1} - \alpha_k}{1 - \alpha_k Z^{-1}} \tag{7.34}$$

with $|\alpha k| < 1$ to ensure stability.

For example, a low-pass filter is transformed into a band-pass filter by:

$$Z^{-1} \to \frac{Z^{-2} - \alpha_1 Z^1 + \alpha_2}{\alpha_2 Z^{-2} - \alpha_1 Z^{-1} + 1} \tag{7.35}$$

It would thus appear that every type of filter can be calculated in a direct way using model functions. However, there are important limitations. Firstly, for example, for elliptic filters, the ripples must be constant in the pass and stop bands. Secondly, the methods described do not allow for possible constraints on the impulse response. To overcome these limitations, optimization techniques have to be employed.

### 7.2.3 Iterative Techniques for Calculating IIR Filter with Frequency

As with FIR filters, optimization methods allow for the calculation of IIR filters with any specification. Nevertheless, the calculation is somewhat more sensitive than for FIR filters as precautions have to be taken to avoid producing an unstable system.

Two methods will be presented. These correspond to two different optimization criteria, the first of which is minimization of the mean square error [6].

#### 7.2.3.1 Minimizing the Mean Square Error

The transfer function of a filter is given in a factorized form by the equation which was introduced earlier:

$$H(Z) = a_0 \prod_{i=1}^{\frac{N}{2}} \frac{1 + a_1^i Z^{-1} + a_2^i Z^{-2}}{1 + b_1^i Z^{-1} + b_2^i Z^{-2}}; a_0 > 0 \tag{7.36}$$

by regarding the numerator and denominator as having the same degree $N$ (even).

Assume $D(f)$ is the function which approximates the frequency response of the filter $H(f)$. The difference between these functions represents an error which can be minimized by least squares for a number of points, $N_0$, on the frequency axis. Thus,

$$E = \sum_{n=0}^{N_0-1} \left( |H(f_n)| - \left|D\left(f_n\right|\right) \right)^2$$

The value $E$ is a function of the set of $2N + 1$ parameters, which are the coefficients of the filter:

$$E = E\left(a_0, a_1^i, a_2^i, b_1^i, b_2^i\right) \quad \text{with} \quad 1 \leqslant i \leqslant \frac{N}{2}$$

The minimum corresponds to the set of $2N + 1$ parameters $x_k$ such that:

$$\frac{\partial E}{\partial x_k} = 0; \quad 1 \leqslant k \leqslant 2N + 1$$

For the parameter $a_0$, one can set $H(Z) = a_0 H_1(Z)$, whence:

$$\frac{\partial E}{\partial a_0} = 0 = 2 \sum_{n=0}^{N_0-1} (a_0 |H_1(f_n)| - |D(f_n)|)|H_1(f_n)|$$

Thus, the value of $a_0$ is:

$$a_0 = \frac{\sum_{n=0}^{N_0-1} |D(f_n)||H_1(f_n)|}{\sum_{n=0}^{N_0-1} H_1(f_n)|^2} \tag{7.37}$$

The optimization is restricted to $2N$ variables.

The procedure consists of taking an initial function $H_1^0(Z)$, which is found, for example, by the direct calculation method given in the preceding section for elliptic filters, and then assuming that it is sufficiently close to the optimum for the function $E$ to be represented by a quadratic function with $2N$ parameters $x_k$. The desired optimum is then obtained through an increment in the parameters represented by the vector $\Delta X$ with $2N$ element such that:

$$E(X + \Delta X) \approx E(X) + \sum_{k=1}^{2N} \frac{\partial E}{\partial x_k} \Delta x_k + \frac{1}{2} \sum_{k=1}^{2N} \sum_{l=1}^{2N} \frac{\partial^2 E}{\partial x_k \partial x_l} \Delta x_k \Delta x_l$$

By using $A$ to denote the matrix with $2N$ rows and $N_0$ columns which has elements:

$$a_{ij} = 2\frac{\partial}{\partial x_i}[a_0|H_1(f_j)|]$$

and by using $\Delta$ to denote the column vector $e_n$ with $N_0$ terms such that:

$$e_n = a_0|H_1(f_n)| - |D(f_n)|$$

the condition of least squares is obtained by requiring $E(X + \Delta X)$ to be an extremum. As in Section 5.4, when calculating the coefficients of FIR filters, we have:

$$\Delta X = -[AA^t]^{-1}A\Delta$$

The calculation is then repeated with the new values for the parameters, which should ultimately lead to the required optimum. The chances of achieving this and the rate of convergence depend on the increments given to the parameters, and one of the best strategies is offered by the Fletcher and Powell algorithm [7].

To ensure stability in the resulting system, either the stability can be controlled at each stage, or the final system can be modified by replacing the poles $P_i$ outside the unit circle by $1/P_i$, which does not modify the modulus of the frequency response except for a constant factor. In the latter case, it is generally necessary to return to the optimization procedure to achieve the optimum.

Mean square error minimization can be applied to other functions as well as the frequency response – for example, the group delay [8].

#### 7.2.3.2 Chebyshev Approximation

This criterion corresponds to a limitation in the amplitude of the ripple in the frequency response of the filter in some frequency bands. One elegant approach consists of applying the Remez algorithm, already used to calculate the coefficients of linear-phase FIR filters. The technique consists of using an initial filtering function $H_0(Z)$ which approximates the required $H(Z)$. This function can, for example, be of the elliptic type calculated by the method shown in Section 7.2.4 using adequate specifications. It is written as:

$$H_0(Z) = \frac{N_0(Z)}{D_0(Z)}$$

As the zeros of the filter functions are generally on the unit circle, the numerator $N(Z)$ can be regarded as the transfer function of a linear-phase FIR filter.

The first stage of the iterative technique is to calculate a new value for the numerator, $N_1(Z)$, using the algorithm for FIR filters. In this calculation, $D_0(f)$ is the function to be approximated in the pass band, and $1/|D_0(f)|$ is used as a weighting factor.

One then looks for a new value for the denominator $D_1(Z)$. A function which approximates $|N_1(f)|$ in the pass band can be sought directly by again using the algorithm for FIR filters. It is

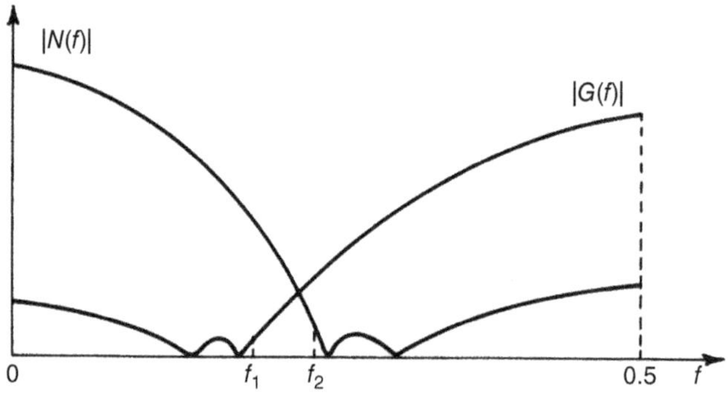


**Figure 7.8** The N(f) and G(f) functions for a low-pass filter.

more satisfactory, however, to use an adaptation of the calculation techniques employed for analog filters.

By assuming that the required function $H(f)$, written as:

$$H(f) = \frac{N(f)}{D(f)}$$

is such that:

$$|H(f)| \leq 1$$

one can write:

$$|G(f)|^2 = |D(f)|^2 - |N(f)|^2$$

and hence:

$$|H(f)|^2 = \frac{1}{1 + \left|\frac{G(f)}{N(f)}\right|^2}$$

Figure 7.8 shows the functions $|G(f)|$ and $|N(f)|$ for a low-pass filter. The zeros of the function $G(Z)$ lie on the unit circle, and they can be calculated using the algorithm for linear-phase FIR filters. The weighting function is determined from $1/|N(f)|$.

By optimizing the stop and pass bands alternately, the required filter function is obtained after several iterations and the filter coefficients are then obtained. The stability of the filter requires that only those poles of $H(z)$ which are inside the unit circle be conserved.

### 7.2.4 Filters Based on Spheroidal Sequences

The filter design criterion can be the maximization of the energy concentration in a given frequency band, instead of a set of specifications on the frequency response.

Let $\lambda$ be a scalar intended to represent the energy concentration and defined by:

$$\lambda = \frac{\int_{-f_c}^{f_c} H(f)\overline{H(f)}df}{\int_{\frac{-1}{2}}^{\frac{1}{2}} H(f)H(f)df} \tag{7.38}$$

where $[-f_c, f_c]$ is the band in which the energy has to be concentrated.

For the response:

$$H(f) = \sum_{n=-P}^{P} a_n e^{-j2\pi fn}$$

a direct calculation yields:

$$\lambda = \frac{\sum_{n=-P}^{P} \sum_{m=-P}^{P} a_n a_m \left[\frac{\sin(n-m)2\pi f_c}{(n-m)\pi}\right]}{\sum_{n=-P}^{P} a_n^2} \tag{7.39}$$

and in matrix form:

$$A^t RA = \lambda A^t A$$

which is an eigenvalue equation. The filter coefficients are the elements of the eigenvector corresponding to the largest eigenvalue of the matrix $R$, whose elements are the terms:

$$\frac{\sin(n-m)2\pi f_c}{(n-m)\pi}$$

The elements of the eigenvectors of the matrix $R$ are called the discrete prolate spheroidal sequences [9].

An FIR filter has been obtained; it is also possible to derive an IIR filter. To that end, consider the following purely recursive function:

$$|H(f)|^2 = \frac{1}{1 + \left|\sum_{n=1}^{N} b_n e^{j2\pi fn}\right|^2}$$

The coefficients can be calculated to minimize the energy of the denominator in the band $[-f_c, f_c]$, under the condition $|H(f_c)|^2 = 0.5$. Then, the same method as above can be applied and the coefficients $b_n$ $(1 \leq n \leq N)$ can be taken as the elements of the eigenvector associated with the smallest eigenvalue of the spheroidal matrix. First, the scaling factor of the eigenvector is chosen such that $|H(f_c)|^2 = \frac{1}{2}$. Then, the poles of the analytic expansion of $|H(f)|^2$ are calculated, and the desired filter transfer function $H(Z)$ is obtained by keeping only those few poles which are inside the unit circle to ensure stability.

The procedure can be made reasonably simple by using iterative techniques and exploiting the structural properties of the spheroidal matrix [9].

*Example*: Assume: $N = 4; f_s = 1; f_c = 0.1$. The minimal eigenvector $V_{\min}$ is:

$$V_{\min}^{t[1.0,-2.773,-2.773,1.0]}$$

If $T$ designates the matrix whose elements are the terms $e^{j2\pi f_c(n-m)}$ with $1 \leq n, m \leq N$, then the scaling factor leading to the equality $V_{\min}^{t_{\min}}$ is 10.46.

After factorization of the analytical expansion $H(Z)H(Z^{-1})$, the transfer function finally obtained is:

$$H(Z) = \frac{0.0704}{(Z - 0.73 + j0.446)(Z - 0.73 - j0.446)(Z - 0.741)} \tag{7.40}$$

The technique presented above for low-pass filtering can be extended to high-pass filtering.

### 7.2.5 Structures Representing the Transfer Function

IIR filters can be produced using circuits which directly perform the operations represented by the expression for their transfer function. The term $Z^{-1}$ corresponds to a delay of one sampling period and is achieved by storage in the memory. The coefficients to be created by the circuits are those of the transfer function, with the same sign for the numerator and the opposite sign for the denominator.

Direct realization of transfer functions may necessitate high-accuracy calculations and it is generally preferable to resort to structures decomposed into sums or products of first- and second-order sections.

Only canonical structures – that is, those which require the minimum number of elementary operators, computing circuits, and memories – will be examined.

Decomposition into a product corresponds to the cascade structure in which the filter is generated from a set of first- and second-order sections:

$$H(Z) = a_0 \frac{\prod_{i=1}^{N}(1 - Z_i Z^{-1})}{\prod_{i=1}^{N}(1 - P_i Z^{-1})} = a_0 \cdots \frac{1 - Z_i Z^{-1}}{1 - P_i Z^{-1}} \cdots \frac{1 - 2Re(Z_j)Z^{-1} + |Z_j|^2 Z^{-2}}{1 - 2Re(P_j)Z^{-1} + |P_j|^2 Z^{-2}} \cdots \tag{7.41}$$

This structure is the one most frequently used because, in addition to its modularity, it presents the useful properties of low sensitivity to coefficient wordlength limitation and to round-off noise.

The function $H(Z)$ can also be decomposed into rational fractions:

$$H(Z) = a_0 + \cdots + \frac{\alpha_i}{1 - P_i Z^{-1}} + \cdots + \frac{\alpha_j + \beta_j Z^{-1}}{1 - 2Re(P_j)Z^{-1} + |P_j|^2 Z^{-2}} + \cdots \tag{7.42}$$

The approach corresponds to connecting the $M$ basic elements in parallel as shown in Figure 7.9. The numbers $y(n)$ are obtained by summing the outputs from the different elements, to which the input numbers $x(n)$ are applied.

IIR filters can also be implemented with the help of phase shifters.

Butterworth, Chebyshev, and elliptic transfer functions can be decomposed into a sum of two phase shifters [10]. For such a function, this gives:

$$H(Z) = \frac{N(Z)}{D(Z)} = \frac{1}{2}[A_1(Z) + A_2(Z)] \tag{7.43}$$

where $A_1(Z)$ and $A_2(Z)$ are the transfer functions of the phase shifters.

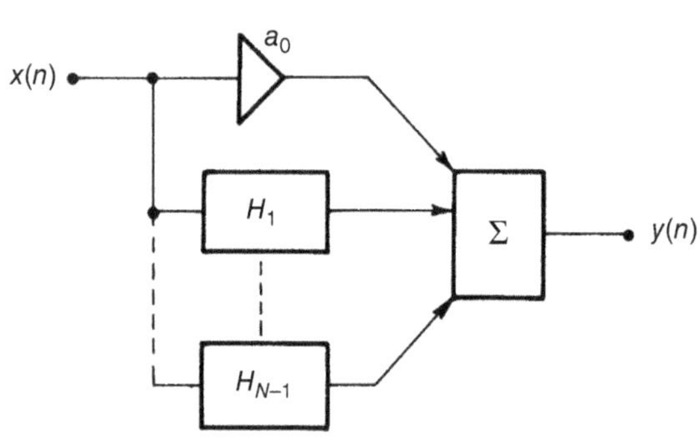


**Figure 7.9** The parallel structure.

Calculation of $A_1(Z)$ and $A_2(Z)$ from $H(Z)$ involves the complementary function $G(Z) = M(Z)/D(Z)$ such that:

$$|G(f)|^2 = 1 - |H(f)|^2 \tag{7.44}$$

It is assumed that the initial function $H(Z)$ is such that $N(Z)$ is a symmetric polynomial and $M(Z)$ is asymmetric – that is:

$$\overline{N}(Z) = Z^N N(Z) \quad \overline{M}(Z) = Z^N M(Z) \tag{7.45}$$

Under these conditions, combining equations (7.44) and (7.45) yields:

$$\overline{N}(Z)N(Z) + \overline{M}(Z)M(Z) = \overline{D}(Z)D(Z) \tag{7.46}$$

and

$$[N(Z) + M(Z)][N(Z) - M(Z)] = Z^{-N}D(Z^{-1})D(Z) \tag{7.47}$$

Note that the zeros of $N(Z) + M(Z)$ and $N(Z) - M(Z)$ are harmonic conjugates and that the zeros of $D(Z)$ are their inverses. Designating the poles of the filters by $P_i(i = 1, \ldots, N)$, and hence the zeros of $D(Z)$, one can write, to within a constant:

$$N(Z) + M(Z) = \prod_{i=1}^{r}(1 - Z^{-1}P_i)\prod_{i=r+1}^{N}(Z^{-1} - P_i) \tag{7.48}$$

and

$$N(Z) - M(Z) = \prod_{i=1}^{r}(Z^{-1} - P_i)\prod_{i=r+1}^{N}(1 - Z^{-1}P_i)$$

where $r$ is the number of zeros of the polynomial $N(Z) + M(Z)$ within the unit circle. Dividing by $D(Z)$, one obtains:

$$H(Z) + G(Z) = \frac{\prod_{i=r+1}^{N}(Z^{-1} - P_i)}{\prod_{i=r+1}^{N}(1 - Z^{-1}P_i)} \tag{7.49}$$

and similarly:

$$H(Z) - G(Z) = \frac{\prod_{i=1}^{r}(Z^{-1} - P_i)}{\prod_{i=1}^{r}(1 - Z^{-1}P_i)} \tag{7.50}$$

The phase shifters $A_1(Z)$ and $A_2(Z)$ have the following expressions:

$$A_1(Z) = \prod_{i=r+1}^{N}\frac{Z^{-1} - P_i}{1 - Z^{-1}P_i} A_2(Z) = \prod_{i=1}^{r}\frac{Z^{-1} - P_i}{1 - Z^{-1}P_i} \tag{7.51}$$

Finally, the filter $H(Z)$ and its complement $G(Z)$ are obtained by the arrangement shown in Figure 7.10.

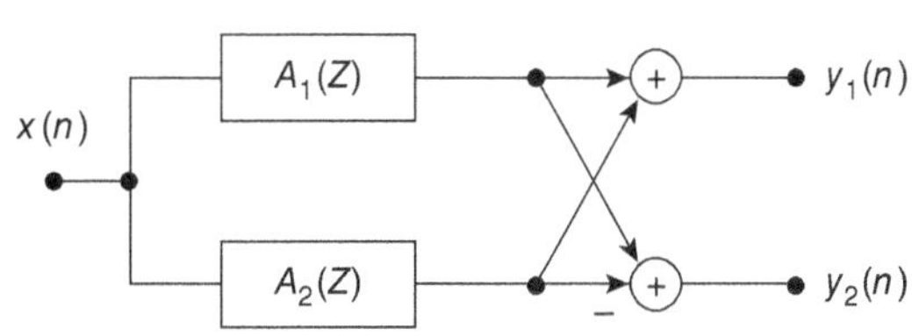


**Figure 7.10** Realization of an IIR filter and the complementary filter using two phase shifters.

The general procedure for designing the phase shifters from an elliptical filter is as follows:

(1) Calculate the transfer function $H(Z) = N(Z)/D(Z)$ of an elliptic filter of odd order $N$.
(2) Calculate the coefficients of the antisymmetric polynomial $M(Z)$ from $N(Z)$ and $D(Z)$ by using equation (7.46).
(3) Determine the inverses of the poles of $H(Z)$ which are the roots of the polynomial $N(Z)+M(Z)$.
(4) Calculate $A_1(Z)$ and $A_2(Z)$ using expression (7.51).

A simplified approach, when the order $N$ is not very high, consists of finding $A_1(Z)$ and $A_2(Z)$ directly by combining poles. Hence, for:

$$H(Z) = 0.0546\left[\frac{1+1.8601Z^{-1}+2.9148Z^{-2}+2.9148Z^{-3}+1.8601Z^{-4}+Z^{-5}}{(1-0.4099Z^{-1})(1-0.06611Z^{-1}+0.4555Z^{-2})(1-0.4993Z^{-1}+0.8448Z^{-2})}\right]$$

this gives:

$$A_1(Z) = \frac{0.4555-0.6611Z^{-1}+Z^{-2}}{1-0.6611Z^{-1}+0.4555Z^{-2}}$$

$$A_2(Z) = \frac{(-0.4099+Z^{-1})(0.8448-0.4993Z^{-1}+Z^{-2})}{(1-0.4099Z^{-1})(1-0.4993Z^{-1}+0.8448Z^{-2})}$$

The basic structure of phase shifters is useful since it provides two complementary filters with the same calculations, which is useful in filter banks, as shown in Chapters 11 and 12. Furthermore, it is less sensitive than other structures to rounding of the coefficients.

Note that filters which can be decomposed as the sum of phase shifters are entirely defined by their poles.

A sum of phase shifters as in Figure 7.10 is the most efficient realization of an elliptic filter since it requires a number of multiplications equal to the order of the filter.

### 7.2.6 Limiting the Coefficient Wordlength

Practical implementation of a filter implies limitations of the number of bits in the representation of filter coefficients, which form one of the multiplication terms. The effect of this on the complexity is important because multiplication is often the most critical factor. It is, therefore, necessary to find the minimum number of bits which allows the constraints imposed on the filtering function to be met.

Limitation of the number of bits of the scale factor $a_0$ appears as a modification in the gain of the filter, but it does not affect the form of the frequency response. As the filter gain is specified with a certain tolerance at a given frequency (for example, 800 Hz or 1000 Hz for a telephone channel), it is necessary to ensure that the binary representation of $a_0$ permits this constraint to be met.

Limitation of the number of bits in the other coefficients modifies the transfer function by introducing error polynomials $e_N(Z)$ and $e_D(Z)$ in the numerator and the denominator. The actual transfer function is $H_R(Z)$:

$$H_R(Z) = \frac{(N(Z)+e_N(Z))}{(D(Z)+e_D(Z))} \tag{7.52}$$

If the rounding errors in the coefficients are denoted by $\delta a_i$ and $\delta b_i$, these error functions can be written as a function of the normalized frequency ($f_s = 1$):

$$e_N(f) = \sum_{i=1}^{N} \delta a_i e^{-j2\pi fi}; \quad e_D(f) = \sum_{i=1}^{N} \delta b_i e^{-j2\pi fi}$$

In fact, these expressions form the Fourier series expansion of functions which are periodic in frequency. The Bessel–Parseval equation (1.7) relating the power of a signal to the power of its components allows the following equation to be written:

$$\int_0^1 \left| e_N(f)^2 df = \sum_{i=1}^{N} |\delta a_i|^2\right.$$

If $q$ denotes the quantization step,

$$|\delta a_i| \leq \frac{q}{2}$$

and an upper limit is obtained for $|e_N(f)|$ by:

$$|e_N(f)| \leq N\frac{q}{2} \tag{7.53}$$

A statistical estimate $\sigma$ of $|e_N(f)|$ can be obtained by regarding the $\delta a_i$ as random variables uniformly distributed over the range $[-q/2, q/2]$. It is evaluated from the effective value of the function $e_N(f)$. Thus,

$$\sigma^2 = \int_0^1 |e_n(f)|^2 df = N\frac{q^2}{12}$$

and hence

$$\sigma = \frac{q}{2}\sqrt{\frac{N}{3}} \tag{7.54}$$

This estimate is valid for both $|e_N(f)|$ and $|e_D(f)|$. It is clearly less than the bound (7.53) given above and it is, in fact, more realistic when $N$ exceeds several units.

The consequences of rounding the coefficients can be analyzed separately for the numerator and denominator of the transfer function by considering the stop band for one and the pass band for the other. Examination of the configuration of the poles and zeros in the $Z$-plane shows that the poles determine the response of the filter in the pass band and the zeros in the stop band.

In the stop band, the denominator coefficient wordlength limitation can be neglected and, in terms of the variable, $\omega = 2\pi f$, one has:

$$H_R(\omega) = \frac{(N(\omega) + e_N(\omega))}{D(\omega)}$$

The error on the response is then estimated by:

$$|H_R(\omega) - H(\omega)| \frac{\approx \sigma}{|D(\omega)|}$$

If the specification requires that the ripples in the stop band be less in modulus than $\delta_2$, then, by separating the tolerance into two equal parts – one for the ripple when there is no rounding error on the coefficients, and the other to allow for the error caused by this rounding – one has:

$$\frac{\sigma}{|D(\omega)|} < \frac{\delta_2}{2} \tag{7.55}$$

In the pass band, the numerator coefficient wordlength limitation can be neglected:

$$H_R(\omega) \approx \frac{N(\omega)}{D(\omega) + e_D(\omega)} \approx \frac{N(\omega)}{D(\omega)}\left[1 - \frac{e_D(\omega)}{D(\omega)}\right] \tag{7.56}$$

If the ripple in the pass band is less in modulus than $\delta_1$, then by again dividing the tolerance into two equal parts,

$$\frac{\sigma}{|D(\omega)|} < \frac{\delta_1}{2} \tag{7.57}$$

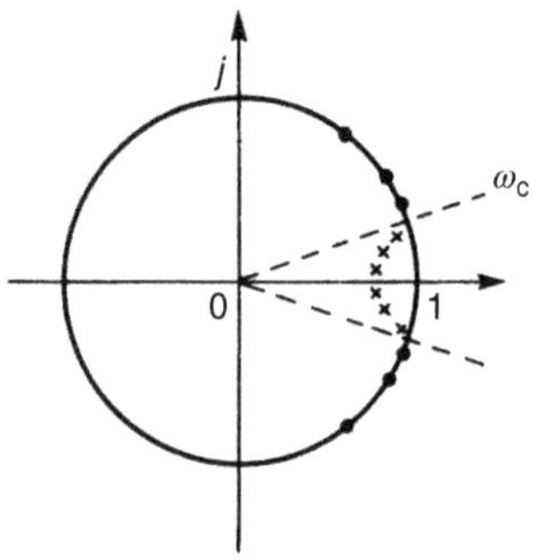


**Figure 7.11** The poles and zeros of a narrow low-pass filter.

This condition is generally much more restrictive than the earlier one because the function $|D(\omega)|$ has very low values in the pass band and is even more restrictive with a more selective filter. Further, when the pass band is narrow, the coefficients can have large values. For a low-pass filter such as that in Figure 7.11, the following equation can be written:

$$D(Z) \approx [1 - Z^{-1}]^N$$

and thus:

$$b_i \approx \frac{N!}{i!(N-i)!} \tag{7.58}$$

Under these conditions, a very large number of bits is required if both large values of the coefficients are to be represented and a very low quantization error is required. For this reason, decomposed structures are used almost exclusively with first- and second-order sections.

Let us first consider the cascade structure which corresponds to the decomposition (7.41) of the transfer function. If the order $N$ of the filter is even, this is written as:

$$H(\omega) = N(\omega)/D(\omega) = \prod_{i=1}^{N/2} N_i(\omega)/D_i(\omega)$$

The polynomials $N_i(\omega)$ and $D_i(\omega)$ are of the second degree.

In the pass band, if rounding of the coefficients of the polynomials $N_i(\omega)$ is neglected, we obtain:

$$H_R(\omega) \approx \prod_{i=1}^{N/2} \frac{N_i(\omega)}{D_i(\omega) + e_i(\omega)}$$

or

$$H_R(\omega) \approx \frac{N(\omega)}{D(\omega)}\left[1 - \sum_{i=1}^{N/2} \frac{e_i(\omega)}{D_i(\omega)}\right] \tag{7.59}$$

Then, using equation (7.53), we obtain:

$$|e_i(\omega)| \leq q$$

and the overall relative error $e(\omega)$ in the frequency response is found to be bounded by:

$$|e(\omega)| \leq q \sum_{i=1}^{N/2} \frac{1}{|D_i(\omega)|} \tag{7.60}$$

This expression demonstrates the benefit of the decomposed structure as the bound of the error is proportional to:

$$\sum_{i=1}^{N/2} \frac{1}{|D_i(\omega)|}$$

instead of (see equation (7.56)):

$$\prod_{i=1}^{N/2} \frac{1}{|D_i(\omega)|}$$

Further, the absolute value of the coefficients of the denominator cannot be greater than 2 if the filter is stable.

With the parallel structure, also in the pass band, the following equation is obtained:

$$H_R(\omega) \approx \sum_{i=1}^{N/2} \frac{N_i'(\omega)}{D_i(\omega) + e_i(\omega)} \approx \sum_{i=1}^{N/2} \frac{N_i'(\omega)}{D_i(\omega)} \left[1 - \frac{e_i(\omega)}{D_i(\omega)}\right]$$

Bearing in mind that the terms $\left|\frac{N_2'(\omega)}{D_i(\omega)}\right|$ are close to unity, the error in this case is approximately the same as that with the cascade structure.

In the stop band, away from the infinite attenuation frequencies, the following is obtained for the cascade structure:

$$H_R(\omega) \approx \prod_{i=2}^{N/2} \frac{N_i(\omega) + e_i(\omega)}{D_i(\omega)} \approx \frac{N(\omega)}{D(\omega)} \left[1 + \sum_{i=1}^{N/2} \frac{e_i(\omega)}{N_i(\omega)}\right] \tag{7.61}$$

For the parallel structure, the following estimate can be made:

$$H_R(\omega) \approx \sum_{i=1}^{N/2} \frac{N_i'(\omega) + e_i(\omega)}{D_i(\omega)} = \sum_{i=1}^{N/2} \frac{N_i'(\omega)}{D_i(\omega)} + \sum_{i=1}^{N/2} \frac{e_i(\omega)}{D_i(\omega)}$$

or,

$$H_R(\omega) \approx \frac{N(\omega)}{D(\omega)} \left[1 + \sum_{i=1}^{N/2} \frac{e_i(\omega)}{N_i(\omega)} \sum_{\substack{i=1 \\ j \neq i}}^{N/2} \frac{D_j(\omega)}{N_j(\omega)}\right] \tag{7.62}$$

By comparing equations (7.61) and (7.62), it appears that the terms:

$$\alpha_j = \prod_{j=1}^{N/2} \frac{D_j(\omega)}{N_j(\omega)}$$

in the stop band can have values much greater than unity, for example, in the vicinity of the zeros of the filter. The result is that in the stop band, the parallel structure is more sensitive to rounding errors than the cascade structure.

Finally, the cascade structure allows the coefficients of the IIR filters to be represented using fewer bits. Thus, this structure is the one most frequently used.

### 7.2.7 Round-Off Noise

Another limitation occurs in the practical realization of IIR filters because of the limits on the capacity of the data memories. This limit is the origin of the round-off noise. It will be analyzed for the *D–N* structure, but the same arguments apply to the *N–D* one. While this structure has specific advantages, mainly related to the calculation of partial sums and to the sequencing of the multiplications, it is the *D–N* structure which is most often used because it is generally easier to design, construct, and test.

In the presence of a signal – that is, for nonzero values of $x(n)$ – rounding before storage in the memory with quantization step $q$ is equivalent to superposition

on the input signal of an error signal $e(n)$ such that $|e(n)| < q/2$, assumed to have uniform spectrum and power so that $\sigma^2 = q^2/12$.

If other roundings are involved (for example, in the multiplications), it is apparent that the error signals produced are added either to the input or to the output signal depending upon whether they correspond to the coefficients of the recursive or non-recursive part. Consequently, to simplify the analysis, only the case of single quantization is considered. (By modifying the power of the injected noise, it is always possible to reach a scenario where only single quantization is required).

The error signal applied to the input of the filter undergoes the filtering function and, by applying equation (4.25), the power of the round-off noise at the output is:

$$N_c = \frac{q^2}{12}\int_0^1 \left|\frac{N(f)}{D(f)}\right|^2 df \tag{7.63}$$

or, as a function of the set $h(k)$, the impulse response of the filter,

$$N_c = \frac{q^2}{12}\sum_{k=0}^{\infty} |h(k)|^2 \tag{7.64}$$

The implementation of the cascade structure presents some possibilities for reducing this noise power [11].

When the filter is realized as a cascade of $N/2$ second-order sections, the round-off noise produced in each section undergoes the filtering function of that section and of the following ones. In this case, it should be noted that the amplitude, or level, at the input of each section, varies with the rank of the section and the frequency of the signal being considered.

If the rounding procedure is the same for all sections, the noise produced is the same and the contributions are added to each other. The total noise at the output of the filter under these conditions has a power of:

$$N_c = \frac{q^2}{12}\sum_{j=1}^{N/2}\left(\int_0^1 \prod_{l=j}^{N/2}\left|\frac{N_i(f)}{D_i(f)}\right|^2 df\right) \tag{7.65}$$

It is important to arrange the cascade of sections in such a way that the total round-off noise is minimized, and the parameters are available:

(1) The pairing of the poles and zeros to form a section.
(2) The order in which the sections are arranged.
(3) The scaling factor applied to each section.

These three parameters will be examined in turn.

(1) The pairing of the poles and zeros. The products

$$P_j(f) = \prod_{l=j}^{N/2}\left|\frac{N_i(f)}{Di(f)}\right|^2$$

have to be minimized, which means that each of the factors is minimized and, in particular, the lowest maximal value for each factor must be obtained. This condition is approximately fulfilled by the very simple procedure of associating the pole nearest to the unit circle with its closest zero, then the next pole with the zero which is then closest, and so on.

(2) *Determining the order of the sections.* The factor making the largest contribution to the total noise is often the one which has the highest maximal value. It can be worthwhile

to place it at the beginning of the chain so that its contribution appears only once in the total sum following equation (7.65) and to connect the sections in decreasing order of their maxima.

(3) *Calculating the scale factors.* These are the parameters which control the scaling of the numbers in the internal data memories. They are calculated for each section so as to maximize the amplitude of the signal while avoiding clipping.

Once the scale factors are known, all the elements involved in the implementation are available and the power of the rounding noise at the output of the filter can be determined for each value of the number of bits in the internal data memories [12–14].

### 7.2.8 Comparison of IIR and FIR Filters

Both IIR and FIR filters allow any given specification to be satisfied, and a systems designer often has to choose between the two approaches. The criterion is the complexity of the circuits to be employed. In practice, as will be seen in a later chapter, the comparison is reduced mainly to evaluating a single parameter: the number of multiplications to be carried out.

Equations (5.32) and (7.29) give estimates for FIR and IRR filters of order $N$, which is needed to satisfy the specifications of low-pass filtering as expressed by the ripple in the pass and stop bands, and by the widths of the pass band and the transition band.

For an FIR linear phase filter, the coefficients are symmetric and, if $N$ is even, $N/2$ have different values. Such a filter, therefore, requires $N/2$ memories for the coefficients and $N$ memories for the internal data. For each output number, $N/2$ multiplications and $N$ additions have to be performed. If phase linearity is not imposed, the order $N$ can be reduced with minimum-phase filters, as shown by equation (5.56). This reduction depends on the ripple in the pass band and is less than 50%. There is a consequent increase in the number of multiplications as the symmetry of the coefficients disappears. Amongst the advantages of FIR filters, it should be stressed that they are always stable, and are not difficult to implement.

IIR filters are more difficult to achieve except for the particular case of the elliptic type, which is the most efficient and the most often used. The estimate of equation (7.29) shows that the order of the filter is a maximum near the frequency $f_s/4$.

Let $n$ be the filter order. Assuming that the filter is formed of second-order sections, each comprising four coefficients and two data memories (Section 6.4), realization of a filter with $n$ data memories requires $2n$ memories for the coefficients, and each output number involves $2n$ additions and $2n$ multiplications.

If comparison between FIR and IIR filters is limited to the number of multiplications to be carried out to produce an output number, then, in the case of the low-pass filter, the IIR type is more advantageous than the FIR one for values of the parameters such that:

$$N > 4n \tag{7.66}$$

Following equations (5.32) and (7.29), it can be seen that the transition band is the most crucial parameter in the comparison, and similar number of multiplications are obtained in the most adverse conditions for the IIR filter if the transition band is given by: $\frac{1}{3}\frac{f_s}{\Delta f} \approx 2\log\left(\frac{f_s}{\Delta f}\right)$ that is, $\Delta f \simeq f_s/3$. It follows that the inequality (7.66) is satisfied as soon as $\Delta f$ is smaller than $f_s/3$. This is the case in the great majority of applications. For example, for parameter values corresponding to Figure 7.11, this inequality is always valid.

Linearity in the phase can be approximated by an IIR filter over a limited frequency band by supplementing, for example, the basic elliptic filter with a group delay equalizer, made of phase-shifter sections whose properties are given in Section 6.3. Experience has shown that the FIR filter, which exhibits perfectly linear phase, always requires fewer calculations [15]. It is also easy to implement.

Finally, it is recommended that FIR filters be used when linear phase is required and that IIR filters be used in other cases.

Nevertheless, the comparison above has been made with the implicit hypothesis that the sampling rate is the same at the input and output of the filter. The bases for the comparison are noticeably modified if this constraint disappears, as will be shown in a later chapter.

## Exercises

**7.1** Using the formulae given in Section 7.1, calculate the frequency and phase response and the group delay of the filter section defined by the relation:

$$y(n) = x(n) + 0.7x(n-1) + 0.9y(n-1)$$

Using the same formulae, calculate the frequency and phase response and the group delay of the second-order section with the $Z$-transfer function:

$$H(Z) = \frac{b_2 + b_1 Z^{-1} + Z^{-2}}{1 + b_1 Z^{-1} + b_2 Z^{-2}}$$

**7.2** It is proposed to use charts of analog filters in order to calculate a digital band-pass filter. What specifications should be used in order for the digital filter to reject the signal components in the bands (0–0.15) and (0.37–0.5) and show no attenuation in the band (0.2–0.33), assuming fs = 1? Study direct calculation using a low-pass transformation and 7.31.

**7.3** Calculate the coefficients of a Butterworth filter of order 4 whose amplitude has the value 2–1/2 at the frequency fc = 0.25. Give the decomposition into second-order sections.

**7.4** Use a frequency transformation to transform the band-pass filter in Section 7.2.4 (Figure 7.6) into a high-pass filter with a pass-band limit of $fH = 0.4$. How are the poles and zeros changed in this operation?

**7.5** A filter has the poles:

$$P1 = 0.9235 \pm j0.189; P2 = -0.806 \pm j0.354; P3 = 0.7326; P4 = -0.4515$$

and zeros: $Z1 = 0.9953 \pm j0.0965$; $Z2 = Z3 = Z4 = Z5 = -1$

Give the frequency response of the decomposition into second-order sections. Calculate the scaling factors of the sections and the internal data wordlength if the round-off noise added by the filter remains below the power of $\frac{1}{10}$ of the noise present at the input and if the numbers at the input have 10 bits.

**7.6** The specification of the above filter is increased by 0.1 dB in order to permit rounding of the filter coefficients. How many bits are required to represent the coefficients in the cascade structure? Estimate how many bits are required to represent the coefficients for the parallel structure. Find an optimum for rounding the coefficients. Can the number of bits found earlier be reduced?

**7.7** Does the filter given in Section 7.2.3 exhibit auto-oscillations? What are the frequencies and the amplitudes? Answer the same question for the filter described in Exercise 5.

**7.8** How many multiplications are required by the filter in Figure 7.5? How many memory locations are necessary? How many coefficients are required by an FIR filter for the same specification? Compare the number of multiplications and the memory capacities.

**7.9** It is desired to achieve the channel filtering function in a PCM transmission terminal by digital methods. The telephone signal is sampled at 32 kHz and coded into 12 bits, and the filtering is carried out by a low-pass IIR filter. The pass band is 3300 Hz, and the stop band begins at 4600 Hz.
The ripples in the pass and stop bands have value of:

$$\delta_1 \leq 0.015;\ \ \delta_2 \leq 0.04$$

A computer program for elliptic filters produces the following results: order of filter: $N = 4$
zeros: $z1 = 0.09896 \pm j0.995$
$z2 = 0.5827 \pm j0.8127$
poles: $P1 = 0.6192 \pm j0.2672$
$P2 = 0.702 \pm j0.589$
Calculate the transfer function of the filter decomposed into second-order sections.
What is the value of the overall scaling factor, knowing that the amplitude at frequency 0 is 0.99?
The coefficients are quantized into 10 bits. Determine the displacement of the infinite attenuation frequencies and evaluate the additional pass-band ripple.
Calculate the scaling factor to be assigned to each section and estimate the round-off noise produced if the data memories have 16 bits.
Give the complete diagram for the filter.
Evaluate the complexity in terms of:
(1) The number of multiplications and additions per second.
(2) The number of memory bits.

**7.10** The following specifications are given for a low-pass filter:

$$\delta_1 = 0.01;\delta_2 = 0.01;f_1 = 1700\text{Hz};f_2 = 2000\text{Hz};\ \text{ and }\ f_s = 8000\text{Hz}$$

Calculate the filter order. By taking the order $N = 6$, a margin on the in-band ripple is available. Determine the coefficient wordlength.
If the signal-to-noise ratio degradation is limited to 0.1 dB, give the increase of the internal data wordlength with respect to the input data wordlength.

## References

1 A. Oppenheim and R. Schafer, *Digital Signal Processing*, Prentice Hall, Englewood Cliffs, NJ, 1974, Chapters 5 and 9.
2 L. Rabiner and B. Gold, *Theory and Application of Digital Signal Processing*, Prentice Hall, Englewood Cliffs, NJ, 1975, Chapters 4 and 5.

3 R. Boite and H. Leich, *Les filtres numériques: analyse et synthèse des filtres unidimensionnels*, Masson, Paris, 1980.
4 B. Gold and C. Rader, *Digital processing of signals*, McGraw-Hill, New York, 1969.
5 A. G. Constantinides, Spectral transformation for digital filters. *Proceedings of the IEEE*, 8, 1970.
6 A. Deczky, Synthesis of recursive digital filters using the minimum p-error criterion. *IEEE Transactions on Audio and Electroacoustics*, 20, 1972.
7 R. Fletcher and M. J. D. Powell, A rapidly convergent descent method for minimization. *Computer Journal*, 6(2), 1963.
8 J. P. Thiran, Equal ripple delay recursive filters. *IEEE Transactions on Circuit Theory*, 1971.
9 T. Durrani and R. Chapman, Optimal all-pole filter design based on discrete prolate spheroidal sequences. *IEEE Transactions*, ASSP32(4), 716–21, 1984.
10 P. P. Vaidyanathan, S. K. Mitra and Y. Neuvo, A new approach to the realisation of low sensitivity IIR filters. *IEEE Transactions*, ASSP34(2), 350–61, 1986.
11 L. B. Jackson, Round-off noise analysis for fixed point digital filters in cascade or parallel form. *IEEE Transactions on Audio and Electroacoustics*, 1970.
12 A. Peled and B. Liu, *Digital Signal Processing: Theory, Design and Implementation*, John Wiley, New York, 1976.
13 Von E. Lueder, H. Hug and W. Wolf, Minimizing the round-off noise in digital filters by dynamic programming. *Frequenz*, 29(7), 211–14, 1975.
14 D. Mitra, Large-amplitude, self-sustained oscillations in difference equations, which describe digital filter sections using saturation arithmetic. *IEEE Transactions*, ASSP25(2), 1977.
15 L. Rabiner, J. F. Kaiser, O. Herrmann and M. Dolan, Some comparison between FIR and IIR digital filters. *Bells Systems Technical Journal*, 53, 1974.

# 8

# Digital Ladder Filters

The filter structures presented in the previous chapters are deduced directly from their $Z$-transfer functions, with the coefficients applied to the multiplying circuits being those of the powers of $Z^{-1}$. More elaborate structures can be developed.

In analog filtering, structures exist which allow filters with very low ripple and excellent selectivity to be constructed using passive components of limited precision. In digital filtering, these properties can be translated into a reduction in the round-off noise and in the number of bits representing the coefficients.

Analog filter networks are based on cascading two port circuits whose properties will be considered first [1].

## 8.1 Properties of Two-Port Circuits

The general two-port circuit terminated by the resistors $R_1$ and $R_2$ is shown in Figure 8.1, together with the currents $I$ and voltages $V$ at ports 1 and 2. This circuit, assumed to be linear, is defined by its impedance matrix $z$, which establishes the relations between the variables, generally written in reduced form as:

$$R = R_1 = R_2; \quad v = \frac{V}{\sqrt{R}}; \quad i = I\sqrt{R}$$

Then,

$$v = zi \tag{8.1}$$

with:

$$z = \begin{bmatrix} z_{11} & z_{12} \\ z_{21} & z_{22} \end{bmatrix}$$

The values $z_{12}$ and $z_{21}$ are the transfer impedances of the two-port circuit. It is reciprocal if $z_{12} = z_{21}$.

If reversing the circuit does not change the external conditions, it is said to be symmetric, and we have $z_{11} = z_{22}$.

*Digital Signal Processing: Theory and Practice*, Tenth Edition. Maurice Bellanger.

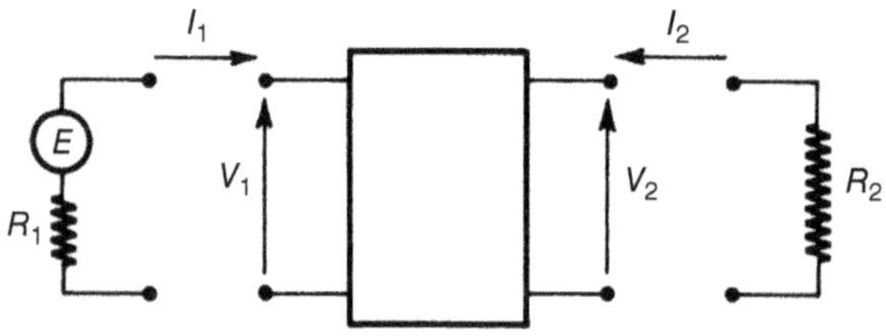


**Figure 8.1** Two-port element with resistive terminations.

The transmission and reflection coefficients of the circuit can be demonstrated if another matrix – the distribution matrix – is introduced. If a reference case is defined with unit terminating resistors, the incident and reflected waves $a$ and $b$ can be written as:

$$a = \frac{1}{2}(v + i) \tag{8.2}$$

$$b = \frac{1}{2}(v - i) \tag{8.3}$$

The relations between variables $a$ and $b$ can be obtained using equation (8.1):

$$a = \frac{1}{2}(z + I_2)i$$

$$\dot{I}_2 = \begin{bmatrix} 1 & 0 \\ 0 & 1 \end{bmatrix}$$

$$b = \frac{1}{2}(z - I_2)i$$

Thus,

$$b = Sa \tag{8.4}$$

where:

$$S = \begin{bmatrix} S_{11} & S_{12} \\ S_{21} & S_{22} \end{bmatrix}$$

and:

$$S = (z - \dot{I}_2)(z + \dot{I}_2)^{-1} \tag{8.5}$$

If the circuit is reciprocal, then:

$$S_{12} = S_{21} = \tau \tag{8.6}$$

where $\tau$ is the transmission coefficient:

$$\tau = \frac{2V_2}{E} \tag{8.7}$$

If the input and output impedances of the circuit are $z_1$ and $z_2$, respectively, one can write:

$$S_{11} = \rho_1 = \frac{z_1 - 1}{z_1 + 1}; \quad S_{22} = \rho_2 = \frac{z_2 - 1}{z_2 + 1} \tag{8.8}$$

The values $\rho_1$ and $\rho_2$ are the reflection coefficients at the input and output of the two-port circuit.

If the circuit is not dissipative, the power that it absorbs is zero, and it can be shown that the distribution matrix of such a reciprocal two-port network has the form [2]:

$$S = \frac{1}{g}\begin{bmatrix} h & f \\ f & \pm h_* \end{bmatrix} \tag{8.9}$$

where $f$, $g$, and $h$ are real polynomials with the following properties:

(1) They are linked by a relation which, on the imaginary axis, corresponds to:

$$|g|^2 = |h|^2 + |f|^2$$

The notation $h_*(p)$ indicates $h(-p)$.

(2) Depending upon whether $f$ is of even or odd degree, the lower or upper sign is taken in equation (8.9).

(3) Each root of g in the complex plane lies in the left-hand half plane.

The polynomials $f$, $g$, and $h$ are the characteristic polynomials of the circuit. The roots of $f(p)$ are generally on the imaginary axis in the stop band and are transmission zeros. The roots of $h(p)$ are attenuation zeros, and for a non-dissipative network, they are generally on the imaginary axis in the pass band.

For the circuit in Figure 8.1, the transmission coefficient is:

$$S_{12} = \frac{2V_2}{E}\sqrt{\frac{R_2}{R_1}} \tag{8.10}$$

The attenuation in decibels is denoted by the function $A_f(\omega)$, where:

$$A_f(\omega) = -10\log|S_{12}(\omega)|^2 = 10\log\left|\frac{g(\omega)}{f(\omega)}\right|^2 = 10\log\left(1 + \frac{|h|^2}{|f|^2}\right)$$

The relation

$$\left|\frac{f(\omega)}{g(\omega)}\right|^2 + \left|\frac{h(\omega)}{g(\omega)}\right|^2 = 1 \tag{8.11}$$

expresses the fact that the non-transmitted power is reflected.

For a cascade arrangement, it is important to pay equal attention to the transfer matrix $t$ defined by:

$$\left[\frac{b_1}{a_1}\right] = t\left[\frac{a_2}{b_2}\right] \tag{8.12}$$

The cascade arrangement is represented by the product of the transfer matrices.

The transfer matrix of non-dissipative two-port circuits takes the form:

$$t = \frac{1}{f}\begin{bmatrix} \pm g_* & h \\ \pm h_* & g \end{bmatrix} \tag{8.13}$$

By way of example, Figure 8.2 gives the transfer matrices of several elementary circuits.

The fact that the circuit element is non-dissipative has important consequences for the attenuation $A_f(\omega)$. In the pass band, $A_f(\omega)$ cannot take negative values. Consequently, at frequencies where $h(\omega)$ is zero, its derivative with respect to any of the parameters must also be zero. In a filter with inductances and capacitances, terminated by resistors, variation of the values of the $L$ and $C$ elements does not affect the attenuation to the first order at frequencies where it is zero.

If the ripple is small, it can be assumed that this property applies over the whole pass band. In practice, it can be taken that, in a ladder filter, for example, the interactions between the different branches are such that a perturbation in one element has repercussions for all the other factors of the attenuation function, having an overall compensating effect which minimizes the incidence of the perturbation.

Given this behavior, it is of interest to find digital filter structures which have similar properties. In effect, in a digital filter where the amplitudes of the ripples in the pass and stop bands are similar,

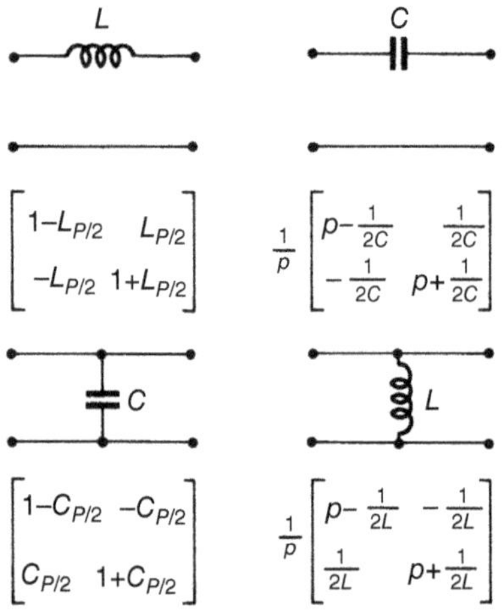


Figure 8.2 Transfer matrices for two-port elements.

the denominator of the transfer function determines the number of bits required to represent the coefficients. Structures derived, for example, from ladder analog filters can therefore be expected to lead to significant gains in the coefficient wordlengths, in the complexity of the multipliers, and also in the power of the round-off noise.

Ladder structures are the most commonly used type in passive analog filtering. The procedure for obtaining the elements of such a structure, using a transfer function, is described in detail in Ref. [2]. It consists of factorizing the overall transfer matrix, defined using the calculated transfer function $H(\omega)$ into partial matrices corresponding to the series and parallel arms of the ladder structure.

The most direct approach for obtaining a digital filter structure from an analog ladder filter is to simulate its voltage–current flow chart.

## 8.2 Simulated Ladder Filters

The representation of ladder filters in terms of their flow chart is used to synthesize active filters from integrator or differentiator circuits. The ladder filter shown in Figure 8.3, and terminated with the resistors $R_1$ and $R_2$, will be used to demonstrate the voltage–current flow chart. Application of Kirchhoff's laws leads to the following relations:

$$I_1 = (E - V_2)R_1^{-1};\quad I_{K-1} = (V_{K-2} - V_K)Z_{K-1}^{-1};\quad I_{N+1} = V_N R_2^{-1}$$
$$V_2 = (I_1 - I_3)Y_2^{-1};\quad V_K = (I_{K-1} - I_{K+1})Y_K^{-1}$$

where the index $K$ takes the values 4, 6, …, $N$.

The flow chart is composed of arcs, each of which is associated with a coefficient representing an impedance or an admittance. With each junction, there is associated either a voltage at the node or the current in a branch. For each arc, there is formed the product of the corresponding coefficient and the magnitude associated with its origin. The magnitude associated with each junction is the sum of the products corresponding to its various associated arcs.

The chart for the ladder filter of Figure 8.3 is shown in Figure 8.4. The currents and voltages are associated with alternate junctions. This topology is defined as "leapfrog."

Simulated ladder digital filters are the structures obtained by simulating each arc in the group, or each branch of the ladder, by an element with an equivalent transfer function.

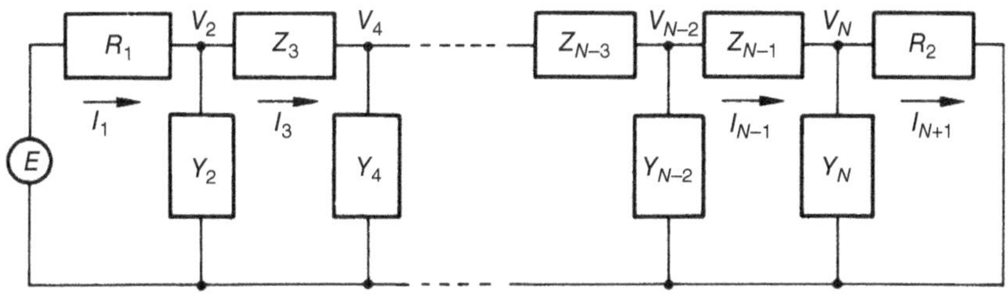


**Figure 8.3** Analog ladder filter.

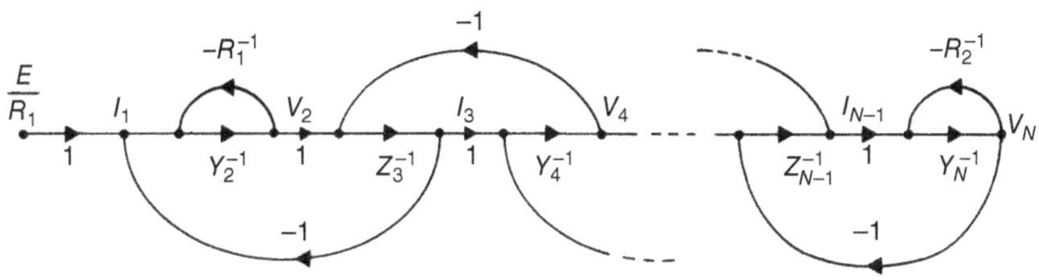


**Figure 8.4** Flow graph of a ladder filter.

A particularly simple case occurs when the series impedors $Z_{K-1}$ are inductors and the parallel branches $y_K$ are capacitors ($K = 4, 6, \ldots, N$). Such filters are purely recursive and are without frequencies of infinite attenuation.

The transfer functions to account for them take the form:

$$Z_{K-1}^{-1} = \frac{R}{sL_{K-1}}; \quad Y_K^{-1} = \frac{1}{sC_K R} \tag{8.14}$$

where $s$ is the Laplace variable and $R$ is a normalization constant. Both cases involve the transfer functions of integrators which are easily produced using operational amplifiers and $R$–$C$ networks. The diagram in Figure 8.5 then appears, being deduced from Figure 8.4. It represents the functions to be implemented and shows the circuit diagram using integrators.

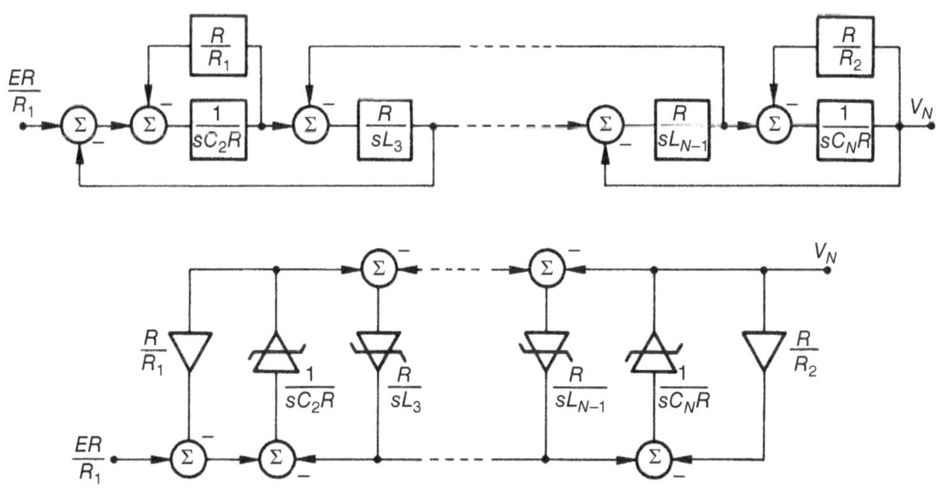


**Figure 8.5** Filter composed of integrators.

The digital realization consists of replacing each integrator with an equivalent function. In Ref. [3], it is shown that the only digital integrator circuit, which is simple to realize, and which is equivalent to an analog integrator, is the one represented by Figure 8.6 and whose $Z$-transfer function, $i(z)$, is written as:

$$I(Z) = \frac{aZ^{-1/2}}{1 - Z^{-1}} \tag{8.15}$$

The equivalence between the analog and digital integrators is obtained in the same way as for any transfer function, by replacing $Z$ with $e^{j\omega T}$. Thus,

$$I(\omega) = \frac{ae^{-j\omega(T/2)}}{1 - e^{-j\omega T}} = \frac{a}{2j} \frac{1}{\sin(\omega(T/2))} \tag{8.16}$$

where $T$ is the sampling period of the digital circuit. This function is equivalent to the analog function $a/j\omega T$ with frequency warping. If $f_A$ denotes the analog frequency and $f_N$ is the digital frequency, we can show:

$$\pi f_A T = \sin(\pi f_N T) \tag{8.17}$$

The frequency warping thus introduced is different from that obtained with the bilinear transform introduced in the previous chapter, as shown in Figure 8.7, and it must be taken into account when calculating a filter from a specification.

The circuit in Figure 8.6 shows the disadvantage of introducing the function $Z^{-1/2}$, which corresponds to an additional memory circuit, but the transfer function of a ladder filter is not altered when the impedances of all branches are multiplied by the same function [3]. This property has already been used to introduce the normalization constant $R$. If the impedances are multiplied by $Z^{-1/2}$, this term is eliminated from all of the $Z$-transfer functions of the integrators in the circuit, which then become:

$$I_i(Z) = \frac{TR}{L_i} \frac{1}{1 - Z^{-1}} \quad \text{for odd } i$$

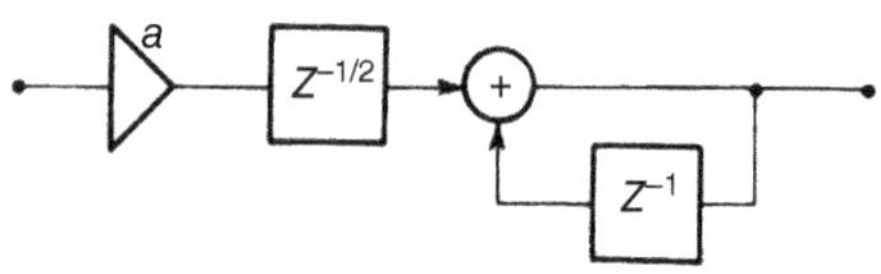


**Figure 8.6** Digital integrator circuit.

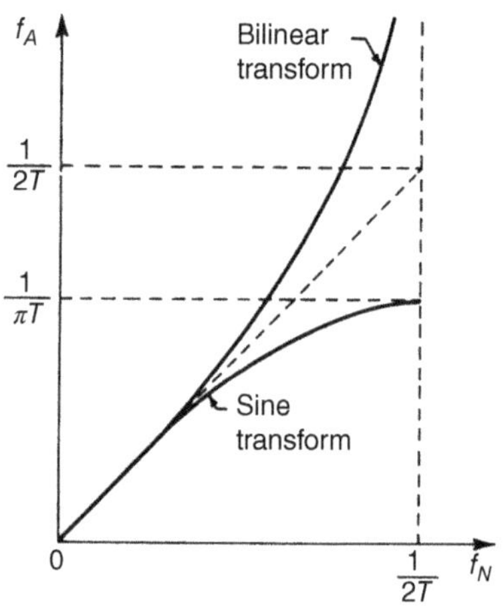


**Figure 8.7** Frequency warping by the sine transform.

and:

$$I_i(Z) = \frac{T}{C_i R}\frac{Z^{-1}}{1 - Z^{-1}} \quad \text{for even } i$$

In contrast, the termination resistances are transformed to $R_1 Z^{-1/2}$ and $R_2 Z^{-1/2}$; the terminations are no longer purely resistive, as they have the transfer functions:

$$R_1 e^{-j\pi f T}; \quad R_2 e^{-j\pi f T}$$

This effect can be neglected when the sampling frequency is large compared to the pass band, as there is an insignificant change in the transfer function of the filter. Further, the resistances $R_1$ and $R_2$ can be chosen as unity, as can the normalization constant $R$. The circuit of the digital filter obtained under these conditions is given in Figure 8.8.

The coefficients have the following values, for odd-order $N$:

$$a_N = \frac{T}{C_{N+1}}; \quad a_{2i-1} = \frac{T}{C_{2i}}; \quad a_{2i} = \frac{T}{L_{2i+1}}; \quad i = 1, 2, \ldots \frac{N-1}{2} \tag{8.18}$$

The filter thus realized involves $N$ multiplications and $N$ memories for a transfer function of order $N$. For these parameters, the structure is canonic. The number of additions is $2N + 1$.

To summarize, the calculation of a simulated ladder digital filter from an imposed specification involves the following stages:

(1) Transposing the specification by modifying the frequency axis using equation (8.17) above.
(2) Calculating the elements of a filter with passive ladder elements $LC$, satisfying the transposed specification.
(3) Using the values of the elements so obtained to calculate the coefficients $a_i$ ($i = 1, 2, \ldots, N$) of the digital filter using equations (8.18).

The chief attraction of the structure obtained in this way is that the coefficients can be represented by a very small number of bits. Also, some multiplications can be replaced by simple additions, and, in certain cases, all the multiplications of the filter can be eliminated, resulting in significant savings in the circuitry. To illustrate this property, let us consider a low-pass filter of order $N = 7$ [3], with elements having the following values (Figure 8.5):

$$R = R_1 = R_9 = 1$$
$$C_2 = 1.2597 = C_8$$
$$L_3 = 1.5195 = L_7$$
$$C_4 = 2.2382 = C_6$$
$$L_5 = 1.6796$$

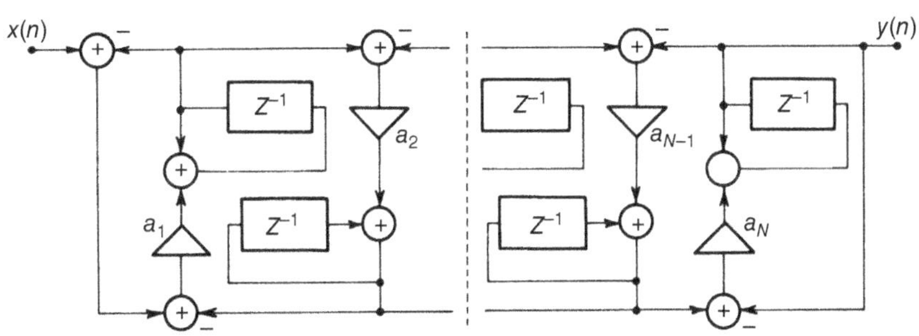


**Figure 8.8** Digital simulated ladder filter.

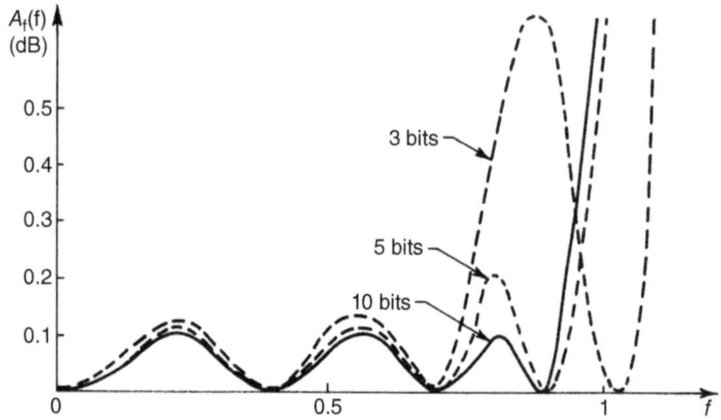


**Figure 8.9** Ripples in the pass band for various coefficient wordlengths.

The coefficients $a_i$ ($i = 1, 2, \ldots, N$) of the corresponding simulated ladder digital filter are calculated with a sampling period $T = 1/\text{fs} = 0.01$, from equation (8.18) above. The ripples of the filter in the pass band are shown in Figure 8.9, where the coefficients are represented by 10, 5, and 3 bits. It is notable that, when represented with 5 bits, the attenuation zeros are conserved. With 3 bits, they are also conserved except for the one closest to the transition band. Thus, the insensitivity to the first order of the attenuation zeros, which was shown in Section 8.1, is demonstrated. In comparison with the cascade structure in the previous chapter, this example shows an estimated gain of 4 or 5 bits for representing the coefficients.

The technique described in this section can be extended to filters other than the purely recursive low-pass filter, but the designs will be more complicated. Also, the need to have a sampling frequency which is large in comparison with the pass band is not conducive to efficient processing. In practice, the simulated ladder structure is primarily used with a different method of performing the calculations – that used in switched-capacitor devices.

## 8.3 Switched-Capacitor Filters

Strictly speaking, because they do not use arithmetic operations, filters using switched-capacitor devices are not digital filters. Nevertheless, they do use the same design methods and are complementary to digital filters. They are frequently used in analog–digital conversion circuits.

The basic principle, which is presented in detail in Reference [4], is as follows. Switching a capacitor $C$ between two voltages $V_1$ and $V_2$ at frequency $f_s$ is equivalent to introducing a resistance $R$ given by:

$$R = \frac{1}{C} f_s$$

between the two potentials. In effect, as is shown in Figure 8.10, the capacitor is charged to voltages $V_1$ and $V_2$ alternately and a charge transfer $C(V_1 - V_2)$ results. If the operations are performed at frequency $f_s$, a current $i$:

$$i = C(V_1 - V_2) f_s$$

flows between the voltages $V_1$ and $V_2$.

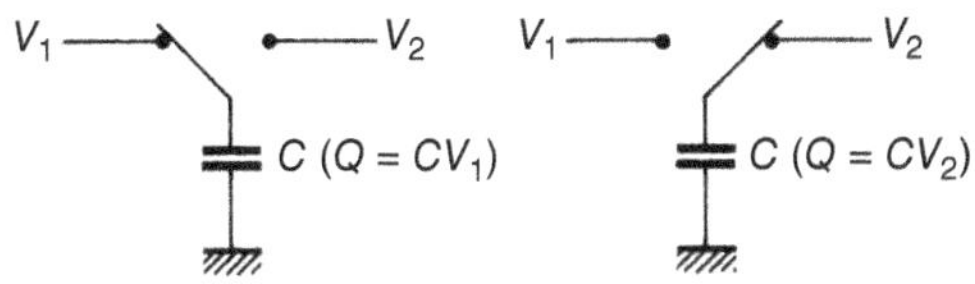


**Figure 8.10** Switching a capacitor between voltages $V_1$ and $V_2$.

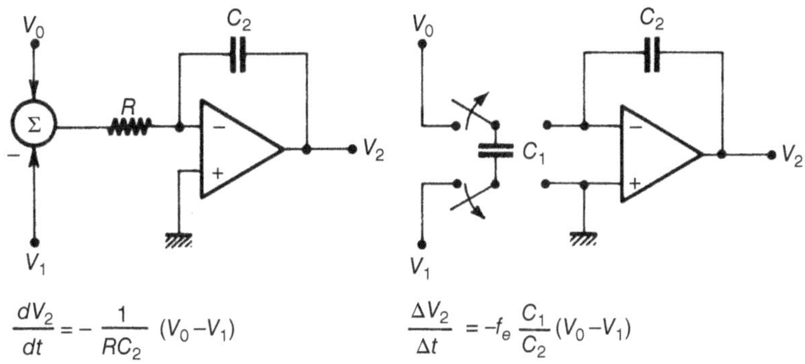


**Figure 8.11** Switched-capacitor integrator.

This equivalent resistor is inserted in an integrator circuit as shown in Figure 8.11. The integrator being considered has an adder at its input, like those in Figure 8.5. The equation describing the operation of the analog integrator is also shown in Figure 8.11. In the switched capacitor version, the capacitor $C_1$ is alternately connected at frequency $f_s$ across the input of the operational amplifier, and between voltages $V_0$ and $V_1$. The equation for the variation $\Delta V_2$ in the output voltage during the interval $\Delta t$, which is assumed to be large in comparison with the period $1/f_s$, is shown in the diagram.

The condition that the two types of integrators are equivalent is:

$$C_1 = \frac{1}{f_s R} \tag{8.19}$$

However, in order to completely analyze the switched-capacitor integrator, it is necessary to take account of the sampling [5] and to calculate its $Z$ transfer function. Assume that $v_e(t)$ is the input signal and that $v_2(t)$ is the output one. The sampling period $T$ is assumed to be divided into two equal parts. The capacitor $C_1$ is connected to the input, to the integrator for time $T/2$, and to the voltage $v_e(t)$ itself for time $T/2$. Let us assume that this is between times $nT$ and $\left(n + \frac{1}{2}T\right)$. The charge transmitted to the integrator is $Q(nT)$ such that:

$$Q(nT) = C_1 v_e\left[\left(n + \frac{1}{2}\right)T\right]$$

Under these conditions, at time $(n+1)T$, the output voltage is:

$$v_2[(n+1)T] = V_2(nT) - \frac{C_1}{C_2} v_e\left[\left(n + \frac{1}{2}\right)T\right]$$

By taking the $Z$-transform of the two components, one has:

$$\frac{V_2(Z)}{V_e(Z)} = H(Z) = -\frac{C_1}{C_2}\frac{Z^{-1/2}}{1 - Z^{-1}} \tag{8.20}$$

One finds the same type of transfer function as was given by equation (8.15) for digital circuits. The switched-capacitor integrator performs in exactly the same way as the digital circuits described in the previous section, and the same warping of the frequency axis is involved. It should be noted that, to ensure that no further delay is introduced and that this function is conserved when cascading two integrators, the capacitors of the two integrators must be switched in antiphase.

A design involving a switched-capacitor device for a simulated ladder filter, like that in Figure 8.5, is obtained by substituting integrator circuits and calculating the value to be given to the switched capacitors in each case.

*Example:* Assume an implementation using switched capacitors for a Butterworth filter of order 4, with an analog circuit as shown in Figure 8.12(a).

The procedure described in the previous section results in the design (Figure 8.12(b)) for producing a filter from integrators, assuming unit terminal resistances. The switched-capacitor design is shown in Figure 8.12(c). The coefficients $a_i$ $(i = 1, 2, 3, 4)$ which define the ratios of the capacitances are given by equation (8.18).

If the filter has a 3 dB attenuation frequency $f_c$ equal to 1 kHz, the analog parameters are as follows:

$$R_1 = R_2 = 1$$

$$C_2 = 121.8 \times 10^{-6}; \quad C_4 = 294.1 \times 10^{-6}$$

$$L_3 = 294.1 \times 10^{-6}; \quad L_5 = 121.8 \times 10^{-6}$$

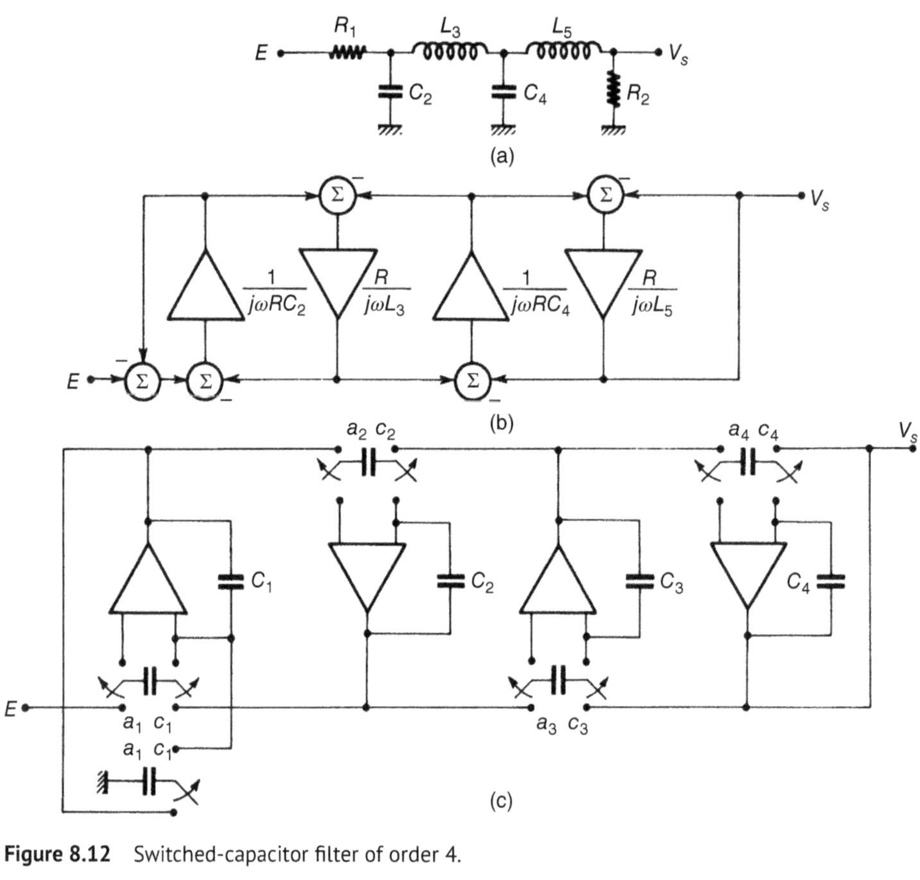


**Figure 8.12** Switched-capacitor filter of order 4.

With a sampling frequency of 40 kHz:

$$a_1 = \frac{1}{4.87} = 0.205 = a_4$$
$$a_2 = \frac{1}{11.76} = 0.085 = a_3$$

Thus, in switched-capacitor devices, the precision and stability of the integrator time constant depends on the external sampling frequency and on the ratio of the capacitances. These devices allow very selective filters to be produced on a silicon chip as a monolithic integrated circuit.

## 8.4 Lattice Filters

The lattice structure occurs in the analysis and synthesis of speech for simulating the vocal tract, and also more generally in systems for linear prediction. It allows the realization of finite impulse response (FIR) and infinite impulse response (IIR) filters [6].

Consider the structure with $M$ sections as represented in Figure 8.13. The outputs $y_1(n)$ and $u_1(n)$ of the first section are related to the input set $x(n)$ by:

$$\begin{aligned} y_1(n) &= x(n) + k_1 x(n-1) \\ u_1(n) &= k_1 x(n) + x(n-1) \end{aligned} \tag{8.21}$$

Similarly, $y_2(n)$ and $u_2(n)$, the outputs of the second element, are related to the inputs by:

$$\begin{aligned} y_2(n) &= x(n) + k_1(1+k_2)x(n-1) + k_2 x(n-2) \\ u_2 = (n) &= k_2 x(n) + k_1(1+k_2)x(n-1) + x(n+2) \end{aligned} \tag{8.22}$$

By iteration, $y_M(n)$ and $u_M(n)$, the outputs produced by the filter are related to the set $x(n)$ by the following equations, which correspond to FIR filtering:

$$y_M(n) = \sum_{i=0}^{M} a_i x(n-i) \tag{8.23}$$

$$u_M(n) = \sum_{i-0}^{M} a_{M-i} x(n-i) \tag{8.24}$$

The two FIR filters thus obtained have the same coefficients but in reverse order. Their $Z$-transfer functions, $H_M(Z)$ and $U_M(Z)$, are image polynomials. Thus,

$$H_M(Z) = \sum_{i=0}^{M} a_i Z^{-i}$$
$$U_M(Z) = \sum_{i=0}^{M} a_{M-i} Z^{-i} = Z^{-M} H_M(Z^{-1}) \tag{8.25}$$

An iterative process is used to determine the coefficients $k_i$ of the lattice filter from the coefficients $a_i$ $(1 \leqslant i \leqslant M)$. Firstly, the coefficient $a_0$ is assumed to be equal to unity. Then it is straightforward, using the equations given earlier (and also directly from Figure 8.13), to prove that:

$$k_M = a_M$$

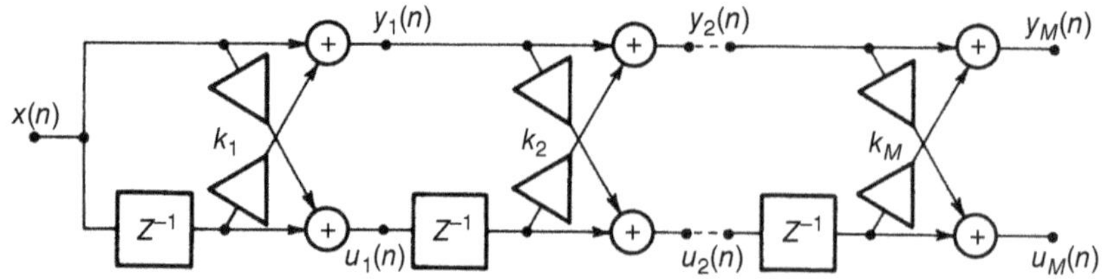


**Figure 8.13** FIR lattice filter.

This point is the basis for the calculation. By using $H_m(Z)$ and $U_m(Z)$ $(1 \leqslant m \leqslant M)$ to denote the corresponding transfer functions at the outputs of the $m$th section, the following matrix equation can be written:

$$\begin{bmatrix} H_m(Z) \\ U_m(Z) \end{bmatrix} = \begin{bmatrix} 1 & k_m Z^{-1} \\ k_m & Z^{-1} \end{bmatrix} \begin{bmatrix} H_{m-1}(Z) \\ U_{m-1}(Z) \end{bmatrix}$$

By assuming $k_m \neq 1$, this equation can also be written as:

$$\begin{bmatrix} H_{m-1}(Z) \\ U_{m-1}(Z) \end{bmatrix} = \frac{1}{1-k_m^2} \begin{bmatrix} 1 & -k_m \\ -k_m Z & Z \end{bmatrix} \begin{bmatrix} H_m(Z) \\ U_m(Z) \end{bmatrix} \tag{8.26}$$

Thus, the polynomials $H_{m-1}(Z)$ and $U_{m-1}(Z)$ are image polynomials of degree $m-1$ whose coefficients $a_{i(m-1)}$ $(1 \leqslant i \leqslant m-1)$ are calculated using the coefficients $a_{im}$ $(1 \leqslant i \leqslant m)$ of the polynomials $H_m(Z)$ and $U_m(Z)$.

Under these conditions, this becomes:

$$a_{mm} = k_m; \quad a_{(m-1)(m-1)} = k_{m-1}$$
$$a_{(m-1)(m-1)} = \frac{1}{1-a_{mm}^2}[a_{(m-1)m} - a_{mm}a_{1m}] \tag{8.27}$$

The coefficients $k_m$ $(1 < m < M)$ are thus calculated in $M$ iterations.

*Example:* Consider the transfer function $H_3(Z)$ such that:

$$H_3(Z) = 1 - 1.990Z^{-1} + 1.572Z^{-2} - 0.4583Z^{-3}$$

Then:

$$k_3 = -0.4583$$

Using equation (8.28), one can write:

$$H_2(Z) = 1 - 1.607Z^{-1} + 0.8355Z^{-2}$$

Thus:

$$k_2 = 0.8355$$

By application of equation (8.29), we have:

$$k_1 = 0.88756 \quad \text{and} \quad H_1(Z) = 1 - 0.8756Z^{-1}$$

The realization of purely recursive IIR-type filters results in a dual structure, as represented in Figure 8.14. The sets $x_1(n)$, $u(n)$, and $y(n)$ are related by the equations:

$$y(n) = x_1(n) - k_1 y(n-1)$$
$$u_1(n) = k_1 y(n) + y(n-1)$$

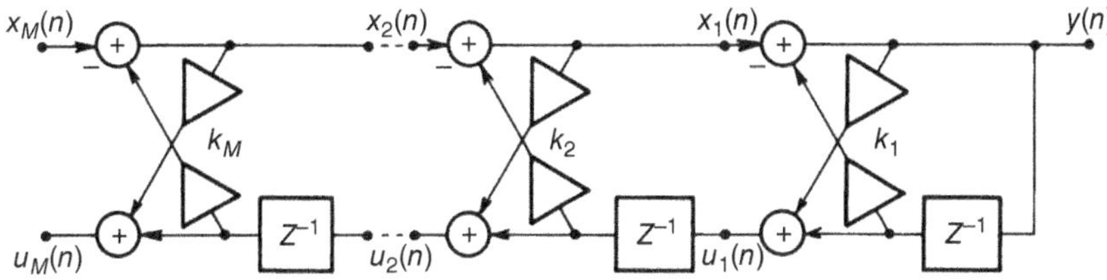


**Figure 8.14** Purely recursive IIR lattice filter.

Similarly, the sets $x_2(n)$, $x_1(n)$, $u_1(n)$, and $u_2(n)$ are related by:

$$x_1(n) = x_2(n) - k_2 u_1(n-1)$$
$$u_2(n) = k_2 x_1(n) + u_1(n-1)$$

This results in the transfer function $H_2(Z)$ between the input $x_2(n)$, and the output $y(n)$ given by:

$$H_2(Z) = \frac{1}{1 + k_1(1+k_2)Z^{-1} + k_2 Z^{-2}}$$

Similarly, the transfer function $U_2(Z)$ relates $u_2(n)$, and $y(n)$ where:

$$U_2(Z) = k_2 + k_1(1+k_2)Z^{-1} + Z^{-2}$$

By iteration, it can be seen that the sets $x_M(n)$ and $y(n)$, on the one hand, and $u_M(n)$, and $y(n)$ on the other, are related by the equations:

$$y(n) = x_M(n) - \sum_{i=1}^{M} b_i y(n-1) \tag{8.28}$$

$$u_M(n) = \sum_{i=0}^{M-1} b_{M-i} y(n-i) + y(n-M) \tag{8.29}$$

Consequently, the transfer functions $H_M(Z)$ and $U_M(Z)$ are:

$$H_M(Z) = \frac{1}{1 + \sum_{i-1}^{M} b_i Z^{-i}} = \frac{1}{D_M(Z)}$$

$$U_M(Z) = \sum_{i=0}^{M-1} b_{M-i} Z^{-i} + Z^{-M} = Z^{-M} D_M(Z^{-1}) \tag{8.30}$$

The coefficients $k_i$ $(1 \leqslant i \leqslant M)$ of the lattice filter are calculated by iteration from the coefficients $b_i$ of the IIR filter, after noting that:

$$k_M = b_M$$

By using $H_m(Z)$ and $U_m(Z)$ to denote the relating functions for the set of $m$ sections $(1 \leqslant m \leqslant M)$, it is possible, using the equations of definition:

$$x_{m-1}(n) = x_m(n) - k_m u_{m-1}(n-1)$$
$$u_m(n) = k_m x_{m-1}(n) + u_{m-1}(n-1)$$

to produce the following matrix equation:

$$\begin{bmatrix} D_m(Z) \\ U_m(Z) \end{bmatrix} = \begin{bmatrix} 1 & k_m Z^{-1} \\ k_m & Z^{-1} \end{bmatrix} \begin{bmatrix} D_{m-1}(Z) \\ U_{m-1}(Z) \end{bmatrix}$$

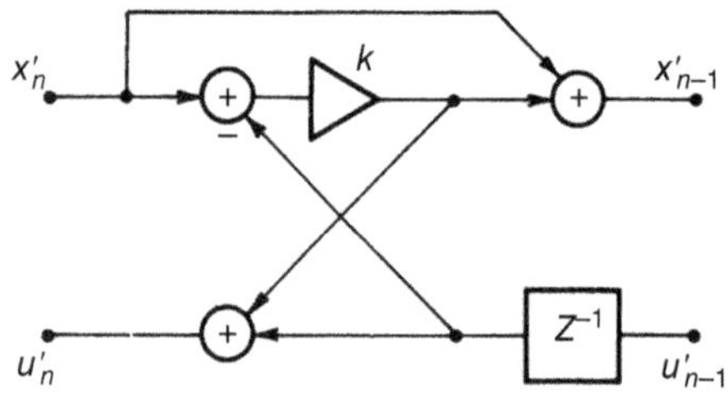


**Figure 8.15** Lattice filter cell with single multiplication.

As in the case of FIR filters, this matrix equation is also written, for $k_m \neq 1$:

$$\begin{bmatrix} D_{m-1}(Z) \\ U_{m-1}(Z) \end{bmatrix} = \frac{1}{1-k_m^2} \begin{bmatrix} 1 & -k_m \\ -k_m Z & Z \end{bmatrix} \begin{bmatrix} D_m(Z) \\ U_m(Z) \end{bmatrix} \tag{8.31}$$

As with FIR filters, this expression allows for the calculation of the coefficients $k_i$ $(1 \leqslant i \leqslant M)$ of the IIR lattice filter in $M$ iterations, using the polynomial $D_M(Z)$, where:

$$D_M(Z) = 1 + \sum_{i=1}^{M} b_i Z^{-i}$$

The lattice structures given in Figures 8.13 and 8.14 are canonical for the data memories but not for the multiplications. They can be made canonical, for an IIR type filter, for example, by using the single-multiplication section represented in Figure 8.15.

However, a further addition is then necessary. The equations of this first-order section are as follows:

$$(1+k)x_1(n) = y(n) + ky(n-1)$$
$$(1+k)u_1(n) = ky(n) + y(n-1)$$

To within a factor of $(1+k)$, they are equivalent to the equations for a two-multiplier lattice.

In contrast with the structures described in the previous sections, lattice filters do not have any particular advantage for the number of bits needed to represent the coefficients. Nevertheless, in practice, they have one interesting property in that a necessary and sufficient condition for an IIR filter to be stable and have its poles inside the unit circle is that the coefficients have a modulus of less than unity:

$$1 \leq i \leq M$$

This property is obvious for $k_1$ in Figure 8.14 if the appropriate section is isolated. It can be extended to the other coefficients by considering the subcircuits and using recurrence.

This results in a control of the stability which can be realized quite simply and is of particular use in systems such as adaptive filters, where the values of the coefficients are constantly changing.

The lattice structures considered above are either non-recursive or purely recursive. Note that the purely recursive structure can be supplemented to make a general filter; it is sufficient to form a weighted sum of the variables $u_m(n)$. That is, the expression:

$$v_M = \gamma_0 y(n) + \sum_{m=1}^{M} \gamma_m u_m(n)$$

defines FIR-type filtering of the signal $y(n)$ by virtue of equation (8.29). As the coefficients $b_i (1 \leq i \leq M)$ are fixed, the coefficients $\gamma_i$ can be determined to obtain any numerator for the general filter.

It is also useful to observe that the purely recursive structure consists of the pure all-pass function. In fact, equations (8.29) and (8.30) yield:

$$H_D(Z) = \frac{U_M(Z)}{X(Z)} = \frac{b_M + b_{M-1}Z^{-1} + \ldots + Z^{-M}}{1 + b_1 Z^{-1} + \ldots + b_M Z^{-M}}$$

This expression shows that, as indicated in Section 6.3, the signal $u_M(n)$ is the output of an all-pass network with $x(n)$ as the input. The transfer function $H_D(Z)$ can be expressed directly as a function of the lattice coefficients by a continued fraction:

$$H_D(Z) = k_M + \cfrac{\left(1 - k_M^2\right) Z^{-1}}{k_M Z^{-1} + \cfrac{1}{k_{m-1} Z^{-1} \cfrac{(1-k_{M-1}^2)Z^{-1}}{k_{M-1}Z^{-1}+}}}$$

$$\ddots \qquad \ddots$$

$$+ \cfrac{1}{k_1 + \cfrac{(1-k_1^2)Z^{-1}}{k_1 Z^{-1}+1}}$$

This observation can be used to calculate the poles of the lattice filter directly [7].

An interesting application of the above results is the implementation of the notch filter introduced in Section 6.3. The notch filter output $y_N(n)$ is obtained simply by incorporating one more adder into Figure 8.14 to carry out the following operation:

$$y_N(n) = x_M(n) + u_M(n) \tag{8.32}$$

For order 2, the transfer function of the all-pass circuit is:

$$H_D(Z) = \frac{k_2 + k_1(1 + k_2)z^{-1} + z^{-2}}{1 + k_1(1 + k_2)z^{-1} + k_2 z^{-2}}$$

A very peculiar property of the approach is that the frequency $\omega_0$ and the 3 dB attenuation band width $B_{3N}$ can be tuned independently [8]. This decoupling effect is due to the relationships:

$$k_1 \simeq -\cos\omega_0$$

$$k_2 \simeq \frac{1 - \tan \pi B_{3N}}{1 + \tan \pi B_{3N}} = (1 - \varepsilon)^2 \tag{8.33}$$

If a subtraction is performed instead of the addition in equation (8.32), the complementary filter is obtained.

## 8.5 Comparison Elements

Having discussed the various structures for digital filters, this is an appropriate point at which to summarize their properties. Reference [9] gives a detailed comparative analysis.

The cascade structure is the simplest to obtain, as the coefficients correspond to a simple factorization of the $Z$-transfer function. It produces a minimum number of multiplications, additions, and memories. However, the representation of the coefficients may require a large number of bits.

The choice between cascade and lattice structures is not normally realistic for filters with fixed coefficients, as the lattice structure only corresponds to particular applications.

The structures derived from analog ladder networks and simulated ladder filters offer the possibility of representing the coefficients with few bits, even for very selective filters. Consequently, elimination of the multipliers can be envisaged. As these are generally the most complex circuits, the saving in materials is appreciable. Nevertheless, some complications do occur. The number of additions is increased, as is the complexity of the sequence of operations. Extra memories are required for storing the intermediate results. Also, multiplexing of the operations between several filters – an important advantage of digital filters – becomes more difficult. In view of these factors, a detailed evaluation is necessary before this type of structure can be used.

## Exercises

**8.1** Give the impedance and distribution matrices for the elementary two port networks of Figure 8.2. Give the impedance, distribution, and transfer matrices when the elements are resonant *LC* circuits.

**8.2** Consider the Butterworth filter of order 4 given in Figure 8.12(a). Draw the corresponding flow chart. Show the circuit of the digital simulated ladder filter and calculate the coefficients from the given values for the analog elements for a sampling frequency of 40 kHz. Study the modification of the transfer function in the pass band introduced by a reduction in the sampling frequency from 40 kHz to 10 kHz.

**8.3** Let us consider the low-pass Chebyshev filter of order 7, whose analog elements are given in Section 8.2. The sampling frequency is taken as 10 kHz. Show the design for the corresponding simulated ladder filter.
What is the number of operations to be carried out in each realization?
Give the frequency response when the coefficients are represented by 5 bits.

**8.4** Calculate the frequency response of the lattice filter given as the example in Section 8.4. How does this response evolve when the parameters are represented by 5 bits? Give the circuit of the filter with single multiplication elements. How should the circuit be modified to produce the filter using an inverse $Z$-transfer function?

## References


1 V. Belevitch, *Classical Network Theory*, Holden-Day, San Francisco, 1968.

2 J. Neirynk and P. van Bastelaer, La synthèse des filtres par factorisation de la matrice de transfert. *Revue MBLE*, 10(1), 1967.

3 L. T. Bruton, Low sensitivity digital ladder filters *IEEE Transactions*, CAS22(3), 1975.

4 B. Hosticka, R. Brodersen and P. Gray, MOS sampled data recursive filters using switched capacitor integrators. *IEEE Journal of Solid-State Circuits*, 12, 1977.

5 R. Brodersen, P. Gray and D. Hodges, MOS switched-capacitor filters. *Proceedings of the IEEE*, 67(1), 1979.

6 S. K. Mitra, P. S. Kamat and D. C. Huey, Cascaded lattice realization of digital filters. *Circuit Theory and Applications*, 5, 1977.

7 W. B. Jones and A. O. Steinhardt, Finding the poles of the lattice filter. *IEEE Transactions*, ASSP33 (4), 1328–31, 1985.

8 T. Saramaki, T. H. Yu and S. K. Mitra, Very low sensitivity realization of IIR digital filters using a cascade of complex all-pass structures *IEEE Transactions*, CAS34, 876–86, 1987.

9 R. E. Crochière and A. V Oppenheim, Analysis of linear digital networks. *Proceedings of the IEEE*, 63(4), 1975.

# 9

# Complex Signals – Quadrature Filters – Interpolators

Complex signals in the form of sets of complex numbers are currently used in digital signal analysis. Some examples of these sets are presented in the chapters on discrete Fourier transforms. In this chapter, analytic signals – a particular category of complex signal – will be studied. Such signals exhibit some interesting properties and occur primarily in modulation and multiplexing. The properties of the Fourier transforms of real causal sets will be examined first [1–3].

## 9.1 The Fourier Transform of a Real and Causal Set

Consider a set of elements $x(n)$ whose $Z$-transform is written as:

$$X(Z) = \sum_{n=-\infty}^{\infty} x(n)Z^{-n}$$

The Fourier transform of this set is obtained by replacing $Z$ with $e^{j2\pi f}$ in $X(Z)$:

$$X(f) = \sum_{n=-\infty}^{\infty} x(n)e^{-j2\pi nf}$$

If the elements $x(n)$ are real numbers, we obtain:

$$X(-f) = X(f) \tag{9.1}$$

The values of $X(f)$ at negative frequencies are complex conjugates of the values at positive frequencies. The supplementary condition of causality can be imposed on the set $x(n)$ and the consequences for $X(f)$ will now be examined.

The function $X(f)$ can be separated into real and imaginary parts:

$$X(f) = X_R(f) + jX_I(f) \tag{9.2}$$

If the set $x(n)$ is real, then using equation (9.1), the function $X_R(f)$ is even. Thus, it is the Fourier transform of an even set $x_p(n)$. The function $X_1(f)$ is the Fourier transform of an odd set $x_i(n)$ such that:

$$\begin{aligned} x_p(n) &= x_p(-n) \\ x_i(n) &= -x_i(-n) \\ x(n) &= x_p(n) + x_i(n) \end{aligned} \tag{9.3}$$

*Digital Signal Processing: Theory and Practice*, Tenth Edition. Maurice Bellanger.

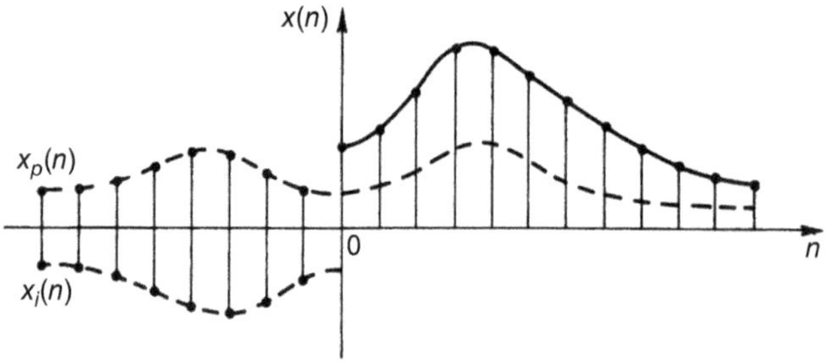


**Figure 9.1** Decomposition of a causal set into even and odd parts.

Under these conditions, we derive:

$$X_R(f) = x_p(0) + 2\sum_{n=1}^{\infty} x_p(n)\cos(2\pi nf) \tag{9.4}$$

$$X_I(f) = -2\sum_{n=1}^{\infty} x_i(n)\sin(2\pi nf) \tag{9.5}$$

If the set $x(n)$ is causal – that is, if:

$$x(n) = 0 \quad \text{for} \quad n < 0$$

the following equations (Figure 9.1) are derived:

$$x_i(n) = x_p(n) = \frac{1}{2}x(n) \quad \text{for} \quad n \geq 1$$
$$x_p(0) = x(0)$$

and hence,

$$X_R(f) - x(0) = \sum_{n=1}^{\infty} x(n)\cos(2\pi nf) \tag{9.6}$$

$$X_I(f) = -\sum_{n=1}^{\infty} x(n)\sin(2\pi nf) \tag{9.7}$$

It can be seen that these two functions are related. To go from one to the other, it is sufficient to change $\cos(2\pi nf)$ to $-\sin(2\pi nf)$ or vice versa. Such an operation is called quadrature, and it will now be expressed analytically.

By definition, a causal set is one which satisfies the equality:

$$x(n) = x(n)Y(n)$$

where the set $Y(n)$ is such that:

$$Y(n) = 0 \ \text{for} \ n < 0$$
$$\qquad 1 \ \text{for} \ n \geq 0$$

This set is a sample of the unit step function $Y(t)$ which, in terms of distributions, has a Fourier transform FY given in Reference [2]:

$$\text{FY} = \frac{1}{j2\pi} vp\left(\frac{1}{f}\right) + \frac{1}{2}\delta(f) \tag{9.8}$$

where $vp(1/f)$ is the distribution defined by:

$$\left\langle vp\left(\frac{1}{f}\right), \phi \right\rangle = VP\int_{-\infty}^{\infty} \frac{\phi(f)}{f} df \tag{9.9}$$

The principal value of the Cauchy integral is itself defined by:

$$\text{VP}\int_{-\infty}^{\infty} \frac{\phi(f)}{f} df = \int_{-\infty}^{\infty} \frac{\phi(f)}{f} df = \lim_{\varepsilon \to 0} \left[ \int_{-\infty}^{-\varepsilon} \frac{\phi(f)}{f} df + \int_{\varepsilon}^{\infty} \frac{\phi(f)}{f} df \right]$$

As the sampling introduces a periodicity to the spectrum, it can be shown that the Fourier transform of the set $Y(n)$ such that:

$$Y(n) = 0 \text{ for } n < 0$$
$$Y(0) = \frac{1}{2}$$
$$Y(n) = 1 \text{ for } n > 0$$

is the distribution $\text{FY}_n$:

$$\text{FY}_n = \frac{1}{2j} vp[\cot \pi \text{f}] + \frac{1}{2}\delta(\text{f}) \text{ for } -\frac{1}{2} \leq \text{f} \leq \frac{1}{2} \tag{9.10}$$

The convolution product of the Fourier transforms corresponds to the product of the two sets. Hence,

$$X(f) = \left[ \frac{1}{2j} vp[\cot \pi f] + \frac{1}{2}\delta(f) \right] * X(f) + \frac{1}{2}x(0)$$

By separating the equation into real and imaginary parts, one obtains:

$$X_R(f) + jX_I(f) = vp[\cot \pi f] * [X_I(f) - jX_R(f)] + x(0)$$

The equations which relate the real and imaginary parts of $X(f)$ are:

$$X_R(f) = x(0) + \overline{\int}_{\frac{-1}{2}}^{\frac{1}{2}} X_I(f') \cot[\pi(f - f')] df' \tag{9.11}$$

$$X_I(f) = -\overline{\int}_{\frac{-1}{2}}^{\frac{1}{2}} X_R(f') \cot[\pi(f - f')] df' \tag{9.12}$$

or further, in a different form, without introducing Cauchy's principal values,

$$X_R(f) = x(0) - \int_{-1/2}^{1/2} [X_I(f) - X_I(f')] \cot[\pi(f - f')] df' \tag{9.13}$$

$$X_I(f) = \int_{-1/2}^{1/2} [X_R(f) - X_R(f')] \cot[\pi(f - f')] df' \tag{9.14}$$

The real and imaginary parts of the Fourier transform of a causal set are related by equations (9.11) and (9.12), which correspond to the Hilbert transform for continuous signals.

*Example:* Consider the set $U_k(n)$ such that:

$$U_k(n) = 0 \quad \text{for } n \neq k$$
$$U_k(k) = 1$$
$$X_R(f) = \cos(2\pi kf); \quad X_I(f) = -\sin(2\pi kf)$$

It can be shown directly that:

$$\int_{-1/2}^{1/2} \cos(2\pi kf')\cot[\pi(f-f')]df' = \int_{-1/2}^{1/2} \cos[2\pi k(f-f')]\cot(\pi f')df' = \sin(2\pi kf)$$

$$\int_{-1/2}^{1/2} \sin(2\pi kf')\cot[\pi(f-f')]df' = \int_{-1/2}^{1/2} \sin[2\pi k(f-f')]\cot(\pi f')df' = -\cos(2\pi kf)$$

On the other hand, using Parseval's equation, we can write:

$$\int_0^{1/2} X_R^2(f)df = \int_0^{1/2} X_I^2(f)df \tag{9.15}$$

The real and imaginary parts of $X(f)$ have the same power.

## 9.2 Analytic Signals

Analytic signals correspond to causal signals in which time and frequency are exchanged. Their spectrum has no negative frequency component, and their name derives from the fact that they represent the restriction to the real axis of an analytic function of a complex variable – that is, they can be expanded in series in a region which contains that axis. The properties of analytic signals are deduced from the properties of causal signals by exchanging time and frequency.

Consider the signal $x(t) = x_R(t) + jx_I(t)$ such that:

$$X(f) = 0 \text{ for } f < 0$$

The functions $x_R(t)$ and $x_I(t)$ are Hilbert transforms of each other:

$$x_R(t) = \frac{1}{\pi}\int_{-\infty}^{\infty} \frac{x_I(t')}{t-t'}dt' \tag{9.16}$$

$$x_I(t) = -\frac{1}{\pi}\int_{-\infty}^{\infty} \frac{x_R(t')}{t-t'}dt' \tag{9.17}$$

The Fourier transform of the real function:

$$x_R(t) = \frac{1}{2}[x(t) + \bar{x}(t)] \tag{9.18}$$

is the function $X_R(f)$ such that:

$$X_R(f) = \frac{1}{2}[X(f) + \bar{X}(-f)] \tag{9.19}$$

that is, $X_R(f) = X(f)/2$ for positive frequencies and $X_R(f) = \frac{\bar{X}(-f)}{2}$ for negative ones.

Similarly:

$$X_I(f) = -j\frac{1}{2}[X(f) - \bar{X}(-f)] \tag{9.20}$$

Figure 9.2 shows the decomposition of the spectrum of a signal into real and imaginary parts.

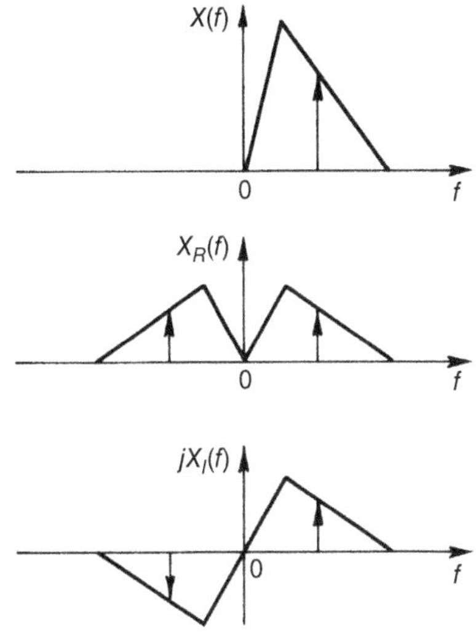


**Figure 9.2** Spectrum of an analytic signal.

*Example:*

$$x(t) = e^{j\omega t}; \quad x_R(t) = \cos\omega t = \frac{1}{2}[e^{j\omega t} + e^{-j\omega t}]$$

$$x_I(t) = \sin\omega t = -j\frac{1}{2}[e^{j\omega t} - e^{-j\omega t}]$$

Finally, it can be seen that the following equations hold between $X_R(f)$ and:

$$X_I(f) = -jX_R(f) \quad \text{for} \quad f > 0$$
$$jX_R(f) \quad \text{for} \quad f < 0$$

That is, $X_I(f)$ is obtained from $X_R(f)$ by a rotation of the components through $\pi/2$. The Hilbert transform consists of orthogonalizing the signal components. This is a filtering operation, and the frequency response $Q(f)$ is represented in Figure 9.3.

*Example:*

$$x_R(t) = \int_0^{\infty} [A(f)\cos(2\pi ft) - B(f)\sin(2\pi ft)]df \tag{9.21}$$

$$x_I(t) \int_0^{\infty} [A(f)\sin(2\pi ft) + B(f)\cos(2\pi ft)]df \tag{9.22}$$

The properties of continuous analytic signals can be transferred to discrete signals after certain modifications.

A discrete signal has a periodic Fourier transform. A discrete analytic signal $x(n)$ deduced from a real signal is a discrete signal whose Fourier transform $X_n(f)$, which has the period $f_s = 1$, is zero for $-\frac{1}{2} < f < 0$ (Figure 9.4).

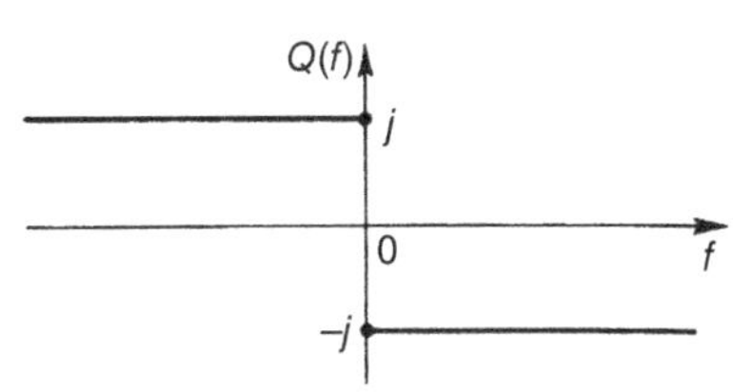


**Figure 9.3** Frequency response of a quadrature filter.

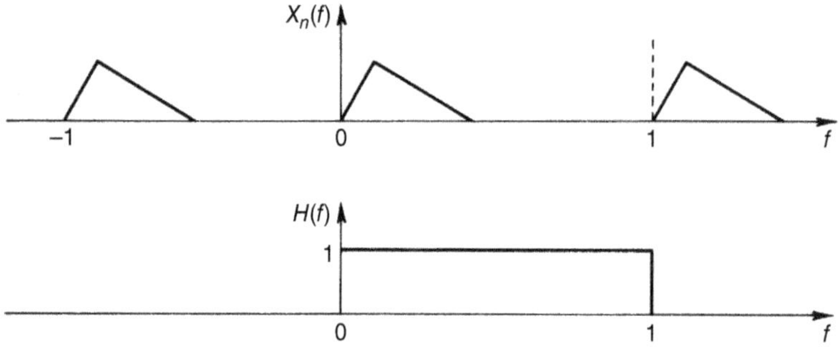


**Figure 9.4** Spectrum of a discrete analytic signal and an interpolation filter.

If a discrete signal $x(n)$ is obtained by sampling a continuous analytic signal $x(t)$ at a frequency $f_s = 1$, it is worth noting that the continuous signal can be reconstructed from the discrete values by using a reconstruction filter which preserves only the signal components contained in the band $(0, f_s)$, as is shown in Figure 9.4. The reconstruction formula is:

$$x(t) = \sum_{n=-8}^{\infty} x(n) \frac{\sin[\pi(t-n)]}{\pi(t-n)} e^{j\pi(t-n)} \tag{9.23}$$

Sampling does not introduce any degradation to an analytic signal $x(t)$ if its spectrum does not contain any components with frequencies greater than or equal to $f_s$. Thus, the sampling theorem for an analytic signal is:

*An analytic signal which does not contain any components with frequencies greater than or equal to $f_m$ is wholly determined by the set of its values sampled at time intervals of $T = /f_m$.*

The set $x(n)$ is decomposed into a real set $x_R(n)$ and an imaginary set $x_I(n)$, such that:

$$x(n) = x_R(n) + jx_I(n)$$

The corresponding Fourier transforms $X_{nR}(f)$ and $X_{nI}(f)$ are obtained from the Fourier transform $X_n(f)$ by equations (9.19) and (9.20) given above.

$$X_{nI}(f) = -jX_{nR}(f) \quad \text{for} \quad 0 < f < \frac{1}{2}$$
$$jX_{nR}(f) \quad \text{for} \quad -\frac{1}{2} < f < 0$$

The relations between the sets $X_R(n)$ and $X_I(n)$ are obtained by considering the quadrature filter whose frequency response is given in Figure 9.5. The impulse response of this filter is the set $h(n)$, such that:

$$h(n) = \int_{-1/2}^{0} je^{j2\pi nf} df + \int_{0}^{1/2} (-j)e^{j2\pi nf} df$$
$$h(n) = \frac{2}{\pi n} \sin^2\left(\frac{n\pi}{2}\right) \text{ for } n \neq 0$$
$$h(0) = 0 \tag{9.24}$$

By applying the set $x_R(n)$ to this filter, we obtain the set $x_I(n)$, thus:

$$x_I(n) = \frac{2}{\pi} \sum_{\substack{m=-\infty \\ m\neq n}}^{\infty} x_R(n-m) \frac{\sin^2[\pi(m/2)]}{m} \tag{9.25}$$

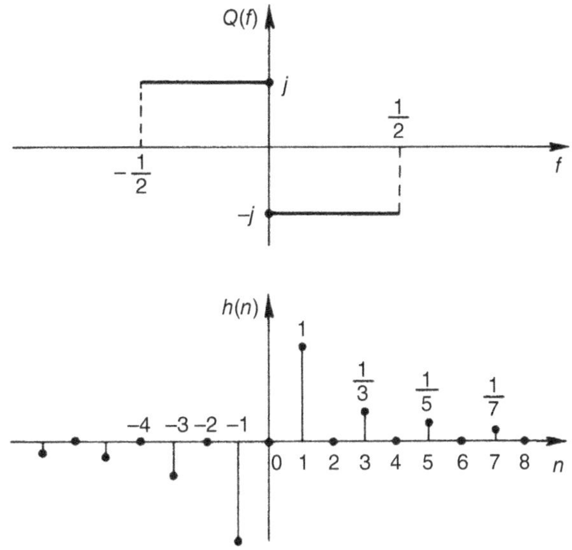


**Figure 9.5** Responses of the quadrature filter.

and similarly:

$$x_R(n) = -\frac{2}{\pi} \sum_{\substack{m=-\infty \\ m \neq n}}^{\infty} x_I(n-m) \frac{\sin^2[\pi(m/2)]}{m} \tag{9.26}$$

The sets $x_R(n)$ and $x_I(n)$ are related by the discrete Hilbert transform [4].

Examination of the elements of the set $h(n)$ leads to several observations. Firstly, the fact that every other element is zero implies that, if the set $x_R(n)$ also has every other element zero, then so must the set $x_I(n)$, and further, the sets $x_R(n)$ and $x_I(n)$ must be interleaved. An example will be given later.

The impulse response of the quadrature filter corresponds to a linearphase FIR filter as described in Section 5.2. Its frequency response is:

$$Q(f) = -j2 \sum_{n=1}^{\infty} h(n) \sin(2\pi nf) \tag{9.27}$$

In order to realize this filter, the number of coefficients must be limited.

## 9.3 Calculating the Coefficients of an FIR Quadrature Filter

A realizable quadrature filter is easily obtained by limiting the number of terms over which summation (9.27) is performed. The frequency response then deviates from the ideal response. In practice, the filter is specified by setting a limit $\delta$ for the ripple in a frequency band $(f_1, f_2)$, as shown in Figure 9.6. A satisfactory FIR filter can be obtained using a low-pass filter, and the results are presented in Chapter 5.

One interesting example is that of the filter whose response is given in Figure 9.7. This filter is called a half-band filter because the pass band represents half of the useful band.

Further, $H$ (0.25) = 0.5, and the response is assumed to be asymmetric about the point $f_c = 0.25$ – that is, $H(0.25 + f) = 1 - H(0.25 - f)$. This filter is specified by the transition band $\Delta f$ and the ripple in the pass and stop bands, which are equal to $\delta_0$. Its coefficients can be

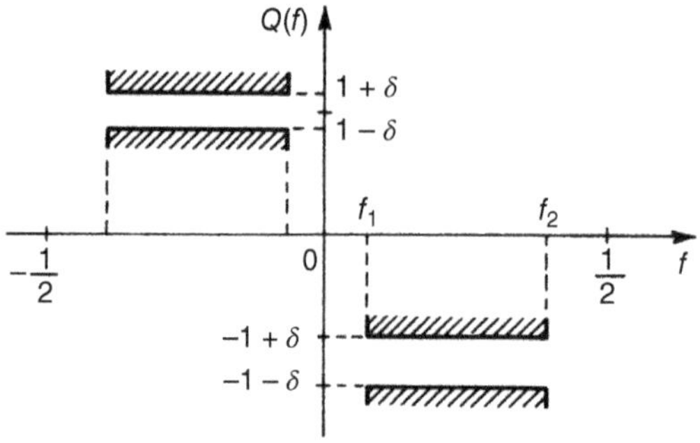


**Figure 9.6** Mask for a quadrature filter.

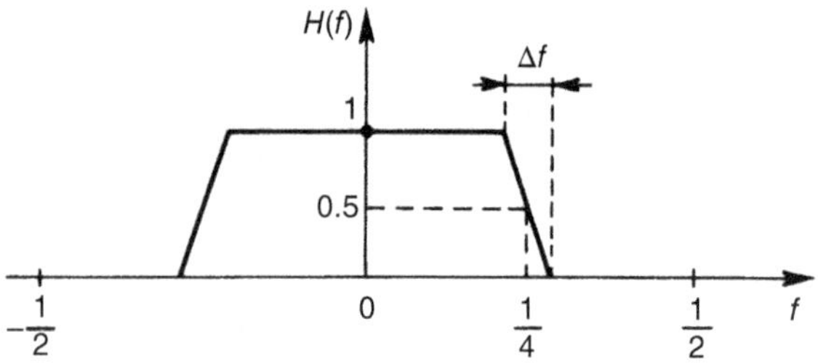


**Figure 9.7** Frequency response of a half-band filter.

calculated using conventional FIR filter design programs. This type of filter is of interest because the symmetry of its frequency response implies that the coefficients $h_n$ are zero for even values of $n$ except for $h_0$. Thus, if $n = 2p$, decomposition of $H(f)$ into the sum of two functions leads to:

$$h_{2p} = \int_0^{0.25} \cos(2\pi 2pf)df + (-1)^p \left[\int_0^{\frac{\Delta f}{2}} H(0.25 + f)\cos 4\pi pf df \right.$$
$$\left. + \int_0^{\Delta f/2} [H(0.25 - f) - 1]\cos 4\pi pf df\right]$$

and hence,

$$h_{2p} = 0$$

For a number of coefficients $N = 4M + 1$, the frequency response is:

$$H(f) = e^{-j2\pi 2Mf}\frac{1}{2}\left[1 + 2\sum_{i=1}^{M} h_{2i-1}\cos[2\pi(2i-1)f]\right] \tag{9.28}$$

Translation of this response through 0.25 on the frequency axis leads to the function $H'(f)$ such that:

$$H'(f) = H(f - 0.25) = e^{-j2\pi 2Mf}\frac{1}{2}\left[1 - 2\sum_{i=1}^{M}(-1)^i h_{2i-1}\sin[2\pi(2i-1)f]\right]$$

The coefficients $h'_n$ of the corresponding filter are:

$$h'_{2i-1} = j(-1)^i h_{2i-1};\quad h'_{-(2i-1)} = -j(-1)^i h_{2i-1};\quad 1 \le i \le M$$

They have imaginary values. By comparing the expression for $H'(f)$ with equation (9.27) for $Q(f)$, it becomes clear that the set of coefficients $a_n$ such that:

$$a_{2i-1} = -2(-1)^i h_{2i-1}; a_{-(2i-1)} = 2(-1)^i h_{2i-1}; 1 \le i \le M$$

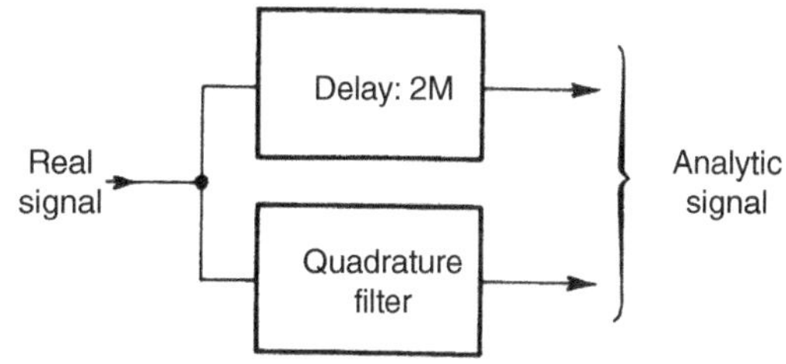


**Figure 9.8** An analytic FIR filter.

represents the coefficients of a quadrature filter whose ripple equals $2\delta_0$ in the band:

$$\left[\Delta f/2, \frac{1}{2} - \Delta f/2\right].$$

*Example.* For the specification $\delta_0 = 0.01$ and $\Delta f = 0.111$, it is found that $M = 5$.

$$a_1 = 0.6283 \quad a_3 = 0.1880$$
$$a_5 = 0.0904 \quad a_7 = 0.0443$$
$$a_9 = 0.0231$$

The expression $H'(f)$ corresponds to a complex filter which contains two parts: a circuit producing a delay of $2M$ elementary periods, and a quadrature filter, as shown in Figure 9.8. The outputs of these two circuits form the real and imaginary parts of the complex signal. The system can be said to comprise two branches – one real and the other imaginary. It allows a real signal to be converted to an analytic signal and is thus an analytic FIR filter.

It should be noted that this property holds true even if the fundamental low-pass filter is not of the half-band type. In this case, the coefficients with even index do not cancel out. In fact, the translation through a frequency of 0.25 corresponds to multiplication of the coefficients by a complex factor, so that the $h'_n$ takes the values:

$$h'_n = e^{-j(\pi/2)n} h_n \tag{9.29}$$

Under these conditions, the real branch of the system is not a simple delay. A filtering function is achieved at the same time as the analytic signal is generated.

Thus, circuits with finite impulse response allow an ideal quadrature filter to be approximated without error in the phase shift but with approximation of the amplitude in the pass band. Circuits with infinite impulse response, or recursive circuits, provide an alternative approach. By using pure phase shifters, they allow for the approximation of the quadrature filter with no error in the amplitude, but with an approximation in the phase.

## 9.4 Recursive 90° Phase Shifters

A recursive phase shifter is characterized by the fact that the numerator and denominator of its $Z$-transfer function are image polynomials. That is, they have the same coefficients but in reverse order. The properties of phase shifters were introduced in Section 6.3.

It is possible to design a pair of phase shifters such that the output signals have a phase difference which approximates 90° with an error of less than $\mathcal{E}$, in a given frequency band $(f_1, f_2)$. The calculation techniques are the same as for IIR filters. The procedure for producing a phase difference with elliptic behavior is as follows [5]:

(1) Determination of the order of the circuit:

$$N = \frac{K(k_1)K(\sqrt{(1-k^2)})}{K(k)K\left(\sqrt{(1-k_1^2)}\right)}$$

with the following values for the parameters:

$$k = \frac{\tan(\pi f_1)}{\tan(\pi f_2)};\ k_1 = \left[\frac{1-\tan(\pi\varepsilon)}{1+\tan(\pi\varepsilon)}\right]^2$$

(2) Determination of the zeros $z_i$ of the $Z$-transfer function:

$$A = Sn\left[\frac{(4i+1)K(\sqrt{(1-k^2)})}{2N}, \sqrt{(1-k^2)}\right]$$

(where $Sn$ is the elliptic function).

$$p_i = -\tan(\pi f_1)\frac{A}{\sqrt{(1-A^2)}}$$

$$z_i = \frac{1+p_i}{1-p_i} \quad \text{for } 0 \le i \le N-1$$

*Example:* Assuming the specification $f_1 = 0.028; f_2 = 0.33$ and $\mathcal{E} = 1°$, it is found that:

$$\tan(\pi f_1) = 0.0875; k = 0.0505; k_1 = 0.9657; N = 4.8$$

By taking $N = 5$, we obtain:

$$p_0 = -0.0395 z_0 = 0.9240$$
$$p_1 = -0.3893 z_1 = 0.4396$$
$$p_2 = -3.8360 z_2 = -0.5864$$
$$p_3 = -1.0039 z_3 = -0.00197$$
$$p_4 = -0.1509 z_4 = 0.7377$$

In forming the circuit, the first three zeros $z_1$ are attributed to one branch and the last two to the other. The variation in the phase difference with frequency is given by the function $\phi(f)$ represented in Figure 9.9.

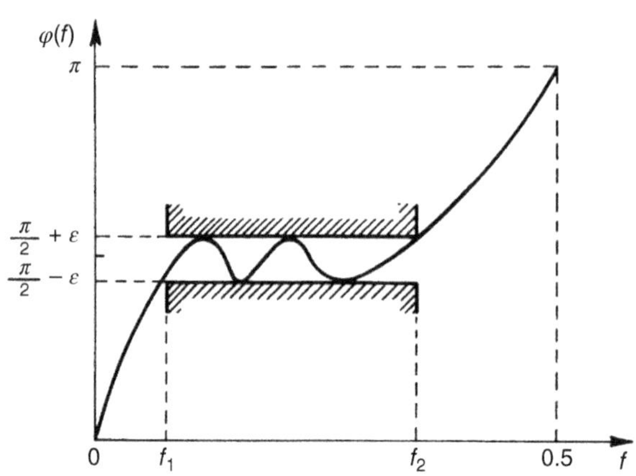


**Figure 9.9** Characteristic of 90° phase shifter.

Recursive phase shifters allow two orthogonal signals to be obtained. It should be noted that in performing this operation, they also introduce phase distortion which is the same for both signals. These circuits can be used in modulation and multiplexing equipment.

## 9.5 Single Side-Band Modulation

Modulation of a signal results in a displacement of the spectrum on the frequency axis. It is single side-band (SSB) modulation if, for a real signal, the part of the spectrum which corresponds to positive frequencies is displaced toward positive frequencies and the part which corresponds to negative ones is displaced toward negative frequencies. Thus, the signal $s(t) = \cos \omega t$ has the modulated signal:

$$s_{\rm m}(t) = \cos(\omega + \omega_0)t$$

This operation can be performed by the following procedure:

(1) Form the analytic signal $s_{\rm a}(n) = s_R(n) + {\rm j}s_{\rm I}(n)$ which corresponds to the real signal represented by the set $s(n)$.
(2) Multiply the set $s_{\rm a}(n)$ by the set of complex numbers:

$$\cos(2\pi n f_0) - {\rm j}\ \sin(2\pi n f_0)$$

and retain the real part $sm(n)$ of the set thus obtained. Then,

$$s_{\rm m}(n) = s_{\rm R}(n)\cos(2\pi n f_0) + s_{\rm I}(n)\sin(2\pi n f_0)$$

The signal spectrum evolves as shown in Figure 9.10 and the corresponding circuits are shown in Figure 9.11.

If the analytic filter is of the FIR type, the set $s_R(n)$ is simply the delayed $s(n)$, which is different from the case of 90° recursive phase shifters. The set which corresponds to the modulated signal $s_m(n)$ can be added to other modulated sets to provide, for example, a frequency-multiplexed signal in telephony.

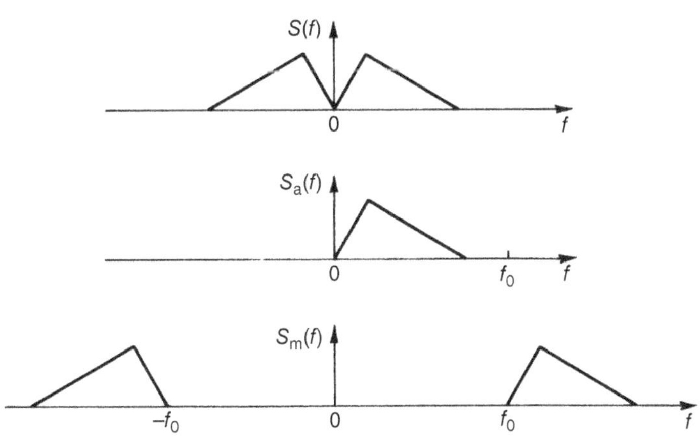


**Figure 9.10** Single side-band modulation.

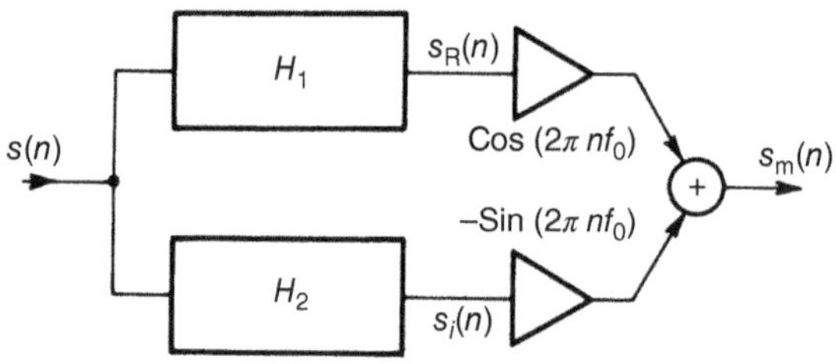


**Figure 9.11** Single side-band modulation circuit.

## 9.6 Minimum-Phase Filters

The properties of causal and analytic signals discussed at the beginning of the chapter help to clarify a point concerning the phase characteristics of filters [3]. The frequency response of a filter $H(f)$ is written as:

$$H(f) = A(f)e^{-j\phi(f)} \text{ with } A(f) = |H(f)|$$

and

$$\phi(f) = -\arg[H(f)]$$

The term $A(f)$ is the attenuation and $\phi(f)$ is the phase shift produced by the filter on a sinusoidal signal, at frequency $f$.

If the filter has real coefficients, its impulse response, the set $h(n)$, is real, and consequently:

$$H(f) = \overline{H(-f)}; A(f) = A(-f) \text{ and } \phi(f) = -\phi(-f)$$

The term $h(n)$ is obtained by:

$$h(n) = 2\int_0^{1/2} A(f)\cos[2\pi nf - \phi(f)]df \tag{9.30}$$

As the response cannot precede the application of the signal to the filter, a realizable filter must be causal, with:

$$h(n) = 0 \quad \text{for } n < 0$$

It follows that if the response is decomposed into a real and imaginary part:

$$H(f) = H_R(f) + jH_I(f)$$

the functions $H_R(f)$ and $H_I(f)$ are related by equations (9.11) and (9.12) given in Section 9.1.

Now, a filter is often only specified by the data for the constraints on the amplitude:

$$A^2(f) = H_R^2(f) + H_I^2(f)$$

and this results in a lack of definition in the phase shift.

To express this generally, a certain length of time, corresponding to the propagation time through the system, is required to process a signal. This is characterized by the variation of the phase shift with frequency. In order to minimize this propagation time, it is necessary to find the minimum phase-shift characteristic, which removes the lack of definition in the calculation of the filter. Another approach is to specify a linear phase shift.

A stable and realizable filter has a $Z$-transfer function, $H(Z)$, whose poles are inside the unit circle and whose zeros may be outside it. Assume $A_0$ is such a zero and $H_1(Z)$ is the function such that:

$$H_1(Z) = \frac{1 - 2\text{Re}(Z_0)Z^{-1} + |Z_0|^2 Z^{-2}}{|Z_0|^2 - 2\text{Re}(Z_0)Z^{-1} + Z^{-2}}$$

This is the transfer function of a phase shifter of the second order, which was introduced in Section 6.3 and whose group delay is given by equation (6.45). One can write:

$$H(Z) = H_1(Z)H_2(Z)$$

where $H_2(Z)$ is a function which is zero at $Z_0^{-1}$ and introduces a smaller phase shift than $H(Z)$. By iteration, it results that minimum phase shift is achieved with the function $H_m(Z)$, which is obtained by replacing all the zeros in $H(Z)$ outside the unit circle with their inverse.

The formulation of the minimum phase condition is that the function:

$$\log[H(Z)] = \log[A(Z)] - j\phi(Z)$$

should not have any poles outside the unit circle.

Under these conditions, the functions $\log[A(f)]$ and $\phi(f)$ are related by equations (9.11) and (9.12), which correspond to the Hilbert transform. They are the Bayard–Bode equations for discrete systems:

$$\log[A(f)] = K - \int_{-1/2}^{1/2} \phi(f') \cot \pi(f - f')\mathrm{d}f' \tag{9.31}$$

$$\phi(f) = \int_{-1/2}^{1/2} \log[A(f')] \cot \pi(f - f')\mathrm{d}f' \tag{9.32}$$

The constant $K$ is a scaling factor for the amplitude.

Similarly, a function $H_M(Z)$ whose zeros are outside the unit circle is said to be maximum phase.

## 9.7 Differentiator

A differentiator is a quadrature filter whose frequency response is proportional to frequency:

$$H_d(\omega) = D(\omega)e^{-\frac{j\pi}{2}} \tag{9.33}$$

$$D(\omega) = \omega \text{ for } 0 \le \omega \le \omega_1 \le \pi$$

If the upper band edge $\omega_1$ equals $\pi$, the filter is said to be full-band.

The frequency response of the corresponding $N$ coefficient digital filter is expressed by:

$$H(\omega) = R(\omega)e^{-j\left(\frac{\pi}{2}+\frac{\omega(N-1)}{2}\right)} \tag{9.34}$$

$R(\omega)$ is the real function:

$$R(\omega) = \sum_{i=1}^{P} h_i \sin(i\omega); N = 2P + 1 \tag{9.35}$$

$$R(\omega) = \sum_{i=1}^{P} h_i \sin\left(\left(i - \frac{1}{2}\right)\omega\right); N = 2P$$

The corresponding impulse responses have been given in Section 5.2.

Conventional design techniques apply to these devices – amongst others, the least-squares method. The full-band case is particularly simple, because $N$ is even, as shown by equation (9.35) above, and the least-squares method leads to the following set of coefficients [6]:

$$h_i = \frac{8}{\pi} \frac{(-1)^{i+1}}{(2i-1)^2} \tag{9.36}$$

If the desired function $D(\omega)$ is proportional to a power of the frequency, the differentiator is said to be of higher order.

The conversion of a real signal into a complex signal often involves an interpolation operation, particularly in analog-to-digital converters in communication transceivers.

## 9.8 Interpolation Using FIR Filters

Interpolation consists of calculating signal samples between known samples. In fact, it is a filtering function which can be efficiently realized by FIR filters [7, 8].

Let us assume, as illustrated in Figure 9.12, that the value $\boldsymbol{x(nT+\tau)}$ has to be calculated from the sample sequence $x(nT)$ with the help of a FIR filter having $N = 2P+1$ coefficients. The delay $\boldsymbol{\tau}$ is such that $|\tau| \leq \boldsymbol{T/2}$.

The filter itself introduces the delay $KT$, with $K$ being an integer, and the output $y(n)$ is required to be:

$$y(n) \approx x[(n-K)T+\tau] \tag{9.37}$$

The delay can be expressed by the $Z$-transfer function as well. Taking $T = 1$:

$$H(Z) = \sum_{i=0}^{N-1} a_i Z^{-i} \approx Z^{-K+\tau} \tag{9.38}$$

and in the frequency domain:

$$\left( \sum_{i=0}^{N-1} a_i e^{-ji\omega} \right) e^{j(K-\tau)\omega} \approx 1 \tag{9.39}$$

Alternatively,

$$e^{-j\tau\omega} \sum_{i=0}^{N-1} e^{-j(i-K)\omega} \approx 1 \tag{9.40}$$

Letting $K=P$, after change of variables:

$$e^{-j\tau\omega} \sum_{i=-P}^{P} h_i\, e^{-ji\omega} \approx 1 \tag{9.41}$$

and finally:

$$G(\omega) = \sum_{i=-P}^{P} h_i\, e^{-j(i+\tau)\omega} \approx 1; |\tau| \leq \frac{1}{2} \tag{9.42}$$

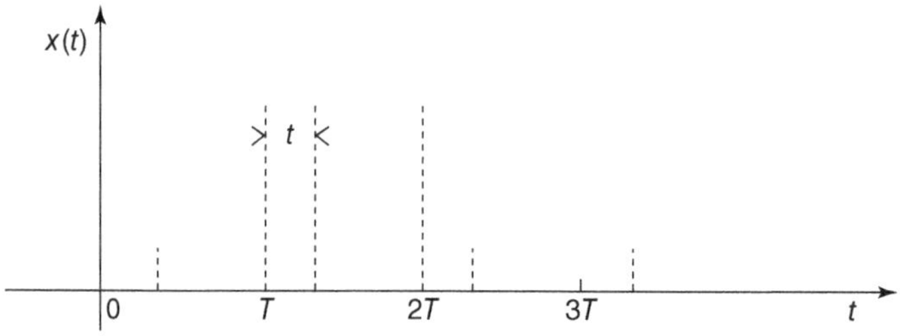


**Figure 9.12** Interpolation with delay $\tau$.

The interpolator coefficients $h_i$ are derived from this expression and conventional approximation techniques such as the least-squares method can be used.

For systems in which the delay may vary in time, such as synchronization loops, it is helpful to be able to relate the coefficient values to the delay values, in order to make the interpolator be able to track the delay evolution. Lagrange interpolation is a suitable approach in this case.

## 9.9 Lagrange Interpolation

In the frequency domain, Lagrange interpolation corresponds to "max-flat" filtering – that is, the derivatives of the frequency response vanish at the origin. Then, the filter coefficients are obtained by solving the following linear system:

$$G(0) = 1; G^{(p)}(0) = 0; 1 \leq p \leq P \tag{9.43}$$

For $P = 1$, one obtains:

$$\begin{aligned} &h_{-1} + h_0 + h_1 = 1 \\ &(\tau - 1)\, h_{-1} + \tau\, h_0 + (\tau + 1)h_1 = 0 \\ &(\tau - 1)^2\, h_{-1} + \tau^2\, h_0 + (\tau + 1)^2 h_1 = 0 \end{aligned} \tag{9.44}$$

which leads to the matrix equation:

$$\begin{bmatrix} 1 & 1 & 1 \\ \tau - 1 & \tau & \tau + 1 \\ (\tau - 1)^2 & \tau^2 & (\tau + 1)^2 \end{bmatrix} \begin{bmatrix} h_{-1} \\ h_0 \\ h_1 \end{bmatrix} = \begin{bmatrix} 1 \\ 0 \\ 0 \end{bmatrix} \tag{9.45}$$

whose solution is:

$$h_{-1} = \frac{\tau(\tau + 1)}{2}; h_0 = 1 - \tau^2; h_1 = \frac{(\tau - 1)\tau}{2}$$

One can observe that the norm of the function $H(\omega)$ can be written as:

$$||H||_2^2 = \sum_{i=-P}^{P} h_i^2 = 1 - \frac{3}{2}\tau^2 + \frac{3}{2}\tau^4 \tag{9.46}$$

This shows that the quadratic interpolation error grows with time $\tau$ and it is maximum for $\tau = 1/2$.

An efficient implementation is obtained by noting that matrix equation (9.45) leads to the following expression of coefficients:

$$h_i = \sum_{j=0}^{N-1} b_{ij}\, \tau^j \tag{9.47}$$

with $b_{00} = 1$ and $b_{i0} = 0; i \neq 0$. Then, one obtains:

$$Z^P H(Z) = \sum_{i=-P}^{P} \left( \sum_{j=0}^{N-1} b_{ij} \tau^j \right) Z^{-i}$$

and, by inverting summations:

$$Z^P H(Z) = \sum_{j=0}^{N-1} C_j(Z) \tau^j \tag{9.48}$$

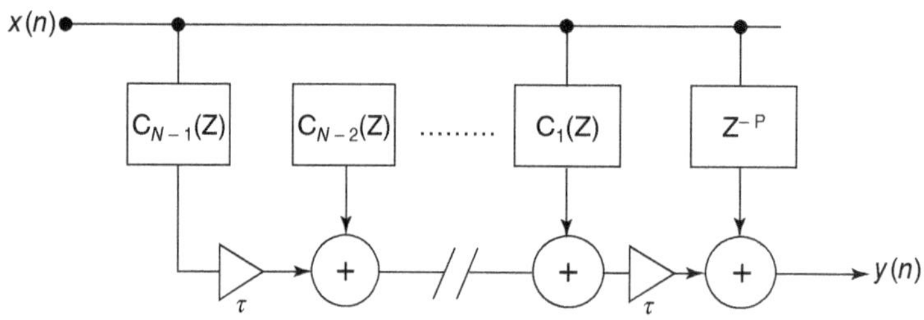


**Figure 9.13** Implementation of the Lagrange interpolator.

with:

$$C_0 = 1; C_j(Z) = \sum_{i=-P}^{P} b_{ij} Z^{-i}$$

The corresponding implementation diagram is provided in Figure 9.13. It can easily be adapted to the evolutions of the delay.

The connection with the interpolation formula mentioned in Section 5.6 can be established from the definition of the filter coefficients.

The general expression of coefficients $h_i$, a solution of (9.45), is obtained by noticing that the square matrix is of Vandermonde type. Determinants and sub-determinants are zero whenever 2 rows or 2 columns are identical. It follows that they can be expressed as products.

Generalizing to arbitrary periods of time Δ, the interpolation filter coefficients are written as:

$$a_i = \prod_{j=0;j\neq i}^{N-1} \frac{\Delta - j}{i - j}; 0 \le i \le N-1 \tag{9.49}$$

and interpolated values:

$$x(nT - \Delta) \approx y(n) = \sum_{i=0}^{N-1} a_i\, x[(n-i)T] \tag{9.50}$$

The quality of the interpolation depends on the number of coefficients and, when this number tends toward infinity, the sampling formula (1.57) can be obtained using the identity:

$$\sin \pi t = \pi t \prod_{n\neq 0} \left(1 - \frac{t}{n}\right) \tag{9.51}$$

which, letting $\boldsymbol{\Delta} = t/\boldsymbol{T}$, leads to:

$$a_i = \prod_{j\neq i} \frac{\Delta - j}{i - j} = \prod_{j\neq 0} \frac{k - (\Delta - i)}{k} = \frac{\sin \pi(\Delta - i)}{\pi(\Delta - i)} \tag{9.52}$$

In certain applications, such as curve fitting or image processing, data are available in sets or blocks, and therefore, the values have to be interpolated in the blocks.

## 9.10 Interpolation by Blocks – Splines

When processing finite sets of data, filters that do not satisfy the Nyquist condition – that is, filters whose impulse response does not vanish at all integer multiples of the sampling period – can be employed. However, then, if known samples have to be preserved, some preprocessing is required, which, in fact, is an inverse filtering operation. Splines are widely used functions of this type [9].

The spline function of degree $m$ is defined by the $m$ following convolutions:

$$B_m(t) = B_0(t) * B_0(t) * \ldots * B_0(t) \tag{9.53}$$

where $\boldsymbol{B_0(t)}$ is the unit length impulse:

$$B_0(t) = \begin{cases} 1; |t| < 1/2 \\ 0.5; |t| = 1/2 \\ 0; |t| > 1/2 \end{cases}$$

Convolution is integration; thus, a higher degree yields a smoother function, because it is necessary to consider higher-order derivatives to find discontinuities. A widely used function is the cubic spline:

$$B_3(t) = \begin{cases} \frac{2}{3} - |t|^2 + \frac{|t|^3}{2}; 0 \leq |t| < 1 \\ \frac{2-|t|^3}{6}; 1 \leq t < 2 \\ 0; 2 \leq |t| \end{cases} \tag{9.54}$$

Given a set of $N$ sample values $\boldsymbol{s(nT), 0 \leq n \leq N-1}$, inverse filtering is carried out first, so that a new set $x(n)$ is obtained, to which the interpolation filter is applied.

With the cubic spline, assuming $T = 1$, the $Z$-transfer function of the samples is:

$$B_3(Z) = \frac{Z + 4 + Z^{-1}}{6} \tag{9.55}$$

The inverse is, in product form:

$$B_3^{-1}(Z) = \frac{6}{2+\sqrt{3}} \quad \frac{1}{1+(2-\sqrt{3})Z^{-1}} \quad \frac{1}{1+(2-\sqrt{3})Z} \tag{9.56}$$

or else:

$$B_3^{-1}(Z) = \frac{6-3\sqrt{3}}{2\sqrt{3}-3} \left[ \frac{1}{1+(2-\sqrt{3})Z^{-1}} + \frac{1}{1+(2-\sqrt{3})Z} - 1 \right] \tag{9.57}$$

Of the 2 factors in (9.56), one is causal, and one is anti-causal. The sequence $s(n)$ can be processed by the first factor and the output $u(n)$ is:

$$u(n) = s(n) - (2-\sqrt{3})\, u(n-1) \tag{9.58}$$

The second factor is applied to $u(n)$ and the output is the desired sequence $x(n)$:

$$x(n) = \frac{6}{2+\sqrt{3}} u(n) - (2-\sqrt{3})\ x(n+1) \tag{9.59}$$

In that case, the calculations must be carried out in reverse order.

The two above recurrences require initial values – namely $u(0)$ and $x(N-1)$ – whose determination depends on limit conditions of the data block. In general, a symmetrical extension outside the data block is retained – that is, $s(-n) = s(n)$ and $s(N-1+k) = s(N-1-k)$. The periodicity is $2N-2$ and the series development of the first factor in (9.56) leads to the following initial value:

$$u(0) = \sum_{n=0}^{\infty} (\sqrt{3}-2)^n\, s(n)$$

which is:

$$u_0 = \frac{1}{1-(\sqrt{3}-2)^{2N-2}} \sum_{n=0}^{2N-3} (\sqrt{3}-2)^n \; s(n) \tag{9.60}$$

For large values of $N$, depending on the level of accuracy required, the summation may be limited to the first terms.

In the other direction, $x(N-1)$ can be calculated directly with the following scheme – start from (9.57), perform series developments of the two terms, use the corresponding expression of $u(n)$ as a function of $s(n)$ for $n = N-1$; then, due to the symmetry of $s(n)$ about $N-1$, the following initialization is obtained:

$$x(N-1) = \frac{6-3\sqrt{3}}{2\sqrt{3}-3} \; [2\,u(N-1) - s(N-1)] \tag{9.61}$$

It is readily verified that the constant signal $s(n) = 1$ leads to:

$$u(0) = \frac{1}{3-\sqrt{3}} = u(n); \quad x(N-1) = 1 = x(n)$$

Once the transformed block has been made available, interpolation is carried out:

$$s(t) = \sum_n x(n)\, B_m(t-nT) \tag{9.62}$$

The summation is limited to the terms for which the spline function is not null. As in the previous section, when the degree $n$ grows, this interpolator tends toward the ideal interpolator.

## 9.11 Interpolations and Signal Restoration

The loss of several samples, adjacent or otherwise, in a signal can be compensated by an interpolation operation exploiting the known characteristics of the signal. In particular, when certain frequencies are missing or are very small, the DFT can be relied upon to estimate the lost temporal values.

Consider a block of $N$ samples $\boldsymbol{x(n)}, \boldsymbol{0 \le n \le N-1}$, in which $P$ samples $\boldsymbol{x(p)}, \boldsymbol{n_0 \le p \le n_0 + P - 1}$, are unknown. The signal is known to have few high-frequency components in its spectrum and the DFT outputs $\boldsymbol{y(k)}, \boldsymbol{N-P \le k \le N-1}$ are assumed to be null. If the input vector $X$ is split into a known vector $X_1$ and an unknown vector $X_2$, the DFT output can be expressed by:

$$Y = T_1 X_1 + T_2 X_2 \tag{9.63}$$

The inputs of the matrices $T_1$ and $T_2$ are those of the DFT properly arranged. Splitting the DFT output vector leads to:

$$\begin{bmatrix} Y_1 \\ Y_2 \end{bmatrix} = \begin{bmatrix} T_{11} & T_{12} \\ T_{21} & T_{22} \end{bmatrix} \begin{bmatrix} X_1 \\ X_2 \end{bmatrix} \tag{9.64}$$

The vector $Y_2$ corresponds to the $P$ frequency components which are either null or negligible. Then, the desired missing input vector $X_2$ is given by:

$$X_2 = -T_{22}^{-1} T_{21} X_1 \tag{9.65}$$

The restoration is perfect if the vector $Y_2$ is null, and approximate if it contains small residual values. In the presence of non-negligible values, further processing, such as filtering, must be introduced.

An important special case is the restoration of a block of pixels in an image from the bordering pixels.

The compression of fixed images or video streams is often based on decomposition into blocks – for example, blocks of $8 \times 8$ or $16 \times 16$ pixels, in combination with the discrete cosine transform (DCT-2D). During transmission or manipulation, some blocks may be lost or damaged and some restoration action is needed to minimize the visual impact. Then, the above technique can apply.

According to the definition given in Section 3.3.4, without the scale factors, considering a $M \times N$ matrix of real data $x(i,j)$, the DCT delivers a matrix with the same dimensions:

$$y(k,l) = \sum_{i=0}^{M+1}\sum_{j=0}^{N+1} \cos\left(\frac{\pi(2i+1)k}{2M}\right)\cos\left(\frac{\pi(2j+1)l}{2N}\right) x(i,j) \tag{9.66}$$

The impact of the discontinuities at the beginning and end of the sequence, mentioned in Section 2.1, is avoided by this transform. A property of most of these images is that they have few high-frequency components. This property can be exploited in the restoration of unknown blocks from their surrounding pixels.

The unknown block is an $M \times N$ rectangle to be determined from the $2(M+N+2)$ surrounding pixels, as illustrated in Figure 9.14a. In the transformed matrix, the $M \times N$ high-frequency elements are set to zero, as indicated in Figure 9.14b.

Once the matrices have been scanned and data are available in vectors, expressions (9.64) and (9.65) apply. Elements of the matrices $\boldsymbol{T}_{ij}$ $(\boldsymbol{1 \le i, j \le 2})$ are the elements of the DCT. The term $\boldsymbol{G} = -\boldsymbol{T}_{22}^{-1}\boldsymbol{T}_{21}$ is called the interpolation mask.

The approach yields particularly simple results because it turns out that every unknown pixel $X$ is obtained from eight neighboring pixels, as indicated in Figure 9.15, by

$$\begin{aligned} X &= aX_1 + (1-a)X_3 + bX_4 + (1-b)X_2 - abX_5 \\ &-a(1-b)X_6 - (1-a)(1-b)X_7 - (1-a)bX_8 \end{aligned} \tag{9.67}$$

The parameters $a$ and $b$ depend on the block dimensions $M$ and $N$. Table 9.1 lists a set of numerical values.

For the details of the calculations leading to expression (9.67), and for examples with test images, see reference [10].

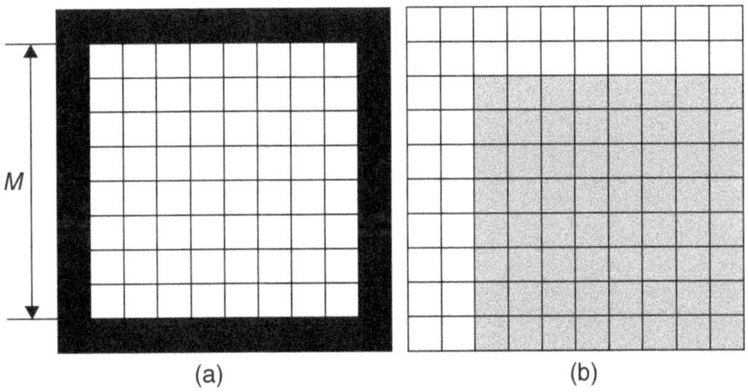


**Figure 9.14** (a) Unknown and known pixels; (b) DCT coefficients.

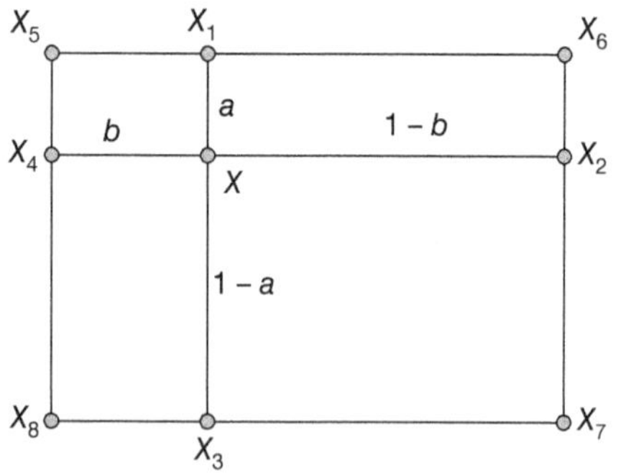


**Figure 9.15** Determination of an unknown pixel from eight neighboring pixels.

**Table 9.1** Parameters of the interpolation mask.

| *M* | | | *a, b* | | |
|---|---|---|---|---|---|
| 2 | 0.7071 | | | | |
| 3 | 0.8090 | 0.5 | | | |
| 4 | 0.8660 | 0.6340 | | | |
| 5 | 0.9010 | 0.7225 | 0.5 | | |
| 6 | 0.9239 | 0.7832 | 0.5995 | | |
| 7 | 0.9397 | 0.8264 | 0.6736 | 0.5 | |
| 8 | 0.9511 | 0.8580 | 0.7298 | 0.5792 | |
| 9 | 0.9595 | 0.8818 | 0.7731 | 0.6423 | 0.5 |
| 10 | 0.9659 | 0.9001 | 0.8070 | 0.6930 | 0.5658 |
| 11 | 0.9709 | 0.9145 | 0.8340 | 0.7341 | 0.6205 |

## 9.12 Conclusion

The transformation of real signals into complex signals is a filtering operation, and the quadrature filter involved can have any frequency response. In particular, when its response is proportional to frequency, it is called a differentiator.

In practice, real–complex conversions are efficiently implemented through interpolation, with a half-band filter. The mask of the filter is defined from the performance objectives in terms of frequency distortion and image–band residuals.

Interpolation is a fundamental operation linked to sampling. In theory, it is defined by the sampling formula which corresponds to a linear-phase infinite impulse response filter. In practice, a linear-phase FIR filter is used, in order to keep known samples untouched. A particularly important case is Lagrange interpolation, which corresponds to a max-flat filter, whose frequency response derivatives vanish at the origin.

Block interpolation, as in image processing, can use filter responses that do not preserve known samples, such as spline functions, which constitute another approximation of the ideal interpolator.

Lost samples or lost blocks of samples in 1D and 2D signal sequences can be efficiently restored by interpolation techniques.

## Exercises

**9.1** Calculate the Fourier transform $X(f)$ of the real causal set $x(n)$ such that:

$$x(n) = 0 \quad \text{for} \quad n < 0$$
$$\qquad a^n \quad \text{for} \quad n \geq 0$$
$$n \geq 0$$

with:

$$|a| < 1$$

Decompose $X(f)$ into real and imaginary parts.

**9.2** Show that the function $X(Z)$ such that:

$$X(Z) = \frac{1}{1 - aZ^{-1}}$$

can be obtained from its real part $X_R(\omega)$ on the unit circle, given by:

$$X_R(\omega) = \frac{1 - a\cos\omega}{1 - 2a\cos\omega + a^2} \quad \text{with } |a| < 1$$

**9.3** Starting from a real signal represented by the set $x(n)$, a complex signal is formed whose real and imaginary parts are given by:

$$x_R(n) = x(n)\cos[2\pi(n/4)]$$
$$x_I(n) = x(n)\sin[2\pi(n/4)]$$

What comments can be made about the sets $x_R(n)$ and $x_I(n)$?
Is the signal obtained an analytic signal?
A half-band filter is applied to each of the sets $x_R(n)$ and $x_I(n)$ and the complex signal filtered in this way is multiplied by the complex set $e^{-j2\pi n/4}$. What operation has been carried out on the real signal $x(n)$? Perform this set of operations on the signal $x(n) = \cos(\pi n/5)$.

**9.4** Study the effect of coefficient wordlength limitation on the quadrature FIR filter. Following the procedure described in Sections 5.7 and 5.10 for linear phase filters, find an expression for the estimate of the coefficient wordlength as a function of the quadrature filter parameters – that is, ripple and transition band.

**9.5** Give a simplified expression for the order of the 90° IIR phase shifter. Study the effect of the coefficient wordlength limitation and find a formula for estimating the coefficient wordlength as a function of the parameters. Check the results in the example given in Section 9.4.

**9.6** Consider a function $H(Z)$ defined by:

$$H(Z) = [(1 - Z_0 Z^{-1})(1 - Z_0 Z^{-1})]^2$$

with

$$Z_0 = 0.5(1 + j)$$

This is a minimum-phase function. Give the expression for the linear-phase function and the maximum-phase function which have the same amplitude characteristic. Compare the impulse responses.

**9.7** A real signal with sampling frequency $2f_s$ is converted into a complex signal with sampling frequency $f_s$ with the following half-band filter:

$$H(Z) = -0.0506 + 0.2954\,Z^{-2} + 0.5\,Z^{-3} + 0.2951\,Z^{-4} - 0.0506\,Z^{-6}$$

Calculate the filter response at frequencies 0 and $f_s/8$. Derive the ripple and transition bandwidth.
The filter is included in an IQ modulator fed with the sequence $x(n) = \sin\left(\frac{n\pi}{4}\right)$. Give the expression of the complex output $y(n)$ and show that it has frequency components at frequencies $f_s/4$ and $3f_s/4$. Give the amplitudes of these components.

**9.8** In a frequency modulation receiver, the discriminator (amplitude/frequency converter) is a five-coefficient differentiator, and the coefficient vector is:

$$h = [-0.1766 \quad 0.9696 \quad 0 - 0.9696 \quad 0.1766]$$

Give the filter response. What is the pass-band width and the ripple of this filter?

**9.9** Quadrature filter based on DFT.
The complex sequence associated with the $N$ samples of the real signal $x_r(n)$ is to be determined. The DFT of the sequence, $X(k)$, is calculated, the values obtained are multiplied by 2, and $X(k)$ is set to zero for $\frac{N}{2} \le k \le N-1$. The inverse DFT yields the desired complex sequence $x_c(n)$.
Give the expression of the impulse response $h(n)$ of the corresponding filter (refer to Section 2.4.1). Compare with the quadrature filter coefficients given by expression (9.24) and explain the differences. A numerical example starts from the sequence:

$$x_r(n) = \cos\left(2\pi(n-1)\frac{3{,}5}{16}\right)\ ;0 \le n \le 15$$

which leads to the following table:

| $n$ | 5 | 6 | 7 | 8 | 9 | 10 | 11 |
|---|---|---|---|---|---|---|---|
| $\boldsymbol{x_r(n)}$ | 0.8315 | −0.3827 | −0.9808 | 0 | 0.9808 | 0.3827 | −0.8315 |
| $\boldsymbol{x_c(n)}$ | 0.9565 | −0.3827 | −0.8558 | 0 | 1.1058 | 0.3827 | 0.7065 |
| | +j0.6040 | +j0.9158 | +j0.1960 | −j1.0047 | −j0.2457 | +j0.9158 | +j0.4371 |

Using the first values of $h(n)$, justify the values $\boldsymbol{x_c(n)}$ in the vicinity of $n = 8$.
Comment on the accuracy of the approach for quadrature filter design.

**9.10** A sequence is made of the following terms:

$$x(n) = e^{j\left(\frac{n\pi}{4}+\frac{\pi}{3}\right)} + 0.5\,e^{jn3\pi/4}; \qquad 0 \le n \le 7$$

Verify the DFT values: $\boldsymbol{y(6) = y(7) = 0}$.
Assuming $\boldsymbol{x(4)}$ and $\boldsymbol{x(5)}$ are unknown, find them using (9.65). Give the expression of the matrix $\boldsymbol{T_{22}}$ and apply (9.65).

## References


1 E. A. Guillemin, *Theory of Linear Physical Systems*. John Wiley, New York, 1963.
2 B. Picinbono, *Principles of Signals and Systems*. Artech House Inc., London, 1988.
3 A. Oppenheim and R. Schafer, *Digital Signal Processing, Chapter 7*. Prentice-Hall, Englewood Cliffs NJ, 1974.
4 B. Gold, A. Oppenheim and C. Rader, Theory and implementation of the discrete Hilbert transform. *Proc. of Symp. Computer Processing in Communications*, Vol. 19. Polytechnic Press, New York, 1970.
5 B. Gold and C. Rader, *Digital Processing of Signals, Chapter 3*. McGraw-Hill, New York, 1969.
6 G. Mollova, Compact formulas for least squares design of digital differentiators. *Electronics Letters*, 35(20), 1695–97, 1999.
7 T. I. Laakso, V. Valimaki, M. Karjalainen and U. K. Laine, Splitting the unit delay—tools for fractional delay filter design. *IEEE Signal Processing Magazine*, 13(1), 30–60, 1996.
8 J. J. Fuchs, B. Delyon, Minimum L1-norm reconstruction function for 0versampled signals: application to time delay estimation. *IEEE Transactions*, 46, 1666–73, 2000.
9 M. Unser, Splines: a perfect fit for signal and image processing. *IEEE Signal Processing Magazine*, 16(6), 22–38, 1999.
10 Z. Al Kachouh, M. Bellanger, Efficient restoration technique for missing blocks in images. *IEEE Transactions on Circuits and Systems for Video Technology*, 13(12), 1182–86, 2003.

# 10

# Multirate Filtering

Multirate filtering is a technique for reducing the calculation rate needed in digital filters and, in particular, the number of multiplications to be performed per second. As will be shown later, this parameter is generally regarded as a reflection of the complexity of the system.

In a filter, the number of multiplications $M_R$ to be performed per second is given by:

$$M_R = Kf_s$$

where $f_s$ is the frequency at which the calculations are made. The parameter $f_s$ generally corresponds to the sampling frequency of the signal represented by the numbers to be processed. The factor $K$ depends on the type of filter and on its performance.

In reducing the value of $M_R$, the factor $K$ can be influenced by choosing the most appropriate type and structure of a filter and by optimizing the order of that filter to suit the constraints and required characteristics. Also, $f_s$ can be influenced by changes in the sampling frequency during processing. In many practical cases, the advantages thus obtained are considerable.

The sampling frequency for a real signal must be more than twice its bandwidth, which can vary during processing. For example, a filtering operation eliminates undesirable components, so the useful bandwidth is reduced. Once the useful bandwidth has been decreased, the sampling frequency of the signal can itself be reduced. As a result, the sampling frequency can be adapted to the bandwidth of the signal at each stage of processing, so as to maximize the filter's computation speed. Before studying the development and implementation of this fundamental principle, it is appropriate first to analyze the effect of a change in the sampling frequency on the signal and its spectrum.

## 10.1 Decimation and $Z$-Transform

Sample rate reduction, or decimation, modifies the $Z$-transform of the signal, as does sample rate increase, or interpolation. The reduction by a factor of two is illustrated in Figure 10.1, which shows the decimation symbol and the signal spectrum. The suppression of every other sample is carried out by adding the initial sequence in which the sign of every other sample has been inverted to the non-modified initial sequence. The impact of such a sign inversion is a spectral shift of about half the sampling frequency. Once the spectra are added, the periodicity in the frequency domain is divided by two.

When the sequence $s(n)$ is split into two interleaved sequences, the $Z$-transforms of these sequences take on the following form:

$$S_0(Z^2) = [S(Z) + S(-Z)]/2; Z^{-1}S_1(Z^2) = [S(Z) - S(-Z)]/2$$

*Digital Signal Processing: Theory and Practice*, Tenth Edition. Maurice Bellanger.

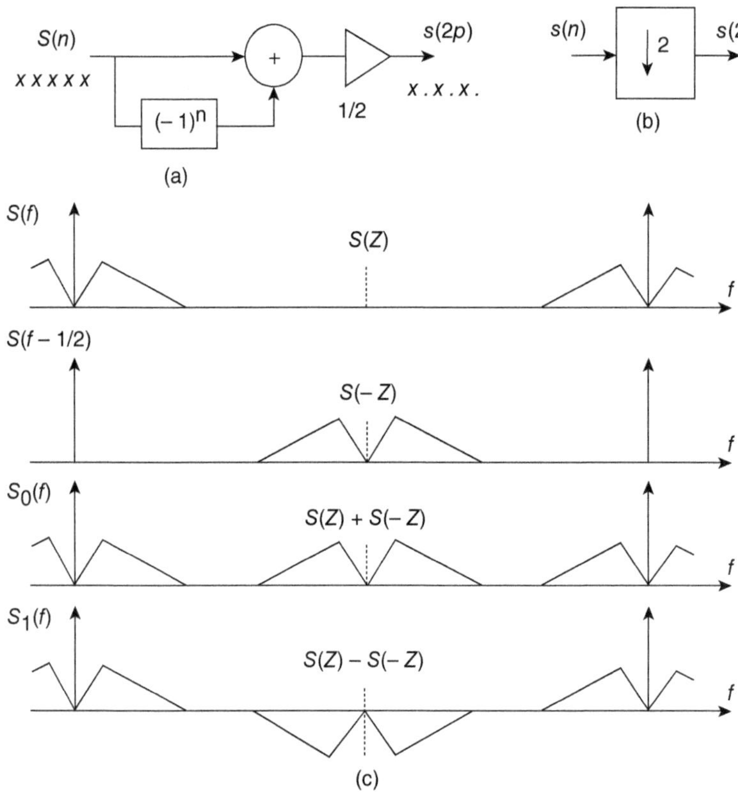


**Figure 10.1** Downsampling by 2. (a) Decimation operations; (b) decimation symbol; (c) spectra and Z-transforms.

The term $Z^2$ characterizes the doubling of the sampling period, while the term $Z^{-1}$ reflects the interleaving operation. The $Z$-transform of $s(n)$ is recovered as:

$$S(Z) = S_0(Z^2) + Z^{-1}\, S_1(Z^2)$$

The decomposition and reconstruction formulas, expressed above for factor two, can be generalized to any whole number $M$, and this generalization is presented below with the help of Fourier transforms.

Given the signal $s(t)$ whose spectrum $S(f)$ has no component with frequency higher than $f_m$, and assuming that the signal is sampled with period $T$ such that:

$$1/MT > 2f_m$$

where $M$ is a whole number, let us examine the relationship between the Fourier transforms $S_i(f)$ of the sets:

$$s\left[\left(n + \frac{i}{M}\right) MT\right]; \quad i = 0,1,2,\ldots, M-1$$

From the results in Section 1.2, the Fourier transform of the distribution $u_0(t)$,

$$u_0(t) = \sum_{n=-\infty}^{\infty} \delta(t - nMT)$$

is the distribution $U_0(f)$ given by:

$$U_0(f) = \sum_{n=-\infty}^{\infty} e^{-j2\pi fnMT} = \frac{1}{MT}\sum_{n=-\infty}^{\infty} \delta\left(f - \frac{n}{MT}\right)$$

The Fourier transform of the distribution $u_i(t)$,

$$u_i(t) = \sum_{n=-\infty}^{\infty} \delta\left[t - \left(n + \frac{i}{M}\right)MT\right]; \quad i = 0,1,\ldots,M-1 \tag{10.1}$$

is the distribution $U_i(f)$ given by:

$$U_i(f) = \sum_{n=-\infty}^{\infty} e^{-j2\pi f(n+i/M)MT} = \frac{1}{MT}e^{-j2\pi fiT}\sum_{n=-\infty}^{\infty} \delta\left(f - \frac{n}{MT}\right)$$

or,

$$U_i(f) = \frac{1}{MT}\sum_{n=-\infty}^{\infty} e^{-j2\pi(in/M)}\delta\left(f - \frac{n}{MT}\right) \tag{10.2}$$

As $S_i(f)$ ($i = 0, 1, \ldots, M-1$) is the convolution product of $S(f)$ with the distribution $U_i(f)$, we have:

$$S_i(f) = \frac{1}{MT}\sum_{n=-\infty}^{\infty} e^{-j2\pi(in/M)}S\left(f - \frac{n}{MT}\right) \tag{10.3}$$

Let us calculate the spectrum $S^M(f)$, where:

$$S^M(f) = \sum_{i=0}^{M-1} S_i(f) = \frac{1}{MT}\sum_{n=-\infty}^{\infty} S\left(f - \frac{n}{MT}\right)\sum_{i=0}^{M-1} e^{-j2\pi(in/M)}$$

As the second summation cancels out except for values of $n$ which are multiples of $M$, this becomes:

$$S^M(f) = \frac{1}{T}\sum_{n=-\infty}^{\infty} S\left(f - \frac{n}{T}\right) \tag{10.4}$$

which is the spectrum corresponding to a signal sampled at frequency $1/T$.

The $S_i(f)$ terms can also be expressed as a function of $S^M(f)$, and, in equation (10.3), the summation can be decomposed as follows:

$$S_i(f) = (1/MT)\sum_{n=-\infty}^{\infty}\sum_{m=0}^{M-1} S\left[f - \left(n + \frac{m}{M}\right)\frac{1}{T}\right] e^{-j2\pi(im/M)}$$

or:

$$S_i(f) = (1/M)\sum_{m=0}^{M-1} e^{-j2\pi(im/M)}\frac{1}{T}\sum_{n=-\infty}^{\infty} S\left[\left(f - \frac{m}{MT}\right) - \frac{n}{T}\right]$$

and finally:

$$S_i(f) = (1/M)\sum_{m=0}^{M-1} e^{-j2\pi(im/M)}S^M\left(f - \frac{m}{MT}\right) \tag{10.5}$$

Figure 10.2 illustrates the case for $M = 4$. The spectra $S_i(f)$ correspond to interleaved samples at frequency $1/MT$ and the spectrum $S^M(f)$ corresponds to a sample at frequency $1/T$. A change in the sampling frequency will exchange these spectra.

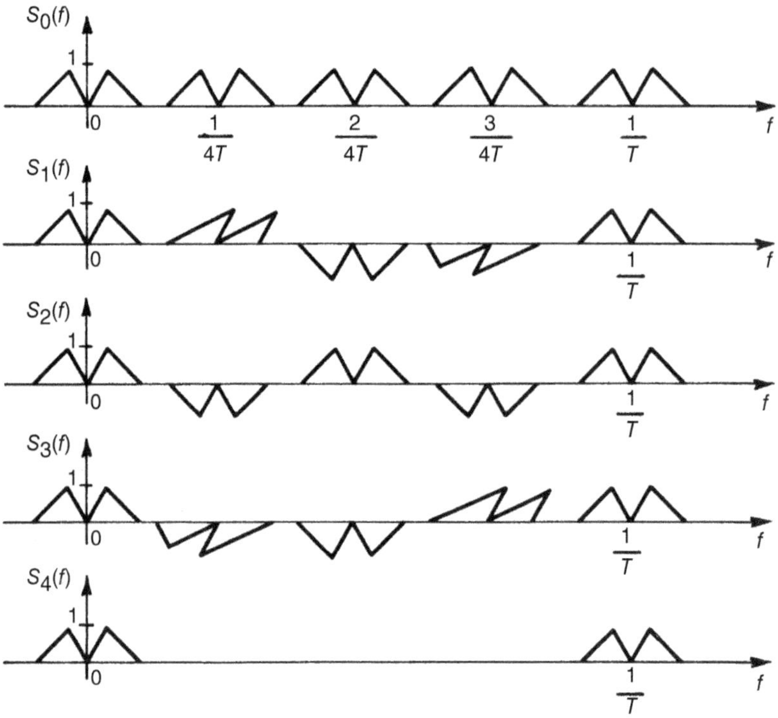


**Figure 10.2** Spectra obtained by interleaved samples.

It is interesting to note, in Figure 10.2, that retarding the set of sampling pulses causes phase rotations through multiples of $2\pi/M$ for image bands around multiples of the sampling frequency $1/MT$.

Addition of all the sets of retarded pulses causes a cancellation of the image bands except around frequencies which are multiples of $1/T$, which becomes the new sampling frequency. This is an application of the linearity properties of the Fourier transform.

It is also useful to establish relations between the $Z$-transfer functions of the sequences involved. Let $S(Z)$ be the $Z$-transform of the set $s(nT)$. By definition:

$$S(Z) = \sum_{n=-\infty}^{\infty} s(nT)Z^{-n} \tag{10.6}$$

The spectrum $S^M(f)$ of the signal sampled with period $T$ is obtained by replacing $Z$ with $(e^{j2\pi fT})$. Then:

$$S^M(f) = S(ej2\pi fT)$$

Decomposing the summation in $S(Z)$ leads to:

$$S(Z) = \sum_{n=-\infty}^{\infty} \sum_{i=0}^{M-1} S(nMT + iT)Z^{-(nM+i)}$$

or:

$$S(Z) = \sum_{i=0}^{M-1} Z^{-i} \sum_{n=-\infty}^{\infty} S(nMT + iT)Z^{-nM}$$

Defining $S_i(Z^M)$ by:

$$S_i(Z^M) = \sum_{n=-\infty}^{\infty} S(nMT + iT)Z^{-nM}$$

we obtain:

$$S(Z) = \sum_{i=0}^{M-1} S_i(Z^M)Z^{-i} \tag{10.7}$$

The terms $s_i(Z^M)$ are $Z$-transforms of the sets $s[(n + i/M)MT]$ for $i = 0, 1, \ldots, M - 1$. The factor $Z^{-i}$ reflects the interleaving of these sets.

Now, it is necessary to express $S_i(Z^M)$ as a function of $S(Z)$. Substituting $Ze^{-j2\pi m/M}$ for $Z$ in equation (10.7), we find:

$$S\left(Ze^{-\frac{j2\pi m}{M}}\right) = \sum_{i=0}^{M-1} e^{j2\pi mi/M}\, Z^{-i}\, S_i(Z^M)$$

and, in matrix form:

$$\begin{bmatrix} S(Z) \\ S\left(Ze^{-\frac{j2\pi}{M}}\right) \\ \cdot \\ \cdot \\ \cdot \\ S(Ze^{-j2\pi(M-1)M}) \end{bmatrix} = T_N^{-1} \begin{bmatrix} S_0(Z^M) \\ Z^{-1}S_1(Z^M) \\ \cdot \\ \cdot \\ \cdot \\ Z^{-(M-1)}S_{M-1}(Z^M) \end{bmatrix}$$

where $T_N^{-1}$ is the matrix of the inverse DFT. Multiplying the two sides of this equation by $T_N$, we get:

$$Z^{-i}S_i(Z^M) = (1/M) \sum_{m=0}^{M-1} e^{-j2\pi(im/M)}\, S(Ze^{-j2\pi(m/M)});\ 0 \le i \le M - 1 \tag{10.8a}$$

which corresponds to equation (10.5) for frequency responses.

Setting $W = e^{-j2\pi/M}$, as we did in Chapter 2, we find:

$$S_i(Z^M) = \frac{1}{M} \sum_{m=0}^{M-1} W^{im} Z^i (S(ZW^m)) \tag{10.8b}$$

Equations (10.7) and (10.8) are fundamental for multirate filtering.

The results obtained, and particularly equations (10.3) and (10.4), are valid for signals $s(t)$ whose spectrum is not limited to the frequency $1/2MT$, but spectrum aliasing then occurs.

## 10.2 Decomposition of a Low-Pass FIR Filter

Multirate filtering will be introduced first for FIR filters, where it appears naturally. Consider a low-pass FIR filter which eliminates components with a frequency greater than or equal to the frequency $f_c$ in a signal sampled at frequency $f_s$. The filtered signal only requires a sampling frequency equal to $2f_c$, and in fact, it is sufficient to provide the output numbers at this frequency.

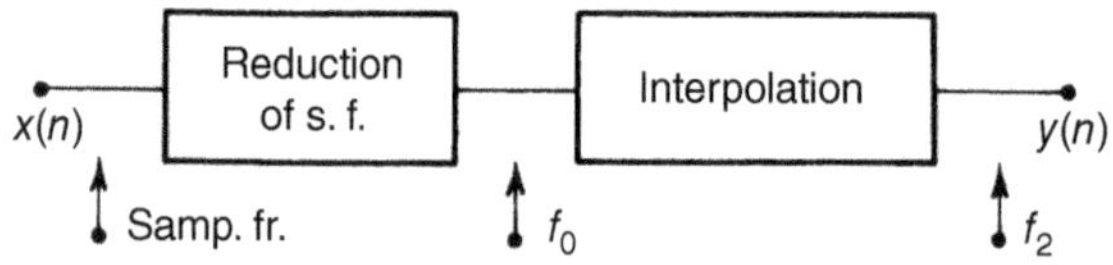


**Figure 10.3** Filter with sample rate reduction and increase.

In an FIR filter of order $N$, the relation which determines the number of the output set $y(n)$ from the set of input numbers $x(n)$ is written as:

$$y(n) = \sum_{i=0}^{N-1} a_i x(n-i) \tag{10.9}$$

Each output number $y(n)$ is calculated from a set of $N$ input numbers by weighted summation with the coefficients $a_i$ $(i = 0, 1, \ldots, N-1)$. Under these conditions, the input and output rates are independent and a decrease in the output sampling rate by a factor $k = f_s/2f_c$, assumed to be an integer, results in a reduction in the computation rate by the same factor.

The same reasoning applies to raising the sampling rate, or interpolation. In this case, the output rate is greater than the input rate. To show the savings in computation, it is sufficient simply to regard the rates as being equal by incorporating a suitable amount of null data into the input set.

The independence of the output from the input in FIR filters can be exploited in narrow band-pass filters, even if the input and output rates are identical, by dividing the filtering operation into two phases [1]:

(1) Reducing the sampling frequency from the value $f_s$ to an intermediate value $f_0$ such that:

$$f_0 \geq 2f_c$$

(2) Raising the sampling frequency or interpolating from $f_0$ to $f_s$.

Figure 10.3 illustrates this decomposition. If the two operations are carried out with similar filters of order $N$, the number of multiplications $M_D$ to be performed per second is given by:

$$M_D = Nf_0 \times 2 \tag{10.10}$$

This value is to be compared with direct realization by a single filter, which results in the value $M_R$ such that:

$$M_R = Nf_s \tag{10.11}$$

Consequently, decomposition is advantageous as long as $k = f_s/2f_c$ is greater than 2 – that is, as long as:

$$f_c < \frac{f_s}{4}$$

This approach, therefore, appears to be well suited to narrow band-pass filters.

It should be noted, however, that the filtering function obtained in the two cases is not exactly the same, and that distortions are introduced by the decomposition, as shown in Figure 10.4. In effect, the intermediate subsampling at frequency $f_0 \geq 2f_c$ has three consequences:

(1) Aliasing of residual components with frequencies greater than $f_0/2$ into the band below $f_0/2$. Harmonic distortion results. Its power $B_R$ depends on the attenuation of the sampling rate reduction filter and is calculated from its transfer function $H(f)$ by using the results given in the previous chapters.

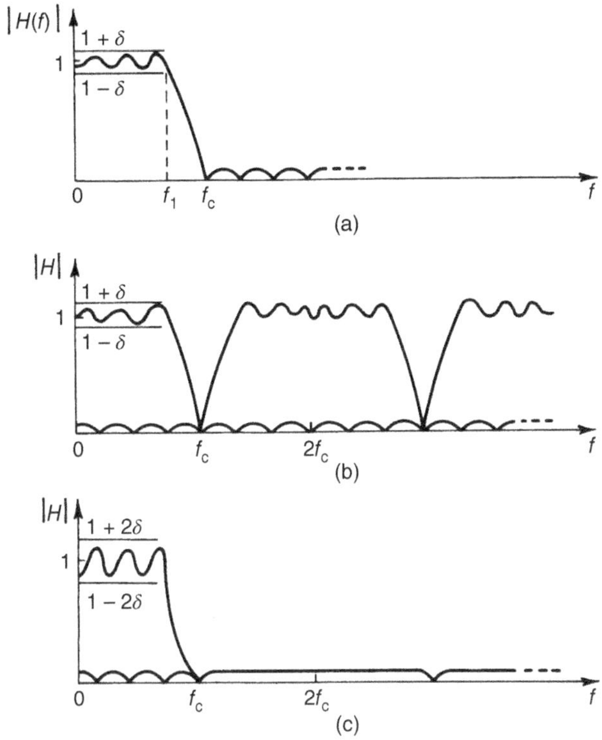


**Figure 10.4** Frequency response of multirate filter, (a) direct filter response; (b) sample rate reduction filter response; (c) multirate filter response.

For example, if the input signal has a uniform spectral distribution and unit power, the total power $B_T$ of the aliased signal is:

$$B_T = \frac{1}{f_s} \int_{f_0/2}^{f_s - f_0/2} |H(f)|^2 df \tag{10.12}$$

If $f_1$ is the filter pass-band edge, an upper bound on $B_T$ is provided by:

$$B_T < \sum_{i=0}^{N-1} |a_i|^2 - \frac{2f_1}{f_s}$$

The distortion can be assumed to have a uniform spectral distribution and only the power in the pass band is considered. In this case, we obtain:

$$B_R < \frac{2f_1}{f_0} \left[ \sum_{i=0}^{N-1} |a_i|^2 - \frac{2f_1}{f_2} \right] \tag{10.13}$$

Allowance must be made for this degradation when calculating the sampling rate reduction filter [2].

(2) The periodicity in frequency of the response of the sampling rate reduction filter, with period $f_0$, introduces a distortion whose power $B_i$ is a function of the attenuation of the interpolation filter.

If this filter is the same as the sampling rate reduction filter, with the same assumptions, we obtain:

$$B_i = (1/f_s)\int_{f_0/2}^{f_s-f_0/2} |H(f)|^2 df$$

This distortion outside the pass band can be troublesome when other signals need to be added to the filtered signal.

(3) Cascading two filters increases the ripple in the pass band. For example, the ripple is doubled if identical filters are used in both operations.

Finally, the subunits of the multirate filter should be designed so that the complete system satisfies the overall specifications imposed on the filter [3].

The circuit in Figure 10.3 is simplified if the sampling frequencies of the signal before and after filtering can be different. The principle can also be applied to high-pass and band-pass filters, provided, for example, modulation and demodulation stages are introduced.

The principle of decomposition can be extended to the sampling rate reduction subunit and to the interpolation subunit, which introduces a further advantage. The FIR half-band filter is a particularly efficient element for implementing these subunits.

## 10.3 Half-Band FIR Filters

The half-band FIR filter was introduced in Section 9.3. This is a linear phase filter, and the frequency response $H(f)$ takes the value 1/2 at the frequency $f_s/4$.

It is antisymmetric about this point – that is, the function $H(f)$ satisfies the equations:

$$H(f_s/4) = 0.5; \quad H\left(\frac{f_s}{4}+f\right) = 1 - H\left(\frac{f_s}{4}-f\right) \tag{10.14}$$

For a number of coefficients $N = 4M + 1$,

$$H(f) = e^{-j2\pi 2Mf} \times \frac{1}{2}\left[1 + 2\sum_{i=1}^{M} h_{2i-1}\cos[2\pi(2i-1)f]\right] \tag{10.15}$$

The coefficients $h_n$ are zero for even value of $n$ except for $h_0$. Figure 10.5 illustrates the properties of this filter. The specification is defined by the ripple ($\delta$) in the pass and stop bands, and by the width $\Delta f$ of the transition band. Given these parameters, the formulae in Section 5.7 can be used to estimate the filter parameters. The filter order is:

$$\left[N \simeq \frac{2}{3}\log(1/10\delta^2)f_s/\Delta f\right]$$

By taking the attenuation $A_f$ as:

$$A_f = 10\log(1/\delta^2)$$

and bearing in mind the particular significance of the frequency $f_s/4$ in this type of filter, one can write:

$$[M \simeq (2/3)(A_f/10 - 1)f_s/4\Delta f]$$

Hence, the relationship between the attenuation (in decibels) and the transition band, for a given number of coefficients, can be expressed simply as:

$$A_f = 10 + 15M(4\Delta f/f_s) \tag{10.16}$$

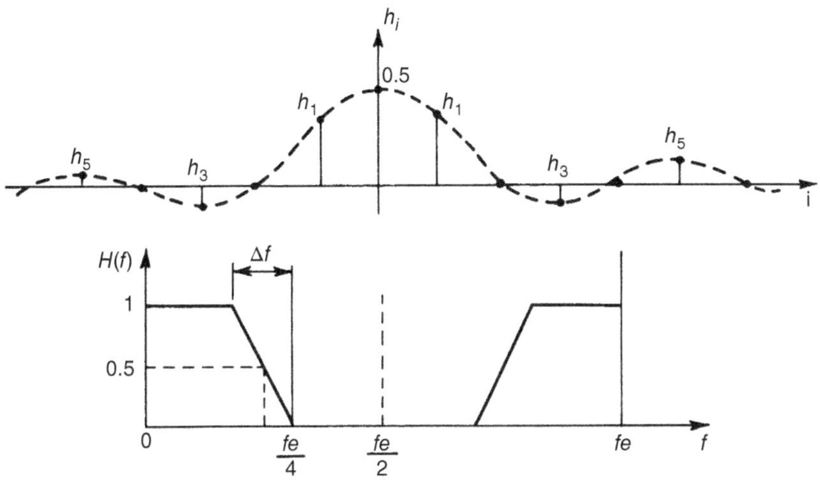


**Figure 10.5** Half-band FIR filter.

This approximation is valid when $M$ exceeds a few units. The coefficients are calculated using the general program for FIR filters, with appropriate data. The relationship between the filter input and output sets is:

$$y(n) = \frac{1}{2}\left[x(n-2M) + \sum_{i=1}^{M} h_{2i-1}[x(n-2M+2i-1) + x(n-2M-2i+1)]\right] \tag{10.17}$$

and $M$ multiplications have to be performed for each element of the output set $y(n)$. It should be noted that these operations are performed only on elements of the input set with an odd index. It follows that if such a filter is used to reduce the sampling frequency from $f_s$ to $f_0 = f_s/2$, the number of multiplications to be performed per second is $Mf_0$. The same is true if the sampling frequency $f_s$ is increased from $f_0$ to $2f_0$, which is achieved simply by calculating a sample between two samples of the input set.

**Table 10.1** Half-band FIR filter.

| Filter | $h_0$ | $h_1$ | $h_3$ | $h_5$ | $h_7$ | $h_9$ |
|---|---|---|---|---|---|---|
| F 1 | 1 | 1 | | | | |
| F 2 | 2 | 1 | | | | |
| F 3 | 16 | 9 | 1 | | | |
| F 4 | 32 | 19 | −3 | | | |
| F 5 | 256 | 180 | −25 | 3 | | |
| F 6 | 346 | 208 | −44 | 9 | | |
| F 7 | 512 | 302 | −53 | 7 | | |
| F 8 | 802 | 490 | −116 | 33 | −6 | |
| F 9 | 8192 | 5042 | −1277 | 429 | −116 | 18 |

Finally, the number of multiplications to be performed per second in a half-band filter with a change of sampling frequency is:

$$M_R = \left[\frac{2}{2}\log\left(\frac{1}{10\delta^2}\right)\frac{f_s}{\Delta f}\frac{1}{4}\right] f_s/2 \tag{10.18}$$

*Example:* A group of half-band filters with useful application characteristics [3] is shown in Table 10.1, which gives the quantized coefficients, the quantization step being taken as unity [4].

The frequency response can be calculated simply from equation (10.15). Filters F4, F6, F8, and F9 correspond to a unit value of the parameter $4\Delta f/f_S$, with ripples of 37, 50, 67, and 79 dB, respectively. Filters F2, F3, and F5 have monotonic responses.

The advantages of the particular filter structure described in this section can be applied to general multirate filtering.

## 10.4 Decomposition with Half-Band Filters

The properties of the elementary half-band filter can be used to produce the multirate filter design shown in Figure 10.6. The intermediate frequency $f_0$ is:

related to the sampling frequency $f_s$ by a power of two:

$$f_s = 2^P f_0 \tag{10.19}$$

The sampling frequency is reduced or increased by a cascade of $P$ half-band filters. The complete unit comprises a basic filter which operates at frequency $f_0$ and has a cascade of half-band filters on either side.

The overall low-pass filter is specified by the following parameters:

(1) Ripple in the pass band: $\delta_1$
(2) Ripple in the stop band: $\delta_2$
(3) Width of the transition band: $\Delta f$
(4) Edge of the pass band: $f_1$
(5) Edge of the stop band: $f_2$

To calculate the half-band filters, their specifications have to be defined. The ripple in the pass band is assumed to be divided between the half-band filters and the basic filter. Also, each filter has to have a ripple in the stop band smaller than $\delta_2$. As a result, for each half-band filter, the ripple $\delta_0$ is given by:

$$\delta_0 = \min\left\{\frac{\delta_1}{4P}, \delta_2\right\} \tag{10.20}$$

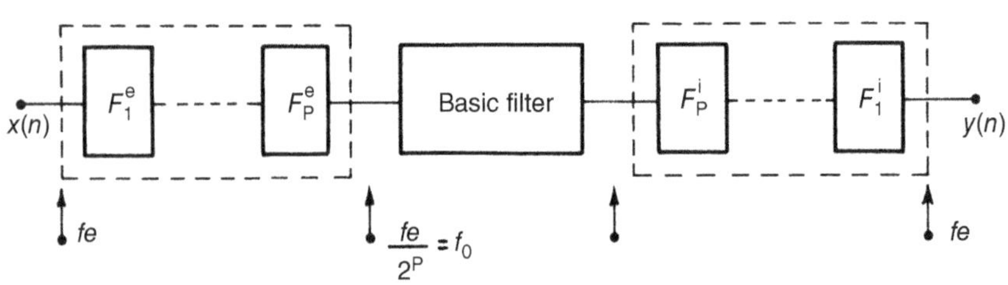


**Figure 10.6** Decomposition with half-band filters.

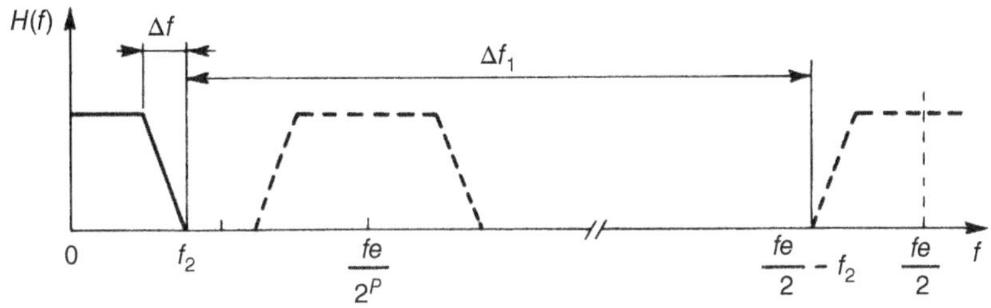


**Figure 10.7** Transition band of first half-band filter.

The first half-band filter $F_1^e$ of the cascade can be determined if its transition band $\Delta f_1$ is fixed (Figure 10.7). To fix $\Delta f_1$, it must be taken that the role of the first filter is to eliminate those components of the signal which can be folded in the useful band after having the sampling frequency. Thus:

$$\Delta f_1 = \frac{f_s}{2} - 2f_2 \tag{10.21}$$

From equation (10.18), the number of multiplications $M_C$ to be performed in the first filter is:

$$M_{C_1} = \frac{2}{3}\log\left(\frac{1}{10\delta_0^2}\right)\left(\frac{f_s}{\Delta f_1}\right)\frac{1}{4} \tag{10.22}$$

The same approach can be taken for the other filters in the cascade, and it can be shown that the total number of multiplications is estimated as:

$$M_c \approx \frac{1}{3}\log\left(\frac{1}{10\delta_0^2}\right)f_s \tag{10.23}$$

To estimate the volume of calculations to be performed per second in the complete filter, the order $N$ of the basic filter must be determined:

$$\left[N \approx D\left(\frac{\delta_1}{2}, \delta_2\right)\frac{1}{\Delta f}\frac{f_s}{2^P}\right] \tag{10.24}$$

with:

$$D\left(\frac{\delta_1}{2}, \delta_2\right) = \frac{2}{3}\log\left(\frac{1}{5\delta_1\delta_2}\right)$$

The values of the parameters showing the complexity of the complete filter are, finally:

$$M_R = f_s\left[D(\delta_0) + \frac{1}{2^{2P+1}}\frac{f_s}{\Delta f}D\left(\frac{\delta_1}{2}, \delta_2\right)\right] \tag{10.25}$$

*Example*: Consider a narrow low-pass filter defined by the following:

$$f_s = 1; \quad f_2 = 0.05; \quad \Delta f = 0.025; \quad \delta_1 = 0.01; \quad \delta_2 = 0.001$$

The parameters have the values:

$$P = 3; \quad \delta_0 = \min\{0.01/12; 0.001\} = 0.00083; \quad D(\delta_0) = 3.3$$
$$D(\delta_1/2, \delta_2) = 2.76; k = 0.8; \quad S(k) = 1.6$$

Thus: $M_R = 6.2$
A direct realization results in a filter of order $N = 110$, which corresponds to the values: $M_R = 55$

From the point of view of generalizing to apply more broadly, it is important, in order to optimize multirate processing in more general cases than that of a single filter, to emphasize the phase-shift function.

## 10.5 Digital Filtering by Polyphase Network

The phase relations between different samples of the same signal were examined in detail in Section 10.1. The results will be used to analyze multirate filtering from this point of view [5].

Assume that the sampling frequency $f_s$ is reduced by a factor $N$. Let $X(Z)$ be the $Z$-transform of the input set $x(n)$; the Fourier transform is obtained by replacing $Z$ with $e^{j2\pi f/f_s}$, forming $\left(e^{j2\pi f/f_s}\right)$. The output set $y(Nn)$ sampled at frequency $f_s/N$ has $Y(Z^N)$, a function of $Z^N$, as its $Z$-transform. Consequently, if phase-shifting circuits are involved in this operation, their transfer function is also a function of the variable $Z^N$ and can be calculated from the overall filter function.

The structure of a phase shifter will be discussed initially for the single element making up the half-band FIR filter, which is defined by equation (10.17). This relation can be rewritten as:

$$y(n) = \frac{1}{2}\left[x(n-2M) + \sum_{i=1}^{2M} a_i x(n-2i+1)\right] \tag{10.26}$$

in which:

$$a_i = h_{(2M-2i+1)} = a_{2M-i+1} \text{ for } 1 \leq i \leq M$$

The corresponding $Z$-transfer function is:

$$H(Z) = \frac{1}{2}\left[Z^{-2M} + Z^{-1}\sum_{i=0}^{2M-1} a_{i+1} Z^{-2i}\right] \tag{10.27}$$

or alternatively:

$$H(Z) = \frac{1}{2}[H_0(Z^2) + Z^{-1}H_1(Z^2)] \tag{10.28}$$

The corresponding frequency response is:

$$H(f) = \frac{1}{2}\left[e^{-j2\pi(f/f_s)2M} + e^{-j2\pi(f/f_s)}H_1(f)\right] \tag{10.29}$$

The function $H_0(f)$ corresponds to a delay, and so is the characteristic of a purely linear phase shifter.

Because the coefficients are symmetrical, $H_1(f)$ is also a linear phase function. Since this part of the filter is operating at the frequency $f_s/2$, $H_1(f)$ displays the periodicity $f_s/2$. As the number of the coefficients is even, we can use the results from Section 5.2, and write:

$$H_1(f) = \exp\left(-j2\pi f\left(M-\frac{1}{2}\right)\frac{2}{f_s}\right)|H_1(f)| \tag{10.30}$$

or:

$$H_1(f) = e^{-j2\pi(f/f_s)2M} e^{-j\phi(f)} |H_1(f)| \tag{10.31}$$

The function $\phi(f)$ is linear and has the periodicity $f_s/2$. Consequently, it is expressed by:

$$\phi(f) = \pi\left(\left[\frac{2f}{f_s} + \frac{1}{2}\right] - \frac{2f}{f_s}\right) \tag{10.32}$$

where [x] represents the largest whole number contained in x.

Regarding the amplitude, using equation (10.8) and with the symmetry of $H(f)$, we get:

$$e^{-\frac{j2\pi f}{f_s}} H_1(f) = H(f) - H\left(\frac{f_2}{2} - f\right) \tag{10.33}$$

Introducing the antisymmetry with respect to frequency $f_s/4$ shown in Section 10.4, we find:

$$|H_1(f)| = 2|H(f)| - 1; 0 \leq f \leq f_s/4$$

The ripple is double with respect to $H(f)$ and the function is null at $f = f_s/4$ – the frequency which corresponds to a change in phase of $\pi$.

Figure 10.8 shows the functions $|H_1(f)|$ and $\phi(f)$ which characterize the filter.

The phase $\Phi(f)$ such that:

$$\Phi(f) = \phi(f) + 2\pi(f/f_s) \tag{10.34}$$

is constant and has the value 0 or $\pi$.

Finally, the circuit whose frequency response is $e^{-j2\pi f/f_s} H_1(f)$ approximates a pure linear phase shifter in the pass and stop bands. Its number of coefficients and its complexity depend on the degree of approximation – that is, on the transition band $\Delta f$ and on the amplitude $\delta$ of the ripple. These results are similar to those of Section 9.3.

The half-band filter appears as a network with two branches, as in Figures 10.9 and 10.10. The overall response corresponds to the sum of the responses of the two branches as shown in Figure 10.11.

If the sign of function $H_1(f)$ is changed, then a high-pass filter is obtained, as shown in Figure 10.11. Thus, a pair of filters is determined by the calculations for one of them. Let $B_0(z)$ and $B_1(z)$ be the transfer functions of these filters. The corresponding system is shown in Figure 10.10.

**Figure 10.8** Function $H_1(f)$ for half-band filter.

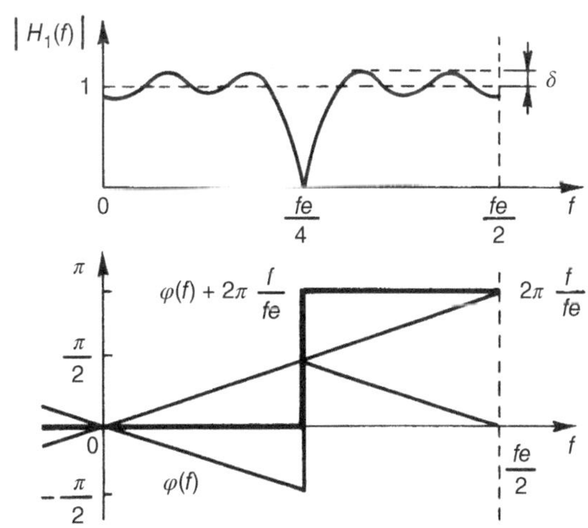


**Figure 10.9** Half-band filter with phase shifters.

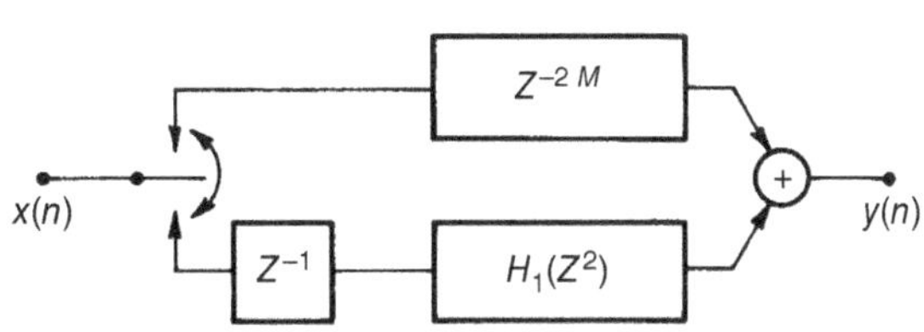

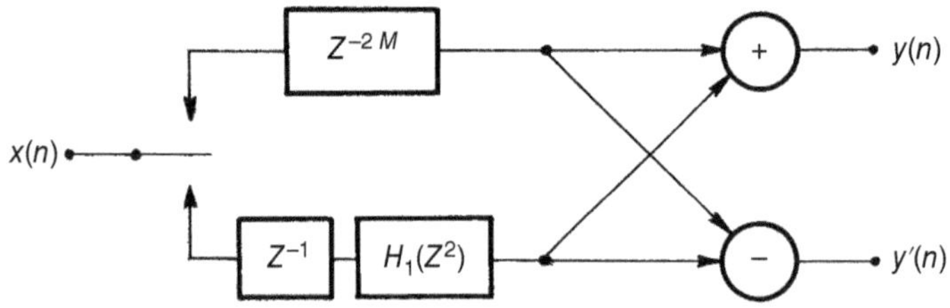


**Figure 10.10** Filter bank with two filters.

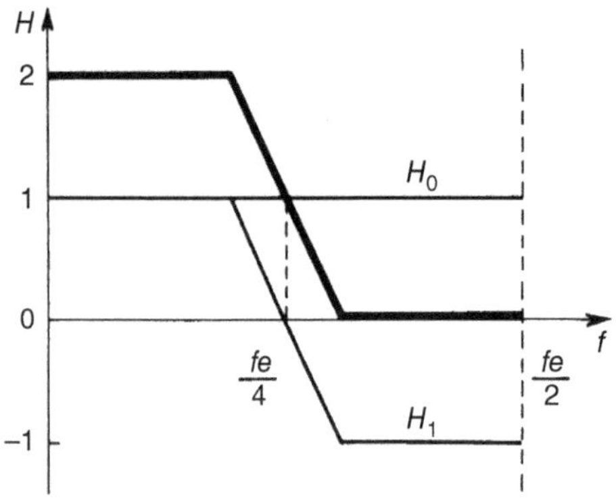


**Figure 10.11** Amplitude of half-band filter frequency response.

It is characterized by the matrix equation:

$$\begin{bmatrix} B_0(Z) \\ B_1(Z) \end{bmatrix} = \begin{bmatrix} 1 & 1 \\ 1 & -1 \end{bmatrix} \begin{bmatrix} Z^{-2M} \\ Z^{-1}H_1(Z^2) \end{bmatrix} \tag{10.35}$$

It is worth noting that the matrix of the discrete Fourier transform of order 2 appears in the expression. The generalization to a bank of $N$ filters leads to a DFT of order $N$. This is presented in Section 10.8.

It should also be noted that by changing the sign of every other coefficient in $H_1(f)$, the frequency response is displaced by $f_s/4$ and the quadrature filter from Section 9.3 is obtained.

The results obtained for the half-band filter can be generalized quite simply by using an FIR filter to reduce or increase the sampling frequency by a factor of $N$. Assume $H(Z)$ is the $Z$-transfer function of such a filter. By assuming that it has $KN$ coefficients, we can write:

$$H(Z) = \sum_{i=1}^{MN} a_i Z^{-n} = \sum_{n=0}^{N-1} Z^{-n} H_n(Z^N) \tag{10.36}$$

with:

$$H_n(Z^N) = \sum_{k=0}^{K-1} a_{kN+n} (Z^{-N})^k$$

This filter can be implemented by a network with $N$ paths which is called a polyphase network because each path has a frequency response which approximates that of a pure phase shifter. The phase shifts are constant in frequency and are whole multiples of $2\pi/N$. When there is a change in sampling frequency by a factor of $N$, the circuits in the different paths of the network operate at the frequency $f_s/N$.

As infinite impulse response filters have greater selectivity than finite impulse response filters, it is important to study multirate filters with recursive elements.

## 10.6 Multirate Filtering with IIR Elements

The basic technique for calculating a multirate filter with IIR elements is to perform the same type of decomposition on the transfer function $H(Z)$ of the overall IIR filter as was used to produce equation (10.36). The function $H(Z)$ is assumed to be a rational fraction in which the denominator and the numerator have the same degree.

Such a decomposition is obtained by finding the poles of $H(Z)$:

$$H(Z) = a_0 \frac{\Pi_{k=1}^{K}(Z - Z_k)}{\Pi_{k=1}^{K}(Z - P_k)} \tag{10.37}$$

From the identity:

$$Z^N - P_k^N = (Z - P_k)\left(Z^{N-1} + Z^{N-2}P_k + \cdots + P_k^{N-1}\right) \tag{10.38}$$

one can write:

$$H(Z) = a_0 \frac{\Pi_{k=1}^{K}(Z - Z_k)\left(Z^{N-1} + P_k Z^{N-2} + \cdots + P_k^{N-1}\right)}{\Pi_{k=1}^{K}\left(Z^N - P_k^N\right)}$$

which, in another form, is:

$$H(Z) = \frac{\sum_{i=0}^{KN} a_i Z^{-i}}{1 + \sum_{k=1}^{K} b_k Z^{-Nk}}$$

Thus:

$$H_n(Z^N) = \frac{\sum_{k=0}^{K} a_{kN+n} Z^{-Nk}}{1 + \sum_{k=1}^{K} b_k Z^{-Nk}} \tag{10.39}$$

$$H_n(Z^N) = \frac{N_n(Z^N)}{D(Z^N)} \tag{10.40}$$

Each path of the polyphase network is thus determined. They all have the same recursive part and are distinguished by the non-recursive one as shown in equation (10.40). In principle, when compared with the previous section, the difference is that the individual IIR phase shifters obtained do not have linear phase.

For realization purposes, it is worth pointing out that the poles are raised to power $N$, which is highly advantageous because they are spread inside the unit circle.

## 10.7 Filter Banks Using Polyphase Networks and DFT

A discrete Fourier transform computer is a bank of filters (Section 2.4) which are suitable for multirate filtering. However, it should be noted that filtering functions achieved in this way have significant overlap. To improve the discrimination between the components of the signal, the numbers are weighted before applying the discrete Fourier transform. The weighting coefficients are samples of functions called spectral analysis windows. The production of filter banks by means of a polyphase network and a DFT represents the generalization of such spectral analysis windows [6, 7].

Assume that a bank is to be created with $N$ filters which cover the band $[0, f_s]$ and are obtained by shifting a basic filter function along the frequency axis by $mf_s/N$, with $1 < m < N - 1$.

If $H(Z)$ is the basic filter $Z$-transfer function, a change in frequency by $mf_s/N$ appears as a change in the variable from $Z$ to $Ze^{j2\pi m/N}$. That is, the filter with index $m$ has a transfer function $b_m(z)$ given by:

$$B_m(Z) = H(Ze^{j2\pi m/N})$$

By applying the decomposition of $H(Z)$ introduced in the earlier sections, this becomes:

$$B_m(Z) = \sum_{n=0}^{N-1} Z^{-n} e^{-j2\pi(mn/N)} H_n(Z^N)$$

By allowing for the fact that the functions $H_n(Z^N)$ are the same for all the filters, a factorization can be introduced which results in the following matrix equation:

$$\begin{bmatrix} B_0 \\ B_1 \\ \vdots \\ B_{N-1} \end{bmatrix} = \begin{bmatrix} 1 & 1 & 1 & \cdots & 1 \\ 1 & W & W^2 & \cdots & W^{N-1} \\ \vdots & \vdots & \vdots & & \vdots \\ 1 & W^{N-1} & W^{2(N-1)} & \cdots & W^{(N-1)^2} \end{bmatrix} \begin{bmatrix} H_0(Z^N) \\ Z^{-1}H_1(Z^N) \\ \vdots \\ Z^{-(N-1)}H_N(Z^N) \end{bmatrix} \tag{10.41}$$

where $W = e^{-j2\pi m/N}$.

The square matrix is the matrix of the discrete Fourier transform. The bank of filters is realized by forming a cascade of the polyphase network described in Figure 10.13 and a discrete Fourier transform.

The operation of this device is illustrated in Figure 10.12 which shows, for the case where $N = 4$, the phase shifts introduced at the various points in the system so as to preserve only the signal in the band $[1/2NT, 3/2NT]$.

The polyphase network has the effect of correcting, in the useful part of the elementary band $[1/2NT, 3/2NT]$, for the interleaving of the numbers at the output of the discrete Fourier transform computer. This prevents overlap between the filters and leads to the filter function in Figure 10.13. This function depends only on the basic filter $H(Z)$, which can be an FIR or IIR filter, and which can be specified so that the filters in the bank have no overlap or have a crossover point, for example, at 3 dB or 6 dB.

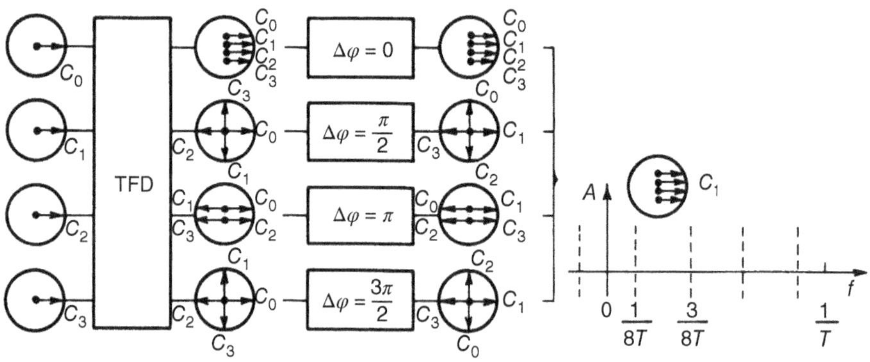


**Figure 10.12** Phase shifts in a filter bank.

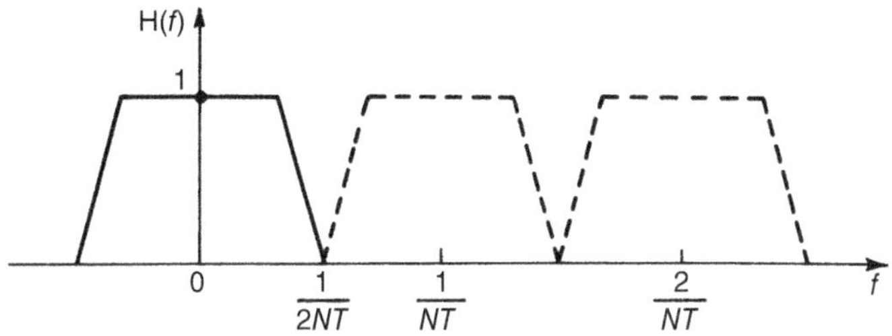


**Figure 10.13** Filtering function of a filter bank.

The impulse response of the basic filter is the spectral analysis window of the system. If the filters are specified to have no frequency overlap, the sampling frequency at the output of the filters, or at the input according to the method of use, can have the value $1/NT$ and the overall calculation can be performed at this frequency.

If $N$ is assumed to be a power of 2, the fast Fourier transform algorithm can be used to calculate the discrete Fourier transform, and the number of real multiplications $M_R$ to be performed during a sampling period in the complete system is:

$$M_R = N \times 2K + 2N \log_2 \frac{N}{2} = 2N \left[K + \log_2 \frac{N}{2}\right] \tag{10.42}$$

This value can be compared with the value $2KN^2$ required by $N$ IIR filters of the same order operating at frequency $1/T$. An application in telecommunications will be discussed in a later chapter.

## 10.8 Conclusion

FIR filters have the property that the sampling frequencies at the input and output are independent. This can be exploited to adapt the computation rate to the bandwidth of the signal during processing. This property can be extended to recursive structures through the use of a suitable transformation. Phase shifting is the basic function involved.

Multirate filtering can be applied to narrow-band filters. It can provide considerable savings in computation when there is a factor of more than an order of magnitude between the sampling frequency and the pass band of the filter, as often occurs in practice.

The use of these techniques requires more detailed analysis of the processing. The limitations on their use result primarily from complications in the computation sequencing produced by the cascading of stages operating at different frequencies. This point should be examined carefully for each potential application of multirate filtering so that an excessive increase in the control unit or in the instruction program does not outweigh the computational gains.

## Exercises

**10.1** Give the number of bits assigned to the coefficients of the half-band filters in Table 10.1. Use the results in Section 5.8 to evaluate the extra ripple introduced in the half-band filters by limiting the number of bits representing the coefficients. Test this evaluation on the filters $F_6$, $F_8$, and $F_9$.

**10.2** Estimate the number of coefficients of the three filters in the cascade in the example in Section 10.4.
As each multiplication result is rounded, analyze the round-off noise produced by the reduction in sampling rate and give an expression for its power. Perform the same analysis for an increase in sampling rate and the basic filter. Compare the results with the direct realization. Give an estimate of the aliasing distortion.

**10.3** A filter for a telephone channel has a pass band which extends from 0 to 3400 Hz with a ripple of less than 0.25 dB. Above 4000 Hz, the attenuation is greater than 35 dB. For a sampling frequency $f_s = 32$ kHz, design a multirate filter as a cascade of half-band filters. Compare the number of multiplications and additions to be performed per second and the size of the memory with the values obtained for direct FIR filtering.
The signal applied to the filter is composed of numbers with 13 bits, and the computations are performed using 16-bit registers. Evaluate the power of the round-off noise with a reduction and an increase in the sampling rate.

**10.4** A discrete Fourier transform computer is a bank of uniform filters with characteristics as shown in Section 2.4. Study the phase shifts introduced in the odd Fourier transforms and in the doubly odd Fourier transforms in Section 3.3. What are the characteristics of the banks of filters so obtained?

**10.5** Consider a bank of two filters to be produced using the IIR filter of order 4 given as an example in Section 7.2. The zeros and the poles of the upper half-plane have coordinates (Figure 7.5):

$$Z_1 = -0.816 + j0.578; \qquad Z_2 = -0.2987 + j0.954$$

$$P_1 = 0.407 + j0.313; \qquad P_2 = 0.335 + j0.776$$

Using the equations in Section 10.6, calculate the $Z$-transfer functions of the polyphase network paths. Give the coordinates of the poles and the zeros in the complex plane.
Use the results of Chapter 7 to determine the effect on the frequency response of limiting the number of bits in the coefficients of the denominator of the transfer function. Compare with the direct realization.
Draw the circuit diagram of this bank of two filters and determine the number of multiplications required, assuming that the sampling frequency at the output of each filter is half the value at the input.

## References


1 M. Bellanger, J. Daguet and G. Lepagnol, Interpolation, extrapolation and reduction of computation speed in digital filters. *IEEE Transactions*, 22(4), 1974.

2 F. Mintzer and B. Liu, Aliasing error in the design of multirate filters. *IEEE Transactions*, 26(2), 1978.

3 R. E. Crochière and L. R. Rabiner, *Chapter 5: Multirate Digital Signal Processing*, Prentice-Hall Inc., Englewood Cliffs, NJ, 1983.

4 D. J. Goodman and M. J. Carey, Nine digital filters for decimation and interpolation. *IEEE Transactions*, 25(4), 1977.

**5** M. Bellanger, G. Bonnerot and M. Coudreuse, Digital filtering by polyphase network: application to sample rate alteration and filter banks. *IEEE Transactions*, 24(2), 1976.

**6** P. P. Vaidyanathan, *Multirate Systems and Filter Banks*, Prentice Hall, 1993.

**7** N. J. Fliege, *Multirate Digital Signal Processing*, John Wiley, Chichester, 1994.

# 11

# QMF Filters and Wavelets

The compression of certain signals, such as speech, sound, or images, involves sub-band decomposition combined with sampling rate reduction and reconstruction from sub-bands after storage or transmission. The simplest approach to these operations is to use banks of two filters [1].

## 11.1 Decomposition into Two Sub-Bands and Reconstruction

The block diagram of the system is given in Figure 11.1. The signal $x(n)$ to be analyzed is fed to a set of two filters – namely, a low-pass filter $H_0(Z)$ and a high-pass filter $H_1(Z)$ whose outputs are decimated by factor two. They are called analysis filters. The reconstruction is performed from two sequences in which every other sample is null, filtered by a low-pass filter $G_0(Z)$ for one and a high-pass filter $G_1(Z)$ for the other, which are called synthesis filters.

As explained in Section 10.1, due to undersampling at analysis filter output, the image bands which are added to the useful signal are expressed by $H_0(-Z)X(-Z)$ and $H_1(-Z)X(-Z)$. These undesired signals disappear at the output of the synthesis filters if the following condition is met:

$$G_0(Z)H_0(-Z) + G_1(Z)H_1(-Z) = 0 \tag{11.1}$$

Then, it is sufficient to have the same filters in both subsets and the relations:

$$G_0(Z) = H_1(-Z); G_1(Z) = -H_0(-Z) \tag{11.2}$$

The reconstruction condition takes the form:

$$H_0(Z)H_1(-Z) - H_1(Z)H_0(-Z) = Z^{-K} \tag{11.3}$$

$K$ is the delay incurred by the signal in the process.

Next, the coefficients have to be calculated, depending on the filter type and the frequency domain specifications. The number of coefficients controls the level of sub-band separation.

The simplest approach, which minimizes computational load, consists of selecting identical linear phase filters.

## 11.2 QMF Filters

As pointed out in the previous chapter, a two-branch polyphase network yields a low-pass filter and a high-pass filter as well, with no additional calculations. Then, decimation can take place at input and the system block diagram is given in Figure 11.2.

*Digital Signal Processing: Theory and Practice*, Tenth Edition. Maurice Bellanger.

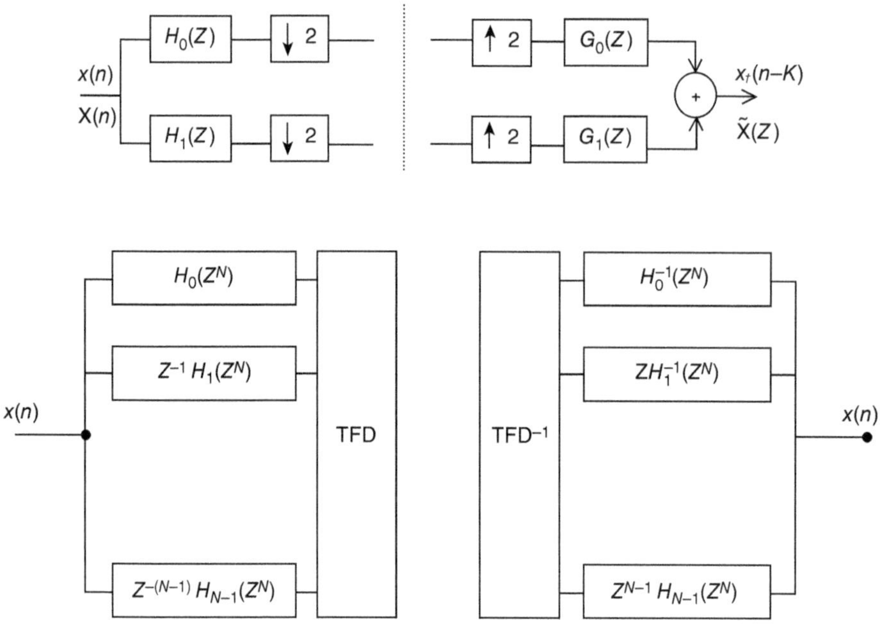


**Figure 11.1** Bank of 2 filters for decomposition and reconstruction.

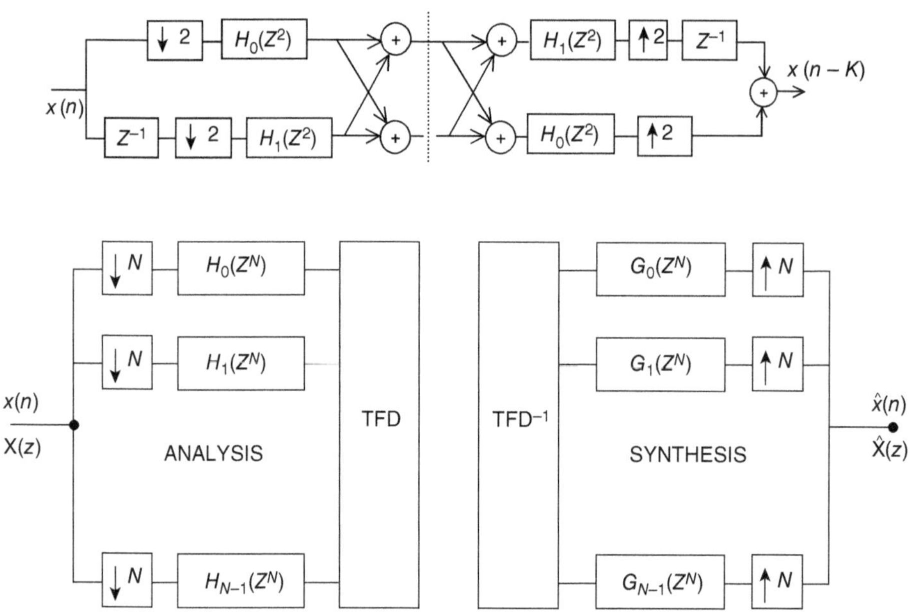


**Figure 11.2** QMF filter.

The transfer functions $H_0(Z^2)$ and $H_1(Z^2)$ represent the polyphase decomposition of prototype filter $H(Z)$ – that is,

$$H(Z) = H_0(Z^2) + Z^{-1}H_1(Z^2) \tag{11.4}$$

Next, it is necessary to determine the conditions to be met by the prototype filter so that the basic relations (11.1) and (11.2) hold. The system transfer function is expressed by:

$$T(Z) = Z^{-1}H_1(Z^2)\,H_0(Z^2) \tag{11.5}$$

The two polyphase components are linked to the prototype filter by (see Section 10.1): $H_0(Z^2) = \frac{1}{2}\,[H\,(Z) + H(-Z)]; Z^{-1}H_1(Z^2) = \frac{1}{2}\,[H(Z) - H(-Z)]$ and the transfer function is:

$$T(Z) = \frac{1}{4}\,[H^2(Z) - H^2(-Z)] \tag{11.6}$$

It can easily be shown that in order to obtain the sign inversion needed for reconstruction in (11.2), it is necessary to choose a prototype filter with an even number of coefficients: $N = 2P$. The frequency response of such a filter is written as (see Section 5.2):

$$H(f) = e^{-j2\pi f(P-1/2)}2H_R(f) \tag{11.7}$$

where $H_R(f)$ is a real even function. Also:

$$H\left(f - \frac{1}{2}\right) = e^{-j2\pi f(P-1/2)}\,e^{j\pi(P-1/2)}\,2\,H_R\left(f - \frac{1}{2}\right) \tag{11.8}$$

In such conditions, on the unit circle, we get:

$$\frac{1}{4}[H^2(Z) - H^2(-Z)]_{Z=e^{j2\pi f}} = e^{-j2\pi f(2P-1)}\left[H_R^2(f) + H_R^2\left(\frac{1}{2} - f\right)\right] \tag{11.9}$$

The reconstruction condition is:

$$H_R^2(f) + H_R^2\left(\frac{1}{2} - f\right) = 1 \tag{11.10}$$

The phase term in (11.9) gives the system delay – that is, the delay for decomposition and reconstruction – namely $K = 2P - 1$.

The term QMF (quadrature mirror image filter) stems from the polyphase decomposition of the even number coefficient prototype filter. In fact, as explained in Chapter 5, it is an interpolator, which implies that it introduces a delay that is an odd multiple of half the sampling period. Hence the decomposition (see Section 10.1):

$$H(Z) = Z^{-1/2}H_{1/2}(Z^2) + Z^{-3/2}H_{3/2}(Z^2)$$

and, in simpler form:

$$H(Z) = Z^{-1/2}[H_0(Z^2) + Z^{-1}H_1(Z^2)] \tag{11.11}$$

With respect to polyphase components, we find:

$$H_0(Z^2) = Z^{1/2}[H(Z) + jH(-Z)];\, H_1(Z^2) = Z^{1/2}[H(Z) - jH(-Z)] \tag{11.12}$$

and the frequency response:

$$|H_0(f)| = \left|H(f) + j\,H\left(\frac{1}{2} - f\right)\right| \tag{11.13}$$

The base band and the image band are in quadrature. By extension, the name QMF applies to all banks of two filters used for decomposition and reconstruction.

Next, the coefficients have to be calculated. Relation (11.10) shows that the transfer function $H^2(Z)$ is a Nyquist odd coefficient number half-band finite impulse response (FIR) filter and $H(Z)$ is a low-pass half-Nyquist. The calculation starts with a mask specifying the pass-band edge $f_1$, the stop-band edge $f_2$, the ripples $\delta_1$ and $\delta_2$, and imposing the amplitude $\sqrt{2}/2$ at frequency 1/4. In order for $H^2(f)$ to meet condition (11.10) with ripple $\delta$, it is sufficient to take:

$$f_2 = \frac{1}{2} - f_1; H(0) = 1; \delta_1 = \frac{\delta}{2}; \delta_2 = \sqrt{\delta}$$

*Example:*

A low-pass filter has the following parameters:

$$\Delta f = 0.24; f_1 = 0.13; f_2 = 0.37; \delta = 0.01$$

Calculated coefficients:

$$h_1 = h_8 = 0.015235; h_2 = h_7 = -0.085187$$
$$h_3 = h_6 = 0.081638; h_4 = h_5 = 0.486502$$

The two branches of the polyphase network obtained $H_0(Z^2)$ and $H_1(Z^2)$ have the same coefficients, but in reverse order.

The even coefficients of the filter $\sum_{i=1}^{15} h_i' Z^{-i}$ should be null, except for $h_8'$.

In fact, we obtain: $\sum_{i=1;\, i\neq 4}^{7} \left(h_{2i}'\right)^2 = 1.7 \times 10^{-5}$

Iterative methods can be employed, if necessary, to complete the calculation and achieve a better approximation of the symmetry.

The quality of the reconstruction depends on the ripple $\delta$ and, therefore, on the number of coefficients. If perfect reconstruction is sought, the option of identical filters must be dropped, and undersampling is necessary at analysis input [2].

## 11.3 Perfect Decomposition and Reconstruction

Letting $P(Z) = H_0(Z)\, H_1(-Z)$, the reconstruction condition (11.3) is written:

$$P(Z) - P(-Z) = Z^{-K} \tag{11.14}$$

This means that $P(Z)$ is a low-pass filter with an odd number of coefficients and all the even index coefficients are null, except for the center coefficient, which is unity. For example, for $M = 2$ different coefficients, we get:

$$P(Z) = h_3 + h_1 Z^{-2} + Z^{-3} + h_1 Z^{-4} + h_3 Z^{-6} \tag{11.15}$$

The coefficients of the low-pass and high-pass filters, $H_0(Z)$ and $H_1(-Z)$, are subject to the constraint that, in their product, the coefficients of the terms $Z^{-1}$ and $Z^{-5}$ are null. Some degrees of freedom remain, which are exploited to obtain specific properties.

In a first example, polynomials of different degrees are chosen, with symmetrical coefficients and zeros at $Z = 1$. Then, taking:

$$H_1(-Z) = \frac{1}{2}(1 + Z^{-1})^2$$
$$H_0(z) = (1 + Z^{-1})(\alpha_0 + \alpha_1 Z^{-1} + \alpha_1 Z^{-2} + \alpha_0 Z^{-3}) \tag{11.16}$$

the perfect reconstruction condition imposes that, in the product $P(Z)$, the coefficient of $Z^{-1}$ is zero and the coefficient of $Z^{-3}$ is unity, which leads to $\alpha_0 = -1/8$ and $\alpha_1 = 3/8$. Finally:

$$H_0(Z) = \frac{1}{8}[-1 + 2Z^{-1} + 6Z^{-2} + 2Z^{-3} - Z^{-4}] \tag{11.17}$$

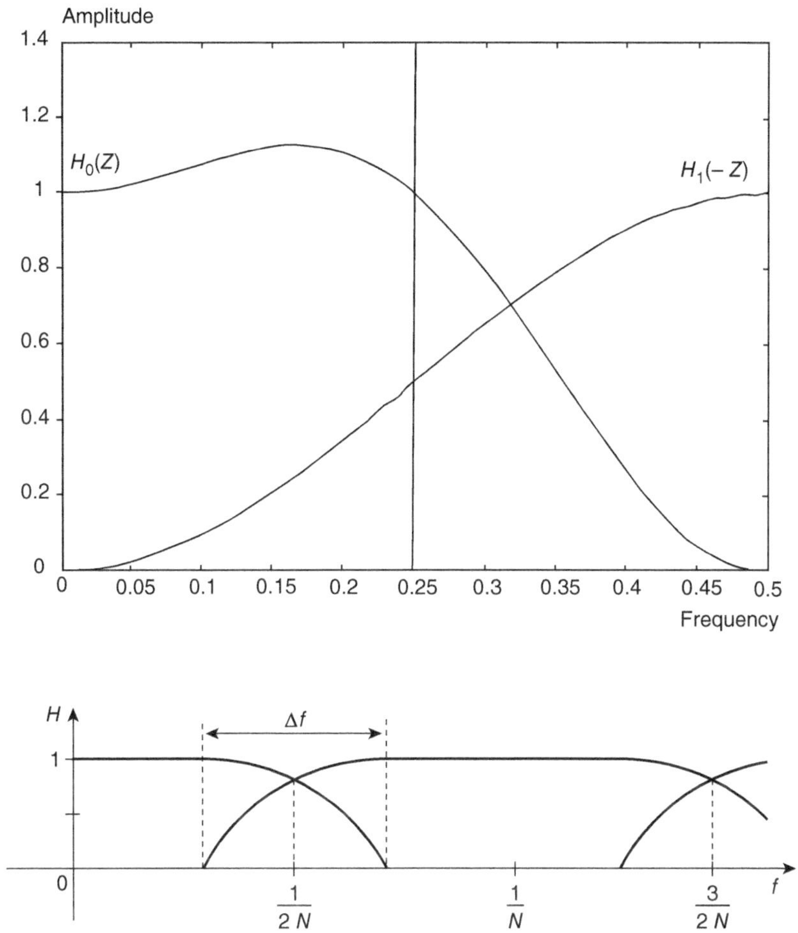


**Figure 11.3** Frequency responses of the filters in the standard JPEG 2000 lossless.

The two filters obtained are at the basis of the reversible transform used in the still image compression standard JPEG 2000 – option lossless. The frequency responses are given in Figure 11.3. It is worth pointing out that the two sub-bands are unbalanced, so the filters are not of the half-Nyquist type.

Decomposition into two equal sub-bands can be achieved provided linear phase constraint is dropped and the following factorization of the half-band filter is performed:

$$P(Z) = H_0(Z)\, Z^{-k} H_0(Z^{-1}) \tag{11.18}$$

The filters $H_0(Z)$ and $H_1(-Z)$ have the same coefficients but in reverse order and the polynomial $P(Z)$ has degree $2K$ and $2K+1$ coefficients.

As a function of $M$, the number of different coefficients, the equality $2K+1 = 4M$ holds, which implies $K = 2M{-}1$. The fact that the whole number $K$ is odd allows relation (11.2) to be satisfied, and with:

$$G_0(Z) = Z^{-k} H_0(Z^{-1}); H_1(Z) = -Z^{-k} H_0(Z^{-1}); G_1(Z) = -H_0(-Z)$$

it can readily be verified that conditions (11.1) and (11.3) are met.

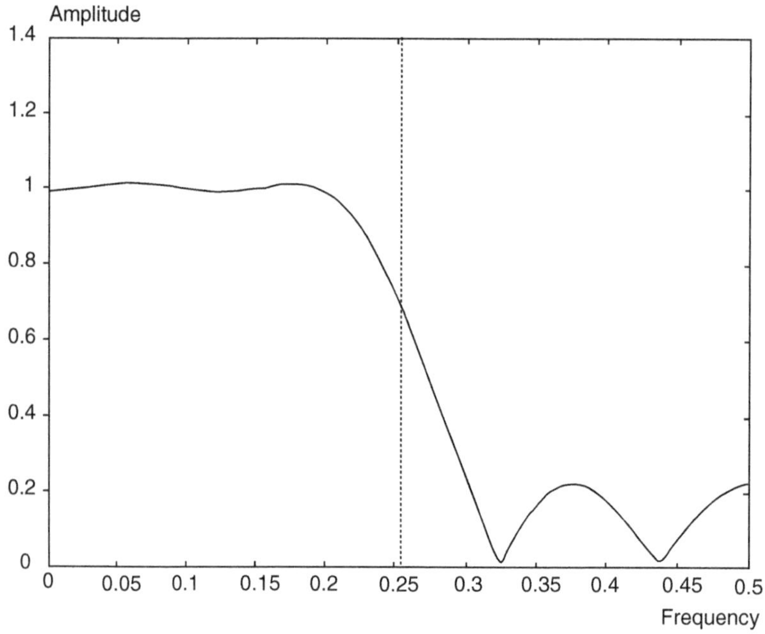


**Figure 11.4** Frequency response of the low-pass analysis filter.

The coefficient calculation procedure is the minimum phase technique for FIR filters described in Section 5.13. The ripple is added to the center coefficient of the half-band filter, which makes the zeros on the unit circle double, and the minimum and maximum phase factors are extracted [3].

As an illustration, let us consider a filter $P(Z)$ with $2K+1 = 15$ coefficients, computed with the specifications $f_1 = \frac{1}{2} - f_2 = 0.2$. The $M = 4$ different coefficients are: $h_1 = 0.62785$; $h_3 = -0.18681$; $h_5 = 0.08822$; $h_7 = -0.05297$.

The ripple value is $\delta = 0.047$ and the central coefficient becomes $h_0 = 1.047$. Taking one of the zeros that are on the unit circle and those that are inside, the first filter is obtained:

$$H_0(Z) = 0.3704 + 0.5111\,Z^{-1} + 0.2715\,Z^{-2} - 0.0885\,Z^{-3} - 0.1346\,Z^{-4} + 0.0338\,Z^{-5}$$
$$+ 0.0973\,Z^{-6} - 0.0703\,Z^{-7}$$

The corresponding frequency response is shown in Figure 11.4. With respect to the initial filter, the ripple in the attenuated band is $\sqrt{\delta}$.

Among the characteristics of the factors of $P(Z)$, the regularity of the frequency response is worth emphasizing, because it is important for the compression of signals and, particularly, of images. In filtering, this property corresponds to the presence of multiple zeros at point $Z = -1$ in the $Z$-transfer function. On the theoretical side, the approach is justified by the wavelet theory.

## 11.4 Wavelets

The objective of the wavelet theory is the representation of signals in the time-frequency domain. It is a representation which is not possible with the Fourier analysis because it assumes periodic or finite duration signals. Thus, in order to localize a signal in both time and frequency with the Fourier transform, a sliding window has to be introduced.

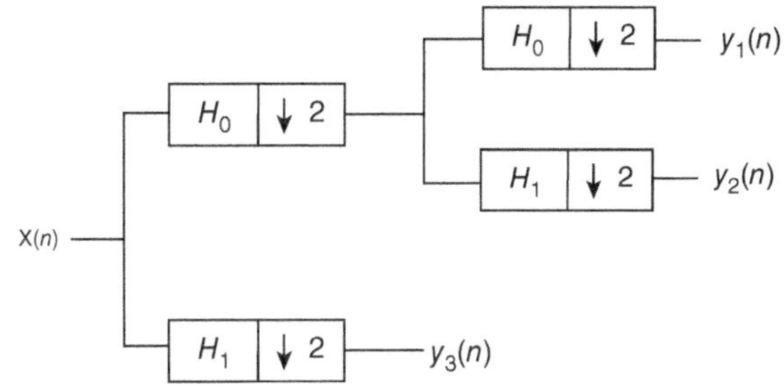


**Figure 11.5** Nonuniform filter bank in tree structure.

As the basis for decomposition, the wavelet transform uses a set of functions called wavelets, deduced from a generating function through translation and dilation. It allows for analysis of signals with arbitrary duration [4, 5].

In practice, the discrete wavelet transform is a nonuniform filter bank and an efficient approach for implementation consists of cascading banks of two filters, such as those described above, with decimation by factor two at each stage and the same coefficients for all the filters. The operations of translation and dilation are performed automatically by the sampling rate changes. The tree structure for analysis is shown in Figure 11.5. By completing the lower part in the figure, a uniform filter bank can be obtained.

The coefficients are calculated with the objective of maximum regularity – that is, function $P(Z)$ featuring the maximum number of zeros at point $Z = -1$.

A half-band FIR filter with $N$ different coefficients has $4N-1$ coefficients and degree $4N-2$. With its $2(N-1)$ null coefficients, the function $P(Z)$ includes a factor of degree $2(N-1)$. Thus, the degree of the remaining factor is $2N$. Then, the analysis and synthesis filters are derived from the factorization of $P(Z)$.

In a minimum phase solution, the filters have $2N$ coefficients and the $Z$-transfer function of the low-pass filter has $N$ zeros at $Z = -1$ in the complex plane. As in the previous sections, numerical values can be determined directly by combining this constraint with the conditions of canceling the coefficients of the odd terms in $P(Z)$. Alternatively, $P(Z)$ can be obtained and factorized.

Table 11.1 provides the coefficients of filters $H_0(Z)$ for the first values of $N$. The coefficients of the other filters involved in analysis and synthesis, $H_1(Z)$, $G_0(Z)$, $G_1(Z)$ as in Figure 11.1, are given by:

$$H_0(Z) = \sum_{i=1}^{2N} h_{0,i} Z^{-i};\, h_{1,i} = (-1)^i h_{0.2N+1-i};\, g_{0,i} = h_{0.2N+1-i};\, g_{1,i} = h_{1.2N+1-i} \tag{11.19}$$

Frequency responses are provided in Figure 11.6. It is worth mentioning the similarity with Butterworth filter responses (see Section 7.2.3), which have the same zeros at $Z = -1$, also have the property of perfect reconstruction when the cutoff frequency is set in the middle of the useful band, and have a monotonic frequency response.

It is also possible to obtain linear phase filters, by giving different odd numbers of coefficients to the low-pass and high-pass filters. For example, Table 11.2 gives the coefficients of the (9, 7) filters used in the JPEG 2000 standard for highrate lossy compression [6].

The accuracy of the reconstruction is evaluated by computing the impulse response of the analysis/synthesis system. Multiplying by $H_1(-Z)$ the polynomial which is obtained by canceling the coefficient with an odd index in $H_0(Z)$ and adding the product by $H_0(-Z)$ of the polynomial obtained by canceling the coefficient with even index in $H_1(Z)$, it can readily be verified that the center coefficient of the resulting polynomial is unity, while other coefficients take on almost

**Table 11.1** Coefficients of minimum-phase wavelets.

| $N = 2$ | $N = 3$ | $N = 4$ | $N = 5$ |
|---|---|---|---|
| 0.482963 | 0.332671 | 0.230378 | 0.160102 |
| 0.836516 | 0.806892 | 0.714847 | 0.603828 |
| 0.224144 | 0.459878 | 0.630881 | 0.724307 |
| −0.129410 | −0.135011 | −0.027984 | 0.138427 |
| | −0.085441 | −0.187035 | −0.242295 |
| | 0.035226 | 0.030841 | −0.032245 |
| | | 0.032883 | 0.077571 |
| | | −0.010597 | −0.006242 |
| | | | −0.012581 |
| | | | 0.003336 |

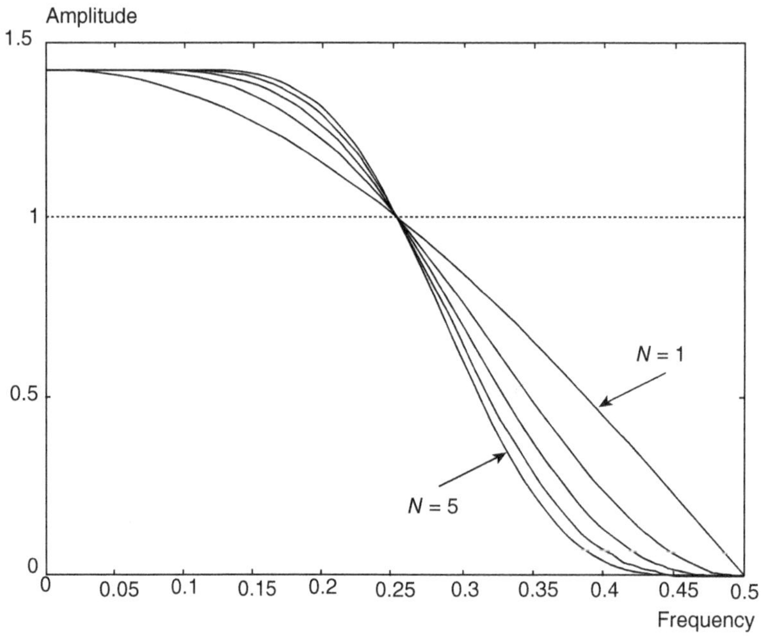


**Figure 11.6** Frequency responses of minimum phase wavelets.

zero values, the error being smaller than $2 \times 10^{-6}$. The discrepancies stem from rounding the coefficients.

Figure 11.7 shows the frequency responses of the filters $H_0(Z)$ and $H_1(Z)/2$. Again, the partition of the signal is unbalanced, but less so than in Figure 11.3, due to the higher number of filter coefficients. These filters have half their zeros at points $Z = \pm 1$ in the complex plane, which brings high regularity to the frequency responses – an important property in image processing.

Regarding arithmetic complexity, it is worth pointing out the small number of multiplications, close to that of polyphase techniques, because it is possible to benefit from the coefficient symmetries for both the analysis and synthesis.

**Table 11.2** Filters for lossy compression in JPEG 2000.

| $Z^i$ | $H_0(Z)$ | $H_1(Z)$ |
|---|---|---|
| $i = 0$ | 0.602949 | 1.115087 |
| $i = \pm 1$ | 0.266864 | −0.591272 |
| $i = \pm 2$ | −0.078223 | −0.057544 |
| $i = \pm 3$ | −0.016864 | 0.091272 |
| $i = \pm 4$ | 0.026749 | – |

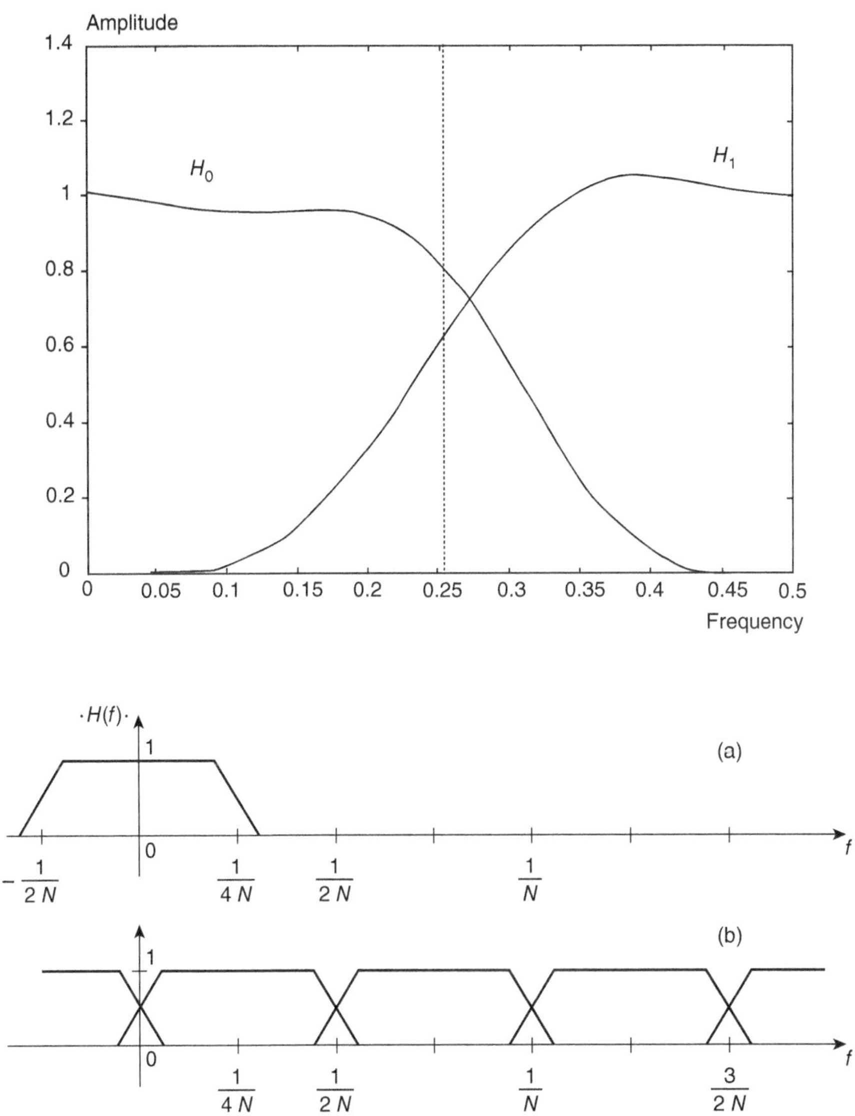


**Figure 11.7** Frequency responses of the filters in the standard lossy JPEG 2000.

## 11.5 Lattice Structures

The factorization (11.8) of the half-band filter can also be realized in lattice representation, canceling every other coefficient. The corresponding modular structure is shown in Figure 11.8.

The transfer functions of analysis and synthesis filters are obtained through the calculations provided in Section 8.5. For example, for three cells, $K = 3$, we find:

$$H_0(Z) = 1 + \alpha_1 Z^{-1} - \alpha_1\alpha_2 Z^{-2} + \alpha_2 Z^{-3}$$
$$H_1(Z) = -\alpha_2 - \alpha_1\alpha_2 Z^{-1} - \alpha_1 Z^{-2} + Z^{-3} \tag{11.20}$$

This is equivalent to the previous relation: $H_1(Z) = -Z^{-K}H_0(-Z^{-1})$. The system transfer function is:

$$T(Z) = Z^{-2(K-1)} \prod_{i=1}^{K} \left(1 + \alpha_i^2\right) \tag{11.21}$$

The delay is $2(K-1)$ sampling periods. The lattice coefficients are computed from filter specifications.

For example, imposing a zero at $Z = -1$ and $H_0(1) = 4$, we obtain $\alpha_1 = 1 + \sqrt{2}; \alpha_2 = 1 - \sqrt{2}$.

Comparison with the previous section reveals that the lattice approach is less efficient than other factorizations; referring to Table 11.1, the filter with four coefficients has two zeros at $Z = -1$. However, the lattice approach has benefits when it comes to implementation, modularity of the structure, and insensitivity to coefficient rounding, because perfect reconstruction is not impacted [7].

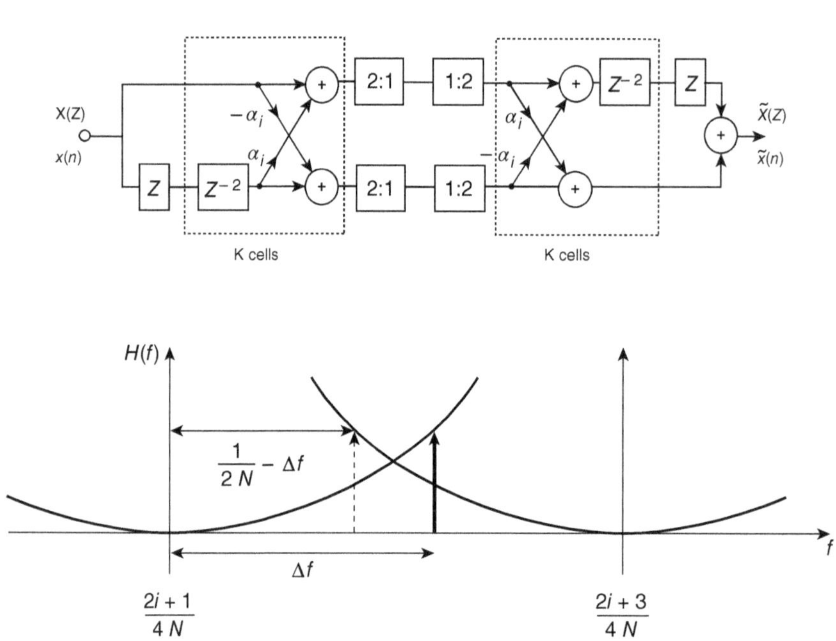


Figure 11.8 Bank of 2 filters in lattice structure.

## Exercises

**11.1** The signal $x(n) = \cos\left(\frac{n\pi}{4}\right)$ is applied to the filter with transfer function $H(Z) = -0.050 + 0.117\,Z^{-1} + 0.452\,Z^{-2} + 0.452\,Z^{-3} + 0.117\,Z^{-4} - 0.050\,Z^{-5}$
What is the delay incurred by the filter? Give the impulse response and express the output sequence.

**11.2** Give the diagram of a QMF bank of two filters, assuming $H(Z)$ is the prototype filter. Provide the expressions of the signals at the output from the analysis. Compute the transfer function of the analysis/synthesis system and give the expression of the reconstructed signal.

**11.3** In the case of perfect decomposition and reconstruction, the prototype $P(Z)$ has degree 6 and it must be factorized with the same degree factors. Looking for maximum regularity, the low-pass filter $H_0(Z)$ must have 3 zeros at $Z = -1$. Compute the coefficients of the high-pass filter $H_1(Z)$ and find its zeros.

**11.4** Compare the frequency responses with those in Figure 11.3. Quantization occurs at analysis output; how is it amplified in the reconstruction phase?

**11.5** Linear phase analysis filters are assumed to have three and five coefficients. Compute the values which give the best possible sub-band separation and perfect reconstruction.

**11.6** When a unit power white noise is applied at the inputs of the analysis filters, what is the power of the output signals?

**11.7** The signal $x(n) = \cos\left(\frac{n\pi}{4}\right)$ is applied to the analysis filters in Table 11.2. Give the expressions of the output signals before and after decimation. In the reconstruction process, give the expressions of the signals before and after the final addition.

**11.8** The sensitivity of the decomposition/reconstruction process to coefficient accuracy is to be investigated.
Considering the filters in Table 11.1, give an upper bound of the reconstruction error due to coefficient rounding. Check the result with the first filter in the table, assuming four coefficients and four-bit representation. Calculate the frequency response of the corresponding filter.
Work through the same questions in reference to the filters in Table 11.2.

## References


1 R. E. Crochière and L. R. Rabiner, *Multirate Digital Signal Processing*, Prentice-Hall Inc., Englewood Cliffs, New Jersey, 1983.
2 M. Vetterli, Filter banks allowing perfect reconstruction. *Signal Processing*, 10(3), 1986, 219–244.

3 M. Smith and T. Barnwell Exact reconstruction techniques for tree structured sub-band coders, *IEEE Transactions*, ASSP-34, 3, 1986, 434–441.

4 I. Daubechies, Orthonormal bases for compactly supported wavelets. *Communications on Pure and Applied Mathematics*, 41, 1988, 909–996.

5 S. Mallat, *A Wavelet Tour of Signal Processing*, 2nd ed., Academic Press, New York, 1999.

6 A. Skodras, C. Christopoulos and T. Ebrahimi, The JPEG 2000 still image compression standard, *IEEE Signal Processing Magazine*, 18(5), 36–58, 2001.

7 P. P. Vaidyanathan, *Multirate Systems and Filter Banks*, Prentice Hall Inc.: Englewood Cliffs, N.J, 1993.

# 12

# Filter Banks

The signal decomposition and reconstruction techniques presented in the previous chapter can be generalized to any number of sub-bands, using banks of more than two filters. In that case, in principle, decimation can be performed at the output of the analysis filters, but, in general, it is preferable to decimate at system input in order to benefit from the combination of a polyphase network and DFT, thus minimizing the computational complexity.

## 12.1 Decomposition and Reconstruction

In the realization of filter banks using polyphase networks and DFT, as explained in Section 10.7, the operations involved are reversible and this leads to the arrangement shown in Figure 12.1 for the decomposition and reconstruction of a signal [1, 2].

The difficulty, in practice, consists of applying the operations associated with the functions $H_i^{-1}(Z^N)$.

The filter $H(Z)$ which serves as a basis for the process, sometimes called the prototype filter, has a polyphase decomposition whose elements satisfy relation (10.8):

$$z^{-i}H_i(Z^N) = \frac{1}{N}\sum_{m=0}^{N-1} e^{-j(2\pi/N)im} H(Ze^{-j(2\pi/N)m}); 0 \leq i \leq -1 \tag{12.1}$$

If the prototype filter $H(Z)$ has a cutoff frequency of less than $f_s/2N$ and infinite attenuation for frequencies greater than or equal to $f_s/2N$ – that is, if aliasing of the spectrum due to sampling at the rate $1/NT$ is negligible – one can write:

$$H[\exp j2\pi(f/f_s - m/N)]H[\exp j2\pi(f/f_s - k/N)] = 0; \quad m \neq k \tag{12.2}$$

Under these conditions, the following equation is satisfied on the unit circle, except for the factor $Z^{-N}$:

$$H_i(Z^N)H_{N-i}(Z^N) = \frac{1}{N^2}\sum_{m=0}^{N-1} H^2(Ze^{-j(2\pi/N)m}) = H_0^2(Z^N)$$

Hence:

$$\frac{H_i(Z^N)}{H_0(Z^N)}\frac{(H)_{N-i}(Z^N)}{H_0(Z^N)} = 1; \quad 0 \leq i \leq N-1 \tag{12.3}$$

These equations simply convey the phase relations illustrated in Figure 10.12.

*Digital Signal Processing: Theory and Practice*, Tenth Edition. Maurice Bellanger.

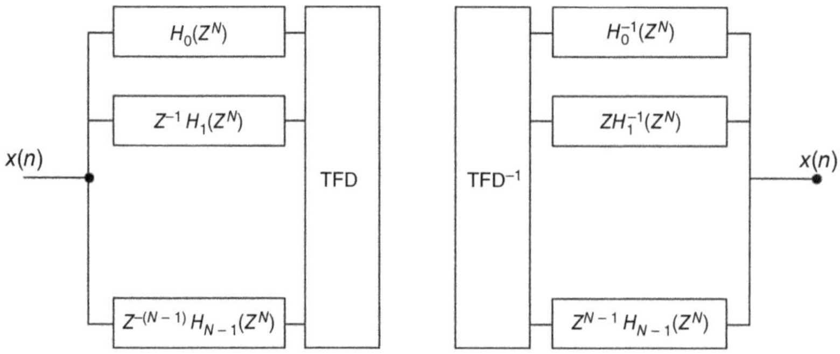


**Figure 12.1** The principle of decomposition and reconstruction.

One can then take $H_{N-i}(Z^N)$ to realize $H_i^{-1}(Z^N)$. Thus, the same filter bank is used for decomposition and reconstruction of the signal; the overall operation corresponds to multiplication by $H_0^2(Z^N)$.

In certain applications, it is not possible to neglect aliasing; this is the case, for example, when it is required to decompose a signal, sample it at the rate $f_s/N$, and then reconstruct it with the greatest possible accuracy over the band $(0, f_s)$. Let $G(Z)$ be the transfer function of the basic filter for the reconstruction. As the product of a discrete Fourier transform and its inverse is equal to unity, the overall operation corresponds to decomposition of the signal $x(n)$ into $N$ interleaved sequences $x(pN + i)$, to which are applied $N$ operators with transfer functions $G_i(Z^N)H_i(Z^N)$.

Figure 12.2 corresponds to a reduction of sampling frequency by $N$ in the decomposition part, sometimes called analysis, and an increase by $N$ in the reconstruction part, sometimes called synthesis. All processing in the corresponding device is performed at a rate $1/N$, which is a particularly effective approach.

The condition for reconstruction with a delay $D$ is written as:

$$G_i(Z^N)H_i(Z^N) = Z^{-D}; \quad 0 \leq i \leq N-1 \tag{12.4}$$

The delay $D$ must be the same in all branches of the polyphase network for which the interleaving corresponds to an increase of sampling frequency by $N$, and this can delay the initial signal, $x(n-D)$.

To determine the inverse functions $G_i(Z^N)$, it is necessary to perform a detailed analysis of the frequency response for the elements of the polyphase network.

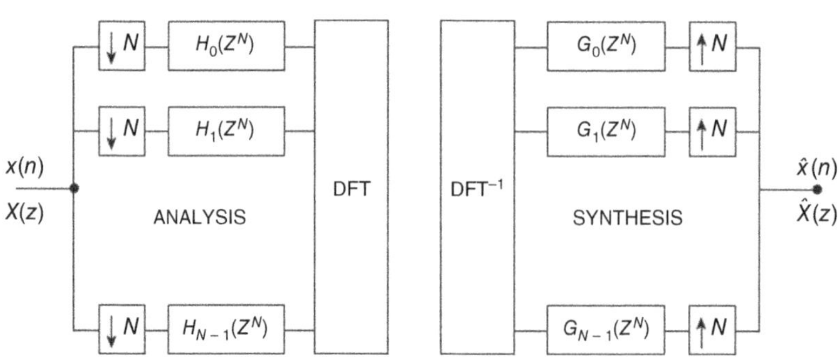


**Figure 12.2** Polyphase filter banks for signal analysis and synthesis.

## 12.2 Analyzing the Elements of the Polyphase Network

The frequency response of the elements $H_i(Z^N)$ of the polyphase network follows directly from equation (12.1). However, simplifications can be made. Indeed, the filters of the bank generally provide limited coverage, as in Figure 12.3. Under these conditions, a given filter is superimposed only on its immediate neighbors if the response of the prototype filter $H(Z)$ is such that $H(f) = 0$ for $|f| > 1/N$.

Then, for the branch of index $i$, one can write:

$$H_i(f) = H(f) + e^{-j(2\pi/N)i} H\left(f - \frac{1}{N}\right) \tag{12.5}$$

Since the periodicity of this response is $1/N$, if the coefficients are real, it is sufficient to consider the response in the interval $0 \leq f \leq 1/2N$, and this gives:

$$H_i(f) = H(f) + e^{-j(2\pi/N)i} \overline{H}\left(\frac{1}{N} - f\right) \tag{12.6}$$

Assuming that the response of the prototype filter is a monotonically decreasing curve in the transition band $\Delta f$, $H_i(f)$ cannot be zero for $f \neq 1/2N$. For the value $1/2N$, this gives:

$$H_i\left(\frac{1}{2N}\right) = H\left(\frac{1}{2N}\right) + (1 + e^{-j(2\pi/N)i}) \tag{12.7}$$

This response is zero for the branch $i = N/2$.

Hence, with the decomposition (12.1), the same result as equation (10.36), the branch of index $N/2$ cannot be inverted since its $Z$-transfer function has a zero in the $Z$-plane at the point $-1$. To obtain a set of invertible branches, it is necessary to use another polyphase decomposition.

In Chapter 5, it was shown that a linear phase FIR filter with an even number of coefficients is an interpolator at mid-sample period. It can be considered to result from an FIR filter with an odd number of coefficients, having the same frequency response by downsampling of a factor two. It is therefore necessary to start with a polyphase decomposition with $2N$ branches and to retain only one branch in two. One then obtains this equation for $H(Z)$:

$$H(z) = \sum_{i=0}^{N-1} Z^{-(i+1/2)} H_{i+1/2}(Z^N) \tag{12.8}$$

With this decomposition, the minimum amplitude $H_{\min}$ in a branch is given by:

$$H_{\min} = \left| H_{(N+1)/2}\left(\frac{2}{2N}\right) \right| = |(1 - e^{j\pi/N})| = 2\sin\frac{\pi}{2N} \tag{12.9}$$

As a result, to have an invertible polyphase network, it is sufficient to require that the linear phase prototype FIR filter have an even number of coefficients.

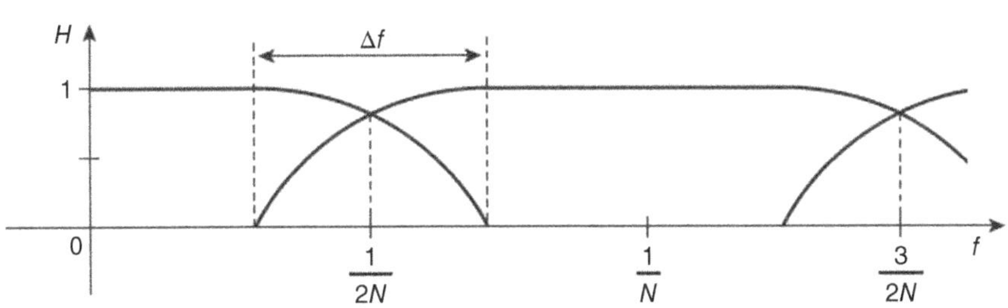


**Figure 12.3** Overlap of adjacent filters.

Note that the zeros of the functions $H_{i+1/2}(Z^N)$ in the $Z^N$-plane with respect to the unit circle are divided equally between the interior and the exterior of the unit circle. The reason is that the elements of $H_{i+1/2}(Z^N)$ are almost of linear phase. Furthermore, the amplitude of the frequency response remains close to unity, so the zeros are far from the unit circle, except for the branches which exhibit high attenuation at the frequency $1/2N$ when $N$ is large.

## 12.3 Determining the Inverse Functions

By placing the transfer functions in the $Z$-plane, determination of the inverse function for the polyphase elements starts with a factorization where the $L_1$ zeros within the unit circle are separated from the $L_2$ zeros which are outside:

$$H_i(Z) = h_{i0}\prod_{h=1}^{L_1}(1 - Z_h Z^{-1})\prod_{l=0}^{L_2}(1 - Z_l Z^{-1}) \tag{12.10}$$

The term $h_{i0}$ is a scaling factor.

For all zeros $Z_l$ outside the unit circle, one can write:

$$\frac{1}{1 - Z_l Z^{-1}} = \frac{-Z}{Z_l}\sum_{l=0}^{\infty}(Z^{-1})^i Z^i \tag{12.11}$$

and, consequently, the inverse of the second factor of (12.10) can be approached with an arbitrary accuracy within a finite number of terms. Consider the function $G_i(Z)$ defined by:

$$G_i(Z) = \frac{\sum_{l=0}^{L_3} a_l Z^{-l}}{\prod_{h=1}^{L_1}(1 - Z_h Z^{-1})} = \frac{\sum_{l=0}^{L_3} a_l Z^{-l}}{1 + \sum_{l=0}^{L_1} a_l Z^{-l}} \tag{12.12}$$

where $L_3$ is an integer.

The condition for inversion is satisfied if:

$$\left(\sum_{l=0}^{L_3} c_l Z^{-l}\right)\left(\sum_{l=0}^{L_3} a_l Z^{-l}\right) = Z^{-(L_2+L_3)} \tag{12.13}$$

The choice of delay $L_2 + L_3$ is justified by the fact that the coefficients of the expansion of (12.11) are decreasing and that the second factor in (12.10) is of maximum phase. The inversion relation is written in matrix form:

$$MA = \begin{bmatrix} C_0 & 0 & 0 & \cdots & 0 \\ C_1 & C_0 & 0 & \dots & 0 \\ \vdots & \vdots & \vdots & & \vdots \\ 0 & 0 & 0 & \cdots & C_{L_3} \end{bmatrix}\begin{bmatrix} a_0 \\ a_l \\ \vdots \\ a_{L_3} \end{bmatrix} = \begin{bmatrix} 0 \\ 0 \\ \vdots \\ 1 \end{bmatrix} \tag{12.14}$$

where the vector $A$ has the unknown coefficients $a_l$ as elements. The system is overdetermined and permits a solution in the least-squares sense, given by the relation:

$$(M^t M)^{-1} M^t \tag{12.15}$$

The polyphase synthesis elements $G_i(Z)$, have the structure of a general IIR filter. As the poles $Z_h$ are far from the unit circle, a realization with a direct structure is possible.

The method of calculating the polyphase synthesis elements described above is general and applies to all analysis filters, with the sole condition that there are no zeros on the unit circle. It is necessary to perform one calculation per branch since different coefficients are obtained. This enables the analysis filter to be specified relatively independently of the synthesis filter. However, it may be useful to sacrifice a little flexibility in order to obtain a simpler and more systematic calculation, as in the previous chapter.

This will be described for a uniform bank of $N$ real filters.

## 12.4 Banks of Pseudo-QMF Filters

The principle relies on the assumption that, for a given filter, the attenuation is such that aliases originate only in the adjacent bands [3]. Let $H(Z)$ be the transfer function of a prototype low-pass linear phase filter having the frequency response represented in Figure 12.4. Consideration of a number of coefficients equal to $LN$ gives:

$$H(Z) = \sum_{k=0}^{LN-1} h_k Z^{-k} \tag{12.16}$$

In the bank, the filter of index $i$, centered on the frequency $(2i + 1)/4N$, has a transfer function $H(Ze^{-j2\pi(-2i+1)/4N})$. During the analysis, a component of the signal at the frequency $(2i + 1)/4N + \Delta f$, with perhaps $1/4N < \Delta f < 3/4N$, will be attenuated by a factor $H(\Delta f)$. Downsampling at a rate $1/N$ will produce a replica of this component at a frequency:

$$\frac{2i+1}{4N} + \frac{3}{4N} - \left(\frac{2i+1}{4N} + \Delta f\right) = \frac{3}{4N} - \Delta f$$

During synthesis, this component, aliased in the band of the filter of index $i$, will be found at the frequency $(2i + 1)/4N + 1/2N - \Delta f$, and it will be subject to attenuation by the synthesis filter of index $i$ – that is, $G(1/2N) - \Delta f$, if $G(Z)$ designates the prototype synthesis filter. Finally, the replicated component will have suffered attenuation:

$$H(\Delta f)G\left(\frac{1}{2N} - \Delta f\right)$$

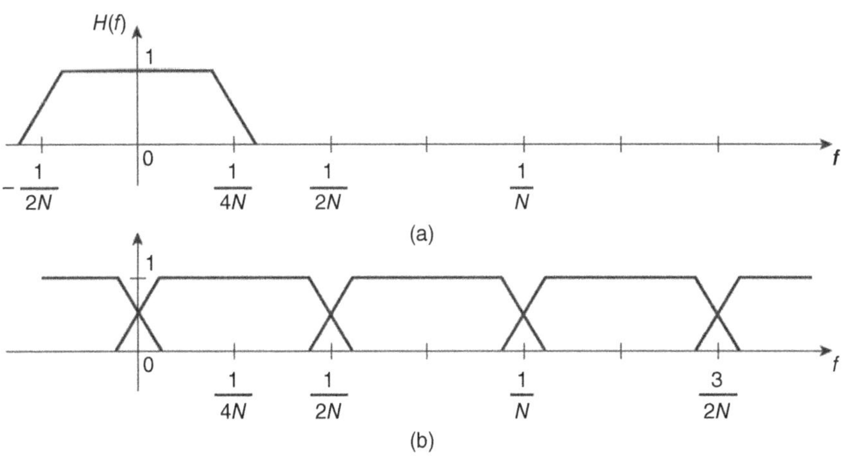


**Figure 12.4** (a) Prototype filter, (b) uniform bank of N real filters.

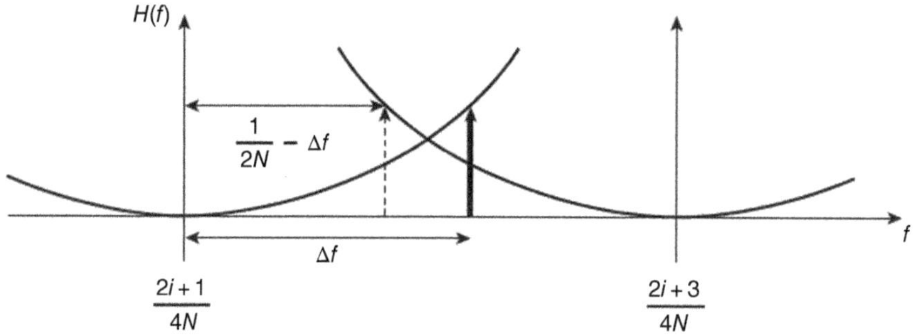


**Figure 12.5** Aliasing of a component in the filter bank.

The same component of the signal will now be processed by the filter of index $i+1$, since it falls in the pass band. Sampling then produces an image component which, during synthesis, will be added to the previous aliased component with attenuation $H(1/2N - \Delta f)G(\Delta f)$. The process is illustrated in Figure 12.5.

Hence, the condition for these components to compensate each other is:

$$\left[H(\Delta f)G\left(\frac{1}{2N} - \Delta f\right)\right]_i + \left[H\left(\frac{1}{2N} - \Delta f\right) G(\Delta f)_{i+1}\right] = 0 \tag{12.17}$$

This condition for the absence of aliasing can be obtained by taking $G(f) = H(f)$ and applying a phase difference of $\pi/2$ between the filters of index $i$ and $i+1$, during analysis and synthesis.

The necessary phase difference can be obtained by introducing phase shifts into the modulation functions – for example, by taking the following values for the coefficients $h_{ik}$ of the analysis filter of index $i$:

$$h_{ik} = 2h_k \cos\left[(2i+1)\frac{\pi}{2N}\left(k - \frac{LN-1}{2}\right) + \theta_i\right] \tag{12.18}$$

with $0 \le i \le N-1$ and $0 \le k \le NL-1$.

For the synthesis filter:

$$g_{ik} = 2h_k \cos\left[(2i+1)\frac{\pi}{2N}\left(k - \frac{LN-1}{2}\right) - \theta_i\right] \tag{12.19}$$

Setting:

$$a_i = e^{j\theta_i} \quad C_i = e^{-j(2i+1)\pi/2N}\frac{LN-1}{2} \tag{12.20}$$

gives the following, for the corresponding transfer functions:

$$H_i(Z) = a_i c_i H(Ze^{-j(2i+1)\pi/2N}) + \bar{a}_i \bar{c}_i H(Ze^{j(2i+1)\pi/2N}) \tag{12.21}$$

$$G_i(Z) = \bar{a}_i c_i H(Ze^{-j(2i+1)\pi/2N}) + a_i \bar{c}_i H(Ze^{j(2i+1)\pi/2N}) \tag{12.22}$$

With the symmetry of the coefficients $h_k$ and alignment of the modulation functions, the following expressions are satisfied:

$$g_{ik} = h_{i(LN-1-k)} \quad G_i(Z) = Z^{-(LN-1)}H_i(Z^{-1}) \tag{12.23}$$

Under these conditions, the total response of the analysis and synthesis systems can be written as:

$$T(Z) = \frac{\hat{X}(Z)}{X(Z)} = \frac{1}{N}\sum_{i=0}^{N-1} H_i(Z)G_i(Z) = \frac{Z^{-(LN-1)}\,N-1}{N}\sum_{i=0}^{N-1} H_i(Z)H_i(Z^{-1}) \tag{12.24}$$

Thus, the global system is of linear phase.

With the assumption that the filters have sufficient attenuation for only the adjacent bands to cause significant aliasing, it is now necessary to determine the values for the angles $\theta_i$ to obtain the desired cancellation.

At the output of the filter $H_i(Z)$, after downsampling, in accordance with equation (10.8), the signal is:

$$X_i(Z^N) = \frac{1}{N}\sum_{m=0}^{N-1} H_i(ZW^m)X(ZW^m) \tag{12.25}$$

The output signal of the system becomes:

$$\hat{X}_i(Z) = \sum_{i=0}^{N-1} G_i(Z)X_i(Z_N) = \frac{1}{N}\sum_{m=0}^{N-1} X(ZW^m)\sum_{i=0}^{N-1} G_i(Z)X_i(ZW^m) \tag{12.26}$$

The condition for perfect reconstruction can then be written as:

$$\sum_{i=0}^{N-1} G_i(Z)H_i(Z) = Z^{-k} \tag{12.27}$$

$$\sum_{i=0}^{N-1} G_i(Z)X_i(ZW^m) = 0; \quad 1 \le m \le N-1 \tag{12.28}$$

To make the aliased components appear to cancel one another out, it is necessary to examine the output signal of each of the synthesis filters. At the output of the filter $G_i(Z)$, the signal $X_i(Z)$ using equation (12.25) can be written as:

$$\hat{X}_i(Z) = G_i(Z)\frac{1}{N}\sum_{m=0}^{N-1} H_i(ZW^m)X(ZW^m) \tag{12.29}$$

However, in light of definition (12.22) and the assumptions concerning attenuation, the filter $G_i(Z)$ allows passage only of the band centered on frequency $(2i+1)/4N$ and the two adjacent bands. Taking account of the distribution of the bands on the frequency axis, the indices $m$ associated with these adjacent bands correspond to a frequency translation such that:

$$\frac{2i+1}{4M} \pm \frac{m}{N} = -\frac{2i+1}{4M} \pm \frac{1}{2N} \tag{12.30}$$

In fact, the downsampling leads to frequency translations which are integral multiples of the frequency $1/N$. Under these conditions, the values of $m$ can be written as:

$$m = \pm i \quad \text{and} \quad m = \pm(i+1)$$

For example, considering the case of Figure 12.5, the aliased component arises from a component at the frequency $-(2i+1)/4N - \Delta f$ shifted by $(i+1)/N$; hence:

$$-\left(\frac{2i+1}{4N} + \Delta f\right) + \frac{i+1}{N} = \frac{2i+3}{4N} - \Delta f$$

Consequently, $\hat{X}_i(Z)$ is limited to the following expansion by using equation (12.21) to define $H_i(Z)$:

$$\begin{aligned}\hat{X}_i(Z) = G_i(Z)\frac{1}{N}[&a_i c_i H(ZW^{-(2i+1)/4})]X(Z) + \bar{a}_i\bar{c}_i H(ZW^{(2i+1)/4})X(Z)\\ &+ a_i c_i H(ZW^{(2i-1)/4})X(ZW^{i}) + \bar{a}_i\bar{c}_i H(ZW^{(1-2i)/4})X(ZW^{-i})\\ &+ a_i c_i H(ZW^{(2i+3)/4})X(ZW^{i+1}) + \bar{a}_i\bar{c}_i H(ZW^{-(2i+3)/4})X(ZW^{-(i+1)})\end{aligned} \tag{12.31}$$

Since:

$$\hat{X}(Z) = \sum_{i=0}^{N-1} \hat{X}_i(Z) \tag{12.32}$$

the aliases in $\hat{X}(Z)$ cancel if the high band of $\hat{X}_i(Z)$ compensates for the low band of $\hat{X}_{i+1}(Z)$. The corresponding condition is obtained by inserting definition (12.22) for $G_i(Z)$ into the expression for $\hat{X}_i(Z)$ and writing the same expression for $\hat{X}_{i+1}(Z)$. The factors of $X[ZW^{i+1}]$ and $X[ZW^{-(i+1)}]$ cancel if the following condition is satisfied:

$$a_i^2 c_i \bar{c}_i H(ZW^{(2i+3)/4}) H(ZW^{(2i+1)/4}) + a_{i+1}^2 c_{i+1} \bar{c}_{i+1} H(ZW^{(2i+1)/4}) H(ZW^{(2i+1)}/4) = 0$$

that is, if:

$$a_{i+1}^2 = -a_i^2 \tag{12.33}$$

It is therefore necessary that the phase shifts satisfy:

$$\theta_{i+1} = \theta_i + \frac{\pi}{2}; \quad 0 \le i \le N-1 \tag{12.34}$$

The first condition for perfect reconstruction (12.27) can thus be written as:

$$\begin{aligned} &\frac{1}{N}\sum_{1=1}^{N-1}[c_i H(ZW^{(2i+1)/4})]^2 + [\bar{c}_i H(ZW^{-(2i+1)/4})] \\ &\qquad + \left(a_0^2 + \bar{a}_0^2\right) H(ZW^{1/4}) H(ZW^{-1/4}) \\ &\qquad + \left(a_{N-1}^2 + \bar{a}_{N-1}^2\right) H(ZW^{(N-1)4}) H(ZW^{-(N-1)4}) = 1 \end{aligned} \tag{12.35}$$

since the cross products do not cancel at the origin and at the sampling half-frequency as the filters are adjacent.

For the cross products to vanish, it is necessary to take:

$$\theta_i = (-1)^i \frac{\pi}{4} \tag{12.36}$$

and the imposed condition for calculation of the frequency response of the prototype filter can finally be written as:

$$\begin{aligned} &|H(f)|^2 + \left|\mathrm{H}\left(\frac{1}{2N} - f\right)\right|^2 = 1 \quad 0 \le f < \frac{1}{2N} \\ &H(f) = 0 \quad \frac{1}{2N} \le f \le \frac{1}{2} \end{aligned} \tag{12.37}$$

Note that the following are also possible for the phase shifts in the analysis and synthesis banks:

(1) $\theta i = (i+1)\pi/2$ – the overall system response cancels at frequencies 0 and 1/2.
(2) $\theta i = i\pi/2$ – the overall response doubles at frequencies 0 and 1/2.

In summary, the design procedure for a bank of $N$ real pseudo-QMF filters consists of the following two operations:

(3) Design of the prototype linear phase filter meeting the specification (12.37), with $LN$ coefficients.
(4) Determination of the transfer functions of the analysis and synthesis filters using the expressions:

$$H_i(Z) = 2\sum_{k=0}^{LN-1} \cos\left[2\pi\left(\frac{2i+1}{4N}\right)\left(k - \frac{LN-1}{2}\right) + (2i+1)\frac{\pi}{4}\right] h_k Z^{-k} \tag{12.38}$$

$$G_i(Z) = 2\sum_{k=0}^{LN-1} \cos\left[2\pi\left(\frac{2i+1}{4N}\right)\left(k - \frac{LN-1}{2}\right) - (2i+1)\frac{\pi}{4}\right] h_k Z^{-k} \tag{12.39}$$

The specification of the prototype filter reflects the required separation between the sub-bands. Several approaches to the design can be envisaged.

## 12.5 Determining the Coefficients of the Prototype Filter

It is the design of a half-Nyquist filter, and the first approach consists of taking a cosine transition band and using (5.37). However, the sampling frequency technique set forth in Section 5.4 may be more efficient. In particular, when the transition band equals sub-band spacing, the coefficients are derived from a simple formula [4].

Let $K$ be a whole number and consider a set of $KN$ samples $H_k$ $(0 \leq k \leq KN - 1)$ in the frequency domain:

$$H_0 = 1 \tag{12.40}$$

$$H_k^2 + H_{K-k}^2 = 1; H_{KN-k} = H_k; 1 \leq k \leq K - 1$$

$$H_k = 0; K \leq k \leq KN - K$$

Generally, the number $N$ of sub-bands is even. The filter coefficients $h_i$ $(0 \leq i \leq KN - 1)$ are obtained with the inverse DFT.

The following constraint is imposed:

$$H_0 + 2\sum_{k=1}^{K-1} (-1)^k H_k = 0 \tag{12.41}$$

Then, the middle coefficient $h_{KN/2}$ is null, and the filter has an odd number of coefficients. There is a double zero at half the sampling frequency, as indicated in Section 5.12, and high attenuation results for high frequencies.

For $K = 3$ and $K = 4$, equations (12.40 and 12.41) define the frequency samples:

$$H_1 = 0.911438; H_2 = 0.411438$$

and

$$H_1 = 0.971960; H_2 = \frac{1}{\sqrt{2}}; H_3 = 0.235147 \tag{12.42}$$

For example, a bank of $N = 16$ filters and $K = 4$ has the coefficients:

$$h_{i+1} = 1 + 2\sum_{k=1}^{K-1} (-1)^k H_k \cos\left(\frac{2\pi ki}{KN}\right); 1 \leq i \leq 63 \tag{12.43}$$

$$h_1 = 0$$

The filter obtained has an odd number of coefficients. The frequency response, whose transition band is equal to 1/16, is shown in Figure 12.6. The attenuation grows with frequency, but it is possible to obtain a constant attenuation using the method indicated in Section 11.2 with appropriate specifications. Iterative techniques may be used in addition to the procedure to better approximate the symmetry condition (12.37).

Once the coefficients are determined, the calculations should be set out and performed so that the number of arithmetic operations is minimal.

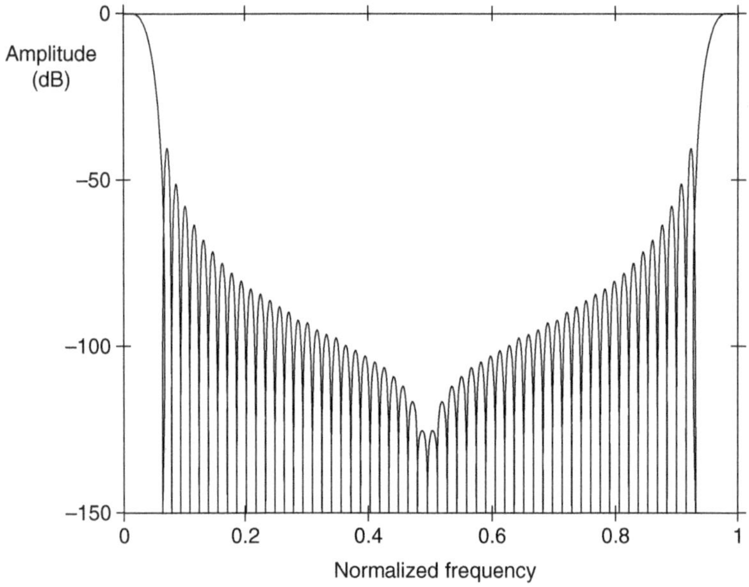


**Figure 12.6** Prototype filter obtained by frequency weighting for a bank of 16 filters and 64 coefficients.

## 12.6 Realizing a Bank of Real Filters

Consider a bank of real filters having the frequency responses shown in Figure 12.4 and an even number of coefficients $2LN$, in which the filter of index $i$ has the coefficients:

$$h_{ik} = h_k \cos\left[\frac{2\pi}{2N}\left(i+\frac{1}{2}\right)\left(k+\frac{1}{2}\right)\right] \tag{12.44}$$

with $0 \leq i \leq N, 0 \leq k \leq 2LN - 1$.

A decomposition into a polyphase network and a discrete Fourier transform can be obtained by setting $k = 2Nl + m$ with $0 \leq l ł \leq L - 1$ and $0 \leq m \leq 2N - 1$.

The output $x_i(n)$ for the filter of index $i$ can be written as:

$$x_i(n) = \sum_{k=0}^{2LN-1} x(n-k)h_k \cos\left[\frac{2\pi}{2N}\left(i+\frac{1}{2}\right)\left(k+\frac{1}{2}\right)\right] \tag{12.45}$$

or by replacing $k$ with $2Nl + m$ and simplifying:

$$x_i(n) = \sum_{m=0}^{2N-1} \cos\frac{2\pi}{2N}\left(i+\frac{1}{2}\right)\left(m+\frac{1}{2}\right)\sum_{l=0}^{L-1}(-1)^l h_{2Nl+m^x}(n-2Nl-m) \tag{12.46}$$

Applying the general transfer function decomposition (10.36) to the prototype filter gives:

$$H(Z) = \sum_{m=0}^{2N-1} Z^{-m}H_m(Z^{2N}) \tag{12.47}$$

and the filters $H_m(Z^{2N})$ are those which arise in the second summation of expression (12.46). To take account of the factor $(-1)^l$, it is sufficient to introduce the functions $H_m(-Z^{2N})$, and the diagram corresponding to the analysis filters is shown in Figure 12.7. The decomposition produces a polyphase network with $2N$ branches, and a cosine transform.

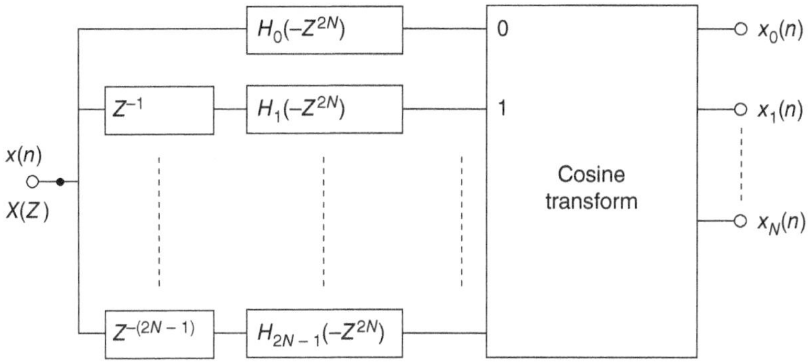


**Figure 12.7** Arrangement of a bank of N real filters.

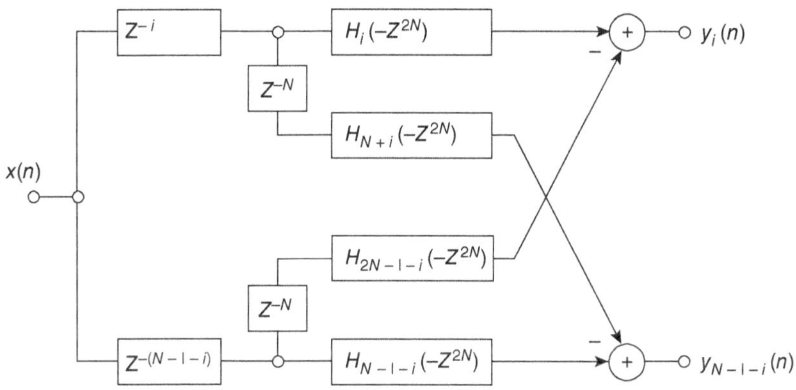


**Figure 12.8** Lattice structure for the polyphase network.

Figure 12.7 can be further simplified since, in the cosine transform considered, two symmetrical inputs are subjected to the same operations, apart from the sign. Factorizing yields an odd-time odd-frequency cosine transform – a special case mentioned in Section 3.3.2. Furthermore, with downsampling, the system operates at a rate $1/N$. As the system consists of $2N$ branches, a given $x(n)$ is processed by the filters $H_i$ and $H_{i+N}$ at two successive instants. Under these conditions, the $2N$ branches of the polyphase network can be regrouped as $N/2$ subgroups having the lattice structure shown in Figure 12.8. The overall configuration is therefore as illustrated in Figure 12.9. In the case of pseudo-QMF filters, this arrangement is applicable with the introduction of phase shifts. In fact, by taking equation (12.45) with the coefficients of the filter given in equation (12.38), the output $x_i(n)$ of the filter of index $i$ with $2LN$ coefficients can be written as:

$$x_i(n) = 2\sum_{k=0}^{2LN-1} x(n-k)h_k \cos\left[2\pi\frac{2i+1}{4N}\left(k - \frac{2LN-1}{2}\right) + (2i+1)\frac{\pi}{4}\right] \tag{12.48}$$

By setting $k = lN + m$ this time, the following double summation is obtained:

$$x_i(n) = 2\sum_{m=0}^{N-1}\sum_{l=0}^{2L-1} \cos\left[(2i+1)(2m+1+N)\frac{\pi}{4N}\cdot\right.$$
$$\left. + (2i+1)(l-L)\pi/2\right] h_{lN+m^x}(n-lN-m) \tag{12.49}$$

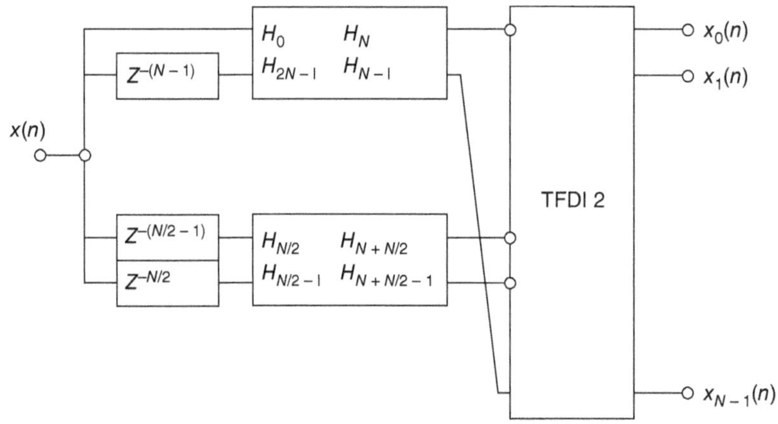


**Figure 12.9** Optimized structure of a bank of N real filters.

A cosine expansion produces these terms:

$$\cos(2i+1)(l-L)\frac{\pi}{2} = \cos(l-L)\frac{\pi}{2}$$

and:

$$\sin(2i+1)(l-L)\frac{\pi}{2} = (-1)^i \sin(l-L)\frac{\pi}{2}$$

Furthermore, the following expression is satisfied:

$$\cos\left[(2i+1)[2(N-1-m)+1]\frac{\pi}{4N}\right] = (-1)^i \sin\left[(2i+1)(2m+1)\frac{\pi}{4N}\right]$$

Finally, to take account of the term $N$ in the cosine argument of (12.49), it is sufficient to scale the data by a factor of $N/2$. By combining all these results, the output $x_i(n)$ is determined as follows:

$$x_i(n) = 2\sum_{m=0}^{N-1} \cos\left[(2i+1)(2m+1)\frac{\pi}{4N}\right] y_m(n) \tag{12.50}$$

with:

$$y_m(n) = [-y_{2,N/2-1-m}(n) + y_{2,N/2+m}(n)]; \quad 0 \le m \le \frac{1}{2}N-1$$

$$y_m(n) = [y_{1,m-N/2}(n) - y_{1,3N/2-M-1}(n)]; \quad N/2 \le m \le N-1$$

and:

$$y_{1,m}(n) = \sum_{m=0}^{2L-1} \cos(l-L)\frac{\pi}{2} h_{lN+m} x(n-lN-m)$$

$$y_{2,m}(n) = \sum_{m=0}^{2L-1} \sin(l-L)\frac{\pi}{2} h_{lN+m} x(n-lN-m)$$

The sequences $y_{1,m}(n)$ and $y_{2,m}(n)$ are interleaved, with sampling frequency $f_s/2N$, and the bank of analysis filters is determined with an odd-time odd-frequency cosine transform. Finally, the phase shifts introduced by the pseudo-QMF technique have been taken into account simply by rearranging the data before the transform.

## Exercises

**12.1** In a bank of four filters, the prototype is linear phase, and it has the coefficients:

$$h = [-0.0218 - 0.0200 - 0.0116\ \ 0.0160\ \ 0.0621\ \ 0.1181\ \ 0.1691\ \ 0.1996]$$

Give the polyphase-FFT (PPN-FFT) decomposition. The sampling frequency is unity. Check that the amplitude of the frequency response in the pass band is the same for all four branches. The phases are linear and the values at frequency 1/8 in radian

$$\text{phase} = [-1.4829 - 1.2459 - 1.0366\ \ 0.8002]$$

Calculate the delay brought by each branch and justify the results.

**12.2** A filter bank can be realized with a FFT whose size equals the number of coefficients of the prototype filter. Justify the approach, using the results in Chapter 2. Compute the coefficients. How are they placed in the analysis and synthesis parts?
As an illustration, the example in Section 12.5 is considered with $K = 4$ and $N = 16$. Give the values of the seven coefficients to apply in front of the summation that yields the output of each filter. The base filter, in the synthesis part, is centered on the FFT output of index 0. At which FFT output are the seven coefficients applied? Answer the same question for the neighboring filter centered on index 4.
Describe the application to the synthesis filter bank.

**12.3** A bank of $N = 16$ filters is used to decompose a signal. The $2N$ coefficients of the prototype filter are the following:

$$h(n) = \sin\left(\frac{\pi}{2N}\left(n + \frac{1}{2}\right)\right); 0 \leq n \leq 2N - 1$$

What is the amplitude $H(f)$ of the frequency response? (Refer to Section 5.8 and find an approximate expression corresponding to the continuous case.)
Compare with the filter associated with the DFT of size $N = 16$, considering the coefficients $\boldsymbol{h_{FFT}(n) = 1; 0 \leq n \leq 15}$.
In a PPN-FFT realization, give the expression of the coefficients in each of the 16 branches.
An alternative realization is based on a FFT of size $2N = 32$. What processing must be introduced at FFT output to realize the prototype filter? Compare the number of multiplications with the previous approach.

## References


1 M. Bellanger and J. Daguet, TDM-FDM transmultiplexer: digital polyphase and FFT. *IEEE Transactions on Communications*, 22(9), 1199–1205, 1974.
2 R. E. Crochière and L. R. Rabiner, *Multirate Digital Signal Processing*, Prentice-Hall Inc., Englewood Cliffs, NJ, 1983.
3 N. J. Fliege, *Multirate Digital Signal Processing*, John Wiley, Chichester, 1994.
4 K. W. Martin, Small side-lobe filter design for multitone data communications. *IEEE Transactions on Circuits and Systems II*, 45(8), 1155–1161, 1998.

# 13

# Signal Analysis and Modeling

The modeling of systems is one of the most important areas of signal processing. Furthermore, modeling is an alternative approach to signal analysis, with properties differing from those of the Fourier transform and those of filters defined in the frequency domain. Linear prediction, in particular, is a simple and efficient tool to characterize some signal types and then compress them. The processing is specified in the time domain, using statistical parameters and, particularly, correlation.

## 13.1 Autocorrelation and Intercorrelation

The degree of similarity between two signals can be described using a correlation coefficient, which should logically take the values +1 for two identical signals, zero for two signals which have no relationship to each other, and −1 for signals in opposition to each other. When time-dependent signals are compared, the correlation coefficient becomes a function of time. It is called the intercorrelation function if the signals are different and the autocorrelation (AC) function if they are the same.

Some definitions and properties will now be restated in order to recapitulate and supplement the discussions in Sections 1.8 and 4.4.

As shown in Section 1.8, the AC function of a random discrete signal $x(n)$ is the set $r_{xx}(n)$ such that:

$$r_{xx}(n) = E[x(i)x(i-n)] \tag{13.1}$$

where $E[x]$ denotes the expectation value of $x$.

With the assumption of ergodicity, this becomes:

$$r_{xx}(n) = \lim_{N\to\infty} \frac{1}{2N+1} \sum_{i=-N}^{N} x(i)x(i-n) \tag{13.2}$$

The function $r_{xx}(n)$ is even. Its value at the origin is the power of the signal and, for any $n$, is:

$$|r_{xx}(n)| \leq r_{xx}(0) \tag{13.3}$$

Consider a set of $N$ coefficients $a_i$ $(1 \leq i \leq N)$. Calculation of the variance of the variable $y(n)$ such that:

$$y(n) = \sum_{i=1}^{N} a_i x(n-i)$$

*Digital Signal Processing: Theory and Practice*, Tenth Edition. Maurice Bellanger.

results in:

$$E[y^2(n)] = \sum_{i=1}^{N}\sum_{j=1}^{N} a_i a_j E[x(n-i)x(n-j)]$$

or,

$$E[y^2(n)] = \sum_{i=1}^{N}\sum_{j=1}^{N} a_i a_j r_{xx}(i-j) \tag{13.4}$$

As this variance is positive or zero, we have:

$$\sum_{i=1}^{N}\sum_{j=1}^{N} a_i a_j r_{xx}(i-j) \geq 0 \tag{13.5}$$

This property characterizes positive functions. If, in the definition of equation (13.1), $x(i-n)$ is replaced by another signal, a function is obtained which allows two different signals to be compared. The intercorrelation function between two discrete signals $x(n)$ and $y(n)$ is the set $r_{xy}(n)$ such that:

$$r_{xy}(n) = E[x(i)y(i-n)] \tag{13.6}$$

With the assumption of ergodicity,

$$r_{xy}(n) \lim_{N\to\infty} \frac{1}{2N+1} \sum_{i=-N}^{N} x(i)y(i-n) \tag{13.7}$$

Similarly,

$$r_{xy}(-n) = E[x(i)y(i+n)] = r_{xy}(n) \tag{13.8}$$

For example, if the signals are the input and output of a filter:

$$y(n) = \sum_{j=0}^{\infty} h_j x(n-j)$$

then, as shown in Section 4.4, one has:

$$r_{xy}(n) = E[y(i)x(i-n)] = \sum_{j=0}^{\infty} h_j r_{xx}(n-j)$$

or,

$$r_{yx}(n) = r_{xx}(n) * h(n) \tag{13.9}$$

Similarly,

$$r_{xy}(n) = r_{xx}(n) * h(-n) \tag{13.10}$$

and also:

$$r_{yy}(n) = r_{xx}(n) * h(n) * h(-n) \tag{13.11}$$

If two random signals are independent, their intercorrelation functions are zero. Further, the following inequality is always valid:

$$|r_{xy}(n)| \leq \frac{1}{2}[r_{xx}(0) + r_{yy}(0)] \tag{13.12}$$

It is worth mentioning that the AC and intercorrelation functions can, in some cases, be computed without multiplication, the signals being replaced by their signs. Thus, if $x(n)$ is a Gaussian signal:

$$r_{xx}(n) = \sqrt{\frac{\pi}{2}}\sqrt{r_{xx}(0)}E[x(i)\text{sing}(x(i-n))] \tag{13.13}$$

$$r_{xx}(n) = r_{xx}(0)\sin\left[\frac{\pi}{2}E[\text{sing}[x(i)x(i-n)]]\right] \tag{13.14}$$

These expressions can greatly simplify the equipment required.

The Fourier transform $\Phi_{xy}(f)$ of the intercorrelation function $r_{xy}(n)$ is called the interspectrum:

$$\Phi_{xy}(f) = X(f)\overline{Y(f)}$$

where $X(f)$ denotes the spectrum of the set $x(n)$ and $\overline{Y(f)}$ is the conjugate spectrum of the set $y(n)$.

If the set $y(n)$ is the output of a filter with transfer function $H(f)$, then:

$$H(f) = \frac{Y(f)}{X(f)} = \frac{Y(f)\overline{X(f)}}{X(f)\overline{X(f)}}$$

Hence,

$$\Phi_{yx}(f) = \Phi_{xx}(f)H(f) \tag{13.15}$$

which corresponds to equation (13.9). Similarly, for equation (13.10), we have:

$$\Phi_{yx}(f) = \Phi_{xx}(f)\overline{H(f)}$$

and finally,

$$\Phi_{yy}(f) = \Phi_{xx}(f)|H(f)|^2 \tag{13.16}$$

These results apply to the spectrum analysis of random signals in general and are useful for the study of adaptive systems.

## 13.2 Correlogram Spectral Analysis

The Fourier transform of the AC function is the spectral power density:

$$S(f) = \sum_{p=-\infty}^{\infty} r(p)e^{-j2\pi pf} \tag{13.17}$$

The AC function of the signal is $r(p)$. In practice, the analysis is performed using a limited number, $N_0$, of signal samples. Therefore, $r(p)$ must be estimated first.

An initial estimation of the AC function is:

$$r_1(p) = \frac{1}{N_0}\sum_{n=p+1}^{N_0} x(n)x(n-p) \tag{13.18}$$

It is biased because its expectation is:

$$E[r_1(p)] = \frac{N_0 - p}{N_0} r(p)$$

The non-biased estimation is given by:

$$r_2(p) = \frac{1}{N_0 - p} \sum_{n=p+1}^{N_0} x(n)x(n-p) \tag{13.19}$$

From P values of the AC function, the so-called correlogram spectral estimation is defined by:

$$S_{CR}(f) = \sum_{p=-(P-1)}^{P-1} r_2(p)e^{-j2\pi pf} \tag{13.20}$$

as a function of the theoretical spectrum:

$$S_{CR}(f) = S(f) * \frac{\sin \pi f(2P-1)}{\sin \pi f} \tag{13.21}$$

The variance is approximated as follows:

$$\mathrm{Var}\{S_{CR}(f)\} \approx \frac{2P-1}{N_0} S^2(f) \tag{13.22}$$

It is apparent that the minimum number of AC values should be used to perform the estimation, which means that only the most significant values of the AC estimation should be retained [1].

In fact, the most direct approach to calculate the signal power spectrum from $N_0$ samples consists of applying the DFT and taking the squared output values $|X(k)|^2$. It can readily be shown that the biased estimation (13.18) is obtained.

An important particular case is the single-frequency signal.

## 13.3 Single-Frequency Estimation

A single frequency $f$ is to be estimated among a set of spurious components which can be seen as an additive white Gaussian noise (AWGN). If $N$ samples $x(n)$ $(0 \le n \le N-1)$ are available, the attainable accuracy depends on $N$ and on the signal-to-noise ratio (SNR). A lower bound of the estimation variance is (see Appendix):

$$\mathrm{BCR}_f = \frac{6}{4\pi^2 \mathrm{SB}\, N^3} \tag{13.23}$$

In practice, the theoretical bound above can be closely approximated, provided the SNR remains greater than a threshold, whose magnitude is roughly:

$$\mathrm{SB}_{\min} \approx \frac{29}{N} \tag{13.24}$$

The simplest estimation technique is based on a DFT of size $N$, and the procedure is made up of two steps. First, a rough estimate is derived, taking the index of the output with maximum magnitude, and then, it is refined through interpolation [2].

Assuming the sampling frequency is unity, the signal is written as:

$$x(n) = Ae^{j(2\pi fn+\varphi)} + b(n) \tag{13.25}$$

$A, f$, and $\varphi$ are the amplitude, frequency, and phase of the component to be identified; the additive noise is $b(n)$. The first step is based on the decomposition:

$$f = \frac{k_0}{N} + \delta f \text{ with } k_0 \text{ integer and } -\frac{1}{2N} < \delta f < \frac{1}{2N} \tag{13.26}$$

The DFT output $y(k)$ which has the maximum amplitude provides $k_0$ in (13.26). After a shift about the origin ($k_0 = 0$), we get:

$$y(k) = Ae^{j\varphi}\frac{\sin\left(N\pi\left(\delta f - \frac{k}{N}\right)\right)}{N\,\sin\left(\pi\left(\delta f - \frac{k}{N}\right)\right)}e^{j\pi\delta f(N-1)} + \beta(k);\quad 0 \le k \le N-1 \tag{13.27}$$

$\beta(k)$ is the noise component in the frequency domain. Since the phase is unknown, in order to assess the frequency deviation, it is necessary to get rid of the phase terms, and one can use the real function:

$$y'(k) = A\frac{\sin\left(N\pi\left(\delta f - \frac{k}{N}\right)\right)}{N\sin\left(\pi\left(\delta f - \frac{k}{N}\right)\right)} + |\beta(k)|e^{j\pi k/N} \tag{13.28}$$

which is obtained by restoring the signs of the $\sin(x)/x$ function in $|y(k)|$. Of course, ambiguity is introduced, and we only know that indices $k$ and $N-k$ are associated with opposite signs. The ambiguity can be solved by observing the sign of the difference abs($y(N-1)$)–abs($y(1)$). From (13.28), it is possible to apply the least squares method to minimize the cost function:

$$J = \sum_{k=-\frac{N}{2}+1}^{\frac{N}{2}-1}\left[y'(k) - \frac{\sin(N\pi(\delta f - k/N))}{N\sin(\pi(\delta f - k/N))}\right]^2 \tag{13.29}$$

The minimum can be reached through iteration – for example, using the procedure set forth in Section 5.4. A direct estimation is obtained by series development of the cardinal sine function. For example, assuming the amplitude A has been estimated and keeping only 3 terms in the cost function (13.29), we obtain:

$$\delta f \approx \frac{1}{N}\frac{\text{abs}(y(1)) + \text{abs}(y(-1))}{[\text{abs}(y(0)) - 1]\,\frac{\pi^2}{3} + 2} \tag{13.30}$$

The estimation is valid in the interval $[-1/2N\ 1/2N]$, and it is degraded in the vicinity of the bounds. For example, for $N = 16$, $1/2N = 0.031$ and $\delta f = 0.01$, we obtain the estimation $\delta f_{estim} = 0.0106$.

Iterative techniques are needed to approach the bound (13.23). A particularly efficient technique is called iterated correlations, according to which the amplitude and the frequency of the sine wave can be estimated independently of the phase. The procedure begins with calculating the AC function of the signal $x(n)$ by:

$$[r_1] = \text{TFD}_{2N}^{-1}\left\{|\text{TFD}_{2N}(x)|^2\right\} \tag{13.31}$$

In order to apply a DFT of size $2N$, $N$ zeros are appended to the sequence $x(n)$. The frequency domain interpolation (see Section 9.10) leads to the AC function because the squaring in the frequency domain amounts to convolution of the signal by its conjugate. Thus, we obtain:

$$r_1(p) = \frac{1}{N}\sum_{n=p}^{N-1}x(n)x^*(n-p) = \frac{N-p}{N}e^{j2\pi pf} + \frac{1}{N}\sum_{n=p}^{N-1}b_1(n);\quad 0 \le p \le N-1 \tag{13.32}$$

The noise component is:

$$b_1(n) = b(n)e^{-j2\pi(n-p)f} + b(n-p)e^{j2\pi pf} + b(n)b^*(n-p) \tag{13.33}$$

Two observations can be made:

- $r_1$ (0) is an estimation of the signal power.
- The sequence $r_1$ ($p$) is a new signal with decreasing SNR as $p$ grows.

In such conditions, it is advisable to keep only the first values for the next step:

$$x_1(n) = r_1(n); \quad x_1\left(\frac{N}{2}+n\right) = 0; \quad 0 \leq n \leq \frac{N}{2} - 1 \tag{13.34}$$

Then, the following calculation is carried out:

$$[r_2] = \text{TFD}_N^{-1}\left\{|\text{TFD}_N(x_1)|^2\right\}$$

The sequence $[r_2]$ has $N$ values, of which only $N/4$ are kept for the next step. The procedure stops at index $q$ such that $[r_q]$ has only four values. Finally, we find:

$$r_1(0) = A^2; \quad \frac{r_q(1)}{r_q(0)} = e^{j2\pi f} \tag{13.35}$$

This algorithm offers very high accuracy, for any frequency to be estimated. For example, if $N = 16, f = 0.1$, and SNR = 100 (20 dB), the variance of the estimated frequency and the bound are:

$$\text{var} = 4.3 \times 10^{-7}; \quad \text{BCR}_f = 3.7 \times 10^{-7}$$

It can be verified that the SNR threshold is close to $\text{SNR}_{\min} = 29/16$.

With regard to the estimation of a real frequency, refer to exercise 1.

## 13.4 Correlation Matrix

The AC matrix of dimension $N$ of a signal is defined by:

$$R_N = \begin{bmatrix} r(0) & r(1) & \cdots & r(N-1) \\ r(1) & r(0) & \cdots & r(N-2) \\ \vdots & \vdots & \vdots & \vdots \\ r(N-1) & r(N-2) & \cdots & r(0) \end{bmatrix} \tag{13.36}$$

with:

$$r(p) = E[x(n)x(n-p)]$$

As the AC function is positive (13.5), the AC matrix is positive and symmetric by definition. In fact, it has double symmetry as it is also symmetric about the second diagonal. This leads to a set of fundamental properties for adaptive systems.

The eigenvalues $\lambda_i$ $(0 \leq i \leq N-1)$ of the AC matrix of order $N$ will be considered first. The characteristic equation:

$$\det(\lambda I_N - R_N) = 0$$

leads to the equations:

$$\det(R_N) = \prod_{i=0}^{N-1} \lambda_i \tag{13.37}$$

$$Nr(0) = \sum_{i=0}^{N-1} \lambda_i = N\sigma_x^2 \tag{13.38}$$

That is, if the determinant of the matrix is nonzero, each eigenvalue is nonzero, and their sum is equal to $N$ times the power of the signal. The positive nature of the matrix $R_N$ further implies that they are all positive:

$$\lambda_i > 0; \quad 0 \leq i \leq N-1 \tag{13.39}$$

To ensure this, it is necessary and sufficient that the following determinants all be positive:

$$r(0);\ \det\begin{bmatrix} r(0) & r(1) \\ r(1) & r(0) \end{bmatrix};\ \ldots;\ \det\begin{bmatrix} r(1) & r(1) & \cdots & r(N-1) \\ r(1) & r(0) & \cdots & \vdots \\ \vdots & \vdots & \vdots & \vdots \\ r(N-1) & \cdots & \cdots & r(0) \end{bmatrix}$$

The corresponding matrices are the AC matrices of order less than or equal to $N$.

Under these conditions, the matrix $R_N$ can be diagonalized so that:

$$R_N = M^t \operatorname{diag}(\lambda_i) M \tag{13.40}$$

where $M$ is a square matrix of dimension $N$, such that $M^t = M^{-1}$, and $\operatorname{diag}(\lambda_i)$ is the diagonal matrix of the eigenvalues. $M^t$ can be equal to $M$ in some cases.

The matrix can be expressed in terms of its normalized eigenvectors $U_i$ $(0 \leq i \leq N-1)$ by:

$$R_N = \sum_{i=0}^{N-1} \lambda_i U_i U_i^t \tag{13.41}$$

It is useful to examine successive powers of the matrices $R_N$ and $R_N^{-1}$. By using both the Cayley–Hamilton theorem (according to which a matrix satisfies its own characteristic equation), and the Lagrange interpolation formula (which has already been used in Section 5.5), the power of a matrix can be expressed as a function of the powers of its eigenvalues:

$$R_N^P = \sum_{i=0}^{N-1} \lambda_i^P \prod_{\substack{j=0 \\ j\neq 0}}^{N-1} \frac{R_N - \lambda_j I_N}{\lambda_i - \lambda_j} \tag{13.42}$$

For large values of the integer $P$, with:

$$\lambda_{\max} = \max_{0 \leq i \leq N-1} (\lambda_i)$$

and, if this maximum corresponds to the value for $i = 0$, one can write:

$$R_N^P \approx \lambda_{\max}^P \prod_{j=1}^{N-1} \frac{R_N - \lambda_j I_N}{\lambda_{\max} \quad \lambda_j} \tag{13.43}$$

Consequently, for large values of $P$, one can make the approximation:

$$R_N^P \approx \lambda_{\max}^P K_N \quad (K_N\text{: Square matrix of dimension } N) \tag{13.44}$$

where $K_N$ is the square matrix of order $N$ of equation (13.43); as the matrix $R_N$ can be diagonalized, and satisfies (13.41), $K_N$ can also be expressed more simply as the product of $M^{-1}$ and a matrix deduced from $M$ and setting all rows to zero except those which correspond to the index of the highest eigenvalue. Similarly, using equation (13.41), with the same conditions, one can write:

$$\left(R_N^{-1}\right)^P \approx \left(\lambda_{\min}^{1}\right)^P K_N' \quad \left(K_N'\text{: Square matrix of dimension } N\right) \tag{13.45}$$

with:

$$\lambda_{\min} = \min_{0 \leq i \leq N-1} (\lambda_i)$$

It will be shown below that these two extreme eigenvalues, $\lambda_{max}$ and $\lambda_{min}$, condition the behavior of adaptive systems.

The physical interpretation of the eigenvalues of the AC matrix is not readily apparent from their definition, but it can be illustrated by comparing them with the spectrum of the signal $x(n)$.

The case where the signal $x(n)$ is periodic and has period $N$ will be considered first. In this case, the set $r(n)$ is also periodic, and is also symmetrical with:

$$r(N-i) = r(i); \quad 0 \leq i \leq N-1$$

Under these conditions, the matrix $R_N$ is a rotating matrix in which each row is derived from the preceding one by shifting. If the set $\Phi_{xx}(n)$ $(0 \leq n \leq N-1)$ denotes the Fourier transform of the set $r(n)$, then it can be shown directly that:

$$R_N T_N = T_N \text{diag}(\Phi_{xx}(n))$$

where $T_N$ is the matrix of the discrete Fourier transform of order $N$. Hence

$$R_N T_N = T_N \text{diag}\left(\Phi_{xx}(n) T_N^{-1}\right) \tag{13.46}$$

Comparison with equation (13.26) shows that, in this case, the eigenvalues of the matrix $R_N$ are the discrete Fourier transform of the AC function – that is, the values of the signal power spectral density. $M$ is the cosine transform matrix.

This relation is also valid for discrete white noise as the spectrum is constant and, since the AC matrix is a unit matrix (to a factor), the eigenvalues are equal.

Real signals generally have a spectral density with nonconstant power, and their AC function $r(p)$ decreases as the index $p$ increases. For sufficiently large $N$, the significant elements of the $N$-dimensional matrix can be regrouped around the principal diagonal. Under these conditions, let $R'_N$ be the AC matrix of a signal $x(n)$ which is assumed to be periodic with period $N$. Its eigenvalues $\Phi_{xx}(n)$ form a sample of the power spectral density. The discrepancy between $R_N$ and $R'_N$ is due to the fact that $R'_N$ is a rotating matrix and the difference appears primarily in the upper right-hand and lower left-hand corners. Thus, $R_N$ can be better approximated by a diagonal matrix than $R'_N$, and consequently, its eigenvalues are less dispersed. In fact, under certain conditions which commonly occur in practice, it can be shown that [3]:

$$\min_{0 \leq n \leq N-1} \Phi_{xx}(n) \leq \lambda_{min} \leq \lambda_{max} \leq \max_{0 \leq n \leq N-1} \Phi_{xx}(n) \tag{13.47}$$

and for sufficiently large $N$:

$$\lambda_{min} \approx \min_{0 \leq f \leq 1} \Phi_{xx}(f); \quad \lambda_{max} \approx \max_{0 \leq f \leq 1} \Phi_{xx}(f) \tag{13.48}$$

In conclusion, it can be considered in practice that, when the dimension of this matrix is sufficiently large, the extreme eigenvalues of the AC matrix approximate the extreme values of the power spectral density of the signal.

## 13.5 Modeling

Digital filters apply to system modeling, as shown in Figure 13.1. The signal $x(n)$ is fed to the system and the model, and the coefficients are computed so that the difference in the outputs is minimized.

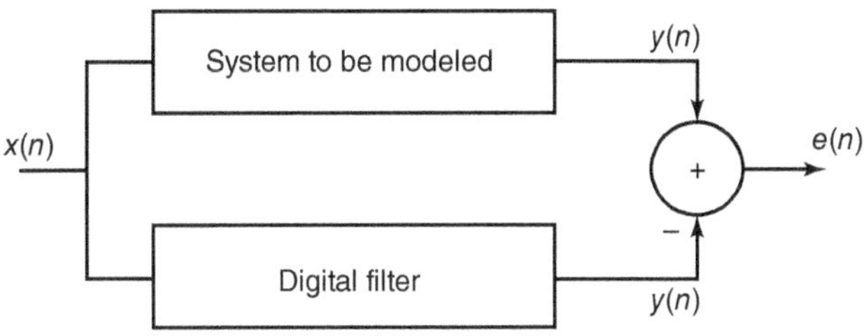


**Figure 13.1** System modeling.

Depending on the knowledge available to begin with, one of a number of filters may be used as a model. However, FIR filters are chosen in general, due to their ease of design and implementation, and the output is:

$$\tilde{y}(n) = \sum_{i=0}^{N-1} h_i x(n-i) = H^t X(n) \tag{13.49}$$

where $X(n)$ is the vector of the $N$ most recent data. The number of coefficients $N$ is chosen in accordance with information available about the model. The output error is defined by:

$$e(n) = y(n) - \tilde{y}(n) \tag{13.50}$$

Next, the coefficients are calculated using the minimum mean square error (MMSE) criterion with cost function:

$$J = E[e^2(n)] \tag{13.51}$$

Canceling the derivatives of this cost function produces the equation $E[e(n)X(n)] = 0$, which defines the decorrelation of the output sequence and the vector of the most recent input data. Then, we get:

$$E[y(n)X(n)] - E[X(n)X^t(n)]H = 0 \tag{13.52}$$

Definition (13.1) of the AC function and (13.36) of the AC matrix show that $E[X(n)X^t(n)] = R_N$ and the coefficients obtained by:

$$H = R_N^{-1} r_{yx} \tag{13.53}$$

Thus, the coefficients of the modeling filter are obtained by multiplying the inverse of the AC matrix by the cross-correlation vector of system input and output defined by:

$$r_{yx} = E\left[ y(n) \begin{bmatrix} x(n) \\ x(n-1) \\ \cdot \\ \cdot \\ \cdot \\ x(n+1-N) \end{bmatrix} \right] \tag{13.54}$$

Combining (13.51) and (13.53) yields the MMSE $E_{\min}$ and the three alternative expressions:

$$\begin{aligned} E_{\min} &= E[y^2(n)] - H^t R_N H \\ E_{\min} &= E[y^2(n)] - H^t r_{yx} \\ E_{\min} &= E[y^2(n)] - r_{yx}^t R_N^{-1} r_{yx} \end{aligned} \tag{13.55}$$

Equalization is a special case, in which the system to be modeled is the inverse of the system which has generated the input sequence $x(n)$. For example, in communications, the transfer function of the equalizer is the inverse of the channel transfer function, in the absence of noise.

*Example:*

The channel output signal $x(n)$ is related to the emitted data $d(n)$, assumed to be un-correlated and of unit power, by:

$$x(n) = d(n) + 0.5d(n-1) + 0.2d(n-2)$$

The first three elements of the AC function are:

$$r(0) = 1.29; \quad r(1) = 0.60; \quad r(2) = 0.20$$

Taking $d(n)$ as the reference signal, $y(n) = d(n)$, the three equalizer coefficients are:

$$H = \begin{bmatrix} 1.29 & 0.60 & 0.20 \\ 0.60 & 1.29 & 0.60 \\ 0.20 & 0.60 & 1.29 \end{bmatrix}^{-1} \begin{bmatrix} 1 \\ 0 \\ 0 \end{bmatrix} = \begin{bmatrix} 0.9953 \\ -0.4971 \\ 00778 \end{bmatrix}$$

The output error is $E_{\min} = 0.0047$ and the $Z$-transfer function $T(Z)$ of the channel-equalizer cascade is:

$$T(Z) = 0.9953 - 0.0015Z^{-1} + 0.0273Z^{-2} - 0.0609Z^{-3} + 0.0156Z^{-4}$$

It is readily verified that the filter $H(Z)$ is a three-coefficient approximation of the inverse of the channel, which is a FIR filter.

## 13.6 Linear Prediction

Linear prediction is a particular case in which the output of the system to be modeled is the signal itself, as shown in Figure 13.2.

The output error, called prediction error, is written as:

$$e(n) = x(n) - \sum_{i=1}^{N} a_i x(n-i) \tag{13.56}$$

The prediction coefficients are:

$$\begin{bmatrix} a_1 \\ \vdots \\ a_N \end{bmatrix} = R_N^{-1} \begin{bmatrix} r(1) \\ \vdots \\ r(N) \end{bmatrix} \tag{13.57}$$

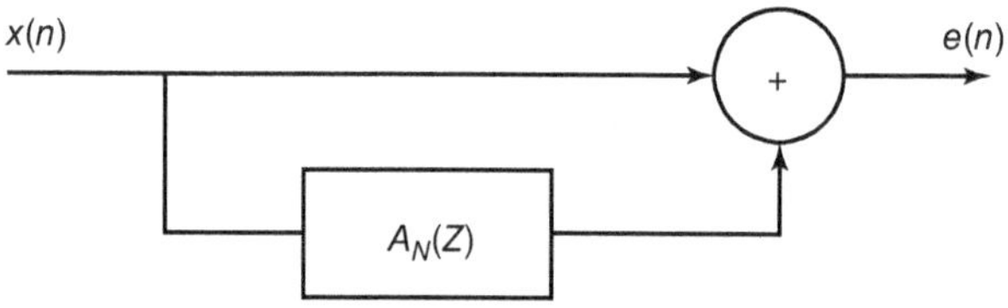


**Figure 13.2** Principle of linear prediction.

The decorrelation of the prediction error and the input signal leads to $N$ relations:

$$r(p) = \sum_{i=1}^{N} a_i r(p-i); \quad 1 \leq p \leq N \tag{13.58}$$

As for the MMSE, we get:

$$E_{aN} = r(0) - \sum_{i=1}^{N} a_i r(i) \tag{13.59}$$

The above expressions can be combined to yield the matrix linear prediction equation:

$$R_{N+1} \begin{bmatrix} 1 \\ -A_N \end{bmatrix} = \begin{bmatrix} E_{aN} \\ 0 \end{bmatrix} \tag{13.60}$$

A signal is said to be "predictable" if the prediction error is null – that is, the following recurrence equation is verified:

$$x(n) = \sum_{i=1}^{N} a_i x(n-i)$$

The predictable signal is made of at most $N/2$ sinusoids, and the filter with transfer function $H(Z) = 1 - A_N(Z)$ has at most $N$ zeros on the unit circle. For example, for $N = 2$ and a sinusoid of frequency $f_0$, the recurrence is:

$$x(n) = 2\cos(2\pi f_0)x(n-1) - x(n-2)$$

In fact, the zeros of $H(Z)$ are either on the unit circle or inside the unit circle because it is minimum phase. The proof is as follows. Assume a zero $z_0$ is outside the unit circle, $|\mathbf{Z_0}| < \mathbf{1}$, then $H'(Z)$ such that:

$$H'(Z) = \frac{H(z)}{(1 - z_0 z^{-1})(1 - \bar{z}_0 z^{-1})} \times \left(1 - \frac{z^{-1}}{z_0}\right)\left(1 - \frac{z^{-1}}{\bar{z}_0}\right)$$

yields a smaller prediction error because:

$$|H'(e^{j\omega})| = \frac{1}{|Z_0|^2} |H(e^{j\omega})|$$

as indicated in Section 9.6. According to expression (4.24) for the signal power at the output of a filter, the power at the output of $H'(Z)$ is lower than that at the output of $H(Z)$ when the same signal is fed to both filters. This contradicts the definition of the prediction filter.

Since the filter $H(Z)$ is minimum phase and invertible, linear prediction is employed to analyze and model signals. In fact, it is possible to retrieve a signal from its prediction error by inverting expression (13.56):

$$x(n) = e(n) - \sum_{i=1}^{N} a_i x(n-i) \tag{13.61}$$

This is so-called autoregressive (AR) modeling. Assuming the prediction error is a white Gaussian noise of power $E_{aN}$, the signal spectrum can be estimated by:

$$S_{AR}(f) = \frac{E_{aN}}{\left|1 - \sum_{i=1}^{N} a_i e^{-j2\pi i f}\right|^2} \tag{13.62}$$

If the filter $H(Z)$ is of IIR type, the model is called autoregressive moving average (ARMA), and the spectrum can again be derived from the model.

## 13.7 Predictor Structures

FIR and IIR filters can be implemented in lattice structures, as indicated in Section 8.5. The approach is particularly advantageous in linear prediction [4].

The prediction coefficients are defined by (13.57) and, due to the properties of the matrix, the inversion may be avoided by an iterative procedure.

The Levinson–Durbin procedure provides a solution for the system of equations (13.58) by recursion over $N$ stages. It begins by taking the power of the error signal as:

$$E_0 = r(0)$$

At the $i$th stage $(1 \leq i \leq N)$, the following calculations are performed:

$$k_i = \frac{1}{E_{i-1}}\left[r(i) - \sum_{j=1}^{i-1} a_j^{i-1} r(i-j)\right]; \quad 1 \leq i \leq N$$
$$a_i^i = k_i \tag{13.63}$$

$$a_j^i = a_j^{i-1} - k_i a_{i-j}^{i-1}; \quad 1 \leq j \leq i-1$$
$$E_i = \left(1 - k_i^2\right) E_{i-1} \tag{13.64}$$

At the $N$th stage, the $N$ coefficients $a_i$ are obtained by:

$$a_i = a_i^N; \quad 1 \leq i \leq N$$

The term $E_i$ corresponds to the power of the residual error with a predictor of order $i$. However, the values of the coefficients $k_i$ obtained at the preceding stages are not changed in the $i$th stage. The procedure is sequential, and the model is improved as the number of stages (and hence the number of coefficients) increases, because, from equation (13.64), the power of the error is reduced at each stage if $|k_i| < 1$.

The coefficients $k_i$ completely define the filter and indicate the method of realization. The error signal at stage $i$ is formed by the set $e_i(n)$, where:

$$e_i(n) = x(n) - \sum_{j=1}^{i} a_j^i x(n-j)$$

The transfer function of the corresponding filter is expressed by $A_i(Z)$ where:

$$A_i(Z) = 1 - \sum_{j-1}^{i} a_j^i Z^{-j}$$

Using equation (13.63), this becomes:

$$A_i(Z) = A_{i=1}(Z) - k_i(Z) - k_i Z^{-i} A_{i-1}(Z^{-1}) \tag{13.65}$$

From the results of Section 8.5, by assuming:

$$B_{i-1}(Z) = Z^{-(i-1)} A_{i-1}(Z^{-1})$$

we obtain:

$$A_i(Z) = A_{i-1}(Z) - k_i Z^{-1} B_{i-1}(Z) \tag{13.66}$$

The function $B_i(Z)$ has the corresponding set $b_i(n)$, with:

$$b_i(n) = b_{i-1}(n-1) - k_i e_{i-1}(n) \tag{13.67}$$

Finally, the coefficients $k_i$ result in a lattice structure as shown in Figure 13.3.

The convergence of the procedure is ensured if:

$$|k_i| < 1; \quad 1 \leq i \leq N \tag{13.68}$$

An alternative decomposition of the prediction filter is provided by the line spectral pair (LSP) method, which consists of splitting the transfer function into two parts with symmetric and anti-symmetric coefficients. Taking recurrence (13.65) at order $N+1$ and designating by $P_N(Z)$ the polynomial obtained with $k_{N+1} = 1$, we get:

$$P_N(Z) = A_N(Z) - Z^{-(N+1)} A_N(Z^{-1}) \tag{13.69}$$

Similarly, designating by $Q_N(Z)$ the polynomial obtained with $k_{N+1} = -1$:

$$Q_N(Z) = A_N(Z) + Z^{-(N+1)} A_N(Z^{-1}) \tag{13.70}$$

Clearly, this is a decomposition of the polynomial $A_N(Z)$, because the sum of the two preceding relations yields:

$$A_N(Z) = \frac{1}{2}[P_N(Z) + Q_N(Z)] \tag{13.71}$$

The coefficients of $P_N(Z)$ and $Q_N(Z)$ exhibit even and odd symmetries, respectively; they are linear phase and, since, as predictors, they are not allowed to have zeros outside the unit circle, all their zeros are on the unit circle. Moreover, if $N$ is even, the relations $P_N(1) = Q_N(-1) = 0$ hold – hence the factorization:

$$P_N(Z) = (1 - Z^{-1}) \sum_{i=1}^{\frac{N}{2}} (1 - 2\cos\theta_i Z^{-1} + Z^{-2}) \tag{13.72}$$

$$Q_N(Z) = (1 + Z^{-1}) \sum_{i=1}^{\frac{N}{2}} (1 - 2\cos\omega_i Z^{-1} + Z^{-2})$$

The two sets of parameters $\theta_i$, $\omega_i$, $1 \leq i \leq N$ provide another representation of prediction parameters.

If $Z_0 = e^{j\omega_0}$ is a zero of $A_N(Z)$ sitting on the unit circle, it is also a zero of $P_N(Z)$ and $Q_N(Z)$. Now, if this zero moves inside the unit circle, the corresponding zeros of $P_N(Z)$ and $Q_N(Z)$ move on the unit circle in opposite directions from $\omega_0$. A necessary and sufficient condition for $1-A(Z)$ to be minimum phase is that the zeros of $P_N(Z)$ and $Q_N(Z)$ be simple and alternate on the unit circle.

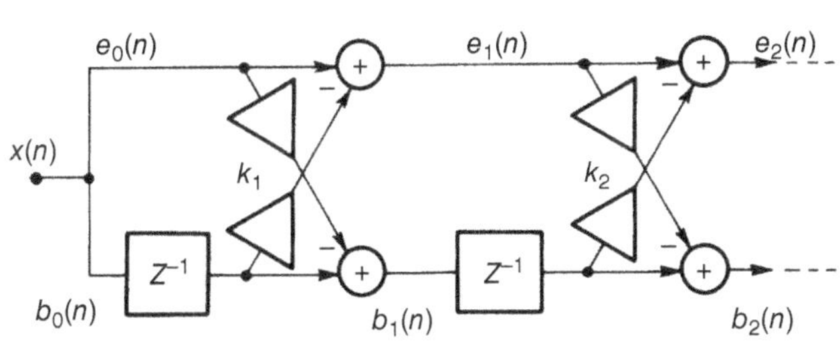


**Figure 13.3** Linear prediction in a lattice structure.

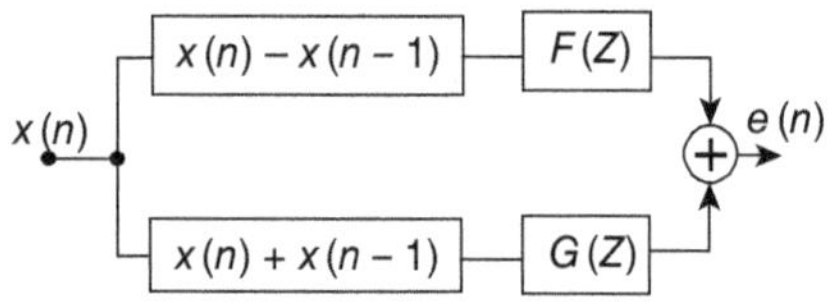


**Figure 13.4** Line spectral pair predictor.

The above approach provides a realization structure for the prediction error filter, as shown in Figure 13.4. The transfer functions $F(Z)$ and $G(Z)$ are the linear phase factors in (13.72). This structure is amenable to implementation as a cascade of second-order sections. The overall minimum phase property is checked by observing the alternation of the $Z^{-1}$ coefficients.

### 13.7.1 Sensor Networks - Antenna Processing

The signal to be processed often comes from a single sensor, but it can also be delivered by a set of sensors, which offers opportunities for improving the performance of existing systems or for introducing new functions. Such a case is illustrated by multiple antenna devices in radio communications. The outputs from the elements of an antenna array are combined to produce a far-field beam pattern which optimizes the reception of a desired signal in a specified direction [5, 6].

A linear network of $N$ elements delivers delayed versions of the signal $s(t)$, as shown in Figure 13.5.

The elementary delay is:

$$\Delta t = \frac{d \sin \theta}{c} \tag{13.73}$$

where $d$ is the distance between two elements, $\theta$ is the incidence angle, and $c$ is the celerity of the signal. The delays can be compensated through $N$ FIR interpolators whose outputs are summed. However, an important simplification occurs if the antenna elements are narrow-band, which is a common case in radiocommunication systems where the frequency band used for information transmission is very low compared to the carrier frequency. Then, the received signals can be viewed as sine waves, and the propagation equations are given by:

$$s(x, t) = S\, e^{j2\pi\left(ft - \frac{x}{\lambda}\right)} \tag{13.74}$$

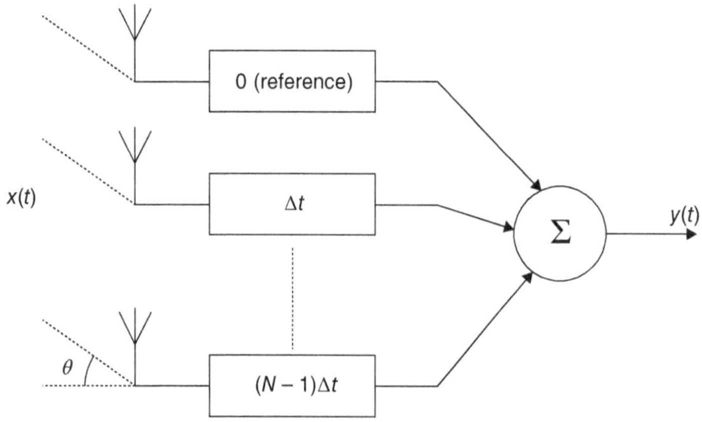


**Figure 13.5** Antenna array.

where $\lambda = c/f$ is the wavelength associated with frequency $f$. The delays translate into phase shifts and, in order to be able to distinguish the arrival angles between $-\pi/2$ and $\pi/2$ and carry out beamforming, the phase differences between elements must be less than $\pi$, which implies:

$$\frac{d \sin\theta}{\lambda} \leq \frac{1}{2}; \quad d \leq \frac{\lambda}{2} \tag{13.75}$$

The value $\lambda/2$ generally retained is the upper bound for the spatial sampling interval. The interpolators boil down to multiplications by weighting coefficients given by:

$$w_i = e^{j\omega\, i\Delta t}; \quad 0 \leq i \leq N-1 \tag{13.76}$$

The calculation of the weighting coefficients is linked to the system environment.

In the presence of Gaussian white noise, a technique similar to linear prediction can be used. One of the antenna elements is taken as the reference, from the corresponding signal $x_0(n)$, a weighted sum of the signals coming from the other elements is subtracted and the following difference is obtained:

$$x_0(n) - \tilde{y}(n) = e(n) = x_0(n) - \sum_{i=1}^{N-1} w_i x_i(n) = x_0(n) - W_{N-1}^t X_{N-1}(n) \tag{13.77}$$

Minimizing the power of this signal yields:

$$R_N \begin{bmatrix} 1 \\ -W_{N-1} \end{bmatrix} = \begin{bmatrix} E \\ 0 \end{bmatrix} \tag{13.78}$$

where $R_N$ is the input covariance matrix, $W_{N-1}$ is the coefficient vector, and $E$ is the output error power. In an evolving environment, adaptive algorithms based on least squares or gradient techniques can be employed to determine the coefficients.

The approach leads to improvements in signal-to-noise ratio. For example, if the noise power is the same for all received signals, the global SNR is improved by a factor close to $N-1$.

Radio links are often plagued by jammers. Undesired signals can be canceled or attenuated if the direction of arrival of the useful signal is known, and a unity response is imposed for that direction. For a known direction $\theta$, with a delay $\Delta t$, the directional vector is expressed by:

$$F = [1, e^{j\omega\Delta t}, \ldots\ldots, e^{j\omega(N-1)\Delta t}]^t \tag{13.79}$$

The coefficients are calculated to minimize the output power subject to the constraint:

$$W_N^t F = 1 \tag{13.80}$$

which yields:

$$W_N = \frac{1}{\overline{F} R_N^{-1} F} R_N^{-1} F \tag{13.81}$$

The corresponding beam provides high attenuation in the direction of jamming signals, which are associated with the zeros of the equivalent $Z$-transfer function. $N$ antenna elements can attenuate $N-1$ undesired signals.

## 13.8 Multiple Sources – MIMO

The environment might convey several useful source signals, which the receiver has to be able to distinguish. The corresponding global system has multiple inputs and multiple outputs, referred to by the abbreviation *MIMO*.

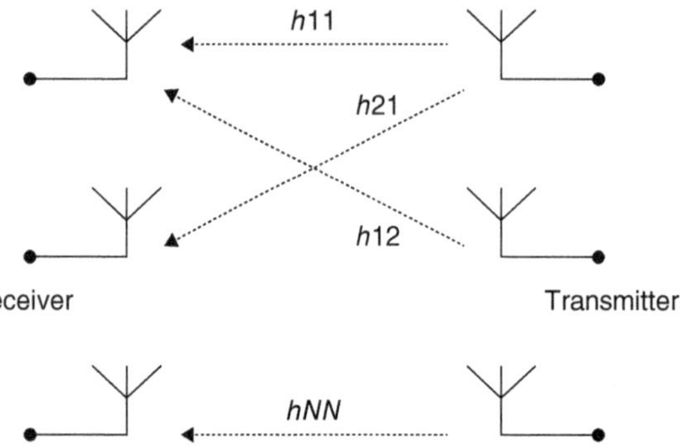


**Figure 13.6** Principle of a MIMO system.

A system with $N$ antennas at the sender and receiver sides is illustrated in Figure 13.6. The $N \times N$ transfer matrix $H$ is written as:

$$H = \begin{bmatrix} h_{11}h_{12}\cdots\cdots\cdots\cdots h_{1N} \\ h_{21}h_{22}\cdots\cdots\cdots\cdots h_{2N} \\ \cdots\cdots\cdots\cdots\cdots\cdots\cdots\cdots \\ h_{N1}h_{N2}\cdots\cdots\cdots h_{NN} \end{bmatrix}$$

where the entries $h_{ij}$ are complex numbers representing the amplitude and the phase of every subchannel linking transmitter and receiver antenna elements.

The analysis of the channel relies on the diagonalization of matrix $H$ and the eigen-decomposition, when it exists:

$$H = M \operatorname{diag}(\lambda_i) M^*; \quad H = \sum_{i=1}^{N} \lambda_i U_i U_i^* \tag{13.82}$$

where $\lambda_i$ and $U_i$ ($i = 1, \ldots, N$) are the eigenvalues and normalized eigenvectors, respectively, and $M$ the rotation matrix:

$$M = [U_1 \cdots\cdots U_N]$$

$$\textit{Example}\text{: } H = \begin{bmatrix} 1re^{-j\theta} \\ -re^{j\theta}1 \end{bmatrix}; \quad \lambda_i = 1 \pm jr; \quad |\lambda_i|^2 = 1 + r^2$$

Eigenvectors:

$$U_1 = \frac{1}{\sqrt{2}} \begin{bmatrix} 1 \\ je^{j\theta} \end{bmatrix}; \quad U_2 = \frac{1}{\sqrt{2}} \begin{bmatrix} 1 \\ -je^{j\theta} \end{bmatrix}$$

Rotation matrix:

$$M = \frac{1}{\sqrt{2}} \begin{bmatrix} 11 \\ je^{j\theta} - je^{j\theta} \end{bmatrix}$$

The decomposition of the transfer matrix can be used to transmit a vector **D** of $N$ data, provided a precoding operation by matrix $M$ is introduced in the transmitter and post-coding by the transpose matrix $M^*$ is introduced in the receiver. The block diagram is given in Figure 13.7 for a Hermitian channel.

The received signal vector is:

$$Y = M^* H M D = \operatorname{diag}(\lambda_i) D = \Lambda D \tag{13.83}$$

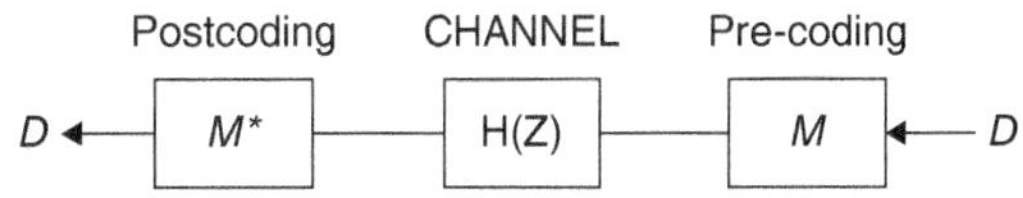


**Figure 13.7** MIMO transmission with Hermitian N × N channel.

The equivalent of a set of $N$ separated channels has been realized and the capacity of the system is the sum of the capacities of the $N$ individual channels. It is worth pointing out that the precoding, which assumes knowledge of the channel at the transmitter side, can be avoided by inverting the channel matrix at the receiver. However, such a scheme might amplify the noise significantly and degrade the transmission performance.

## 13.9 Conclusion

The AC matrix is directly involved in modeling with the MMSE criterion. Although it does not appear explicitly in the most common adaptive filter algorithms, its eigenvalues – in particular, the minimum eigenvalue – control the system operation.

The AC matrix is instrumental in high-resolution spectral analysis of signals, through the harmonic decomposition method. In that method, the signal is modeled by a set of sinusoids in noise, whose frequencies are linked to the zeros of the minimum eigen filter – that is, the filter coefficient vector is the eigenvector associated with the minimum eigenvalue. Compared with linear prediction, harmonic decomposition avoids the bias introduced by the least squares criterion.

Linear prediction allows for real-time signal analysis and can lead to simple and efficient compression methods.

The techniques developed for one-dimensional signals can be extended to multidimensional signals with real-time matrix processing in broad diffusion applications – in particular, mobile radiocommunications.

## Appendix: Estimation Bounds

When estimating a frequency from a set of $N$ samples, three parameters are taken into account – namely amplitude $A$, angular frequency $\omega$, and phase $\varphi$.

Consider the signal:

$$x(n) = Ae^{j(n\omega+\phi)} + b(n); \quad 0 \le n \le N-1 \tag{13.A1}$$

At optimum, and in the absence of bias, the covariance matrix of the parameter estimation vector $\theta$ is related to the power $\sigma_b^2$ of the noise $b(n)$ by:

$$\text{var}\{\theta\} = \left[M_G\overline{M}_G^t\right]^{-1}\sigma_b^2 \tag{13.A2}$$

where $M_G$ is the matrix of the derivatives of the measurements with respect to the parameters. Taking the derivative of $x(n)$, as defined in (13.A1), with respect to the parameters, we obtain:

$$M_G\overline{M}_G^t \begin{bmatrix} N & -jA\frac{N(N-1)}{2} & -jAN \\ jA\frac{N(N-1)}{2} & A^2\frac{N(N-1)(2N-1)}{6} & A^2\frac{N(N-1)}{2} \\ jAN & A^2\frac{N(N-1)}{2} & A^2N \end{bmatrix} \tag{13.A3}$$

Since the three parameters to be estimated are real, their variances are obtained from the real part of the above matrix $M_G \overline{M}_G^t$ only. Assuming the noise is complex, the inverse of the real part of the matrix $M_G \overline{M}_G^t$ must be multiplied by $\sigma_b^2/2$, which yields:

$$\text{var}\{A\} = \frac{\sigma_b^2}{2N}; \quad \text{var}\{\omega\} = \frac{1}{\text{SB}} \frac{6}{N(N^2-1)}; \quad \text{var}\{\phi\} = \frac{1}{\text{SB}} \frac{(N-1)}{N(4N^2-3N+1)} \tag{13.A4}$$

These bounds are called Cramer–Rao bounds.

## Exercises

**13.1** Real signal estimation. Consider the sequence:

$$x(n) = \sqrt{2}\cos(2\pi nf); \quad 0 \le n \le N$$

Apply definition (13.18) to find the AC function. For $f = 1/8$ and $N = 16$, verify that the value $r_1 = 0.618$ is obtained.

Clearly, the AC function does not provide an accurate estimation of the frequency. In order to use the iterated correlation technique, relation (13.31) is modified by canceling the terms with index greater than $N-1$ in the sequence $y(n) = \text{TFD}_{2N}(x)$. Refer to Section 9.2 for justification.

For $f = 0.1$, we find $f_{estim} = 0.1014$, and if a noise such that SNR = 10 (10 dB) is added to the signal, the variance becomes var $= 4.7 \times 10^{-6}$. Compare with the estimation bound (13.23) for complex signals.

**13.2** The signal $x(n) = m + e(n)$, where $m$ is a constant and $e(n)$ is a white noise of power $\sigma_e^2$, is applied to a recursive estimator whose output is:

$$y(n) = (1-b)y(n-1) + bx(n)$$

Assuming $x(n) = 0$ for $n < 0$, compute $y(n)$. If $b = 0.8$, how many samples are needed for $y(n)$ to approach $m$, in the mean, within 1%?

Compute the output mean square error, $E[[y(n) - m]^2]$, for $n > 0$. What is the limit when $n$ tends toward infinity? Study the evolution and the choice of coefficient $b$ for the three cases: $\sigma_e \approx m$; $\sigma_e > m$; and $\sigma_e < m$.

Compare the performance of the recursive estimator with that of the non-recursive estimator defined by:

$$y(n) = \frac{1}{n+1}\sum_{i=0}^{n} x(i)$$

**13.3** Consider the signal sequence $x(n) = \sqrt{2}\sin\left(\frac{n\pi}{4}\right)$. Compute the first three terms of the AC function. Compute the eigenvalues and eigenvectors of the $3 \times 3$ AC matrix. Verify expression (13.40) for the decomposition and reconstruction of the matrix.

**13.4** A second-order predictor with transfer function:

$$H(Z) = 1 - a_1 Z^{-1} - a_2 Z^{-2}$$

is applied to the signal:

$$x(n) = \sqrt{2}\sin(n\omega) + b(n)$$

where $b(n)$ is a Gaussian white noise with power $\sigma_b^2$. Give the power of the predictor output signal. Through derivation, find the expressions of the prediction coefficients. How do these values evolve when the noise power grows?

**13.5** Consider the signal:

$$x(n) = \sin\left(\frac{n\pi}{4}\right) + \cos\left(\frac{n\pi}{3}\right)$$

Compute the first three terms of the AC function. Give the coefficients of the second-order predictor. Place the zeros of the prediction filter in the $Z$-plane.

Give the expression of the prediction error signal and compute its power.

**13.6** Give the polynomial decomposition with even and odd symmetry for the following prediction filter:

$$1 - A_N(Z) = (1 - 1.6\,Z^{-1} + 0.9\,Z^{-2})\ (1 - Z^{-1} + Z^{-2})$$

Give the zeros of polynomials obtained and compare them with those of the initial filter.

**13.7** In a radio transmission link, in order to dispose of large jamming signals, a three-antenna network is employed, and the following sequences are available:

$$x_1(n) = d(n) + \sqrt{2}\sin(n\pi/4 - \pi/6); \quad x_2(n) = d(n) + \sqrt{2}\sin(n\pi/4);$$

$$x_3(n) = d(n) + \sqrt{2}\sin(n\pi/4 + \pi/6)$$

The data $d(n)$ are independent and have unit power. Compute the $3 \times 3$ covariance matrix $R3$ of the input signals.

A weighted summation of these three signals is performed. Write the matrix equation needed to compute the weighting coefficient values such that the summation yields the useful sequence $d(n)$. Deduce that the first and third coefficients are equal and provide the coefficient values.

**13.8** A MIMO system with two antennas at the transmitter side and the receiver side has the channel coefficient matrix:

$$H = \begin{bmatrix} 0.8 & 0.5e^{j\pi/3} \\ 0.7e^{j\pi/4} & 0.6e^{-j\pi/8} \end{bmatrix}$$

Compute the eigenvalues of that matrix.

The total emitted power is unity, and it is uniformly distributed between the two antennas. The additive white noise at each receiver antenna is $\sigma_b^2 = 0.1$. Compute the theoretical capacity of the system expressed in bit/s/Hz.

## References

1 L. Marple, *Digital Spectrum Analysis with Applications*, Prentice-Hall, New York, 1987.

2 E. Aboutanios and B. Mulgrew, Iterative frequency estimation by interpolation on Fourier coefficients, *IEEE Transactions on Signal Processing*, 53(4), 1237–1242, 2005.

3 S.S. Reddy, Eigenvector Properties of Toeplitz Matrices and Their Applications to Spectral Analysis of Time Series, Signal Processing, Vol. 7, North Holland, 1984, pp. 46–56.

4 J. Makhoul, Linear prediction: a tutorial review, *Proceedings of the IEEE*, Vol. 63, pp. 561–580, 1975.

5 A. Paulraj, R. Nabar and D. Gore, *Introduction to Space-Time Wireless Communications*, Cambridge University Press, USA, 2003.

6 D. Tse and P. Viswanath, *Fundamentals of Wireless Communication*, Cambridge University Press, USA, 2005.

# 14

# Adaptive Filtering

Adaptive filtering is used when we need to realize, simulate, or model a system whose characteristics develop over time. It leads to the use of filters whose coefficients change with time. The variations in the coefficients are defined by an optimization criterion and are realized according to an adaptation algorithm, both of which are determined depending on the application. There are many different possible criteria and algorithms [1-4]. This chapter examines the simple but, in practice, most important case in which the criterion of mean square error minimization is associated with the gradient algorithm.

While fixed coefficient filtering is generally associated with specifications in the frequency domain, adaptive filtering corresponds to specifications in time. It is natural to introduce this subject by considering the calculation of filter coefficients in these conditions. We begin by examining FIR filters.

## 14.1 Principle of Adaptive Filtering

The principle of adaptive filtering is illustrated in Figure 14.1. It consists of processing the input signal $x(n)$ to produce an output $\tilde{y}(n)$, whose difference with the reference $y(n)$ is minimized. For every new set of data, reference, and input signal, the coefficients of the filter, assumed to be of FIR type and represented by the vector $H(n)$, are updated.

At time $n$, assuming $n$ data have been received, the cost function for the optimization procedure is chosen as:

$$J(n) = \sum_{p=1}^{n} [y(p) - H^t(n)X(p)]^2 \tag{14.1}$$

where $X(p)$ now designates the column vector of the $N$ most recent input samples at time $p$:

$$X^t(p) = [x(p), x(p-1) \cdots x(p+1-N)]$$

Applying the results from Section 13.5, we obtain the following expression for the filter coefficients:

$$H(n) = R_N^{-1}(n) r_{yx}(n) \tag{14.2}$$

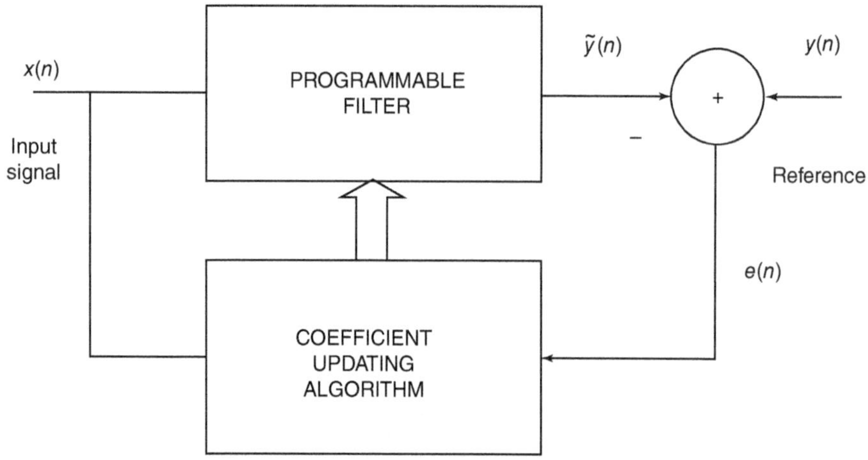


**Figure 14.1** Adaptive filter principle.

The input signal AC matrix can be conveniently expressed as:

$$R_N(n) = \sum_{p=1}^{n} X(p)X^t(p) = \sum_{p=1}^{n} \begin{bmatrix} x(p) \\ x(p-1) \\ \vdots \\ x(p+1-N) \end{bmatrix} [x(p), x(p-1), \ldots, x(p+1-N)] \tag{14.3}$$

Similarly, the estimation of the cross-correlation vector between input and reference is:

$$r_{yx}(n) = \sum_{p=1}^{n} y(p)X(p) \tag{14.4}$$

Whenever a new data set $\{x(n+1), y(n+1)\}$ becomes available, the new coefficient vector $H(n+1)$ can be recursively computed from $H(n)$. The definition equations (14.3) and (14.4) lead to the recursions:

$$\begin{aligned} R_N(n+1) &= R_N(n) + X(n+1)X^t(n+1) \\ r_{yx}(n+1) &= r_{yx}(n) + X(n+1)y(n+1) \end{aligned} \tag{14.5}$$

Now:

$$R_N(n+1)H(n+1) = r_{yx}(n+1) = r_{yx}(n) + X(n+1)y(n+1)$$

and:

$$R_N(n+1)H(n+1) = R_N(n)H(n) + X(n+1)y(n+1)$$

and finally:

$$R_N(n+1)H(n+1) = [R_N(n+1) - X(n+1)X^t\,(n+1)]H(n) + X(n+1)y(n+1)$$

Hence the recursion:

$$H(n+1) = H(n) + R_N^{-1}X(n+1)[y(n+1) - H^t(n)X(n+1)] \tag{14.6}$$

It is interesting to observe that the term:

$$e(n+1) = y(n+1) - H^t(n)\,X(n+1) \tag{14.7}$$

is the error at system output, computed at time $(n+1)$ with the coefficients $H(n)$ available at time $n$. This is called the *a priori* error, while the same calculation with $H(n+1)$ is called the *a posteriori* error.

The algorithms in which the coefficients are computed each time, according to equation (14.6), are the least squares algorithms.

Simplified but very useful algorithms are obtained when the matrix $R_N^{-1}(n)$ is replaced by the diagonal matrix $\delta I_N$, where $\delta$ is a scalar called the adaptation step size. Accordingly, the coefficients are updated by:

$$H(n+1) = H(n) + \delta\, X(n+1)\, e(n+1) \tag{14.8}$$

The algorithm is called the gradient algorithm because the vector $-X(n+1)e(n+1)$ is actually the gradient of the function $\frac{1}{2}e^2(n+1)$, which is the instantaneous value of the quadratic error [4]. Thus, the change in the coefficient values is carried out in the direction of the error gradient, but with the opposite sign, so as to move toward the minimum. The procedure is similar to steepest descent in optimization.

The above procedure can easily be extended to complex signals. Computing the gradient of the squared norm of the complex output error leads to the updating equation for complex coefficients:

$$H(n+1) = H(n) + \delta\, \overline{X}(n+1)\, e(n+1) \tag{14.8bis}$$

Indeed, the computational complexity increases four times.

In a stationary environment, the coefficient vector converges, in the mean, to the theoretical solution. Substituting equation (14.7) into equation (14.8) yields:

$$H(n+1) = [\,I_N - \delta\, X(n+1)X^t(n+1)]H(n) + \delta\, X(n+1)y(n+1) \tag{14.9}$$

Now, defining:

$$R_N = E[X(n)X^t(n)]; \quad r_{yx} = E[y(n)X(n)] \tag{14.10}$$

The optimum set of coefficients is reached when $n$ tends toward infinity:

$$E[H(\infty)] = H_{opt} = R_N^{-1} r_{yx} \tag{14.11}$$

The matrix $R_N$ is the $N \times N$ signal autocorrelation matrix and $r_{yx}$ is the vector of the $N$ first elements of the inter-correlation between input and reference. Finally, the gradient algorithm converges in the mean to the optimal solution $H_{\rm opt}$; hence also the denomination of stochastic gradient, which is sometimes used. The minimization criterion is the least mean squares (LMS) criterion.

Once the convergence has been obtained, the optimum coefficient values are given by (14.11). The minimum value $E_{\rm min}$ of the quadratic error corresponding to the optimum coefficient values can be expressed in terms of the signals $x(n)$, $y(n)$, and their cross-correlation, as indicated in Section 13.5.

The circuit for the resulting adaptive filter is shown in Figure 14.2. The variations of the coefficients are calculated by multiplying for each value of the difference $e(n)$ and summing.

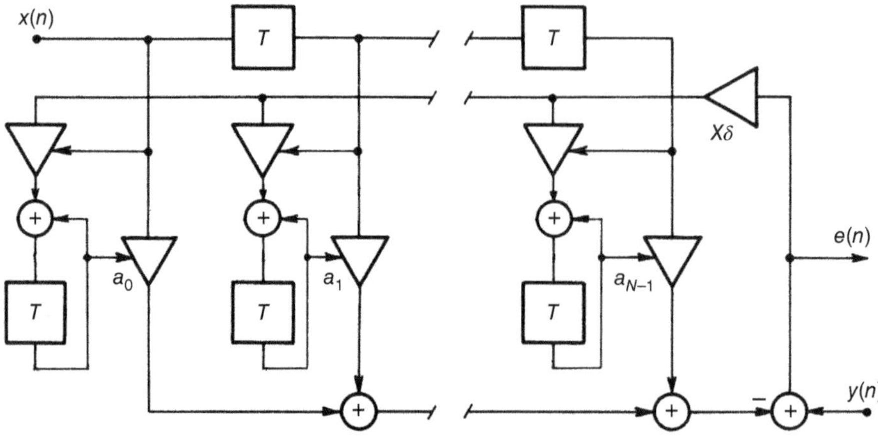


**Figure 14.2** Adaptive FIR filter in direct form.

The choice of the value $\delta$ in equation (14.8) leads to a compromise between the adaptation speed and the value of the residual error when the adaptation is obtained. These two properties will be studied later, but it is first necessary to study the convergence conditions.

## 14.2 Convergence Conditions

The case of a well-dimensioned system without noise is considered first, which means that the residual error is null after convergence. Designating the coefficient deviation by $\Delta H(n) = H_{opt} - H(n)$, we can write the following at time $n$:

$$\Delta H(n+1) = \Delta H(n) + \delta\, X(n+1)\, e(n+1) \tag{14.12}$$

and also:

$$y(n+1) = H^t_{opt} X(n+1); \quad e(n+1) = \Delta H^t(n) X(n+1) \tag{14.13}$$

The norm of the coefficient deviation vector is:

$$\|\Delta H(n+1)\|_2^2 = \|\Delta H(n)\|_2^2 + \delta^2 e^2\,(n+1) X^t\,(n+1) X(n+1) - 2\,\delta e(n+1) \Delta H^t(n) X(n+1)$$

and, using (14.13):

$$\|\Delta H(n+1)\|_2^2 = \|\Delta H(n)\|_2^2 + \delta\, e^2\,(n+1)\,[\delta\, X^t\,(n+1) X(n+1) - 2] \tag{14.14}$$

A monotonic decreasing sequence is obtained, and convergence is guaranteed if the following conditions are satisfied:

$$0 < \delta < \frac{2}{X^t(n+1) X(n+1)} \tag{14.15}$$

The constraints for selecting the adaptation step can be deduced:

$$0 < \delta < \frac{2}{N \max\,[x^2(n)]} \tag{14.16}$$

For signals with high peak factor and a number of coefficients $N$ exceeding several units, the upper bound might be too restrictive and may lead to slow adaptation.

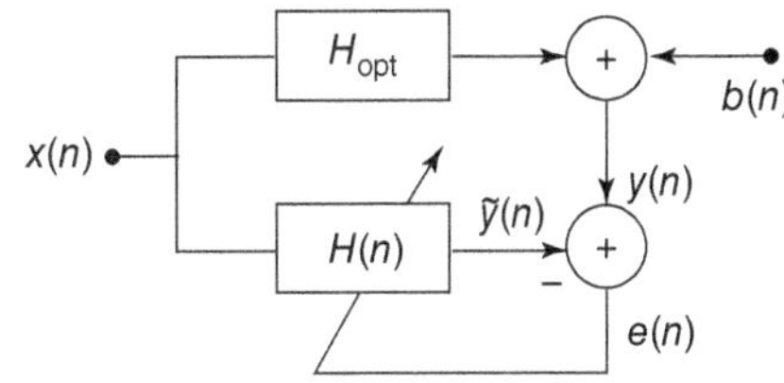


**Figure 14.3** Adaptive filter with measurement noise.

Considering deviation means, taking the expectations of the two terms in (14.14), and assuming independence of $e^2(n+1)$ and $X^t(n+1)\,X(n+1)$, it appears that convergence cannot occur if the following conditions are not satisfied:

$$0 < \delta < \frac{2}{N\sigma_x^2} \tag{14.17}$$

where $\sigma_x^2$ is the input signal power.

The two upper bounds (14.16) and (14.17) are linked by the peak factor $F_c$ of the input signal and, in applications, an intermediate value may be selected – for example (14.17), with some margin. In the case of Gaussian signals, it can be shown that convergence is guaranteed as soon as the adaptation step is smaller than:

$$0 < \delta < \frac{1}{3}\frac{2}{N\sigma_x^2} \tag{14.18}$$

An alternative way of deriving (14.17) is worth mentioning. A convergence condition can be derived by considering the *a posteriori* error defined by:

$$\varepsilon(n+1) = y(n+1) - H^t(n+1)X(n+1) \tag{14.7bis}$$

The adaptive filter works properly if the adaptation is efficient – that is, if the *a posteriori* error is smaller, on average, than the *a priori* one:

$$E[|\varepsilon(n+1)|] < E[|e\,(n+1)|]$$

Substituting the coefficient updating equation (14.8) into the definition of $\varepsilon(n+1)$ yields:

$$\varepsilon(n+1) = [1 - \delta X^t(N+1)X(n+1)]e(n+1)$$

The two errors are, in fact, proportional, and condition (14.15) follows.

If a measurement noise is introduced, the block diagram is as shown in Figure 14.3.

The output error is:

$$e(n+1) = \Delta H^t(n)X(n+1) + b(n+1) \tag{14.19}$$

The noise $b(n)$ is zero mean and its power is $\sigma_b^2$. Then, the quadratic coefficient deviation satisfies the recurrence:

$$\begin{aligned}\|\Delta H(n+1)\|_2^2 = \|\Delta H(n)\|_2^2 &+ \delta\, e^2\,(n+1)\,[\delta\, X^t\,(n+1)X(n+1) - 2\,]\\ &+ 2\delta b^2(n+1) + 2\delta b(n+1)\Delta H^t(n)X(n+1)\end{aligned} \tag{14.20}$$

Since the noise is uncorrelated with the signal, the last term in the right-hand side of the equation is zero in the mean and the convergence condition is derived by taking the expectations of the two terms in (14.20):

$$\left[\delta N\sigma_x^2 - 2\right] E_R + 2\sigma_b^2 < 0 \tag{14.21}$$

where $E_R$ is the mean square error. Then, the convergence stops as soon as the following equality is satisfied:

$$E_R = \frac{\sigma_b^2}{1 - \frac{\delta}{2} N \sigma_x^2} \tag{14.22}$$

Thus, after convergence, a residual error with power $E_R$ remains, and the noise power corresponds to the minimum mean square error $E_{min}$, which is reached when the coefficients take on their optimal value $H_{opt}$. This residual error is analyzed in Section 14.4.

With the stability ensured, it is interesting to evaluate the adaptation speed and to determine the time constant of the adaptive filter.

## 14.3 Time Constant

Let us first consider a filter with a single coefficient $h_1(n)$. In these conditions, equation (14.9) can be written:

$$h_1(n+1) = [1 - \delta x^2(n+1)]\, h_1(n) + \delta x(n+1) y(n+1) \tag{14.23}$$

The coefficient of this filter has the mean value $b = 1 - \delta\,\sigma_x^2$. For a global assessment of the characteristics, that filter can be viewed as a first-order IIR section with a fixed coefficient equal to $b$. Then, for small values of $\delta$, expression (6.5) yields the time constant:

$$\tau \approx \frac{1}{\delta\,\sigma_x^2}$$

The result can be generalized to a filter with $N$ coefficients, under certain conditions.

Assume $[\alpha(n)]$ is the column vector:

$$[\alpha(n)] = M\,[H_{opt} - H(n)] \tag{14.24}$$

where $M$ is the square matrix of order $N$, which appears in the decomposition of $R_N$ given by equation (13.26).

Using (14.11), it can be verified that equation (14.9), which defines the evolution of the coefficients, can be written as follows, in the transformed space and in the mean:

$$E[\alpha(n+1)] = [I_N - \delta\,\mathrm{diag}(\lambda_i)]\, E[\alpha(n)] \tag{14.25}$$

The system corresponds to $N$ time constants:

$$\tau_i = \frac{1}{\delta \lambda_i} \tag{14.26}$$

and it shows that the smallest eigenvalue of the AC matrix, $\lambda_{min}$, determines the convergence time of the adaptive filter. The most favorable case occurs when the input signal is a white noise, because all the eigenvalues are equal to the signal power:

$$\tau_e = \frac{1}{\delta \sigma_x^2} \tag{14.27}$$

This expression provides an estimate of the time constant, which turns out to be proportional to the inverse of the adaptation step. The equations of the coefficient evolution can be simplified in that case, as can that of the mean square error evolution which is:

$$E(n) = \left(1 - \delta\,\sigma_x^2\right)^{2n} \sigma_y^2 \tag{14.28}$$

Assuming the initial coefficient values are null, $\sigma_y^2$ is the power of the reference signal.

In concrete applications, maximum adaptation speed is often sought and, thus, the maximum adaptation step size is contemplated.

The convergence condition (14.17) provides an upper bound for the adaptation step, and if $\delta$ exceeds that bound then the output error will grow. A geometric illustration, based on the error surface representation, assumed to be symmetric, shows that the fastest adaptation is obtained with half that bound, $\delta = \frac{1}{N\sigma_x^2}$. That result can be proven analytically using the approach described in the next section. In these conditions, the time constant satisfies the following inequality:

$$\tau_e \geq N \tag{14.29}$$

In order to complete the examination of this adaptive filter, the residual error after adaptation still has to be evaluated in the general case – that is, with imperfect dimensioning and measurement noise.

## 14.4 Residual Error

After a transition phase corresponding to the convergence, the coefficients of the adaptive filter continually vary about their optimal value, because the adaptation step $\delta$ remains constant. This, incidentally, is the condition for permanent adaptation of the system. As a result, the residual error $E_R$, defined as the limit of the expectation of the quadratic error $E(n)$ as $n$ tends toward infinity, is greater than the minimum value $E_{\min}$.

The residual error $E_R$ is evaluated by considering the vector $[\alpha(n)]$ defined by equation (12.59) and its evolution, described by the following recursion:

$$[\alpha(n+1)] = [\alpha(n)] - \delta MX(n+1)e(n+1) \tag{14.30}$$

In order to estimate the variances of the elements of the vector $[\alpha(n)]$, it is convenient to consider the matrix $[\alpha(n)]\,[\alpha(n)]^t$ whose main diagonal is made of those elements squared:

$$\begin{aligned}[\alpha(n+1)][\alpha(n+1)]^t &= [\alpha(n)[\alpha(n)]^t - 2\delta MX(n+1)e(n+1)[\alpha(n)]^t \\ &\quad + \delta^2 e^2(n+1)MX(n+1)X^t(n+1)M^t\end{aligned} \tag{14.31}$$

In terms of $[\alpha(n)]$, the error is expressed by:

$$e(n+1) = y(n+1) - H_{\text{opt}}^t X(n+1) + X^t(n+1)M^t[\alpha(n)]$$

The evolution of the adaptive system is governed by the above two equations. In order to obtain useful results, it is necessary to make several simplifying assumptions.

The following variables are assumed to be independent:

(1) The output error when the coefficients take on their optimal values.
(2) The data vector: $X(n+1)$.
(3) The coefficient deviation: $H(n) - H_{\text{opt}}$.

The consequence is the following equality:

$$E\left[\left[y(n+1) - H_{\text{opt}}^t\, X(n+1)\right] X^t(n+1)M^t\,[\alpha(n)]\right] = 0 \tag{14.32}$$

Taking the expectation of both sides of equation (14.31) yields:

$$\begin{aligned}E\{[\alpha(n+1)][\alpha(n+1)]^t\} &= [I_N - 2\delta \text{diag}(\lambda_i)]E\{[\alpha(n)][\alpha(n)]^t\} \\ &\quad + \delta^2 E(n)\text{diag}(\lambda_i)\end{aligned} \tag{14.33}$$

After the transient phase, when $n$ tends toward infinity:

$$E\{[\alpha(\infty)][\alpha(\infty)]^t\} = \frac{\delta}{2}E(\infty)I_N \tag{14.34}$$

Because of the definition (14.24) of $[\alpha(n)]$, we also have:

$$E[\,[H_{\text{opt}} - H(\infty)]\,[H_{\text{opt}} - H(\infty)]^t\,] = \frac{\delta}{2}\,E(\infty)\,I_N \tag{14.35}$$

Therefore, after convergence, the coefficient deviations are independent, and they all have the same variance.

Next, the residual error power must be computed.

For the coefficient deviation $\Delta H(n)$, the residual error is given by:

$$E(n) = E_{\min} + \Delta H^t(n)\,R_N \Delta H(n) \tag{14.36}$$

Alternatively, using (14.24):

$$E(n) = E_{\min} + [\alpha(n)]^t\,\text{diag}(\lambda_i\,)\,[\alpha(n)] \tag{14.37}$$

Computing the products, we find:

$$E(n) = E_{\min} + \sum_{i=0}^{N-1} \lambda_i\,\alpha_i^2(n)] \tag{14.38}$$

Since the coefficient deviations have the same variance, factorization can take place and, using (14.35), it appears that the residual error at infinity, $E_{\text{R}} = E(\infty)$, is given by:

$$E_{\text{R}} = \frac{E_{\min}}{1 - (\delta/2)N\sigma_x^2} \tag{14.39}$$

It is worth pointing out that the above equation leads to the stability condition (14.17) derived by a different approach.

In practice, due to the margin generally taken on the step size $\delta$, the following approximation holds:

$$E_{\text{R}} \simeq E_{\min}\left(1 + \frac{\delta}{2}N\sigma_x^2\right) \tag{14.40}$$

In terms of the time constant, with equation (14.27), we have:

$$E_{\text{R}} \simeq E_{\min}\left(1 + \frac{NT}{2\tau_e}\right) \tag{14.41}$$

Thus, the relation between the time constant and the residual divergence is clearly seen. $T$ is the sampling period, taken here as being equal to unity.

The increase in residual error due to the step size $\delta$ can be viewed as a gradient noise.

## 14.5 Complexity Parameters

The complexity parameters of adaptive filters are the same as those for filters with fixed coefficients. The most important are the multiplication rate, the number of bits of the coefficients, and the internal memories. The limitations on the number of bits of the coefficients and internal data increase the residual error $E_{\text{RT}}$. The specifications are generally given in terms of a minimum gain of the system. That is, the ratio of the power of the reference signal $\sigma_y^2$ to the total residual error $E_{\text{RT}}$

should exceed a specified value $G^2$:

$$\frac{\sigma_y^2}{E_{\mathrm{RT}}} \leq G^2 \tag{14.42}$$

The time constant $\tau_e$, if it is imposed, should be compatible with the minimum gain of the system so that the system can be realized using the gradient technique.

The values of the parameters $G$ and $\tau_e$ of the adaptive filter can be used to calculate the number of bits of the coefficients and internal data for each structure. The order of the filter is chosen to be sufficiently large for the minimum divergence $E_{\min}$ to be acceptably small and for equation (14.42) to be satisfied.

In the case of the FIR filter realized as a direct structure as in Figure 14.2, rounding is generally carried out on the output of the multipliers. The noise generated by rounding the internal data with the quantization step $q_2$ corresponds to the addition of the power $N\left(\frac{q_2^2}{12}\right)$ to the minimal error power $E_{\min}$.

If the coefficients are quantized with the step $q_1$, the round-off errors make a vector which has to be incorporated into the coefficient evolution equation. Assuming that round-off errors are independent of other signals, the additional term $\left(\frac{q_1^2}{12}\right) I_N$ has to be introduced in equations (14.33) and equation (14.34) becomes:

$$E\{[\alpha(\infty)][\alpha(\infty)]^t\} = \frac{\delta}{2} E(\infty) I_N + \frac{1}{2\delta}\frac{q_1^2}{12}\mathrm{diag}\left(\frac{1}{\lambda_i}\right) \tag{14.43}$$

Finally, the total residual error $E_{\mathrm{RT}}$ is given by:

$$E_{\mathrm{RT}} = \frac{1}{1-(\delta/2)N\sigma^2{}_x}\left[E_{\min} N\frac{q_2^2}{12}\frac{N}{2\delta}\frac{q_1^2}{12}\right] \tag{14.44}$$

which, for a small step size $\delta$, can be approximated by:

$$E_{\mathrm{RT}} \simeq E_{\min}\left(1+\frac{\delta.N\sigma_x^2}{2}\right) + \frac{N}{2\delta}\frac{q_1^2}{12} + N\frac{q_2^2}{12} \tag{14.45}$$

The relative values of the four terms in this expression can be chosen to suit each application. One common option is to consider $E_{\min}$ as the most important term and to assume that the additional residual error caused by the adaptation step $\delta$ is equal to the noise introduced by the internal rounding – that is,

$$\frac{1}{2}E\frac{\delta.N\sigma_x^2}{2}\frac{N}{2\delta}\frac{q_1^2}{12}\frac{q_2^2}{12}{}_{\min} \tag{14.46}$$

If $b_c$ is the number of bits of the coefficients and $a_{\max}$ is the amplitude of the largest coefficient, then:

$$q_1 = h_{\max}\ 2^{1-b_c}$$

Under these conditions,

$$2^{2b_c} = \frac{2}{3}\ \frac{h_{\max}^2}{\delta^2\ E_{\min}\,\sigma_x^2} \tag{14.47}$$

With the assumption that $E_{\min}$ is the dominant term in equation (14.45) – that is,

$$G^2 \cdot E_{\min} \approx \sigma_y^2$$

and by introducing the time constant, one has approximately:

$$b_c \approx \log_2(\tau_e) + \log_2(G) + \log_2\left(h_{\max}\frac{\sigma_x}{\sigma_y}\right) \tag{14.48}$$

The term $h_{max}\frac{\sigma_x}{\sigma_y}$ depends on the system gain, the signals, and the filter order; it must be determined for every application.

Equation (14.48) shows that the number of bits of the coefficients is directly related to the time constant and to the gain of the system. It can assist designers in selecting from the available options.

## 14.6 Normalized Algorithms and Sign Algorithms

The time constant of an adaptive filter and its residual error are related to the power of the input signal $x(n)$. When this power can vary greatly, the adaptation can be modified as follows:

$$H(n+1) = H(n) + \frac{\delta}{X^t(n+1)X(n+1)} X(n+1)\, e(n+1) \tag{14.49}$$

This is described as a normalized algorithm. It can be verified that the *a posteriori* error is zero if $\delta = 1$. In practice, rather than calculate the scalar product $X^t(n+1)X(n+1)$, a recursive estimate of the power can be made, and this leads to:

$$P_x(n+1) = (1-\varepsilon)\, P_x(n) + \varepsilon\, x^2\,(n+1) \tag{14.50}$$

$$H(n+1) = H(n) + \frac{\delta\, X(n+1)e(n+1)}{P_x\ (n+1)} \tag{14.51}$$

The parameter $\varepsilon$ of the recursive estimate is chosen as a function of the variations of the signal power. It must be at least of order $1/N$, where $N$ is the number of coefficients for the filter.

In certain applications, it is important to minimize the operations, and simplified algorithms are then used in which the variations of the coefficients are a function of the sign of the terms $e(n)$ and $x(n)$, or even the products $e(n)x(n-i)$; these are sign algorithms. The reduction in complexity thus obtained comes at the cost of a degradation in some aspects of system performance [2].

Consider the following coefficient adaptation algorithm:

$$h_i(n+1) = h_i(n) + \Delta e(n+1)\,\mathrm{sign}(x(n+1-i)) \tag{14.52}$$

For nonzero $x$:

$$\mathrm{sign}[x] = \frac{x}{|x|}$$

If the amplitude distribution of $x(n)$ is symmetrical, in a rough approximation, $|x(n)|$ can be replaced by $\sigma_x$. Therefore, expression (14.52) is similar to equation (14.8) with:

$$\delta \approx \frac{\Delta}{\sigma_x}$$

Proceeding further in that direction, the coefficient variations can be limited to constant values:

$$h_i(n+1) = h_i(n) + \Delta\ \ \mathrm{sign}(e(n+1)\,x(n+1-i)) \tag{14.53}$$

Then, as an initial approximation, the coefficient variations in equation (14.53) are comparable with those in equation (14.8), with:

$$\delta = \frac{\Delta}{\sigma_e \sigma_x} \tag{14.54}$$

Starting from zero values for the coefficients, in the convergence phase of the filter, it can be taken that $\sigma_e \simeq \sigma_y$ and the time constant $\tau_s$ for the sign algorithm can be expressed by:

$$\tau_s \approx \frac{\sigma_y}{\Delta \sigma_x} \tag{14.55}$$

After convergence, one can take $\sigma_e^2 = E_{\min}$. If the variation step is sufficiently small, the residual error $E_{RS}$ in the sign algorithm is:

$$E_{RS} \approx E_{\min} \left(1 + \frac{N\Delta}{2} \frac{\sigma_x}{\sqrt{E_{\min}}}\right) \tag{14.56}$$

The residual error is thus found to be larger than with the gradient algorithm. This leads to low values for $\Delta$. It should also be noted that, remembering equation (14.54), the stability condition (14.17) is represented by the inequality:

$$\Delta \leq \frac{2}{N} \frac{\sqrt{E_{\min}}}{\sigma_x} \tag{14.57}$$

which can be taken as a convergence condition and can result in very small values of $\Delta$. In practice, equation (14.52) is usually modified to become:

$$h_i(n+1) = (1-\varepsilon)\, h_i(n) + \Delta \ \operatorname{sign}(e(n+1)\, x(n+1-i)) \tag{14.58}$$

The constant $\varepsilon$, positive and small, introduces a leakage function, which is needed, for example, in transmission systems which must tolerate a certain error rate. Under these conditions, the coefficients are bounded by:

$$|h_i(n)| \leq \frac{\Delta}{\varepsilon}; \quad 0 \leq i \leq N-1 \tag{14.59}$$

This modification leads to an increase in the residual error. The coefficients are biased, and instead of (14.11) for small values of $\varepsilon$ and $\Delta$, we can write:

$$E[H(\infty)] = \left[\frac{\varepsilon}{\Delta} I + R_N\right]^{-1} E[y(n)X(n)] \tag{14.60}$$

The corresponding increase in residual error can be calculated by an expression similar to equation (14.36).

The constants $\varepsilon$ and $\Delta$ are chosen on the basis of the performance to be achieved in each case.

The adaptive filters considered above are of the direct structure FIR type. It is a simple and robust approach and is commonly used. However, as with fixed coefficient filters, other structures can be employed.

## 14.7 Adaptive FIR Filtering in Cascade Form

In certain modeling problems, particularly in automatic control, it is important to know the roots of the $Z$-transfer function of the adaptive filter. It is then convenient to use a cascade arrangement of $L$ second-order sections $H_i \cdot (Z)$, $1 \leq i \leq L$, such that:

$$H_i(Z) = 1 + a_1^i Z^{-1} + a_2^i Z^{-2}$$

From the results of Chapter 6, if the roots $Z_1^i$ and $Z_2^i$ are complex, then:

$$Z_2^i = \overline{Z_1^i}; \quad a_2^i = \left|Z_1^i\right|^2; \quad a_1^i = -2\mathrm{Re}\left(Z_1^i\right)$$

Consider an adaptive filter whose transfer function $H(Z)$ is:

$$H(Z) = \prod_{i=1}^{L} \left(1 + a_1^i Z^{-1} + a_2^i Z^{-2}\right) \tag{14.60}$$

Beginning with a given set of values of the coefficients, variations proportional to the gradient of the error function $E(A)$ must be applied so as to minimize the mean square error. This leads to:

$$\frac{\partial E}{\partial a_k^i} = -\frac{2}{N_0} \sum_{n=0}^{N_0-1} [y(n) - \tilde{y}(n)] \frac{\partial \tilde{y}(n)}{\partial a_k^i}; \quad k = 1.2; \quad 1 \leq i \leq L \tag{14.61}$$

In order to calculate the term $g_k^i$ such that:

$$g_k^i(n) = \frac{\partial \tilde{y}(n)}{\partial a_k^i}$$

one can use the expression for $y(n)$ obtained by an inverse $Z$-transform on the transform $X(Z)$ of the set $x(n)$ (see Section 4.2). Hence:

$$\tilde{y}(n) = \frac{1}{2\pi j} \int_r Z^{n-1} \prod_{i=1}^{L} \left(1 + a_1^i Z^{-1} + a_2^i Z^{-2}\right) X(Z) dZ$$

where $\Gamma$ is a suitable integration contour. Hence:

$$\frac{\partial \tilde{y}(n)}{\partial a_k^i} = \frac{1}{2\pi j} \int_\Gamma Z^{n-1} Z^{-k} \prod_{\substack{l=1 \\ l \neq i}}^{L} \left(1 + a_1^l Z^{1} + a_2^l Z^{-2}\right) X(Z) dZ$$

or,

$$\frac{\partial \tilde{y}(n)}{\partial a_k^i} = \frac{1}{2\pi j} \int_r Z^{n-1} Z^{-k} \frac{H(Z)}{1 + a_1^i Z^{-1} + a_2^i Z^{-2}} X(Z) dZ \tag{14.62}$$

Thus, to form the term $g_k^i(n)$, it is sufficient to apply the set $\tilde{y}(n)$ to a recursive section whose transfer function is the inverse of that of the initial section of order $i$. This recursive section has the same coefficients, but with the opposite sign. The corresponding circuit is given in Figure 14.4.

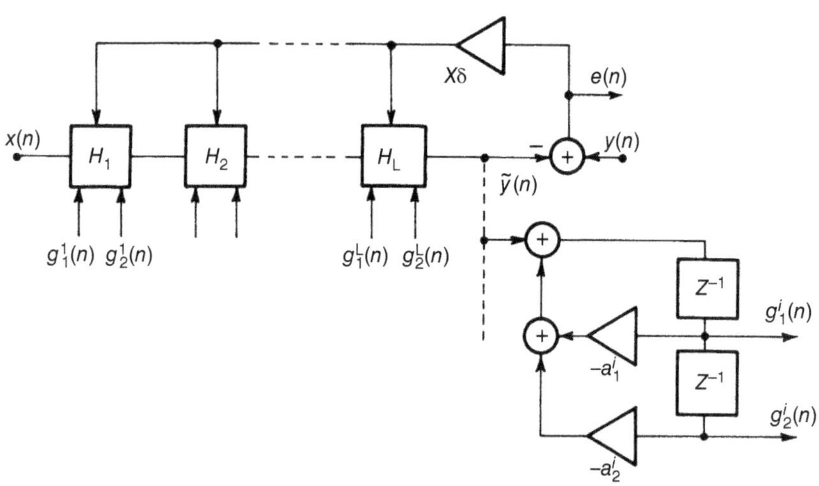


**Figure 14.4** Adaptive FIR filter in cascade form.

The variations in the coefficients are calculated by the expressions:

$$da_k^i(n) = \delta\, g_k^i(n)[y(n) - \tilde{y}(n)]; \quad k = 1.2; \quad 1 \le i \le L \tag{14.63}$$

The filter obtained in this way is more complicated than that in the previous section, but it offers a very simple method of finding the roots, which, due to the presence of a recursive part, should be inside the unit circle in the $Z$-plane to ensure the stability of the system. The techniques derived for FIR filters also apply to IIR filters.

## 14.8 Adaptive IIR Filtering

The coefficients of an IIR filter can be calculated for time specifications by using the least mean squares technique in an iterative procedure, as was done in Section 7.3. Algorithms for adapting the coefficients to the time evolution of the system can also be deduced.

A linear system can be modeled by a purely recursive IIR filter with a $Z$-transfer function, $G(Z)$, such that:

$$G(Z) = a_0 \frac{1}{1 + \sum_{k=1}^{K} b_k Z^{-k}} \tag{14.64}$$

In this case, the model is said to be autoregressive (AR). This important and convenient approach is also appropriate if the best representation of the system corresponds to a $Z$-transfer function, $H(Z)$, which is the quotient of two polynomials:

$$H(Z) = \frac{N(Z)}{D(Z)}$$

where $N(Z)$ has all its roots inside the unit circle and thus has minimum phase. In this case, for a suitable integer $M$, we can write:

$$\frac{1}{N(Z)} \simeq 1 + \sum_{i=1}^{M} c_i Z^{-i}$$

It is then sufficient to let the degree $K$ of the denominator of thc function $G(Z)$ take a value sufficient for representing $H(Z)$. The presence of roots inside the unit circle of the system results in an increase in the number of poles of the model [1].

The general IIR filter corresponds to an autoregressive moving average (ARMA) model. This is the most widely used approach for modeling a linear system. For the IIR filter whose coefficients must be calculated over a set of $N_0$ indices in order to approximate a set $y(n)$, the output is written as:

$$\tilde{y}(n) = \sum_{l=0}^{L} a_l x(n-l) - \sum_{k=1}^{K} b_k \tilde{y}(n-k) \tag{14.65}$$

The error function $E(A, B)$ is expressed by:

$$E(A, B) = \frac{1}{N_0} \sum_{n=0}^{N_0-1} [y(n) - \tilde{y}(n)]^2 \tag{14.66}$$

Starting from a set of values for the coefficients, this function can be minimized using the gradient algorithm if the coefficients are given increments proportional to the gradient of $E(A, B)$ and of the opposite sign. The presence of a recursive part causes complications. Calculation of the gradient leads to the following expressions:

$$\frac{\partial E}{\partial a_l} = -\frac{2}{N_0}\sum_{n=0}^{N_0-1}[y(n) - \tilde{y}(n)]\frac{\partial \tilde{y}(n)}{\partial a_l}; \quad 0 \leq l \leq L$$

$$\frac{\partial E}{\partial b_k} = -\frac{2}{N_0}\sum_{n=0}^{N_0-1}[y(n) - \tilde{y}(n)]\frac{\partial \tilde{y}(n)}{\partial b_k}; \quad 1 \leq k \leq K$$

with:

$$\frac{\partial \tilde{y}(n)}{\partial a_l} = x(n-l) - \sum_{k=1}^{K} b_k \frac{\partial \tilde{y}(n-k)}{\partial a_l} \tag{14.67}$$

$$\frac{\partial \tilde{y}(n)}{\partial b_k} = -\tilde{y}(n-k) - \sum_{j=1}^{K} b_j \frac{\partial \tilde{y}(n-j)}{\partial b_k} \tag{14.68}$$

To show the method of realizing equations (14.67) and (14.68), we can write:

$$H(Z) = \frac{\sum_{l=0}^{L} a_l Z^{-l}}{1 + \sum_{k=1}^{K} b_k Z^{-k}} = \frac{N(Z)}{D(Z)}$$

Then:

$$\tilde{y}(n) = \frac{1}{2\pi j}\int_{\Gamma} Z^{n-1} H(Z) X(Z) dZ$$

and consequently:

$$\frac{\partial \tilde{y}(n)}{\partial a_l} = \frac{1}{2\pi j}\int_{\Gamma} Z^{n-1} Z^{-l} \frac{1}{D(Z)} X(Z) dZ \tag{14.69}$$

$$\frac{\partial \tilde{y}(n)}{\partial b_k} = \frac{1}{2\pi j}\int_{r} Z^{n-1} Z^{-k} \frac{(-1)}{D(Z)} H(Z) X(Z) dZ \tag{14.70}$$

The gradient is thus calculated from the set obtained by applying $x(n)$ and $\tilde{y}(n)$ to the circuits corresponding to the transfer function $1/D(Z)$.

To simplify the implementation, the second terms in equations (14.67) and (14.68) can be ignored, which leads to the following increments:

$$da_l(n) = \delta\,[y(n) - \tilde{y}(n)]x(n-l); \quad 0 \leq l \leq L \tag{14.71}$$

$$db_k(n) = -\delta\,[y(n) - \tilde{y}(n)]\tilde{y}(n-k); \quad 1 \leq k \leq K$$

For each value of the index $n$, the coefficients $a_l$ and $b_k$ are incremented by a quantity which is proportional to the product of the error $e(n) = y(n) - \tilde{y}(n)$ with $x(n-l)$ and $\tilde{y}(n-k)$, respectively.

The stability and parameters of this type of filter can be studied in the same manner as for FIR filters [5]. However, it is not difficult to control the stability of an IIR filter when it is constructed as a cascade of second-order sections, which, as indicated in previous chapters, also offers other advantages.

Consider a filter constructed as a cascade of second-order elements and with a transfer function $G(Z)$. In the AR case, we have:

$$G(z) = a_0 \prod_{i=1}^{L} \frac{1}{1 + b_1^i Z^{-1} + b_2^i Z^{-2}} \tag{14.72}$$

To control the stability of such a filter, it is sufficient, as set forth in Section 6.7, to ensure that the following conditions are fulfilled:

$$|b_2^i| < 1; \quad |b_1^i| < 1 + b_2^i; \quad 1 \leqslant i \leqslant L \tag{14.73}$$

As before, the calculation of the gradient of the error function requires knowledge of the term $g_k^i$, where:

$$g_k^i(n) = \frac{\partial \tilde{y}(n)}{\partial b_k^i}$$

Since,

$$\tilde{y}(n) = \frac{1}{2\pi j} \int_\Gamma Z^{n-1} a_0 \prod_{i=1}^{L} \frac{1}{1 + b_1^i Z^{-1} + b_2^i Z^{-2}} X(Z) dZ$$

we obtain:

$$\frac{\partial \tilde{y}(n)}{\partial b_k^i} = -\frac{1}{2\pi j} \int_\Gamma Z^{n-1} \frac{Z^{-k}}{1 + b_1^i Z^{-1} + b_2^i Z^{-2}} G(Z) X(Z) dZ$$

This expression indicates that the terms $g_k^i(n)$, with $k = 1, 2$ and $1 \leq i \leq L$, are obtained by applying the set $\tilde{y}(n)$ to the $i$th recursive section. The stability of the system is tested by equation (14.73) for each value of the index $n$.

The method which has been discussed is also applicable to an ARMA model, but in this case, the circuits are rather more complicated.

The techniques used in the previous sections involve overall minimization of the mean square error. It is also possible to achieve minimization step by step, using lattice structures.

The lattice filter in Figure 13.3 can also be adapted by a gradient algorithm for each value of the index. Indeed, using the equations:

$$\begin{aligned} e_i(n) &= e_{i-1}(n) - k_i b_{i-1}(n-1) \\ b_i(n) &= b_{i-1}(n-1) - k_i e_{i-1}(n) \end{aligned} \tag{14.74}$$

one can write the gradients as:

$$\begin{aligned} \frac{\partial e_i^2(n)}{\partial k_i} &= -2e_i(n) b_{i-1}(n-1) \\ \frac{\partial b_i^2(n)}{\partial k_i} &= -2b_i(n) e_{i-1}(n) \end{aligned} \tag{14.75}$$

and the following variations can be applied to the coefficients by assuming that the functions $E\left[e_i^2(n) + b_i^2(n)\right]$ for $1 \leqslant i \leqslant N$ are to be minimized:

$$k_i(n+1) = k_i(n) + \delta_i(e_i(n) b_{i-1}(n-1) + b_i(n) e_{i-1}(n)) \tag{14.76}$$

As the power of the signals $e_i(n)$ and $b_i(n)$ decreases with the index $i$, the variation step $\delta_i$ must be related to this power in order to obtain a certain degree of homogeneity of the time constants.

## 14.9 Conclusion

Several techniques for designing and producing adaptive filters have been presented in this chapter. They are based on the gradient algorithm, which is the simplest and most robust approach for changing the coefficients. The direct FIR structure has been studied in detail by developing the adaptation parameters (the time constant and the residual error) and the complexity parameters (multiplication rate and the number of bits of the coefficients and internal data). This is the structure most commonly used in practice. In some specific cases, different structures, such as IIR, mixed FIR–IIR, or lattice structures, can offer significant advantages. Analysis of the stability conditions in these structures and the study of the adaptation and complexity parameters can be performed by a method similar to that given for the FIR structure.

The gradient algorithm results in a relatively slow change in the values of the coefficients of the filter, especially when a low residual error is required and when it is used in its most reduced form: the sign algorithm. In order to find the most rapid rate of adaptation, all the coefficients can be recalculated periodically by using fast iterative procedures. The lattice structure is well suited to this approach and allows for real-time analysis or modeling of signals such as speech with circuits of moderate complexity.

The gradient algorithm can be improved, for example, by using different adaptation steps for the coefficients, which are obtained from statistical estimates of the signal characteristics.

It is possible to envisage criteria which, for certain applications, are more appropriate than the minimization of the mean square error, and algorithms which are more efficient than the gradient algorithm can be developed [1]. However, these algorithms are generally more complicated to put into operation, and problems of sensitivity to any imperfection in the realization may arise.

In conclusion, FIR and IIR structures operating according to the least mean square error criterion and using the gradient algorithm, or its simplest form, the sign algorithm, offer a simple and effective compromise for adaptive filtering applications.

## Exercises

**14.1** A signal $x(n)$ can be modeled by the output of a filter $H(Z)$ when the input is a unit power white noise. The Z-transfer function is:

$$H(Z) = \frac{\frac{1}{2}(1 + Z^{-1})}{1 - 0.5Z^{-1}}$$

Compute the impulse response and show that the three first terms of the AC function take on the values: $r_0 = 1$; $r_1 = 0.75$; $r_2 = 0.375$.

For compression, a two-coefficient predictor $P(Z)$ is used, and the output sequence is:

$$y(n) = x(n) - a_1 x(n-1) - a_2 x(n-2)$$

Compute the optimum coefficient values – those which minimize the output power. Give the output power value and the prediction gain $G_p$.

Compute the impulse response of the cascade $H(Z)P(Z)$.

In an adaptive realization with the gradient algorithm, give the coefficient updating equations and the maximum adaptation step value $\delta_m$. Compute the residual error for adaptation step $\delta = \delta_m/4$ and give the time constant of the adaptive predictor.

**14.2** A sinusoidal signal $x(n)$ is applied to a second-order FIR prediction filter:

$$x(n) = \sin\left(2\pi \frac{3n}{8}\right)$$

Calculate the coefficients $a_1$ and $a_2$ of this filter and locate the zeros in the $Z$-plane.

Starting from zero initial coefficients, examine the trajectory of the zeros of this filter for an adaptation step $\delta = 0.1$.

Give the new values of the prediction coefficients when discrete white noise of power $\sigma^2$ is added to the signal.

**14.3** An FIR filter is used to equalize a channel with transfer function:

$$C(Z) = \frac{0.5}{1 - 0.5\,Z^{-1}}$$

Calculate the power of the received signal $x(n)$, assuming uncorrelated unit power data. For $N = 3$ coefficients, provide the values.

A white noise with power $\sigma_b^2 = 0.1$ is added to the signal $x(n)$. Calculate the new coefficient values and explain the difference. Give the noise amplification factor.

Using the channel + equalizer impulse response, compute the power of the inter-symbol interference.

Compute the eigenvalues of the $3 \times 3$ AC matrix of the signal $x(n)$. In an adaptive realization, the step size is $\delta = 0.05$. Give the time constant of the equalizer and check by simulation.

**14.4** Consider Figure 14.3 with the following filters:

$$H_{opt}(Z) = \frac{1 + 0.5Z^{-1}}{1 - 0.5Z^{-1}}; \quad H(Z) = \frac{a_0 + a_1 Z^{-1}}{1 - bZ^{-1}}$$

The input is a unit power uncorrelated sequence, and the power of the noise $b(n)$ is $\sigma_b^2 = 0.1$.

Compute the optimum coefficient values of the adaptive filter.

In an adaptive realization, give the coefficient updating equations and the bounds of the adaptation step $\boldsymbol{\delta}$. Draw the curves of the evolution of the coefficients for $\boldsymbol{\delta} = \mathbf{0.1}$ and verify the time constant and residual error power after convergence.

## References


**1** B. Widrow and S. D. Stearns, *Adaptive Signal Processing*, Prentice-Hall, Englewood Cliffs, NJ, 1985.

**2** M. Bellanger, *Adaptative Digital Filters*, 2nd Edition, Marcel Dekker Inc., New York, 2001.

**3** O. Macchi, *Adaptative Processing, the LMS approach with Application in Transmission,* Wiley, New York, 1995.

**4** S. Haykin, *Adaptative Filter Theory*, 4th Edition, Prentice-Hall, Englewood Cliffs, New Jersey, 2000.

**5** P. A. Regalia, *Adaptative IIR Filtering in Signal Processing and Control*, Marcel Dekker Inc., New York, 1995.

# 15

# Neural Networks

Adaptive systems are at the heart of digital communication networks. They are particularly critical for data transmission efficiency, which is achieved through channel equalization. This operation involves an initial learning phase, an operational phase, and subsequent decisions and improvements throughout the duration of the communication. Thanks to artificial intelligence, these concepts can be extended to all technical fields, with operational devices that are ever more complex and sophisticated. Signal processing and adaptive techniques, as presented above, are profoundly involved in one such device – namely, the neural network.

The present chapter describes how neural networks operate, and how signal-processing techniques, and specifically adaptive techniques, are exploited. As a starting point, we look at a simple classification operation.

## 15.1 Classification

Classification is a basic operation in shape recognition. Its complexity depends on the space in which it is carried out. In a two-dimensional space, objects defined by their coordinates can be grouped together, and the different groups can be separated by curves. Whenever a new object appears, it is assigned to an existing group, depending on its position with respect to the separation curves. An illustration is provided in Figure 15.1.

N objects are separated into two groups (a and b) by a line with equation:

$$h_0 + h_1X_1 + h_2X_2 = 0 \tag{15.1}$$

The structure of the corresponding classifier is shown in Figure 15.2.

Assuming the objects are represented by points in the plane, during the learning phase, the coordinates of the two categories of points are fed to the input, the output is subtracted from a reference $d$, and the error $e$ obtained is used to determine the coefficients of the separation line.

An iterative procedure can be employed to obtain the coefficient vector $H$, starting from an initial vector. As in the previous chapter, we can write:

$$H(n+1) = H(n) + \frac{\delta}{X(n+1)X^t(n+1)}e(n+1)X(n+1); \quad 0 \leq n \leq N-1 \tag{15.2}$$

with:

$$e(n+1) = d(n+1) - X^t(n+1)H(n) \tag{15.3}$$

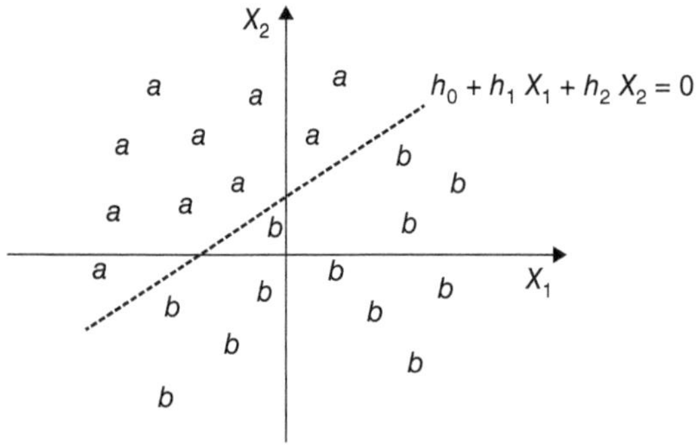


**Figure 15.1** Using a 3-coefficient filter for classification.

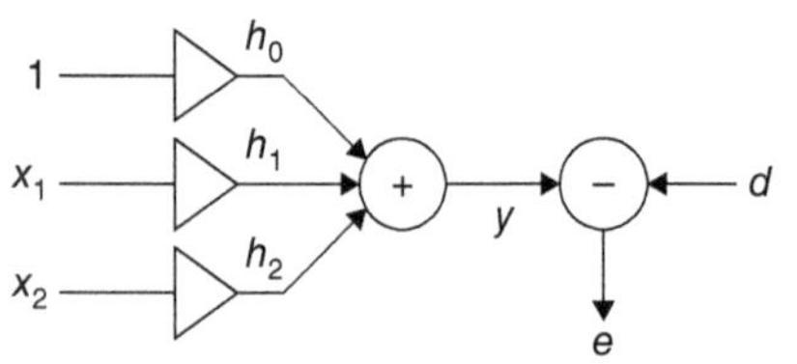


**Figure 15.2** Filter for distinguishing two groups of objects.

The procedure and the reference must be adjusted to the two types of data ($a$ and $b$) to achieve the desired separation. In the absence of prior information, the reference is null ($d(n)=0$). The adaptation step $\delta = 1$ leads to cancellation of the *a posteriori* error as shown in Section 14.6. Hence the need to select $0 < \delta < 1$.

Once the learning phase is completed, the sign of the output $y(n)$ is used for the classification operation. In fact, a decision device is inserted at filter output and the whole structure is called a "neuron".

Now, if two other groups of objects ($c$ and $d$) are introduced, two lines are needed to divide the plane into four parts and potentially achieve separation, which leads to the combination of two neurons, as shown in Figure 15.3. The system has two inputs fed by the coordinates of the objects and two outputs which, if binary $\pm 1$, allow for an object to be assigned to one of the four groups. Nonbinary outputs can also be exploited; they are interpreted as estimations of the probability for the objects to belong to the groups.

The approach can be generalized – a neuron can have a large number of inputs and related coefficients, and several such neurons can be connected to make a system with multiple inputs and multiple outputs, along with their decision devices. Such a system is called a "perceptron" [1].

Now, if the curve needed to separate the groups of objects in the plane is no longer a line, nonlinear processing is required, which is achieved through the combination of several perceptrons. Then, it can be shown that 2 perceptrons are able to distinguish convex domains and

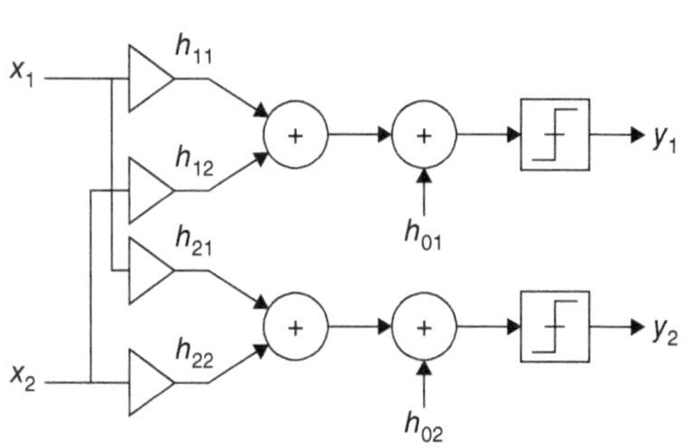


**Figure 15.3** Classification with two neurons.

3 perceptrons can cope with the combination of convex domains – that is, domains of any form. Such systems are called multilayer perceptrons.

Besides object classification, neural networks can serve to approximate or estimate nonlinear functions. This stems from the partitioning capability provided by their decision, or activation, of nonlinear functions. In theory, any function can be approximated with arbitrary precision, although some constraints may have to be introduced for actual realization [2]. This aspect is dealt with at the end of the chapter.

## 15.2 Multilayer Perceptron

A system with two intermediary layers, called hidden layers, is shown in Figure 15.4. If all the connections exist, the network is said to be fully connected.

By definition, in a network made of L+1 layers, the input and output correspond to layers 0 and 1, respectively, and L-1 hidden layers are present in between. The system may be parameterized as follows:

- Number of neurons per layer: $n_l$; $1 \leq l \leq L$
- So-called synaptic coefficients: $h_{ij,\,l}$; $1 \leq i \leq n_{l-1}$; $1 \leq j \leq n_l$
- Nonlinear activation function: $x_{j,\,l} = f(u_{j,\,l})$
- Synaptic weightings:

$$u_{j,l} = \sum_{i=1}^{n_{l-1}} h_{ij,l} x_{i,l-1} + h_{0j,l} \tag{15.4}$$

The term $h_{0j,\,l}$ in the above convolution represents the introduction of a bias, linked to the statistics of the data and the type of activation function. It may be null for zero-mean data.

The nonlinear activation function $f(.)$ may be the sign, as in Figure 15.3, on which a decision is based. However, if the synaptic coefficients in the various branches have to be updated in supervised learning, the activation function must be derivable, so that gradient techniques can be employed. For example, the sigmoid:

$$f(u) = \frac{e^u}{1+e^u}; \quad f'(u) = \frac{e^u}{(1+e^u)^2} = f(u)(1-f(u)) \tag{15.5}$$

or the hyperbolic tangent:

$$f(u) = \tanh(u); \quad f'(u) = 1 - \tan h^2(u) \tag{15.6}$$

are frequently retained.

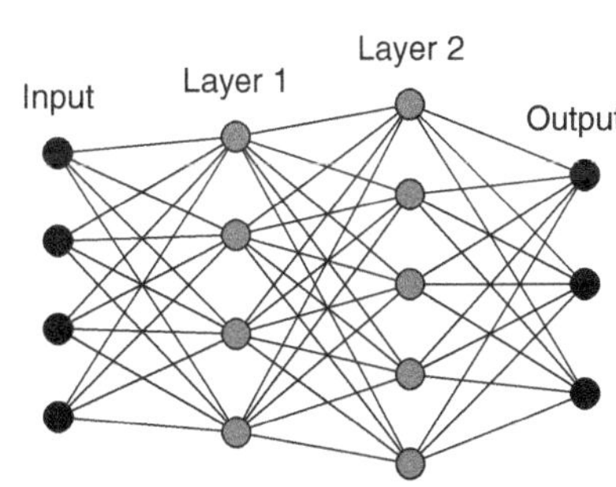


**Figure 15.4** Neural network with 2 hidden layers.

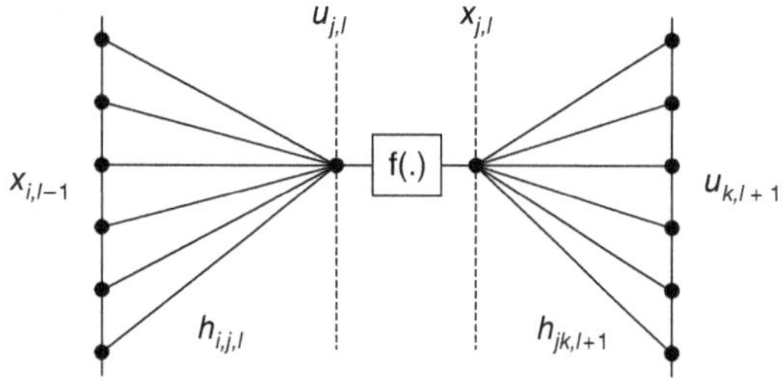


**Figure 15.5** Connecting two stages in the network.

Expression (15.4) and the details of the connections between two stages in the network are illustrated in Figure 15.5.

In the supervised-learning phase, a reference sequence is available, and an error signal is derived from the network output, through subtraction, as in the previous chapter. The presentation of the learning algorithm is simplified if all the layers have the same width $N$ and the output does not include a nonlinear function, as in estimation for example.

## 15.3 The Backpropagation Algorithm

Let $X_0(n)$ be the input vector at time $n$, $Y(n)$ the output vector, and $D(n)$ the reference vector.

In the different layers, the state vectors $U_l(n), X_l(n)\,(1 \leq l \leq L-1)$ are available, as are the synaptic coefficients $h_{ij,\,l}(n)$, which need to be updated at time n+1 with the gradient technique.

Applying the vector $X_0(n+1)$ at network input, the output is $Y(n+1)$, and the following error signal is derived:

$$e_{j,L}(n+1) = d_j(n+1) - y_j(n+1); \quad 1 \leq j \leq N \tag{15.7}$$

The output is expressed as a function of the last state vector $X_{L-1}(n+1)$ by:

$$y_j(n+1) = \sum_{i=1}^{N} h_{ij,L}(n) x_{i,L-1}(n+1) = H_{j,L}^t(n)\, X_{L-1}(n+1); \quad 1 \leq j \leq N \tag{15.8}$$

As in Section 14.1, the coefficient updating equation is:

$$H_{j,L}(n+1) = H_{j,L}(n) + \delta_L\, X_{L-1}(n+1)\, e_{j,L}(n+1); \quad 1 \leq j \leq N \tag{15.9}$$

$\delta_L$ is the adaptation step size.

As for stage $L-1$, the coefficients impact the whole error signal, which implies that the global quadratic cost function is involved – that is:

$$J(n+1) = \frac{1}{2} \sum_{k=1}^{N} e_{k,L}^2(n+1) \tag{15.10}$$

Since the coefficients evolve in the opposite direction to the gradient, the updating equation should be:

$$H_{j,L-1}(n+1) = H_{j,L-1}(n) + \delta_{L-1} X_{L-2}(n+1)\, e_{j,L-1}(n+1); \quad 1 \leq j \leq N \tag{15.11}$$

Then, it is necessary to determine the "error" at stage $L-1$. Let us assume a deviation $\varepsilon$ is introduced on the coefficient $h_{ij,\,L-1}$. Since, by definition, we have:

$$u_{j,L-1} = \sum_{i=1}^{N} h_{ij,L-1} x_{i,L-2} \tag{15.12}$$

the variation of $u_{j,L-1}$ will be $\varepsilon\ x_{i,L-2}$ and, for the output with index $k$, $y_k$, it will be $f'(u_{j,L-1})\varepsilon\, x_{i,L-2}\, h_{jk,L}$, which leads to the following derivative:

$$\frac{\partial y_k}{\partial h_{ij,L-1}} = f'(u_{j,L-1})\, x_{i,L-2}\, h_{jk,L} \tag{15.13}$$

Summing for all the output errors, the gradient of the global cost function is obtained:

$$\frac{\partial J}{\partial h_{ij,L-1}} = -\sum_{k=1}^{N} \frac{\partial y_k}{\partial h_{ij,L-1}} e_{k,L} \tag{15.14}$$

Worth noting also is the following general expression:

$$\frac{\partial J}{\partial h_{ij,L-1}} = \frac{\partial J}{\partial u_{j,L-1}} \frac{\partial u_{j,L-1}}{\partial h_{ij,L-1}} \tag{15.14-bis}$$

Due to (15.13) and (15.14), the error to be used for the updating of the coefficients of stage $L-1$ at time $n+1$ is given by:

$$e_{j,L-1}(n+1) = f'(u_{j,L-1}(n+1)) \sum_{k=1}^{N} h_{jk,L}(n) e_k(n+1) \tag{15.15}$$

As concerns other stages, referring to Figure 15.5, the following relation holds:

$$\frac{\partial J}{\partial u_{j,l}} = f'(u_{j,l}) \sum_{k=1}^{N} h_{jk,l+1} \frac{\partial J}{\partial u_{k,l+1}} \tag{15.16}$$

It follows that relation (15.15) can be extended to stages lower than $L-1$ and a recurrence equation is established for the "errors":

$$e_{j,l-1}(n+1) = f'(u_{j,l-1}(n+1)) \sum_{k=1}^{N} h_{jk,l}(n) e_{k,l}(n+1) \tag{15.17}$$

Therefore, synaptic coefficients can be updated at time $n+1$, from stage $L-1$ to stage 1, which is the backpropagation algorithm.

The algorithm must be initialized so that backpropagation can occur. Unless otherwise specified, the coefficients of the last stage are set to zero. In order for the algorithm to start, the last state vector must differ from zero, which is obtained by setting the coefficients of the stages smaller than $L$ to values such that the first input vector propagates up to stage $L-1$. If a bias is present, the set of updating equations is supplemented by:

$$h_{0j,l}(n+1) = h_{0j,l}(n) + \delta_l\, e_{j,l}(n+1);\quad 1 \le j \le N;\quad 1 \le l \le L-1 \tag{15.11-bis}$$

As shown in the previous chapter, the adaptation step controls the stability, the convergence speed, and the residual error after convergence. However, the results cannot be applied directly due to interdependence of coefficients; some specific analysis is required.

The simplest case is the last stage because each coefficient vector is updated from the related output error. A key observation is that the values in the state vector are bounded by the activation function, assumed to be a sigmoid or hyperbolic tangent. If that bound is unity, an initial determination of the range for the adaptation step in the last stage is:

$$0 < \delta_L < \frac{2}{N} \tag{15.18}$$

Further in the analysis, hidden layers must be accounted for. To begin with, a system with a single hidden layer, $L = 2$, and a single output with index $k$, is considered. Then, a simple approach is to refer to the *a posteriori* error obtained after coefficient updating by:

$$\varepsilon_{k,2}(n+1) = d_k(n+1) - H^t_{k,2}(n+1)X_1(n+1) \tag{15.19}$$

As mentioned in the previous chapter, the stability condition implies that the mean of the amplitude of the *a posteriori* error must be less than the mean of the amplitude of the *a priori* error:

$$E[\,|\varepsilon_{k,2}(n+1)|] < E[|e_{k,2}(n+1)|] \tag{15.20}$$

Once the coefficients of layers 1 and 2 have been updated, the *a posteriori* error becomes as follows, neglecting second-order terms:

$$\varepsilon_{k,2}(n+1) \approx e_{k,2}(n+1)\left[1 - \delta_2 \sum_{j=1}^{N} x^2_{j,1}(n+1) - \delta_1 \sum_{j=1}^{N}\sum_{i=1}^{N}\left[\frac{\partial y_k}{\partial h_{ij,1}}\right]^2\right] \tag{15.21}$$

Introducing the activation function and for a single adaptation step, $\delta$, the stability condition becomes:

$$0 < \delta < \frac{2}{A+B} \tag{15.22}$$

$$A = \sum_{j=1}^{N} [f(u_{j,1}(n+1))]^2$$

$$B = \sum_{i=1}^{N} x^2_{i,0}(n+1) \sum_{j=1}^{N} [f'(u_{j,1}(n+1))\, h_{jk,2}(n)\,]^2$$

The importance of the hyperbolic tangent, bounded as well as its derivative, clearly appears in the expressions above.

Now, when all the outputs are active, equation (15.21) becomes a matrix equation. Letting $H_1$ and $H_2$ designate the $N \times N$ coefficient matrices, it can be shown that the *a priori* and the *a posteriori* output vectors are related by:

$$\in(n+1) = \left[1 - \delta_2 |f(U_1(n+1)|^2 I_N - \delta_1\, |X_0(n+1)|^2\, H^t_2\,(n)\, Du\, H_2(n)\right]\; E(n+1) \tag{15.23}$$

$Du$ is a diagonal matrix whose $N$ entries are: $[f'(u_{j1}(n+1))]^2$. In order to determine the stability limit, the maximum eigenvalue $\lambda_{max}$ of the matrix $H^t_2\, Du\, H_2$ must be considered and relation (15.22) becomes:

$$0 < \delta < \frac{2}{[f(U_1(n+1))|^2 + |X_0(n+1)\,|^2\, \lambda_{max}} \tag{15.24}$$

Note that the maximum eigenvalue of a square matrix is bounded by the sum of the absolute values of the elements of a row or a column.

The above analysis can be extended to systems having several hidden layers. However, in practice, and as a first attempt, one can simply rely on relation (15.18).

The system time constant impacts the length of the learning sequence. It is the sum of the time constants of the different stages, which are proportional to the inverses of the adaptation steps and, thus, are proportional to the number of coefficients, as indicated in Section 14.3. As concerns the cascade of the different hidden layers, one can refer to the beginning of Section 6.2. Overall, a high number of coefficients leads to long learning sequences.

Regarding the residual error after convergence, the results of the previous chapter apply – in particular, the fact that the residual error grows near the stability limit.

The above processing also applies to nonzero mean signals. It suffices to introduce, in each neuron, the branch corresponding to coefficient $h_0$ and update, for example, according to relation (15.11-bis).

It is worth mentioning that, in principle, alternative optimization techniques, which are more complicated and harder to understand and implement than the gradient, may be employed [3].

## 15.4 Examples of Application

1. ***Nonlinear function***. In order to illustrate how the adaptation algorithm works, it is applied to the elementary network shown in Figure 15.6, which corresponds to two inputs, two outputs and a single hidden layer.

   To begin with, it is proposed to approximate a nonlinear function with one output only and the following reference sequence:

$$d(n) = x_2(n) + 2\,x_1(n) - x_1^3(n) \tag{15.25}$$

   The data are assumed to have a zero mean. At time $n+1$ we get:

$$e(n+1) = d(n+1) - y(n+1) \tag{15.26}$$

$$y(n+1) = h_{112}(n)\tanh(u_1(n+1)) + h_{212}(n)\ \tanh(u_2(n+1))$$

$$u_1(n+1) = h_{111}(n)x_1(n+1) + h_{211}(n)x_2(n+1)$$

$$u_2(n+1) = h_{221}(n)x_2(n+1) + h_{121}(n)x_1(n+1)$$

   For coefficient updating:

$$h_{j12}(n+1) = h_{j12}(n) + \delta\, e(n+1)\tanh(u_j(n+1));\ \ j = 1{,}2$$

$$e_{j,1}(n+1) = \left(1 - u_j^2(n+1)\right) h_{j12}(n)e(n+1);\quad j = 1{,}2 \tag{15.27}$$

$$h_{ij1}(n+1) = h_{ij1}(n) + \delta\, e_{j,1}(n+1)\,x_i(n+1);\quad i,j = 1{,}2$$

   Initial coefficient values are set to zero, except $h_{111}$ and $h_{221}$, which are set to 1. Using N = 1000 pairs of random data in the interval [−1 1] and the adaptation step $\delta = 0.1$, we obtain:

| $h_{111}$ | $h_{121}$ | $h_{211}$ | $h_{221}$ | $h_{112}$ | $h_{212}$ |
|---|---|---|---|---|---|
| 1.67 | −0.05 | 0.07 | 1.07 | 1.27 | 1.07 |

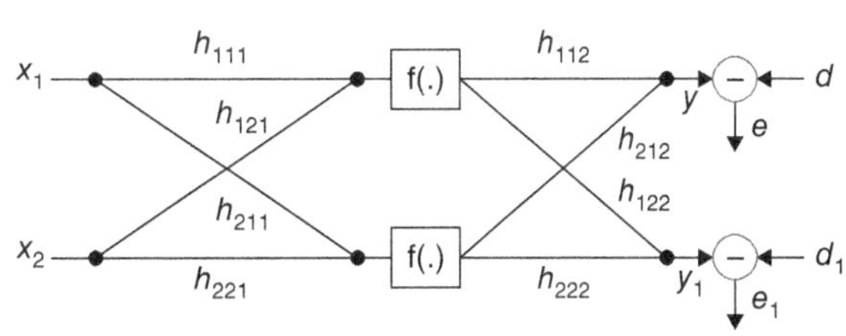


**Figure 15.6** Approximation of a nonlinear function.

The standard deviation after convergence is 5%. Invoking the series expansion:

$$\tanh(u) = u - \frac{1}{3}u^3 + \frac{2}{15}u^5 + \dots \tag{15.28}$$

Estimations $a$, $b$ of the coefficients of the terms $x_2, x_1$ in (15.25) are given by:

$$a = h_{211}\, h_{112} + h_{221} h_{212} = 1.23; \quad b = h_{111} h_{112} + h_{121} h_{212} = 2.07 \tag{15.29}$$

Comparing to the slope of function $x_2 = f(x_1)$ at the origin, which equals 2, we find $b/a = 1.7$. Note that the function has a maximum and, in contrast, a monotonic increasing function would yield a more accurate estimation, as illustrated in Exercise 5.
As for the coefficient of the term $-x_1^3$, setting the factor of $x_2$ to unity, we obtain:

$$c = \left( \frac{1}{3} h_{111}^3 h_{112} \right) / a^3 = 1{,}06$$

In general, accurate estimation of nonlinear functions, with the above activation functions and a single hidden layer, requires large numbers of neurons [2].

2. ***Classification.*** The network shown in Figure 15.6 is able to perform a classification operation in the plane with its 2 outputs and the curves:

$$x_2 + 2x_1 - x_1^3 = 0; \quad x_2 - 2x_1 + x_1^3 = 0 \tag{15.30}$$

Four areas are distinguished, and the input data pairs can be assigned to one of those. In the process, equations (15.26) and (15.27) are supplemented by:

$$e_1(n+1) = d_1(n+1) - y_1(n+1) \tag{15.31}$$

$$y_1(n+1) = h_{122}(n) \tanh(u_1(n+1)) + h_{222}(n)\ \tanh(u_2(n+1))$$

and coefficient updating:

$$h_{j22}(n+1) = h_{j22}(n) + \delta\, e_1(n+1) \tanh(u_j(n+1)); \ \ j = 1.2 \tag{15.32}$$

$e_{j,1}(n+1) = \left(1 - u_j^2(n+1)\right)(h_{j12}(n)e(n+1) + h_{j22}(n)e_1(n+1)); \quad j = 1.2$. Using $N = 1000$ pairs of random data in the interval $[-1\ 1]$ and the adaptation step $\delta = 0.1$, we find:

| $h_{111}$ | $h_{121}$ | $h_{211}$ | $h_{221}$ | $h_{112}$ | $h_{212}$ | $h_{122}$ | $h_{222}$ |
|---|---|---|---|---|---|---|---|
| 1.82 | −0.01 | −0.015 | 1.02 | 1.22 | 1.16 | −1.18 | 1.16 |

Worth noting is the impact of the second output on the coefficients of the first part of the circuit. After learning, the classification error is less than 2%.

3. ***Graphic symbol recognition.*** Image recognition is among the most important applications of neural networks. In order to provide a simple yet convincing illustration, digit recognition is considered.

The 10 digits can be represented in a grid of 5×3 pixels, as shown in Figure 15.7.

Weights −1 and +1 are assigned to white and black pixels, respectively. The digits can be identified through 10 sets of 15 coefficients each. These coefficients take values $\pm 1/15$ and the summation outputs are included in the interval $[-1\ +1]$. Note that digits may differ by just a single pixel. Indeed, it is not necessary to use a network with hidden layers in that case.

However, for recognition of handwritten digits, redundancy must be introduced to account for variations in handwriting. A network with 2 hidden layers is represented in Figure 15.8. The input grid contains $N1$ pixels.

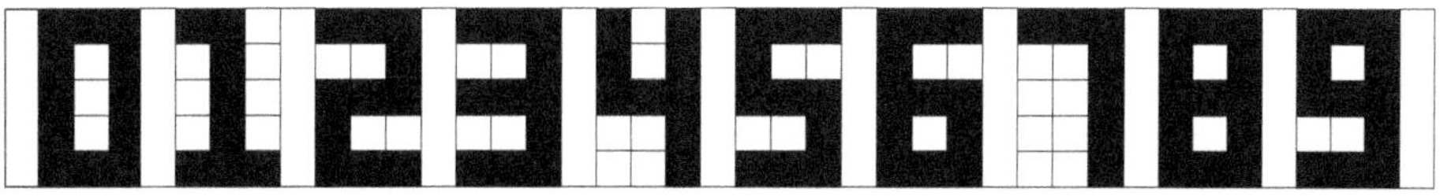

**Figure 15.7** Representing the 10 digits in a grid of 5 × 3 pixels.

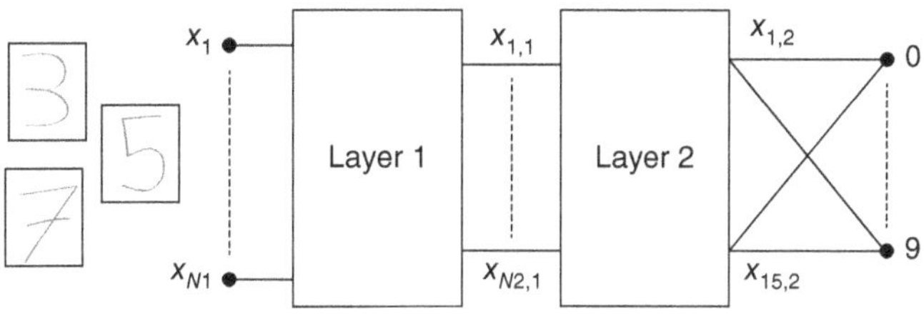


**Figure 15.8** Recognition of handwritten digits.

From $N1$ input values, after learning, layers 1 and 2 yield a set of discriminating values that are used by the 10 sets of output coefficients to provide 10 final values, to decide which digit was applied at input. Learning can be performed digit by digit, using only one output error and scanning all the digits at each iteration.

As an illustration, we take $N1 = 8\times8$, $N2 = 32$ and consider three digits. Learning is carried out with 3, 5 and 7, step $\delta = 0.01$ and 150 iterations. In the coefficient matrices of layers 1 and 2, the diagonal entries of the square central matrices are set to 1 and a single output error is used for coefficient updating. The following discriminations are obtained for the digits, in the absence of noise (a), with SNR = 5 dB (b):

(a)

| | 3 | 5 | 7 |
|---|---|---|---|
| 3 | 1 | 0.11 | −0.37 |
| 5 | 0.11 | 1 | −0.22 |
| 7 | −0.36 | −0.22 | 1 |

(b)

| | 3 | 5 | 7 |
|---|---|---|---|
| 3 | 1 | 0.14 | −0.49 |
| 5 | 0.11 | 1 | −0.36 |
| 7 | −0.43 | −0.30 | 1 |

Clearly, recognition is easily achieved, even in the presence of a high level of noise. The importance of initialization is worth pointing out – a reduction of initial coefficient values leads to an increase of the system time constant and then, it might be necessary to include all the output errors in the coefficient updating process.

In order to cope with multiple ways of writing the same digit, with significant differences, the size of the input grid must be increased, and the learning process must be adjusted accordingly, if a recognition rate objective is prescribed.

The above example illustrates the role of the hidden layers – in one or more steps, they condense the input grid onto the grid employed by the output stage. Of course, the computational cost grows with the size of the input grid and might become prohibitive.

## 15.5 Convolution Neural Networks

In certain fields, such as image processing or vision, the amount of data to be processed simultaneously might be enormous, leading to layers of excessive dimensions in the networks. For example, take a black-and-white image of size $32 \times 32$. The dimension of the input vector is 1024 and the number of coefficients in the layers may reach the millions, which is unrealistic.

However, it is known that compression techniques are particularly efficient with images and certain categories of signals, which is evidence that these signals are highly redundant. In such conditions, it appears well advised – for example, in recognition or classification operations – to introduce one or several compression stages before the neural network. These extra stages can extract characteristics which may considerably reduce the dimensions of the hidden layers and facilitate the mastering of networks. Generally, the corresponding devices are banks of FIR filters, which perform convolutions or correlations, to cancel out redundancies while preserving the specificities of signals [4–6]. The principle is shown in Figure 15.9.

The coefficients in the filter banks depend on the characteristics that are to be extracted. For example, the filters mentioned in Section 5.14 for the extraction of contours in images are frequently used. At the output, a table of characteristics are available. Several such devices may be cascaded, in combination with sample rate reductions, so as to end up with a neural network input vector of significantly reduced size.

In the approach, taking the characteristics of images into account, it might be well advised to use activation functions other than sigmoid or hyperbolic tangents. In fact, in classification for example, we can consider that the desired output is obtained by aligning the input data vector and a coefficient vector, such that the scalar product is maximized. In a multilayer network, the alignment is achieved step by step. The outputs of the first layer correspond to the scalar products of the input vector and the coefficient vectors associated with these outputs. A negative output implies opposite directions for the vectors involved and, thus, it is useless to keep this coefficient vector in the chain leading to the desired global alignment, and this output can be dropped. Hence, the nonlinear activation function ReLU (rectified linear unit), which is zero for negative values of the variable and linear for positive values. In order to make the procedure even more selective, a threshold can be introduced which cancels the small positive values. In fact, it can be argued that the operation of a perceptron layer is similar to a data-clustering operation.

In the implementation of a ReLU-based CNN system, it is worth emphasizing that the derivative of the activation function is not continuous, and the signal amplitude is not bounded, which requires stability control. To that end, an upper bound may be imposed on the activation function output and the step size may be chosen to satisfy the conditions given in Section 15.3. Moreover, initialization may be challenging and, instead of assigning random values to the coefficients, one can employ data-clustering techniques like "$k$-means" [6].

Reference [4] describes an application of CNN to handwritten character recognition.

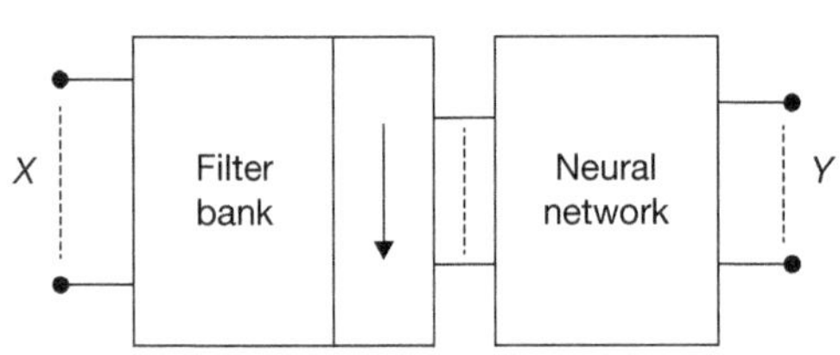


**Figure 15.9** Convolutional neural network (CNN).

## 15.6 Recurrent/Recursive Neural Networks

The techniques described so far apply to blocks of data and they do not account for the evolution of signals over time. Whenever streams, such as video, audio, natural language, or temporal series delivered by sensors, have to be processed, it is necessary to add, into the system, the evolution at a given time, the previous states and, as in Section 14.8 for adaptive filters, we end up with an infinite impulse response [5, 7].

A so-called recurrent neural network with a single hidden layer is depicted in Figure 15.10. Using the terminology of the previous chapter, actually, it is a recursive structure.

The system equations are as follows:

$$X_1(n) = \tanh[AX(n) + BX_1(n-1) + b_1] \tag{15.33}$$

and, if a nonlinear activation function is introduced at output:

$$Y(n) = f[H\,X_1(n) + b_2] \tag{15.34}$$

The terms $b_1$ and $b_2$ are bias coefficients related to the activation functions. The 3 matrices $A$, $B$, and $H$ must have their coefficients updated.

Assuming the vectors $X$, $X_1$, and $Y$ have dimension $N$, the coefficients of matrix $H$ can be updated, as above, using the output errors:

$$H_j(n+1) = H_j(n) + \delta\,X_1(n+1)\,e_j(n+1); \quad 1 \leq j \leq N \tag{15.35}$$

As for the matrices $A$ and $B$, the derivative of the output error with respect to the coefficients is more difficult to determine because of the recursive structure and the nonlinearity in the loop. Taking a simplified approach as in Section 14.8, we can write:

$$u_i(n+1) = \sum_{k=1}^{N} a_{ki}(n)\,x_k(n+1) + \sum_{k=1}^{N} b_{ki}(n)\,x_{k,1}(n+1) \tag{15.36}$$

$$e_{i,1}(n+1) = [1 - \tan h^2 u_i(n+1)] \sum_{j=1}^{N} h_{ij}(n)\,e_j(n+1)$$

$$A_i(n+1) = A_i(n) + \delta_1\,e_{i,1}(n+1)\,X(n+1)$$

$$B_i(n+1) = B_i(n) + \delta_2\,e_{i,1}(n+1)\,X_1(n+1)$$

It must be pointed out that, for matrix $B$, the data are bounded by the nonlinearity, while the input data do not necessarily have the same bound, which can lead to different adaptation steps $\delta_1$ and $\delta_2$ for the two matrices.

A critical issue for recursive systems is stability. In the long run, the evolution of the coefficients must be controlled. If matrix $B$ is diagonal, or near diagonal, the recursive part becomes a set of first-order cells that evolve separately, and the modulus of each single coefficient must remain smaller than unity.

An important aspect of neural networks in general, and the recursive structure in particular, is overdimensioning. In fact, the number of coefficients generally exceeds the number of variables

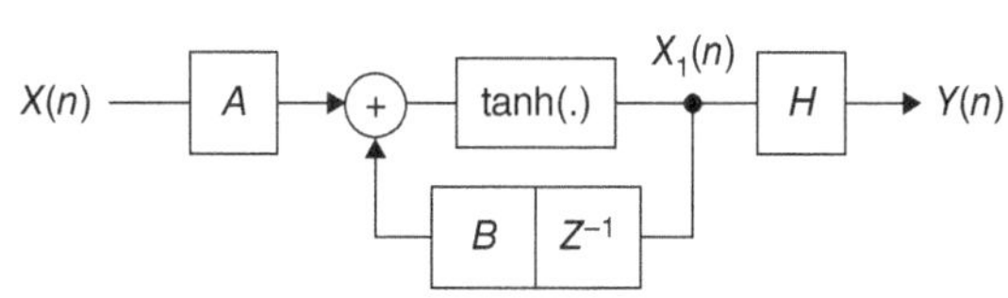


**Figure 15.10** Recursive neural network with a single hidden layer.

of the system to be modelled and, as a result, we see a drift of the adaptive network coefficients linked to the adaptation parameters. That drift can raise problems in long learning phases and in steady-state operation. Then, it may be advisable, in the updating equations of some of the coefficients, to introduce a regularization term $\varepsilon$ which, as described in Section 14.6, limits the amplitude of the coefficients to $\frac{\delta}{\varepsilon}$.

## 15.7 Neural Network and Signal Processing

Thanks to their learning capability, neural networks are able to perform all sorts of functions and, in particular, signal-processing functions, provided a sufficient number of input sequences is available, as well as the corresponding output sequences. There is no need to know how the inside of the system works; this system is viewed as a black box [8].

Such an approach might require considerable amounts of calculations, but the same hardware or software can implement different tasks with or without reconfiguration.

As an illustration, let us consider a discrete cosine transform (DCT) in the presence of an interference signal. With a neural network, there is no need to identify the jammer to enhance performance with respect to the direct calculation.

The DCT outlined in Section 3.3.3 is implemented using a fully connected neural network with $N = 16$ inputs, 16 outputs, and 2 hidden layers also of width $N$. The input vector $X$ is a binary zero-mean random sequence, and the reference $Y$ is obtained by applying relations (3.32). The tan activation functions are used and coefficient updating is performed with step $\delta = 0.02$. The coefficients $h_{ij1}$ for layer 1, $h_{ij2}$ for layer 2, and $g_{ij}$ for the output layer are initialized as follows:

$$h_{ii1} = 1; \quad h_{ii2} = 1; \quad h_{ij1} = h_{ij2} = 0\, i \neq j; \quad g_{ij} = 0; \quad 1 \leq i,j \leq 16$$

The jammer is the sinusoid:

$$s(n) = 0.1 \sin\left(2.25\, n \frac{\pi}{16}\right); \quad 1 \leq n \leq 16 \tag{15.37}$$

The evolution of the output error power Perr (m) for a sequence of $M = 800$ learning blocks is shown in Figure 15.11, for the network with and without hidden layers.

The absence of hidden layers corresponds to iterative calculation, as in the previous chapter.

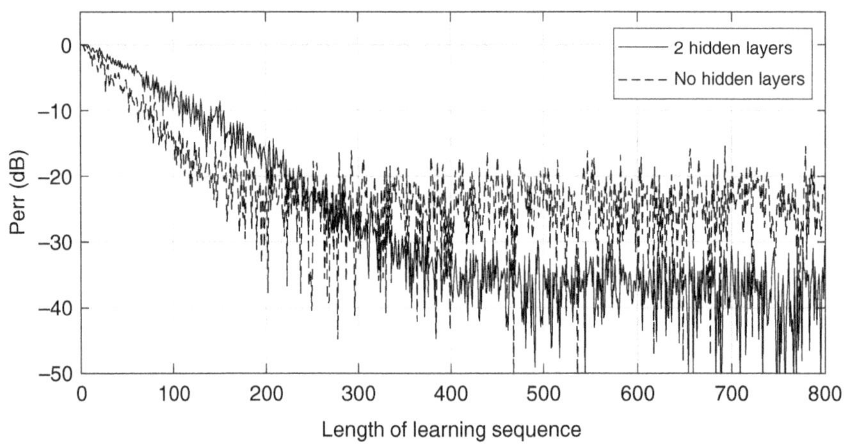


**Figure 15.11** Computation of a DCT in the presence of a jammer.

It can be observed that the two hidden layers double the time constant and bring a gain of about 12 dB in computation accuracy after convergence. In the absence of the jammer, the output error power decreases to −65 dB, this threshold being due to the cascading of the two hidden layers, with their nonlinearities.

The stability limit for the adaptation step is 1/8 without the hidden layers, and it turns out to be slightly smaller with the two layers.

As concerns numerical complexity, the system has 768 coefficients, which implies that, disregarding learning operations, 768 multiplications are carried out per DCT, while direct calculation necessitates fewer than 64 multiplications.

The DCT is employed in image and video compression; a neural network can reduce the degradations arising from imperfect sensors.

## 15.8 On Activation Functions

The nonlinear activation functions used in practice result from a compromise between contradictory requirements. On the one hand, decisions must be made; on the other hand, adaptive techniques (such as learning, in particular) must be implemented. Therefore, a stepwise characteristic is needed for the decision, while a linear characteristic is preferred for adaptivity.

The sigmoid (15.5) is an initial compromise, which provides smooth decision and allows for the direct use of gradient techniques. However, the output is bounded, amplitude distortions occur in the vicinity of the bounds, and derivatives tend toward zero.

The ReLU function is an approach that keeps linearity on a part of the domain and cancels the other part. Adaptive techniques can be applied in the retained part, on the condition that stability conditions are met. If the cancelled part is unimportant in view of the objectives, ReLU can be considered optimal. Thus, applied to the digit recognition task set out in Section 15.4 and with the same parameters, it yields discrimination deviations smaller than those of the hyperbolic tangent.

The ReLU function is appropriate for function approximation because it has the capability to divide a domain into nonoverlapping elementary parts. As an illustration, let us consider the approximation of a half-sinusoid on the interval [01] by 4 segments bounded by the 4 lines whose equations in the plane are:

$$y = 2\sqrt{2}x; \quad y = (4 - 2\sqrt{2})x + \sqrt{2} - 1; \tag{15.38}$$

$$y = (2\sqrt{2} - 4)x + 3 - \sqrt{2}; \quad y = -2\sqrt{2}x + 2\sqrt{2}$$

The block diagram of the network with two hidden layers is shown in Figure 15.12. In fact, the input domain is divided into four parts with linear interpolation in each part.

The coefficients $a = -\frac{\sqrt{2}}{2}$ and $b = 1 - \frac{\sqrt{2}}{2}$ are related to the slopes of the lines and $= \frac{\sqrt{2}}{2}$.

At the output of the first hidden layer, the ReLU functions, represented by rectangles, keep only positive values and, thus, set the negative values to zero and, in that case, also inhibit the addition of the terms $\pm 1$. The output $y$ yields the approximation of $\sin(\pi x)$.

In order to improve the approximation, the approach can be extended by dividing the interval [0 1] into $2^N$ parts with a network of $N$ hidden layers.

A general method to approximate continuous functions on a closed interval using deep ReLU networks is presented in Reference [9].

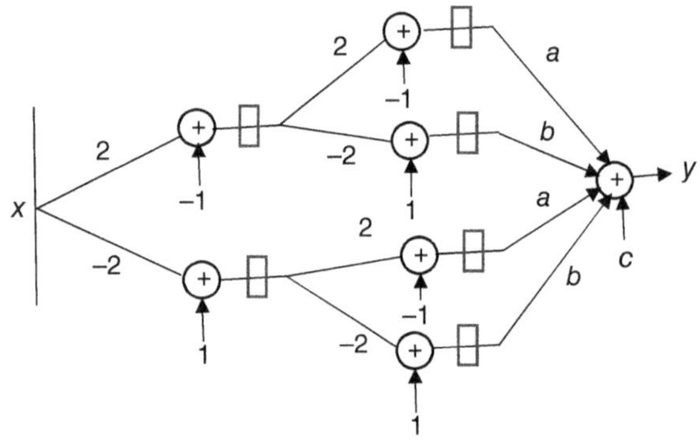


**Figure 15.12** Approximation of a half-sinusoid by a network with ReLU and two hidden layers.

## 15.9 Conclusion

In the field of artificial intelligence, neural networks are a widely used tool which can be leveraged in all technical areas and can produce spectacular results. They apply, among other areas, to the approximation of functions, object classification, and shape recognition. The neural network offers an alternative to specific techniques which exploit *a priori* knowledge on a given topic in cooperation with dedicated algorithms. In fact, the two approaches can be combined, as they are in convolutional neural networks.

On the operational front, the practical design and implementation of neural networks is a challenge. Firstly, in a given situation, the structure must be decided upon. One might think that the more neurons are included in the system, the better will the performance be. However, then, besides the computation requirements, multiple issues manifest themselves in relation to initialization, response time, and overdimensioning, with related drifts and stability problems. Next, one must get to grips with a system comprising large numbers of nonlinearities and be able to validate the results. Finally, though the neural network is a potentially powerful tool, its exploitation requires great care in order to be efficient and possibly competitive in terms of performance and technological resources.

## Exercises

**15.1** Consider 2 groups of points in the plane, with coordinates:

A1 = [1,2}; A2 = [−1,1]; A3 = [−2,0]; B1 = [2,1]; B2 = [−1,−2]; B3 = [2,−2]

Give the equation of a separation line and compute the distances from the points to that line.

Use an iterative method to find a separation line and check that the separation is achieved.

**15.2** Three groups of points in the plane have the coordinates:

$A$ $(x_1, x_2)$: (1,2) (0,4) (−3,3) (−2,−1) (−4,0); $B$: (1,−2) (2,−1) (4,−1) (4,1) (−3,−3) (5,−2)

$C$: (4,5) (6,2) (6,6) (4,6)

Through observation of the points, give the equations of 2 lines which separate the 3 groups.
Propose an iterative method to reach that objective and check the results.

**15.3** An adaptive system is defined by the relations:

$$y(n) = h_1(n-1)\,x_1(n) + h_2(n-1)x_2(n); \qquad e(n) = d(n) - y(n)$$

The input signals consist of real random variables uniformly distributed in the interval $[-1\ 1]$ and the reference signal is expressed by: $y(n) = x_2(n) + 2x_1(n) - x_1^3(n)$.

Give the values of the optimum coefficients $h_1$ and $h_2$ and the corresponding error power. Compare with the estimations in Section 15.4 and justify the differences.

For adaptation step $\delta = 0.1$, compute the excess residual error.

**15.4** A neuron is added to the network in Figure 15.6, so as to obtain a network with 2 hidden layers. Give the block diagram. Keeping the reference sequence, perform a simulation and determine the new estimations $a$, $b$, and $c$. Compare with the case of a single hidden layer. White noise of amplitude 0.01 is added to the reference signal. Verify the impact on the estimations and on the drift of the coefficients.

**15.5** In the first example in Section 15.4, the reference sequence is changed and replaced by the following nonlinear function:

$$d(n) = x_2(n) - \sin\left(\frac{\pi}{2}\,x_1(n)\right)$$

Simulations yield the values: $a = 1.0744$; $b = 1.6154$, and the standard deviation of the error after convergence is 2%. Justify these values and the accuracy achieved.

**15.6** In the third example in Section 15.4, digit 5 leads to the following values of the second layer output:

$$X2 = 0.7\,[-1\ -1\ -1\ -1\ -1\ 1\ 1\ -1\ -1\ -1\ 1\ 1\ 1\ 1\ 1\ ]$$

Give the values of coefficients $g(i, 5)$ $(1 \leq i \leq 15)$ in the output layer.

Compute the number of multiplications per iteration in the learning phase.

Based on the possibilities of the 5×3 grid, estimate the number of different forms of writing that could be taken into account by the 8×8 grid.

**15.7** A source delivers 2 independent signals $d_1(n)$ and $d_2(n)$, uniformly distributed in the interval $[-1,1]$, which are fed to a device performing the processing:

$$d(n) = d_1(n) + jd_2(n); \quad s(n) = d(n) + C\,d(n-1); \quad C = 0.5\,(1+j)$$

followed by a distortion of the amplitude $a$ which becomes: $f(a) = \sin\left(\frac{\pi a}{4}\right) 4/\pi$.

To retrieve the source signal, a recursive neural network with 2 inputs, 2 outputs, and 2 nonlinearities is employed. Write the equations of the network.

The coefficients of the input matrix, the recursive part, and the output matrix are designated by $h_{ij,1}$, $h_{ij,2}$, and $h_{ij,3}$, respectively. A learning sequence yields the following coefficients:

$$h_{ij,2} = [-0.57 - 0.57\ 0.56 - 0.56]; \quad h_{11,1} = 1.05; \quad h_{11,3} = 1.25$$

Justify these values and the output error power $E = 0.006$.

**15.8** The impact of the initial conditions in the example provided in Section 15.7 is investigated. Give the first value of the variables, $x_{i1}$ and $x_{i2}$, at the output of the activation functions of the first and the second hidden layers respectively. Answer the same question for the errors $e_{i1}$ and $e_{i2}$ which are involved in updating at the next step. Deduce the impact of dividing the initial coefficient values by 2.

Justify the values given for the limit of stability. Do likewise for the system time constant.

**15.9** It is proposed to extend the block diagram in Figure 15.12 to obtain an approximation of the half-sinusoid with 8 segments. Provide the corresponding diagram and the coefficient values. Give an estimate of the maximum value of the approximation error.

## References


1 F. Rosenblatt, "The perceptron: a probabilistic model for information storage and organization in the brain", *Psychological Review*, vol.65, n°6, pp. 386–408, 1957.

2 K. Hornik, M. Stinchcombe and H. White, "Multilayer feedforward networks are universal approximators", *Neural Networks*, vol.2, n°5, pp. 359–356, 1989.

3 B.M. Wilamowski, "Neural networks architectures and learning algorithms", *IEEE Industrial Electronics Magazine*, vol.3, n°4, pp. 56–63, 2009.

4 Y. LeCun, L. Bottou, Y. Bengio, and P. Haffner, "Gradient-based learning applied to document recognition," *Proceedings of the IEEE*, vol. 86, no. 11, pp. 2278–2324, 1998.

5 I. Goodfellow, Y. Bengio and A. Courville, "*Deep Learning*", Cambridge, MA, MIT-Press, 2016. http://www.deeplearningbook.org

6 C.C. Jay Kuo," The CNN as a guided multilayer RECOS transform, *IEEE Signal Processing Magazine*, 34 2017, pp.81–89.

7 M. Schuster and K.K. Paliwal, "Bidirectional recurrent neural networks," *IEEE Transactions on Signal Processing*, vol. 45, no. 11, pp. 2673–2681, 1997.

8 M. Narwaria, "The transition from white box to black box", *IEEE Signal Processing Magazine*, 38 2021, pp. 163–173.

9 D. Elbrächter, D. Perekrestenkp, P. Grohs and H. Bölcskei, "Deep neural network approximation theory", *IEEE Transactions on Information Theory*, vol.67, n°5, 2021, pp. 2581–2623.

# 16

# Error-Correcting Codes

Systems for information processing and transmission constitute a major field of application for signal-processing techniques. Error-detection and correction techniques are also widely used in these systems and, therefore, they coexist and even interact with signal processing in communication equipment.

Generally, coding is presented and taught with a mathematical approach [1]. However, some of the most commonly used coding techniques exploit signal-processing concepts, results, and algorithms [2]. For example, Reed–Solomon coding uses the discrete Fourier transform and linear prediction, convolutional coding is FIR filtering, and turbo codes are related to IIR filtering.

This chapter provides an introduction to some important error-correcting codes from a signal-processing perspective, allowing readers to gain an understanding of these codes and assess their strengths and weaknesses. Moreover, a unified view of communication techniques may result.

## 16.1 Reed–Solomon Codes

Reed–Solomon codes are extensions of the BCH (Bose–Chaudhury–Hocquenghem) codes to multi-bit symbols [3, 4]. They exploit predictable signals generated by line errors to identify these errors and subtract them from the received signal so as to recover the transmitted signal.

Before the codes are described, some addition information about linear prediction is provided.

### 16.1.1 Predictable Signals

A signal is said to be predictable if it satisfies a linear recurrence equation. For example, the real sequence:

$$x(n) = A\cos(n\omega t + \phi) \tag{16.1}$$

satisfies equation:

$$x(n) - 2\cos\omega x(n-1) + x(n-2) = 0 \tag{16.2}$$

In fact, to cancel $x(n)$, it is sufficient to apply the following FIR filter transfer function:

$$H(Z) = (1 - e^{j\omega}Z^{-1})(1 - e^{-j\omega}Z^{-1}) = 1 - 2\cos\omega Z^{-1} + Z^{-2} \tag{16.3}$$

The signal is predictable, since, as long as two consecutive samples are known, the entire sequence can be calculated.

Similarly, the complex signal:

$$x(n) = \sum_{i=1}^{P} A_i e^{jn\omega_i} \tag{16.4}$$

is canceled by the so-called prediction FIR filter:

$$H(Z) = 1 - \sum_{i=1}^{P} H(z) = \prod_{i=1}^{P} \left(1 - e^{j\omega_i} z^{-1}\right) a_i z^{-i} \tag{16.5}$$

and it satisfies the recurrence equation:

$$x(n) = \sum_{i=1}^{P} a_i x(n-i) = 0 \tag{16.6}$$

When signal samples are available, the $P$ elementary signals making up $x(n)$ can be extracted in a few steps. First, compute the prediction coefficients $a_i (1 \leq i \leq P)$ from $2P$ consecutive samples with the help of the following matrix, obtained by running the recurrence equation (16.6) $P$ times:

$$\begin{bmatrix} x(P)x(P-1)\dots x(1) \\ x(P+1)x(P)\dots x(2) \\ \cdot \\ \cdot \\ \cdot \\ x(2P-1)x(2P-2)\dots x(P) \end{bmatrix} \begin{bmatrix} a_1 \\ a_2 \\ \cdot \\ \cdot \\ \cdot \\ a_P \end{bmatrix} = \begin{bmatrix} x(P+1) \\ x(P+2) \\ \cdot \\ \cdot \\ \cdot \\ x(2P) \end{bmatrix} \tag{16.7}$$

This linear system is solved efficiently using an order iterative algorithm, beginning at order 1 and finishing at order $P$. As for the computation load, $2(P+1)(P+2)$ multiplications and $P$ divisions are needed. For a given order $i$ of the prediction filter, the algorithm computes the prediction error $e_{i+1}$ at order $i+1$ and stops when the error is zero. For $i = P$, if the error $e_{P+1}$ is not zero, the number of components in $x(n)$ is greater than $P$.

Once the prediction coefficients have been obtained, the frequencies are determined by computing the roots $Z_i = e^{j\omega_i}$ of the prediction filter transfer function $H(Z)$ given by (16.5).

Finally, returning to (16.4), the amplitudes $A_i$ are obtained by expressing $P$ values of the signal $x(n)$, which yields the equations:

$$\begin{bmatrix} Z_1 Z_2 \dots Z_P \\ Z_1^2 Z_2^2 \dots\dots Z_P^2 \\ \cdot \qquad \cdot \\ \cdot \qquad \cdot \\ Z_1^P \dots Z_P^P \end{bmatrix} \begin{bmatrix} A_1 \\ A_2 \\ \cdot \\ \cdot \\ A_P \end{bmatrix} = \begin{bmatrix} x(1) \\ x(2) \\ \cdot \\ \cdot \\ x(P) \end{bmatrix} \tag{16.8}$$

It can be observed that, if the frequencies $\omega_i$ are known, $2P$ samples enable us to compute $2P$ amplitudes.

As above for (16.7), efficient techniques can be used to solve the matrix equation (16.8) and determine the amplitudes $A_i (1 \leq i \leq P)$ without inverting the matrix.

For example, we can apply the signal sequence $x(n)$ to the filter:

$$H_j(z) = \prod_{\substack{i=1 \\ i \neq j}}^{P} \left(1 - e^{j\omega_i} z^{-1}\right) \tag{16.9}$$

in order to get amplitude $A_j$.

Overall, it is verified that $2P$ signal samples are required to identify $P$ components.

From the above development, it appears that two methods exist to generate samples of predictable signals – one in the frequency domain and one in the time domain, which leads to two variants of the codes.

### 16.1.2 Reed–Solomon Codes in the Frequency Domain

In order to protect a set of $K$ samples, or symbols, $x(n)$, $2P$ zeros are appended, which makes a set of $N = K + 2P$ symbols. A DFT of size $N$ yields $N$ values $X(k)$ $(0 \le k \le N-1)$ that are fed to the transmission channel. At the receiver side, the inverse DFT restores the initial symbols, in the absence of error. However, if errors occur, the $2P$ appended samples are no longer zeros: they are predictable signals, whose frequencies are related to the errors.

The identification of the error signal consists of computing the prediction coefficients $a_i (1 \le i \le P)$ and, by recurrence, extending the error signal to the complete set of samples. The $K$ initial symbols are restored by subtraction. The $2P$ samples of the received signal which are not zero in the presence of errors are called the "syndrome". The procedure is illustrated in Figure 16.1.

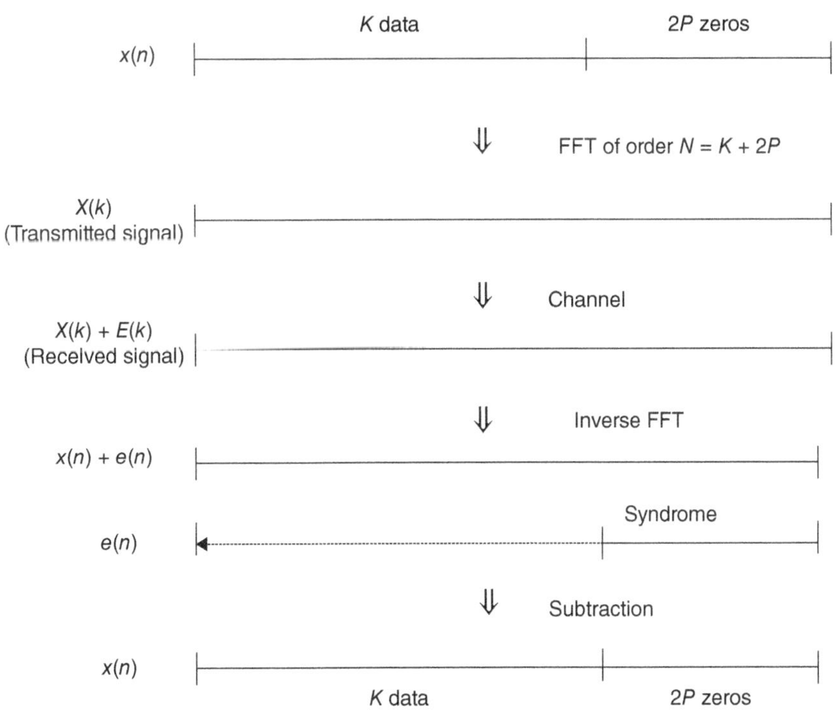


**Figure 16.1** Reed–Solomon coding in the frequency domain.

A strong point of the approach is the simplicity of the decoder, since there is no need to separate the components of the error signal. However, it requires DFT calculations and, more importantly, the coding is not systematic. In systematic coding, the useful data are transmitted as such, and they are just supplemented with protecting data. In fact, the procedure can be modified to obtain that property.

### 16.1.3 Reed–Solomon Codes in the Time Domain

The procedure is shown in Figure 16.2.

The $K$ data symbols are supplemented by $2P$ values, computed in such a manner that the inverse DFT of the total set includes a subset of $2P$ zeros. At the receiver side, an inverse DFT is calculated and the samples which should be zeros make the syndrome. The $2P$ samples of the syndrome allow for the determination of $P$ frequencies, along with the amplitudes of the $P$ corresponding components. Next, the error signal is identified and subtracted from the received signal to recover the initial data.

In general, in order to limit the loss in transmission rate, the length of the syndrome is chosen such that $2P \ll K$. In such conditions, it is advantageous to replace the inverse DFT with a set of $2P$ filters, fed by the sequence $[x(N-1), x(N-2), \ldots, x(0)]$ as shown in Figure 16.3, when the syndrome corresponds to the first $2P$ values of the inverse DFT.

It can readily be verified that $X(k) = \sum_{n=0}^{N-1} x(n)W^{nk}$ with $W = e^{j2\pi/N}$, once the $N$ input values have been processed.

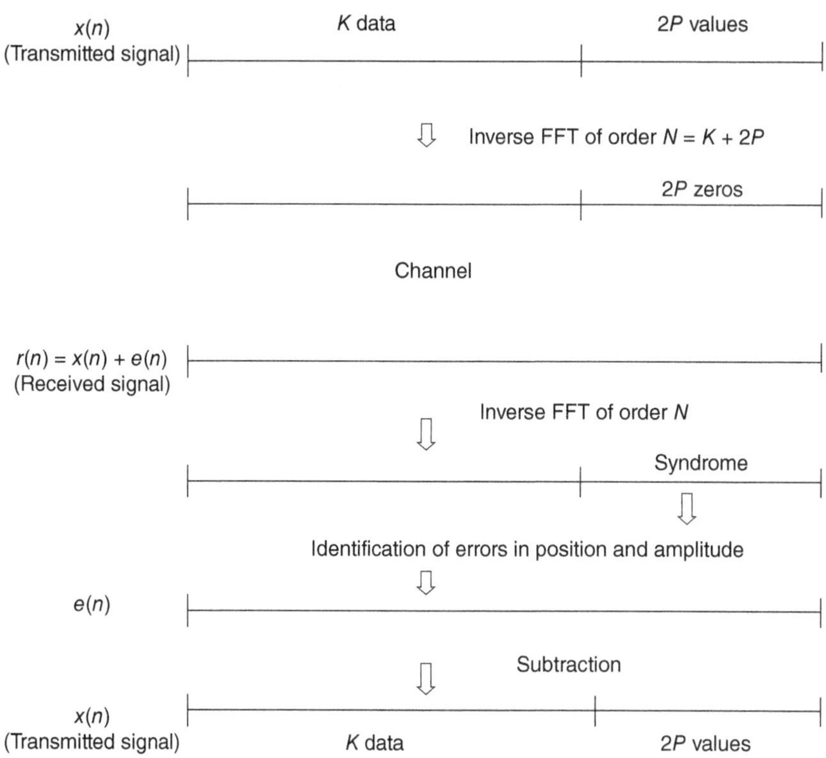


**Figure 16.2** Reed–Solomon coding in the time domain.

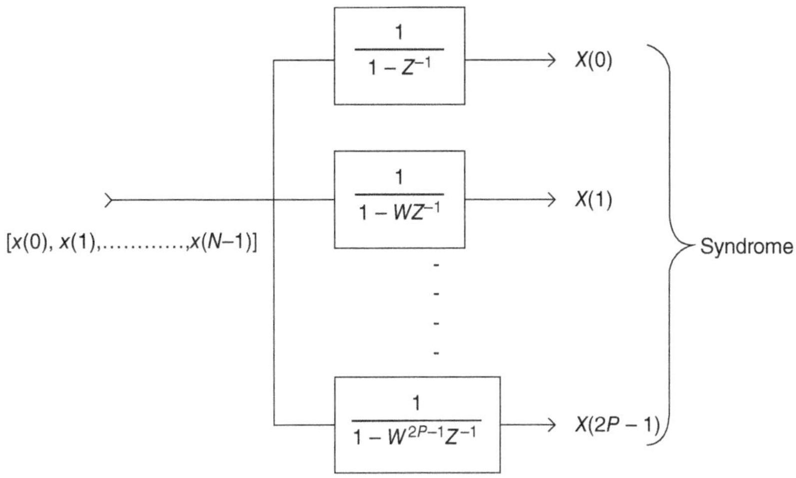


**Figure 16.3** Computing the syndrome by filtering.

Finally, the processing in the receiver comprises the following operations:

- Compute the syndrome by filtering.
- Compute the prediction coefficients.
- Extract the zeros of the prediction filter.
- Compute the amplitudes of the components
- Subtract the components of the error signal.

On the transmitter side, the processing can also be performed by filtering. Assuming the data to be transmitted are $[x(2P), \ldots, x(N-1)]$, the $2P$ complementary values $[x(0), \ldots, x(2P-1)]$ must be computed. The filter with transfer function:

$$H(z) = \frac{1}{\prod_{i=0}^{2P-1}(1 - w^i z^{-1})} = \frac{1}{1 + \sum_{i=1}^{2P} a_i z^{-i}} \tag{16.10}$$

has the following input–output relationship:

$$y(n) = x(n) - u(n); u(n) = \sum_{i=1}^{2P} a_i\, y(n-i) \tag{16.11}$$

Taking the index inversion into account, the sequence $x(2P-i) = u[N-1-(2P-i)]$, $(1 \leq i \leq 2P)$ is fed to the filter and the last $2P$ outputs – that is, $[y(N-2P, \ldots, y(N-1]$ – are zero. The operation can be checked by taking $P = 1$, for example.

The above coding and decoding process has been presented assuming the arithmetic operations are carried out in the complex field. However, the data which must be protected are generally binary – that is, the signals $x(n)$ are $B$-bit numbers, and the same applies to the $2P$ protecting values. Then, in order for the numerical calculations to be exact, arithmetic operations must be carried out in finite fields with $2^B$ elements, as indicated in Section 3.7.

### 16.1.4 Computing in a Finite Field

Galois fields, $GF(2^B)$, possess the required properties; they are algebraic extensions of the field $GF(2) = [0, 1]$. They are defined from a degree-B polynomial $g(x)$ irreducible on the field $[0, 1]$.

**Table 16.1** GALOIS FIELD GF($2^4$)

| Polar representation | Polynomial representation | Binary representation |
|---|---|---|
| 0 | 0 | 0000 |
| $\alpha^0$ | 1 | 0001 |
| $\alpha^1$ | $\alpha$ | 0010 |
| $\alpha^2$ | $\alpha^2$ | 0100 |
| $\alpha^3$ | $\alpha^3$ | 1000 |
| $\alpha^4$ | $\alpha + 1$ | 0011 |
| $\alpha^5$ | $\alpha^2 + \alpha$ | 0110 |
| $\alpha^6$ | $\alpha^3 + \alpha^2$ | 1100 |
| $\alpha^7$ | $\alpha^3 + \alpha + 1$ | 1011 |
| $\alpha^8$ | $\alpha^2 + 1$ | 0101 |
| $\alpha^9$ | $\alpha^3 + \alpha$ | 1010 |
| $\alpha^{10}$ | $\alpha^2 + \alpha + 1$ | 0111 |
| $\alpha^{11}$ | $\alpha^3 + \alpha^2 + \alpha$ | 1110 |
| $\alpha^{12}$ | $\alpha^3 + \alpha^2 + \alpha + 1$ | 1111 |
| $\alpha^{13}$ | $\alpha^3 + \alpha^2 + 1$ | 1101 |
| $\alpha^{14}$ | $\alpha^3 + 1$ | 1001 |

The elements of the field are the successive powers of a primitive element $\alpha$ such that $\alpha^M = 1$ and $M = 2^B - 1$. The number $N$ of code values must be less than or equal to $M$.

Example: $B = 4$; $g(x) = x^4 + x + 1$; $M = 15$.

Letting $\alpha^4 + \alpha + 1 = 0$, the 15 elements of the code are given in Table 16.1.

Coding and decoding consist of running the algorithms described in the above sections, using the code table for the arithmetic operations. Appropriate and optimized algorithms have been developed; they are generally presented with the help of the polynomial terminology [1]. Accordingly, a typical temporal decoding corresponds to the following sequence:

- Compute the syndrome (polynomial $s(x)$ of degree $2P$).
- Compute the localizing polynomial using the Berlekamp–Massey algorithm (iterative calculation of the prediction filter coefficients); extract the localizing zeros by the Chien method (systematic search among the elements of the field).
- Compute the amplitudes of the error by the Forney algorithm.
- Subtract the errors.

Frequency decoding necessitates the definition of a Fourier transform in the field GF($2^B$), using successive powers of a primitive element $\alpha$.

### 16.1.5 Performance of Reed–Solomon Codes

A code C($N$,$K$) is defined by the number $K$ of data to be protected and the total number $N$ of values in the block. Such a code is able to detect $N - K$ errors and correct $P \leq \frac{N-K}{2}$ errors. The number of bits $B$ of each value must be such that $N < 2^B$. In total, each block contains $KB$ useful bits protected by $(N - K)B$ redundancy bits. The performance of the code is measured by the probability error per bit $P_b$ after decoding, as a function of the error probability per symbol $P_s$ before decoding. In case of scattered errors in the channel, and if the bit error rate is $P$, the probability for a $B$-bit symbol to be error free is: $(1 - p)^B$ and we get:

$$p_s = 1 - (1 - p)^B \tag{16.12}$$

After decoding, there is no output error when the number of erroneous symbols in the block of $N$ symbols is less than the correction capability $P$. For $P + 1$ erroneous symbols, the number of bit errors is at most $(P + 1)B$ and the probability is:

$$P_{p+1} = C_N^{P+1}(p_s)^{p+1}\,(1 - p_s)^{N-P-1} \tag{16.13}$$

The corresponding bit error rate is written as:

$$BER \leq \frac{P+1}{N}\,C_N^{P+1}\,(p_s)^{P+1}\,(1 - p_s)^{N-P-1} \tag{16.14}$$

In total, the case of $(P + i)$ erroneous symbols $(1 \leq i \leq N - P)$ must be taken into account and the probabilities must be summed. However, where $P_s$ is small, it might be sufficient to consider only $i = 1$ and stick to expression (16.14).

*Example*:
$N = 204$; $K = 188$; $P = 8$; $B = 8$ bits.
With the line error probability $p = 10^{-3}$, we obtain: $p_s = 1 - (1 - 10^{-3})^8 \approx 0.008$ and

$$BER \leq \frac{9}{204}\frac{204!}{9!\,195!}(0.008)^9(0.992)^{195}$$

which is:

$$BER \leq 1.6\,10^{-6}$$

Referring to the signal-to-noise ratio, and assuming Gaussian white noise, probability $p = 10^{-3}$ corresponds to a SNR of about 10 dB, while 1.6 $10^{-6}$ corresponds to about 14 dB. Thus, the gain brought by the code for this error probability, the coding gain, is about 4 dB.

Now, $p = 10^{-4}$ leads to $BER \leq 10^{-14}$ and the code cancels the line errors, almost completely, in this example, for $p \leq 10^{-4}$.

Clearly, Reed–Solomon codes are recommended for applications requiring very low error rates.

The above derivations assume scattered errors. When errors occur in packets, we can return to the case of scattered errors by introducing symbol interleaving.

## 16.2 Convolutional Codes

Convolutional coding of a sequence of symbols consists of introducing redundancy into the sequence and correlation between symbols [5, 6]. Correlation is introduced through digital filtering – thus, convolution at the transmitter side. At the receiver side, the inverse operation is carried out. The convolution–deconvolution cascade provides the opportunity to approach the conditions for maximum rate transmission in the channel. These conditions need to be explicitly stated.

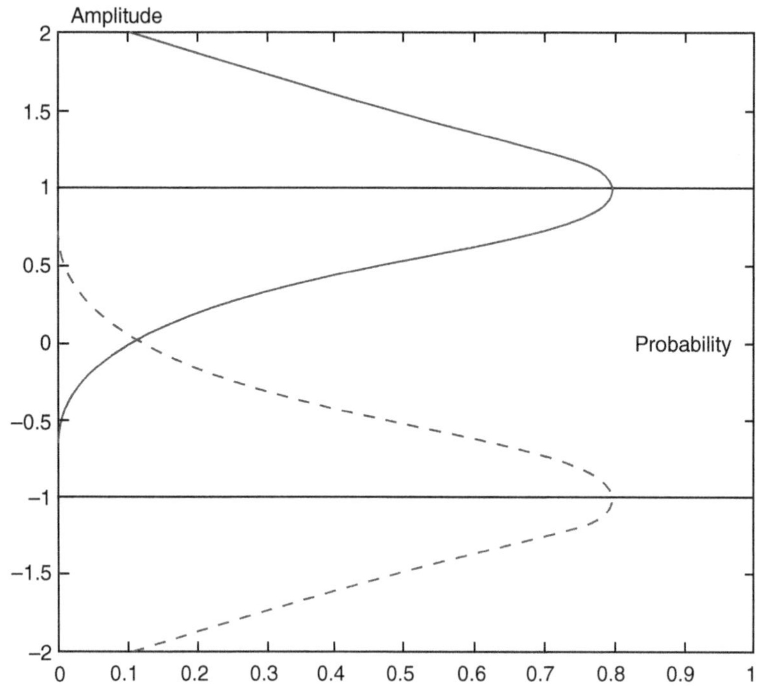


**Figure 16.4** Two-level detection.

### 16.2.1 Channel Capacity

The number of bits per symbol that can be transmitted by a channel, assumed to be free of distortion but noisy, with a specified error probability, depends on the signal-to-noise ratio and on the probability distribution law of the amplitudes of the noise.

As an illustration, consider real equiprobable symbols $d(n) = \pm 1$, and a Gaussian law as shown in Figure 16.4. Then, the error probability $(P_e)_{\pm 1}$ is expressed by:

$$(P_e)_{\pm 1} = \int_1^{\infty} \frac{1}{\sigma_b \sqrt{2\pi}} e^{-\frac{x^2}{2\sigma_b^2}} \mathrm{d}x \tag{16.15}$$

where $\sigma_b^2$ stands for the noise power. After a change of variable, we find:

$$(P_e)_{\pm 1} = \frac{1}{\sqrt{2\pi}} \int_{1/\sigma_b}^{\infty} e^{-\frac{x^2}{2}} \mathrm{d}x \tag{16.16}$$

Referring to Annex 2 in Chapter 1, the error probability $P_e = 10^{-5}$ requires $\frac{1}{\sigma_b} = 4.4$ and SNR = 12.9 dB. For symbols having more than 2 levels, (16.16) must be modified to account for neighboring levels:

$$P_e(\Delta) = \frac{2}{\sqrt{2\pi}} \int_{\Delta}^{\infty} e^{-\frac{x^2}{2}} \mathrm{d}x \tag{16.17}$$

where $\Delta = \frac{1}{\sigma_b}$

Next, the error probability must be linked to the theoretical channel capacity presented in Section 1.13. The upper bound of the number of bits that can be transmitted without error, for a given SNR, is expressed by:

$$C = \frac{1}{2}\log_2\left(1 + \frac{S}{B}\right) \tag{16.18}$$

Thus, to transmit 1 bit, we must have SNR = 3 – that is, 4.77 dB. The difference of 8.1 dB with the $10^{-5}$ error probability above represents the loss which can possibly be compensated for by coding. For symbols with $2^N$ equiprobable symbols, uniformly spread in the amplitude range $\pm 2^N$, the signal power is:

$$S = \frac{1}{3}(2^{2N} - 1) \approx \frac{1}{3}2^{2N} \tag{16.19}$$

The error probability can be introduced into the expression of the number of bits $C_N$ that symbols can carry. Referring to the peak signal power, $3S$, we write:

$$C_N = \frac{1}{2}\log_2\left(1 + \frac{3}{\Delta^2}\frac{S}{B}\right) \tag{16.20}$$

where the parameter $\boldsymbol{\Delta}$, introduced in (16.17), represents the error probability.

Worth noting is the fact that the capacity limit (16.18) corresponds to $\Delta = \sqrt{3}$, which means that the greatest noise standard deviation that allows $C_N$ bits to be transmitted without error is expressed by:

$$(\sigma_b)_{\text{lim}} = \frac{1}{\sqrt{3}} = 0.58 \tag{16.21}$$

This should be compared with half the distance between neighboring levels, which is unity. Applying relation (16.17), we note that, in the absence of coding, this noise level would lead to the error probability of 0.0833.

### 16.2.2 Approaching the Capacity Limit

Before studying the conditions for approaching the limit, it is necessary to revisit the derivation in Section 1.13.

Instead of decoding symbols independently as above, the global decoding of a set of $M$ symbols of $N$ bits each, in the presence of Gaussian noise power $B = \sigma_b^2$, is contemplated. Then, an $M$ component noise vector is added to the useful signal, the energy being $E_b = M\sigma_b^2$.

When $M$ tends toward infinity, the end of the noise vector sits on the hypersphere with radius $\sqrt{M}\,\sigma_b$. Each set of $M$ symbols can be represented by a point in a $M$-dimensional hyperspace. In this hyperspace, the received signal samples must occupy a volume greater than $2^{MN}$ times the volume of the noise hypersphere, because each of the potential $2^{MN}$ sets of symbols is accompanied by a noise vector. In order to minimize energy, the symbol volume $V_s$ must be a hypersphere whose radius $R$ must be greater than $2^N$ times the radius of the noise sphere:

$$R > 2^N\sqrt{M}\sigma_b \tag{16.22}$$

The volume $V_M$ of the hypersphere is calculated as a function of the radius $R$ by:

$$V_M = \int_0^R r^{M-1}\mathrm{d}r \int_{i=1,..,M-1} \cdots \int f(\theta_i)\,\mathrm{d}\theta_i = \frac{R^M}{M}F_\theta \tag{16.23}$$

where $F_\theta$ is a function of $\pi$. For example, for $M = 3$, we have $F_\theta = 4\pi$.

Assuming uniform distribution of symbols in the hypersphere, the corresponding signal has energy:

$$E_S = \frac{1}{V_M} \int_0^R r^2 r^{M-1} \mathrm{d}r \int_{i=1} \cdots \int_{M-1} f(\theta_i) \mathrm{d}\theta_i = R^2 \frac{M}{M+2} \tag{16.24}$$

When $M$ tends toward infinity, $R^2$ represents the total energy of the received samples and $R^2/M$ represents the sum of the signal and noise powers. From (16.22), we obtain:

$$S + B > 2^{2N} B \tag{16.25}$$

Hence, the channel capacity limit is defined by (16.18).

Now, in order to come close to this limit, the points representing the transmitted signal must be spread in the hypersphere, which implies that the projections on the coordinate axes are quantized with more than $N$ bits and that relations exist between these projections.

It is worth pointing out that tending toward infinity for the number of symbols of a block means infinite transmission delay.

In practice, the number of symbols in a block at decoding is limited to $M$ and the loss in performance must be assessed. The error probability $P_e$ must be related to the ratio of noise standard deviation $\sigma_b$ and the value $(\sigma_b)_{\lim}$ associated with the limit capacity. Letting $\alpha = \frac{(\sigma_b)_{\lim}}{\sigma_b}$ with $\alpha > 1$, the error probability is expressed as a function of the probability distribution $P(r)$ of the radius $r$ of the noise hypersphere by:

$$P_e = \int_{\alpha\sqrt{M}}^{\infty} P(r) \mathrm{d}r \tag{16.26}$$

To find $P(r)$, consider M unit-variance Gaussian independent random variables $b_i$. The variable:

$$r = \left( \sum_{i=1}^{M} b_i^{\,2} \right)^{1/2} \tag{16.27}$$

has the distribution:

$$P(r) = \frac{2}{2^{\frac{M}{2}} \left( \frac{M}{2} - 1 \right)!} r^{M-1} e^{-\frac{r^2}{2}} \tag{16.28}$$

with $M$ being even. For example, $M = 2$ is the Rayleigh distribution.

The function $P(r)$ is maximum for $r = \sqrt{M-1}$. Figure 16.5 provides an illustration for $M = 256$.

Finally, the 3 decoding parameters are the deviation with respect to the limit SNR $\alpha$, the tolerated error probability $P_e$, and the dimension $M$ of the decoded block. For example, to approach the limit at 1 dB – that is, $\alpha = 1.1$, with $M = 250$, we get $P_e = 10^{-2}$, while with $M = 950$, we get $P_e = 10^{-5}$.

In the case of binary symbols $N = 1$, the above derivation must be slightly modified. In fact, each set of $M$ symbols can be represented by a point of the hypersphere of radius $\sqrt{M}$ in the $M$-dimension hyperspace. The noise hypersphere must be within the cone associated with the solid angle corresponding to $1/2^M$ of the total angle. It can be shown that the following approximation holds:

$$M.[10 \log(\alpha)]^2 \geq 29.10 \log \left( \frac{1}{P_e} \right) \tag{16.29}$$

Thus, for $\alpha = 1.1$, which is a 1 dB deviation with respect to the limit, and $P_e = 10^{-5}$, the block length must be $M = 1450$ symbols.

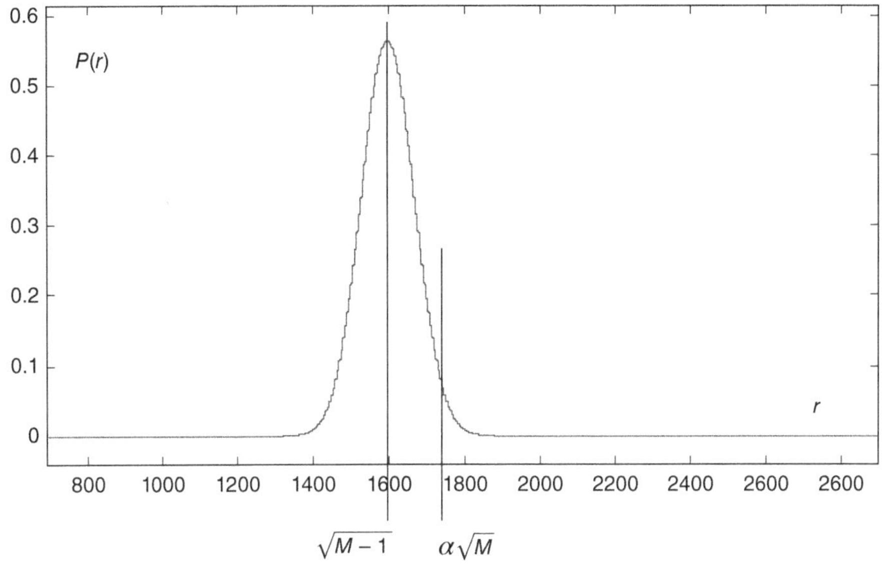


**Figure 16.5** Probability distribution of the noise vector modulus.

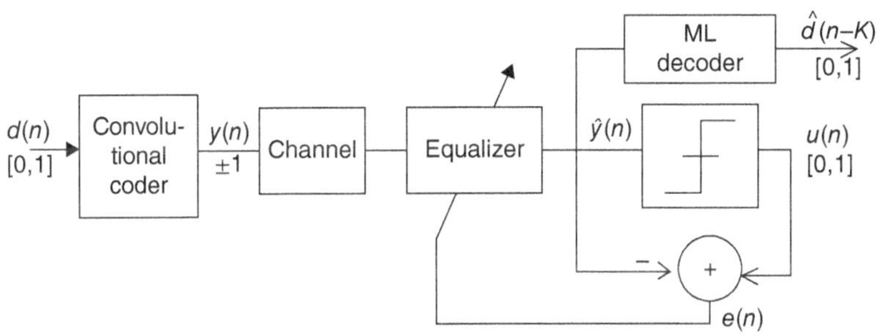


**Figure 16.6** Transmission system with convolutional coding.

Convolutional codes are designed to draw near to the conditions leading to the limit, and a simple scheme is described, to begin with. As for large blocks, iterative techniques are required, with turbo codes.

### 16.2.3 A Simple Convolutional Code

Convolutional codes are based on FIR filtering. A simple system consists of two filters, binary data and doubling the output rate; it is shown in Figure 16.6.

The binary symbols $y(n)$ delivered by the coder are fed to the channel whose distortion is compensated by an equalizer at receiver input. The equalizer output $\tilde{y}(n)$ is applied to a maximum likelihood (ML) decoder, which provides an estimation $\tilde{d}(n-K)$ of the input binary data. The system delay is $K$. The ML decoder has multilevel inputs. The binary symbols $u(n)$ are used to obtain the error signal $e(n)$ needed for equalization. Assuming perfect equalization, the system can be modeled as shown in Figure 16.7.

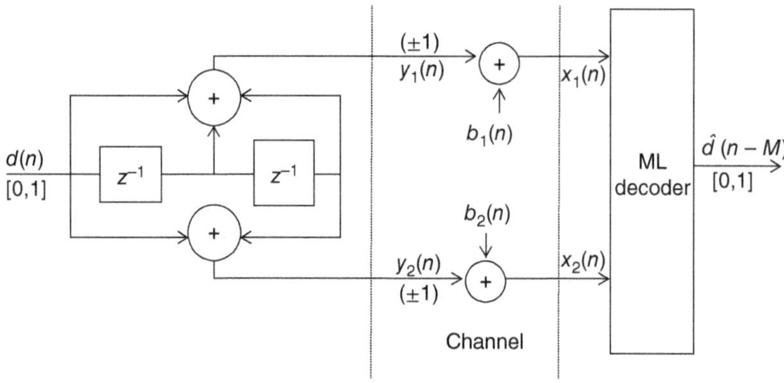


**Figure 16.7** System model with convolutional decoding.

The filter transfer functions are:

$$H_1(Z) = 1 + Z^{-1} + Z^{-2}$$
$$H_2(Z) = 1 + Z^{-2} \tag{16.30}$$

The number of coefficients $L = 3$ is called the "constraint length" of the code. Since input and output data are binary, the arithmetic operations are carried out modulo 2, in the field [0, 1]. The coder output rate is doubled since each input generates 2 outputs; the code rate is $R = 1/2$.

The channel is assumed to be distortion free but noisy. Noise samples are added to the transmitted useful samples, and the received signals $x_1(n)$ and $x_2(n)$ are fed to the *ML* decoder which retrieves the data with the delay $M > L$.

In the decoding process, at time $n = M - 1$, the data sample to be decoded $d(0) = \pm 1$ appears in at most $2L = 6$ received values, as shown in the sequence:

$$\begin{aligned}
x_1(0) &= d(0) \oplus d(-1) \oplus d(-2) + b_1(0)\\
x_2(0) &= d(0) \oplus d(-2) + b_2(0)\\
x_1(1) &= d(1) \oplus d(0) \oplus d(-1) + b_1(1)\\
x_2(1) &= d(1) \oplus d(-1) + b_2(1)\\
x_1(2) &= d(2) \oplus d(1) \oplus d(0) + b_1(2)\\
x_2(2) &= d(2) \oplus d(0) + b_2(2)\\
x_1(3) &= d(3) \oplus d(2) \oplus d(1) + b_1(3)\\
x_2(3) &= d(3) \oplus d(1) + b_2(3)
\end{aligned} \tag{16.31}$$

An error vector of $M$ terms can be built, taking the differences between the received and emitted values. The norm $E_{\min}$ of this vector is taken as the cost function, and it is minimum at time $n = M - 1$ if $d(0)$ is the transmitted value in (16.31). Next, the deviations produced by false $d(0)$ values in (16.31) must be calculated. The impact of modulo additions is worth pointing out because error compensation may occur which necessitates the examination of cases of multiple errors. Then, if the exact value is $d(0) = 0$, we find:

1. Single error

$$E_1 = \sum_{n=0}^{M-1} [x_1(n) - y_1(n)]^2 + [x_2(n) - y_2(n)]^2 \tag{16.32}$$

$$(d(0) = 1) : E_1 - E_{\min} = \Delta_1 = 20\left[1 + \frac{b_1(0) + b_2(0) + b_1(1) + b_1(2) + b_2(2)}{5}\right]$$

2. Double error

$$d(0) \text{ and } d(1) : \Delta_2 = 24\left[1 + \frac{b_1(0) + b_2(0) + b_2(1) + b_2(2) + b_1(3) + b_2(3)}{6}\right]$$

$$d(0) \text{ and } d(2) : \Delta_2' = 24\left[1 + \frac{b_1(0) + b_2(0) + b_1(1) + b_1(3) + b_1(4) + b_2(4)}{6}\right]$$

The average of noise samples appears in the cost function, with factor 5 for the single-error case and 6 for the double-error case. It can readily be verified that the averaging factor is greater than 6 for other cases. In the absence of coding, the averaging factor is 2. Thus, with coding, considering single-error cases only, the coding gain is approximately $G_c = 2.5$, or 4 dB.

It can be observed that the averaging factor corresponds to the number "1" at the filter outputs when the error sequence is fed at the inputs. Then, a single error leads to the impulse response. This number is called the weight of the sequence.

The curves giving the error probability per bit as a function of $E_b/N_0$, with and without coding, are shown in Figure 16.8. The term $E_b$ is the energy per bit and $N_0$ is the noise power spectral density. The ratio $E_b/N_0$ is a theoretical parameter that allows us to obtain generic curves. To get the practical SNR and determine the rates, it is necessary to apply a factor 2 to account for the symbol rate equal to twice the channel bandwidth and multiply by the number of bits in each symbol.

With $L = 3$, the coding gain is close to 4 dB for $P_e = 10^{-6}$. In contrast, it is less than 3 dB for $P_e = 10^{-3}$. In fact, it is necessary to consider the vectors in the vicinity of the ideal vector and add up the error probabilities. Overall, the vectors associated with multiple errors have averaging factors greater than that of the single-error case and their impact is weak for high-SNR situations

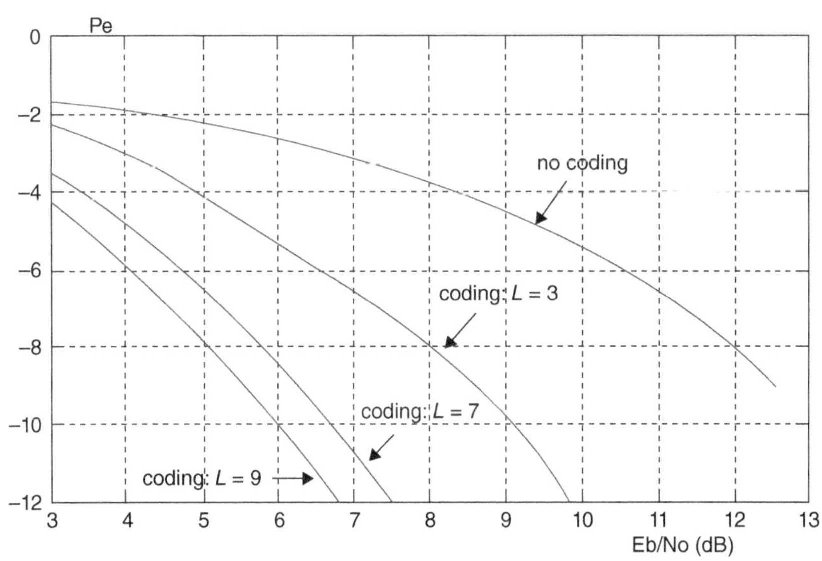


**Figure 16.8** Bit error probability for convolutional coding with rate R = 1/2.

and small error probabilities. The error probability is expressed as a function of the averaging factor $k$ by:

$$p_k = \frac{1}{\sqrt{2\pi}} \int_{\frac{\sqrt{k}}{\sigma_b}}^{\infty} e^{-\frac{x^2}{2}} \mathrm{d}x \approx 0.4\, e^{-\frac{k}{2\sigma_b^2}} \tag{16.33}$$

and for $k+1$:

$$p_{k+1} \approx 0.4 \exp\left(-\frac{k+1}{2\sigma_b^2}\right) \approx p_k \exp\left(-\frac{1}{2\sigma_b^2}\right) \tag{16.34}$$

The vectors in the vicinity of the ideal vector are obtained from the trellis diagram and the graph of the transitions between associated states, which constitute an efficient realization of the ML principle. The generating function of the convolution code is expressed by:

$$T(D, L, N) = D^5L^3N + D^6L^4(1+L)N^2 + D^7L^5(1+L)^2N^3 + D^8L^6(1+L)^3N^4 \tag{16.35}$$

where $D$ is the averaging factor, the exponent of $N$ is the number of errors, and the terms of the polynomial in $L$ represent the multiple error configurations. For example, the averaging factor 7 occurs for 3 errors in the following 4 configurations:

$[d(0), d(1), d(2)]$, $[d(0), d(1), d(3)]$, $[d(0), d(2), d(3)]$, $[d(0), d(2), d(4)]$.

The averaging factor is called "free distance" in coding theory.

### 16.2.4 Coding Gain and Error Probability

The codes are determined by a given constraint length and by the search for the maximum value of the minimum averaging factor. In the single error case, this factor equals the number of nonzero coefficients $N_1$ in the set of polynomials $H_1(Z)$ and $H_2(Z)$. For 2 consecutive errors, it equals the number of alternations $N_2$ of zero and nonzero coefficients. Generally, these two cases account for the smallest averaging factor values and a good initialization strategy is to start from 2 polynomials that lead to equivalent factors for the cases of single error and two consecutive errors. If the ones and zeros alternate in $H_1(Z)$ and $H_2(Z)$, then, $N_2 = 2L$ and $N_1 = \mathrm{L}$.

If a zero is replaced by a one, the number of alternations $N_2$ is reduced by two units and $N_1$ is increased by one unit. Equality is reached for $N_1 = N_2 = 4L/3$ and the coding gain for small error probability is bounded by:

$$G_{\max} = 10 \log\left[\frac{1}{2}\,\frac{4}{3}\,L\right] \tag{16.36}$$

For example, for $L = 7$, a widely used code has coefficients [1111001] and [1011011] and $N_1 = N_2 = 10$. The bound (16.36) is $G_{\max} = 6.7$ dB, while simulations yield a coding gain of 6.3 dB.

Similarly, for $L = 9$, the best code has coefficients [111101011] and [101110001] and $N_1 = N_2 = 12 = 9{\times}4/3$. As shown by the curve associated with that code in Figure 16.8, for $P_e = 10^{-8}$ the coding gain is 7 dB, while bound (16.36) gives $G_{\max} = 7.8$ dB. That gain is obtained for $\frac{E_b}{N_0} = 5$ dB. Note that, for a rate $R = 1/2$ code, the channel capacity is $C = 1/2 = 1/2 \log_2(1{+}1)$ which corresponds to SNR $= 1 = 2\,E_b/N_0$, or $E_b/N_0 = -3$ dB.

The systematic codes keep the initial data sequence, and they are associated with $H_1(Z) = Z^{-P}$. The delay is $P$ samples and the gain, which rests on $H_2(Z)$, is reduced.

The global error probability of a code is computed by considering all the data vectors in the vicinity of the ideal vector and summing the corresponding probabilities. For the averaging factor $k$, the number of error configurations $a_k$ is obtained from (16.35) rewritten as:

$$T(D, 1, 1) = \sum_{k_{\min}}^{\infty} a_k D^k \tag{16.37}$$

For each configuration, the error probability $P_k$ is determined so that the average of the $k$ noise samples exceeds unity as shown by relations (16.32). The error probability is bounded by:

$$P_E < \sum_{k_{\min}}^{\infty} a_k P_k \tag{16.38}$$

The inequality reflects the fact that noise samples are involved in several averaging operations. Approximation (16.33) leads to:

$$P_E < 0.4 \exp\left(-\frac{k_{\min}}{2\sigma_b^2}\right) \sum_{k=k_{\min}}^{\infty} a_k \exp\left(-\frac{k - k_{\min}}{2\sigma_b^2}\right) \tag{16.39}$$

This represents the probability that an error will occur. However, an important feature of convolutional codes is that the output errors often occur in packets, due to the decoder.

In summary, convolutional codes have the following properties:

- The coding gain grows with the constraint length.
- The coding gain is reduced for low SNR values.
- The decoder may generate error packets.

### 16.2.5 Decoding and Output Signals

The ML decoder retrieves the data sample $d(n)$ assuming the $L-1$ previous data are known and searching for the minimum norm of the M-term error vector associated with indices: $n, n+1, \ldots, n+M-1$. If an error occurs at time $n$, it is propagated in the decoder memory, and it can generate new errors. An illustration is provided in Figure 16.9, with $L = 3$, a block of 1000 bits and 3 dB SNR.

To investigate the generation of these errors, it is helpful to consider the signal which represents the noise averaging, according to (16.32). This signal $s_q(n)$ is obtained by the following operations at time $n$:

- Assuming $d(n) = 0$, search for the minimum $E_0(n)$ of the error vector norm.
- Assuming $d(n) = 1$, search for the minimum $E_1(n)$ of the error vector norm.
- Take $s_q(n) = E_1(n) - E_0(n)$.

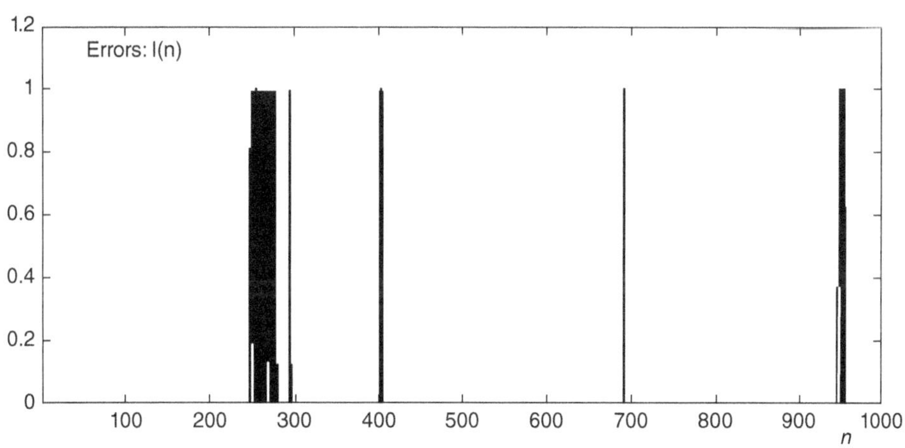


**Figure 16.9** Errors at the output of a convolutional decoder.

The binary data are given the sign of the sequence $s_q(n)$. In fact, the sign of $s_q$ $(n)$ represents the sum of the original data and the error signal and $|s_q(n)|$ is the input noise after averaging, with amplitude shift and aliasing about the origin in the presence of errors. Low values of $|s_q(n)|$ reflect the low reliability of the recovered data at corresponding times.

A decoder that delivers the sequence $s_q(n)$ is called a soft output decoder. This signal may be leveraged in decoders in cascades to extract errors, as in turbo codes.

An approach to cope with error packets consists of cascading the convolutional decoder with an interleaver and a Reed–Solomon decoder. The interleaver performs a permutation which spreads the errors over several blocks of the RS code, which allows for correction.

### 16.2.6 Recursive Systematic Coding (RSC)

An IIR filter can be employed to provide systematic coding which is equivalent to the non-systematic coding provided by a FIR filter. It suffices to replace $H_1$ $(Z)$ with 1 and $H_2$ $(Z)$ with $H_1(Z)/H_2(Z)$, which leads to the diagram shown in Figure 16.10 for $L = 3$ [7].

The output of the decoder is the same as the output of FIR filter-based coder when the input signal is filtered by $H_1(Z)$. Therefore, the minimum averaging factor in ML decoding is the same, but it is obtained for 3 consecutive errors, for $L = 3$. However, the impulse response is infinite, and, as a consequence, a data sample to be decoded is involved in all the following samples that are taken into account in the computation of the error vector norm. Modification of one bit in the input sequence entails a change of the whole following output sequence. The computation of the soft output sequence is carried out as in the FIR coding.

The recursive code can be made circular and transformed into a long block code by terminating the block in such a manner that the memory returns to the initial state. Then, the impulse response is at most equal to the block length. Now, an input sequence with $L - 1$ zeros appended leads to a return to zero with FIR filtering, while the probability is $1/2^{L-1}$ with IIR filtering. Then, due to the ML calculation, an input sequence that does not bring about a return to zero has a low probability of being incorrectly recovered by the decoder.

In short, RSC allows for a long block to be processed with a small constraint length and reduced decoder complexity.

In order to come near to the channel capacity limit mentioned in Section 16.2.2, it is necessary to combine RSC with an interleaving operation, which spreads the error packets, and an iterative process [7, 8].

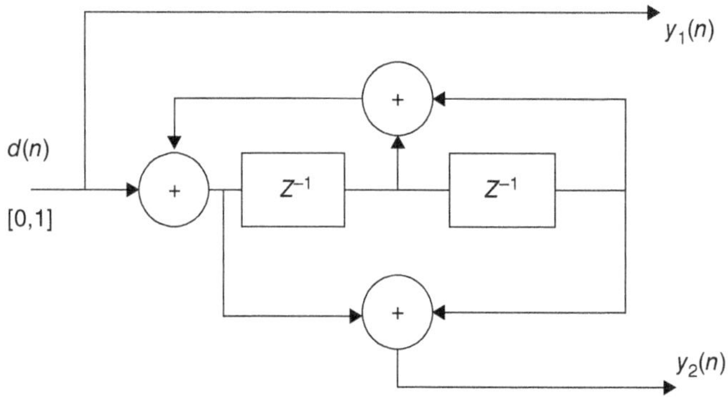


**Figure 16.10** Recursive systematic coder.

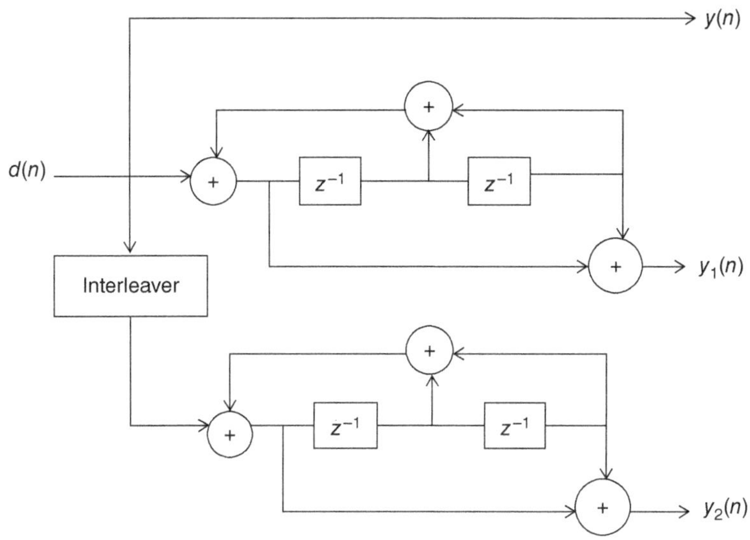


**Figure 16.11** Rate R = 1/3 parallel coder.

### 16.2.7 Principle of Turbo Codes

The data are processed in blocks. Two recursive systematic convolutional coders are deployed in parallel with an interleaver in between. At the decoder, soft outputs are obtained, and the output of the second decoder is fed to the input of the first decoder after deinterleaving. The diagram of the rate $R = 1/3$ coder, based on the filters described in the previous section, is given in Figure 16.11.

The principle of the decoder is shown in Figure 16.12. Each decoder delivers a soft output. The decoded data sample is the sign of the output of decoder 2, $s_p(n)$. The decoder outputs can be calculated through a technique similar to that used to derive $s_q(n)$ in the previous sections or using probability computation.

Decoder 1 receives the signal $x(n)$ of the emitted data and the signal $x_1(n)$ which is the output of the first coder with channel noise added. The soft output is fed to decoder 2, after interleaving, in order to synchronize with the signal $x_2(n)$ from coder 2. Then, the output of decoder 2 is deinterleaved to be fed to the input of decoder 1 after combination with the received signal $x(n)$. The SNR of the sequence $s_p(n)$ is improved at each iteration and the combination with the noisy input signal is performed in such a manner that the SNRs add up. In order for the addition to be efficient, the

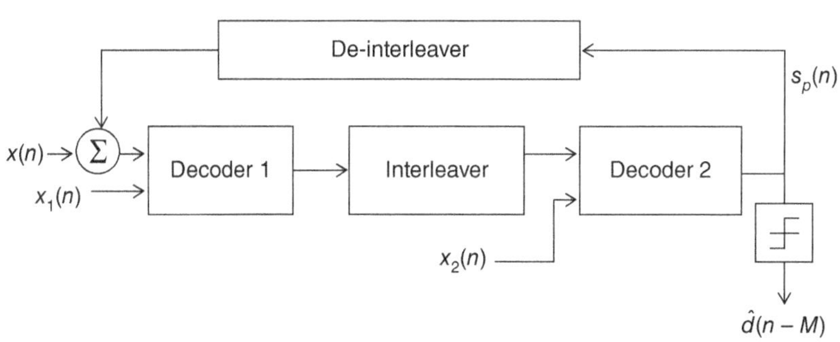


**Figure 16.12** Principle of the iterative decoder.

noise signals must be independent – hence the importance of the interleaving involved in recursive coding. Moreover, it is obvious that using systematic coding and keeping the input signal is essential for the realization.

For low input SNR, many errors occur at first decoding. They are corrected gradually by successive iterations. The procedure is stopped when the output signal variance is not decreasing any more and the retrieved data are obtained as the sign of this signal.

The transmission efficiency is improved with rate $R = 1/2$ and alternatively transmitting the outputs of coders 1 and 2.

*Example:* Let us consider a coder with $L = 5$ and the IIR filter transfer function:

$$\frac{H_2(Z)}{H_1(Z)} = \frac{1 + Z^{-4}}{1 + Z^{-1} + Z^{-2} + Z^{-3} + Z^{-4}}$$

a permutation matrix of size $256 \times 256$ as interleaver, and the block length 65,536 bits. After 18 iterations, the error probability per bit has become smaller than $10^{-5}$ for SNR = 0.7 dB. The SNR limit value, with rate $R = 1/2$, equals 0 dB. In this example, the turbo code comes as close as 0.7 dB to the theoretical limit.

### 16.2.8 Trellis-Coded Modulations

In the previous sections, the data and emitted symbols are binary: $d(n) = [0, 1]$ and $x(n) = \pm 1$. Coding techniques can be extended to multilevel symbols. The principle consists of protecting the low-weight bits with a convolutional code and distributing the amplitudes in such a way that intervals corresponding to the non-protected bits are maximized [9].

An illustration is given in Figure 16.13. Amplitudes of 4- and 8-level symbols are meant to carry 2 information bits with the same power of the emitted signal.

In the absence of coding, 4 levels are needed and the distance between neighboring levels is 2q. Applying a rate $R = 1/2$ convolutional code to one of the 2 bits, as shown in Figure 16.14, a 3-bit symbol is obtained. The amplitude distribution is such that the noncoded bit is associated with the distance 4q, as apparent in Figure 16.13. In order to determine the system coding gain, the

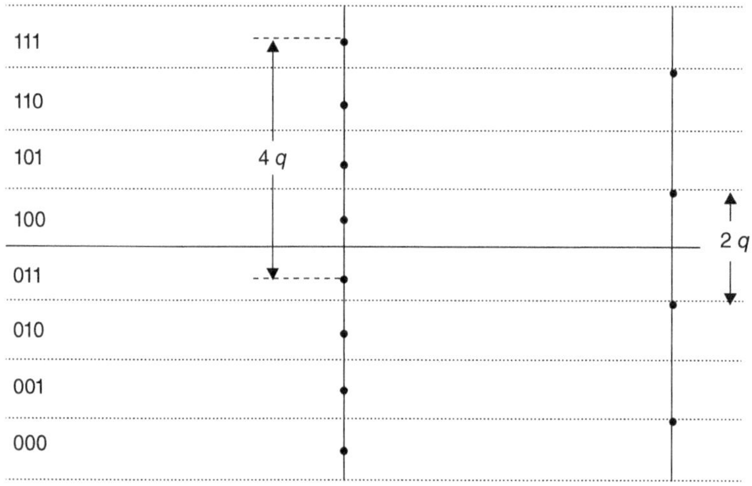


**Figure 16.13** Amplitudes of 4- and 8-level symbols.

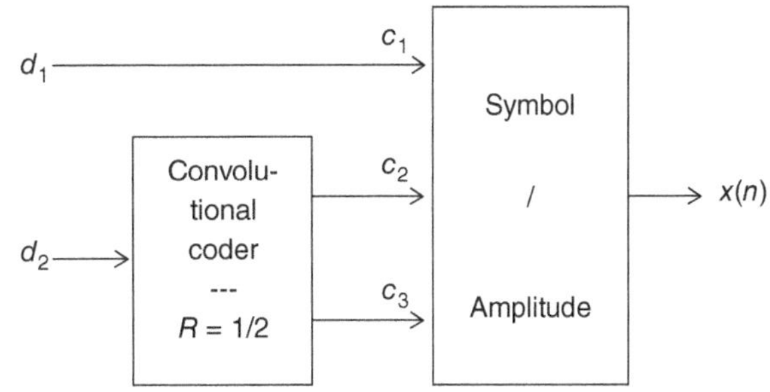


**Figure 16.14** Transmitter with coded modulation.

minimum averaging factor $k_{min}$ of the convolutional code can be associated with a multiplication of the distance between levels by $\sqrt{k_{min}}$. Then, for $k_{min} > 16$, the distance 4q corresponding to the noncoded bit is dominant and the system coding gain, with respect to the absence of coding, can come close to factor 2 – that is, 6 dB.

At decoding, the noncoded bit is introduced in the trellis in the form of parallel paths, which means that the weighting computations are carried out for the 2 possibilities associated with this bit. The assignment of the amplitudes for the 2 coded bits can also be optimized – the greatest available distance is associated with transitions in the trellis that start from the same state or reach the same state. The objective is to maximize the distance between paths.

When the free distance of the code is preponderant, the error probability $P_e$ per symbol is estimated using (16.38) or the simplified version:

$$P_e \approx 0.4\, a_0 \exp\left(-\frac{k_{min}}{2\sigma_b^2}\right) \tag{16.40}$$

where $a_0$ is the number of error configurations leading to the minimum averaging factor $k_{min}$ as explained in Section 16.2.3.

## 16.3 Conclusion

Signal-processing techniques can apply to error detection and correction. Reed–Solomon codes based on the DFT, in combination with linear prediction, are able to correct errors in blocks so that extremely low error rates can be reached.

Convolutional codes use digital filters, and they bring the equivalent of SNR improvements in transmission. The decoder is simple in principle and easy to implement. It exploits the ML technique and the Viterbi algorithm whose complexity depends on the filter order and the length of the code. The delay is proportional to the code length, multiplied by a factor of a few units. Convolutional coding can be extended to multi-bit symbols, with the coded modulation technique, in which one or several low-weight bits are protected.

Turbo codes combine recursive filtering, interleaving, and an iterative procedure. Their performance can come close to the theoretical limit of channel capacity. They process data blocks of large or very large size, which increases complexity and transmission delay.

The combination of a convolutional coder (inner code) and a Reed–Solomon coder (outer code) is a very powerful setup which yields extremely low error rates in transmission and radio broadcasting systems.

## Exercises

**16.1** A Reed–Solomon code uses a DFT of size 32 and the syndrome consists of the 4 values:
$S = [2 + j - 1{,}69 + 2{,}23\,j - 2{,}12(1 + j)2{,}23 - 1{,}67\,j]$
Compute the prediction coefficients and determine the signal which must be subtracted in the frequency domain to retrieve the initial signal.

**16.2** Cancelling an impulse noise. To protect a block of samples $X = [x(0), \ldots, x(5)]$, a suffix $Sx = [x(6), x(7)]$ is added, such that, in the DFT of sequence, the terms $Y(6)$ and $Y(7)$ are null. Give the expression of $Sx$ as a function of $X$.
After addition of an impulsive noise, we have $Y(6) = 2\,e^{-j\pi/2}$, $Y(7) = 2\,e^{j3\pi/4}$. Compute the amplitude of the pulse and its index in the input block. What happens if the spurious pulse falls between two index values and then, how can its impact be reduced?

**16.3** A rate $R = 1/2$ convolutional code has the coefficients [1111] and [1101]. Give the realization diagram and show that the free distance is 6. What is the maximum coding gain?

**16.4** Assuming binary data, $d(n) = \pm 1$, apply the formula that gives the theoretical capacity of a channel, in order to obtain the maximum SNR value which allows for error-free transmission.
The system SNR is assumed to be equal to 8 dB and a rate $R = 1/2$ convolutional code is used. If a bit error rate less than $10^{-8}$ is targeted, give the coding gain needed. Propose a code which is able to meet the objective.

**16.5** A systematic code is defined by:

$$H_1(Z) = 1; H_2(Z) = 1 + Z^{-1}$$

Give the free distance of the code.
Perform the decoding of the sequence: 011010110001101111
The code is made recursive – that is, $H_1(Z) = 1;\quad H_2(Z) = \frac{1}{1+Z^{-1}}$
To use this code, an extra bit is appended to each set of 3 information bits, in order to secure a return to zero. Give the corresponding rate. What is the free distance?

## References


1 W.W. Peterson and E.J. Weldon, *Error Correcting Codes*, MIT Press, Cambridge, MA, 1972.
2 R. Blahut, *Algebraic Methods for Signal Processing and Communications Coding*, Springer-Verlag, New York, 1992.
3 A. Hocquenghem, *Codes Correcteurs d'Erreurs*, revue *Chiffres*, Vol. 2, pp. 147–156, 1959.
4 R.C. Bose and D.K. Ray-Chaudhuri, On a class of error correcting binary group codes, *Information and Control*, Vol. 3, 1960, pp. 68–79.
5 A.J. Viterbi and J.K. Omura, *Principles of Digital Communications and Coding*, McGraw-Hill, New York, 1979.
6 R. Ziemer and R. Peterson, *Introduction to Digital Communication*, Chapter 7: Fundamentals of convolutional coding Prentice Hall, N.J., 2001.

**7** C. Berrou and A. Glavieux, Near optimum error correcting coding and decoding: turbo-codes, *IEEE Transactions* Vol. 44, N° 10, 1996, pp. 1261–71.

**8** C. Berrou, The ten-year-old turbo codes are entering into service, *IEEE Communications Magazine*, 2003, pp. 111–116.

**9** E. Biglieri, D. Divsalar, P.J. McLane and M.K. Simon, *Introduction to Trellis-Coded Modulation*, Macmillan, New York, 1991.

# 17

# Applications

Signal processing is instrumental in the generalization of electronics to all technical fields. A few examples of applications are presented in this chapter, mainly in the field of communications.

## 17.1 Frequency Detection

Assume that the amplitude of a signal component with frequency $f_0$ is to be determined when the signal is sampled at frequency $f_s > 2f_0$. Figure 17.1 represents the set of operations to be performed.

The signal is applied to a narrow band-pass filter centered on the frequency $f_0$. Rectification is then performed by taking the absolute value of the numbers obtained. This set of absolute values is applied to a low-pass filter which provides the desired value of the amplitude. If the frequency component $f_0$ which is to be detected is present, threshold logic provides the logic information.

This process can be analyzed as follows. Assume $s_0(t)$ is the signal to be detected, with:

$$s_0(t) = A\sin(\omega_0 t)$$

Taking the absolute value of the numbers which represent samples of this signal is equivalent to multiplying by a square wave $i_p(t)$ in phase with $s_0$, and of unit amplitude. Using equation (1.6), we can write:

$$i_p(t) = 2\sum_{n=0}^{\infty} h_{2n+1}\sin[(2n+1)\omega_0 t] \tag{17.1}$$

in which:

$$h_{2n+1} = (-1)^n \frac{\sin[\pi(2n+1)/2]}{\pi(2n+1)/2} = \frac{1}{\pi(2n+1)/2}$$

The signal $s_0^*(t)$ obtained after rectification is:

$$s_0^*(t) = 2A\sum_{n=0}^{\infty} h_{2n+1}\sin[(2n+1)\omega_0 t]\sin(\omega_0 t)$$

or:

$$s_0^*(t) = Ah_1 + A\sum_{n=1}^{\infty}(h_{2n+1} - h_{2n-1})\cos(2n\omega_0 t) \tag{17.2}$$

To obtain the amplitude $A$, terms of the infinite sum have to be eliminated. Above a certain order, the parasitic products have frequencies greater than half the sampling frequency $f_s/2$, and

*Digital Signal Processing: Theory and Practice*, Tenth Edition. Maurice Bellanger.

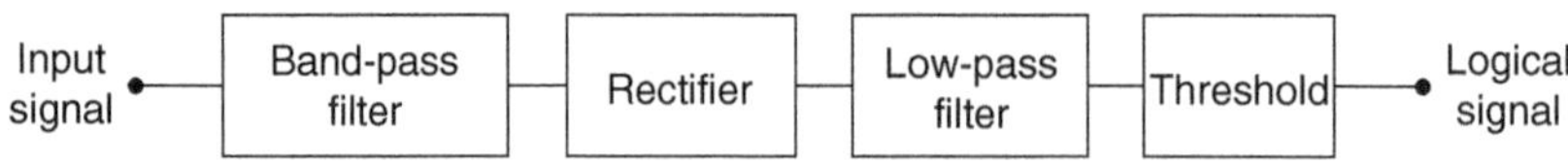


**Figure 17.1** Frequency detection by band-pass filtering.

are aliased in the useful band. The specifications of the low-pass filter, and, in particular, the stop-band edge, have to be chosen so as to eliminate the largest parasites. Those occurring in the pass band result in fluctuations in the measurement of $A$.

In this approach, it is advantageous to use an IIR band-pass filter and an FIR low-pass filter, because the amplitude can be measured at a frequency less than $f_s$. There is another method available, which involves only:

multirate filters. This is based on modulation by two carriers in quadrature at the frequency $f_0$ and is shown in Figure 17.2.

The component to be detected is:

$$s(t) = A\sin(\omega_0 t + \varphi)$$

where $\varphi$ represents the phase of the component relative to the carrier. After low-pass filtering in the two branches to eliminate the unwanted modulation products, the following signals are obtained:

$$S_R = \frac{A}{2}; \quad \sin\varphi; \quad S_I = \frac{A}{2}\cos\varphi \tag{17.3}$$

The required amplitude is:

$$A = 2\sqrt{(S_R^2 + S_1^2)}$$

Accurate evaluation of $X = \sqrt{(S_R^2 + S_1^2)}$ is difficult, and one is generally satisfied with an approximation $X'$, which depends on the phase $\varphi$.

Table 17.1 gives various approximations and the corresponding relative errors. These errors can be reduced by multiplication by a scaling factor $C$ – that is, by calculating the value $X_c'$:

$$X_C' = C\sqrt{(S_R^2 + S_I^2)}$$

Detection of a modulated frequency generally requires less calculation than the method which uses a band-pass filter, but it does require the availability of suitable carrier signals. The operation of detecting a frequency is used in signal transmission systems and forms the basis of receivers of multifrequency codes.

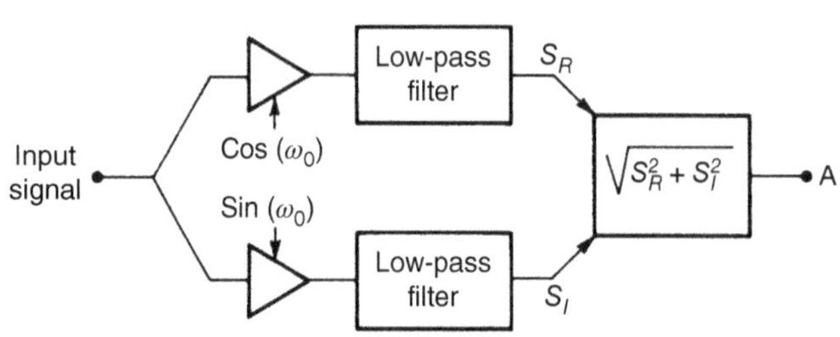


**Figure 17.2** Frequency detection by complex filtering.

**Table 17.1** *Approximation of* $x = \sqrt{(S_R^2 + S_1^2)}$

| $X'$ | max $\left(\frac{X'-X}{X}\right)$ | C | max $\left(\frac{X'_c-X}{X}\right)$ |
|---|---|---|---|
| $\lvert S_R \rvert + \lvert S_I \rvert$ | 0.41421 | 0.20711 | 0.17157 |
| $\max(\lvert S_R \rvert, \lvert S_I \rvert)$ | 0.29289 | 0.85355 | 0.17157 |
| $\max(\lvert S_R \rvert, \lvert S_I \rvert) + \frac{1}{2}\min(\lvert S_R \rvert, \lvert S_I \rvert)$ | 0.11803 | 1.05803 | 0.05573 |

## 17.2 Phase-locked Loop

Phase-locked loops are used for clock recovery in terminals and receivers [1, 2]. The principle is illustrated in Figure 17.3. When the loop is in equilibrium, the frequency produced by the voltage-controlled oscillator is equal to the frequency of the input signal and the phase detector produces a signal whose continuous component is extracted by the narrowband low-pass filter. The phase detector can be a modulator which forms the product of the oscillator output and the signal input. If the nominal frequency of the oscillator is equal to the input frequency, the signals are in quadrature, and the continuous component at the output of the modulator is zero. If not, then the phase difference with respect to the quadrature signal produces a continuous component which shifts the oscillator frequency by the required amount for the frequencies to become equal. The bandwidth of the loop filter determines the capture range, the response time, and the residual noise level.

This operation can be replicated entirely digitally. However, there is additional flexibility with respect to where the phase calculation is performed. The digital oscillator can be realized by means of a phase accumulator connected to a memory which provides samples of the sinusoid. Thus, the input phase values can be directly processed at the input and output of the loop, and the phase difference can be obtained by simple subtraction. A model corresponding to a second-order loop is shown in Figure 17.4. It is a control loop with two coefficients, $K_1$ and $K_2$, corresponding to the proportional and integral control terms, respectively.

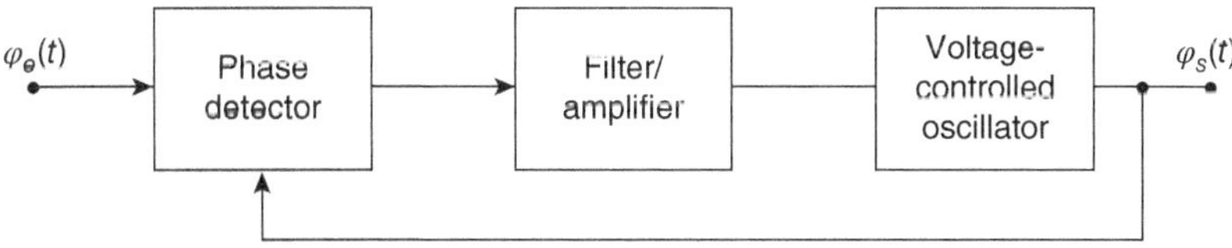


**Figure 17.3** The principle of a phase-locked loop.

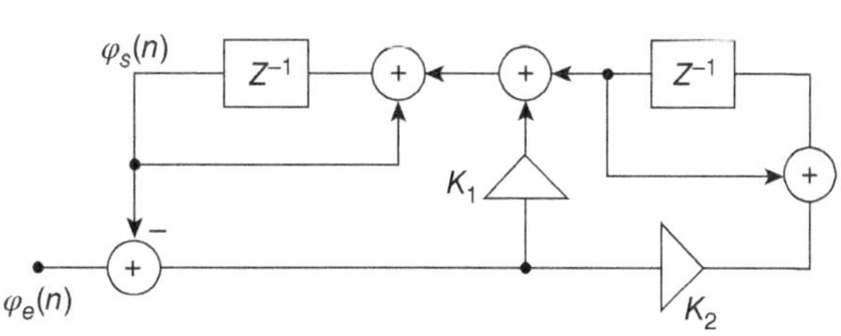


**Figure 17.4** Model of a second-order loop.

The voltage-controlled oscillator is represented by the integrator which provides the output phase $\varphi_s(n)$. The transfer function between the output and the input can be written as:

$$H(Z) = \frac{\Phi_S}{\Phi_e} = \frac{K_1 Z^{-1} + (K_2 - K_1)Z^{-2}}{1 - (2 - K_1)Z^{-1} + (1 - K_1 + K_2)Z^{-2}} \tag{17.4}$$

This is the transfer function of a low-pass filter with a value of 1 at zero frequency. The characteristics of the filter are determined by the two parameters $K_x$ and $K_2$. The region of stability is examined by using the results given in Section 6.7 and setting $b_1 = K_1 = 2$ and $b_2 = 1 - K_1 + K_2$. This gives $1 - K_1 + K_2 < 1$ and $|K_1 - 2| < 2 - K_1 + K_2$. In the plane of the coefficients $K_1$ and $K_2$, the stability domain is a triangle.

The transfer function between the phase shift and the input can be written as:

$$\frac{\Phi_e(Z) - \Phi_s(Z)}{\Phi_e(Z)} = \frac{(1 - Z^{-1})^2}{1 - (2 - K_1)Z^{-1} + (1 - K_1 + K_2)Z^{-2}} \tag{17.5}$$

The presence of the term $(1 - Z^{-1})^2$ in the numerator shows that such a loop is capable of tracking a phase variation.

## 17.3 Differential Coding of Speech

Differential coding of speech leads to reduced data rates, with no additional delay and low computational complexity.

The speech signal has a spectral density which, in the long term, decreases rapidly with frequency from a value of less than 1 kHz. Under these conditions, considering the 8 kHz sampling frequency which is standard in communications, significant prediction gains can be expected [3].

Figure 17.5 shows the principle of differential coding based on linear prediction. At transmission, the prediction error $e(n)$ is quantized by coder C, and, at the receiver, decoder D delivers the signal which is fed to the inverse filter to retrieve the speech signal. The set $e'(n)$ is the result of adding the quantization error to $e(n)$ and $x'(n)$ is the set output by the decoder. The signal $e(n)$ is expressed by:

$$e(n) = x(n) - \tilde{x}(n) = x(n) - \sum_{i=1}^{N} a_i x(n-i)$$

The prediction filter $P$ has a transfer function $P(z)$ such that:

$$P(Z) = \sum_{i=1}^{N} a_i Z^{-i} \tag{17.6}$$

The order $N$ of the filter and the coefficients $a_i$ $(1 \le i \le N)$ should be chosen to minimize the power of the signal $e(n)$. Under these conditions, for a given value of $N$, the coefficients are calculated as

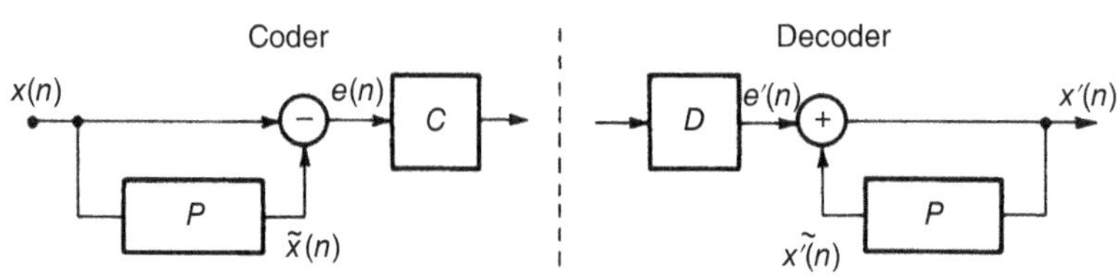


**Figure 17.5** Principle of differential coding.

indicated in Section 13.57, from the elements $r(k)$ $(0 \leq k \leq N)$ of the autocorrelation function of $x(n)$. The following normalized values have been suggested for speech signals:

$$R(0) = 1; \quad R(1) = 0.8644; \quad R(2) = 0.5570; \quad R(3) = 0.2274$$

They show a strong correlation between neighboring samples. The corresponding coefficients have values of:

$$a_1 = 1.936; \quad a_2 = -1.553; \quad a_3 = 0.4972$$

The eigenvalues of the autocorrelation matrix $R_3$ are:

$$\lambda_1 = 2.532; \quad \lambda_2 = 0.443; \quad \lambda_3 = 0.025$$

and we have:

$$A^t_{\text{opt}} R_3 A_{\text{opt}} = 0.947$$

Thus, the corresponding prediction gain is close to 13 dB.

Improvements can be made to the basic principle, in order to achieve a high level of performance:

(1) The prediction is carried out using the sequence $e(n)$ transmitted after quantization, which brings a reduction in quantization distortion power. Moreover, the transmitter and receiver then operate on the same source of information and nothing else has to be transmitted if some adaptive procedures are introduced.
(2) The quantizer is made adaptive by relating the quantization step to an evaluation of the signal power – that is, to take advantage of the fact that speech signals are nonstationary but can be considered almost stationary over short periods of time (of the order of 10 ms).
(3) The prediction is made adaptive to follow the short-term variations of the speech spectrum.

With an adaptive prediction filter of adequate order (for example, 10th-order FIR), the prediction gain for speech can range from 6 dB for unvoiced sounds to 16 dB for voiced ones, with an overall subjective value of about 13 dB.

These techniques are used in communication networks and a system named adaptive differential pulse code modulation (ADPCM) has been standardized by the ITU (International Telecommunication Union) under recommendation G721 [4].

## 17.4 Coding of Sound

Filter banks are the basis of digital sound compression, since they are capable of exploiting the characteristics of the human ear – particularly, the effect of masking. The algorithm represented in Figure 17.6, and standardized as ITU-T/G722, permits transmission of sound on the telephone channel at 64 kbit/s. The audio signal has a bandwidth extending to 7 kHz; it is sampled at 16 kHz and coded in 14 bits. A bank of two QMF filters enables two sub-bands to be obtained, sampled at 8 kHz, which are then coded in ADPCM at rates of 48 and 16 kbit/s for the low- and high-bands, respectively. One multiplexing operation, with possible insertion of data, provides a transmission rate of 64 kbit/s.

Compression of high-quality sound for digital broadcasting or recording is defined by the ISO/IEC 1 1172-2 standard [5, 6]. It is based on a bank of 32 filters with 512 coefficients of the pseudo-QMF type. The signals thus obtained are quantized separately, with a number of bits for each sub-band such that the quantization noise remains at a level below the masking threshold. This threshold, illustrated in Figure 17.7, is defined for each sub-band on the basis of a DFT analysis at 1024 points by applying psycho-acoustic results. The method enables a rate of 128 kbit/s to be achieved for a high-quality monophonic sound channel (MPEG-layer 3/MP3) [6].

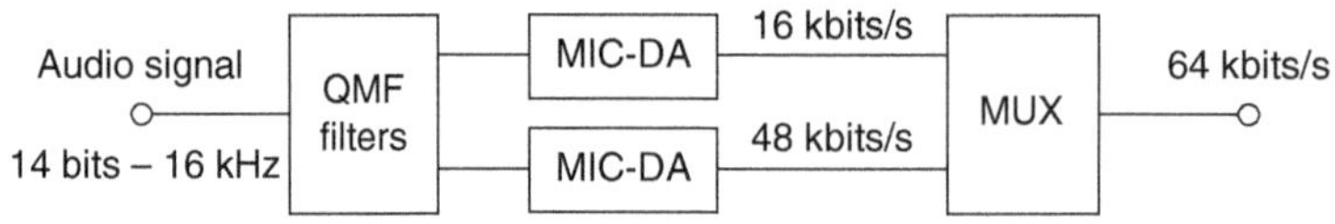


**Figure 17.6** Coding of an audio signal in two sub-bands.

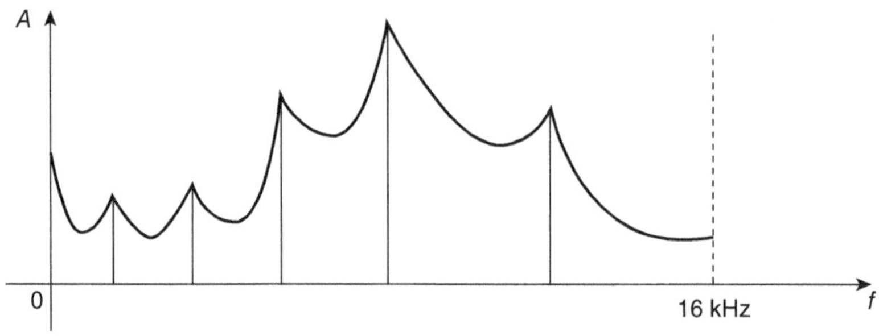


**Figure 17.7** Example of a masking curve for a sound signal.

## 17.5 Echo Cancelation

In communication networks, echoes are created when delayed and attenuated replicas of the signal transmitted by a local terminal toward a distant terminal reach the local receiver.

Specifically, electrical echoes are produced on the transmission lines in the form of reflected signals due to impedance mismatching and imperfections in the hybrid transformers which perform two-wire to four-wire conversions. In the case of speech, these signals are reflected back to the subscriber who is speaking, and they become problematic as the distance between the subscribers increases. There are also echoes arising from acoustic coupling between the microphone and loudspeaker in a telephone set, in which case an adaptive echo canceller can be used to provide a hands-free set.

Echo canceling consists of modeling the echo path and subtracting the generated synthetic echo from the real echo [7].

Two different cases can be distinguished, depending on the type of signals involved – namely speech and data. To begin with the simpler case, data modems are considered.

### 17.5.1 Data Echo Canceller

The most efficient exploitation of two-wire lines is achieved when signals are transmitted simultaneously in both directions and in the same frequency band. The transmission is called full-duplex. The principle is illustrated in Figure 17.8.

#### 17.5.1.1 Two-wire Line

The signal $x_A(n)$ is transmitted from terminal A to terminal B through a two-wire line. At the input of the receiver of terminal A, the signal $y(n)$ has 2 components, the useful data signal $y_B(n)$ issued from terminal B and the reflected signal which is the disturbing echo produced by $x_A(n)$ and designated by $r_A(n)$. The function of the filter $H(Z)$ is to generate a synthetic echo $y(n)$ as close to $r_A(n)$ as possible, so that, after subtraction, the error signal $e(n)$ is close to $y_B(n)$ and the quality of the data transmission from terminal $A$ to terminal $B$ is of satisfactory quality.

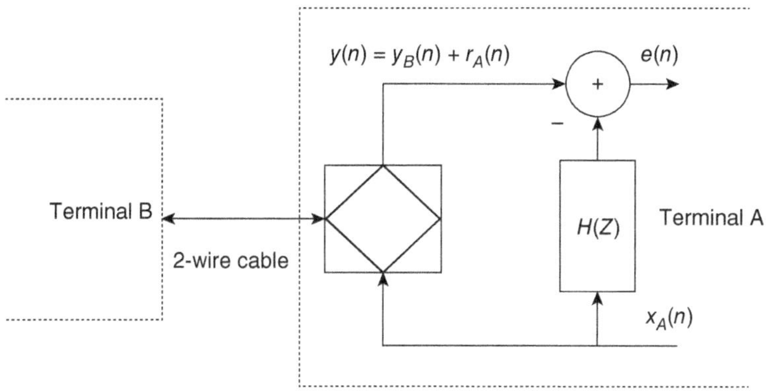


**Figure 17.8** Principle of echo canceling.

The selection of adaptive filter parameters is driven by the context. The number N of coefficients is derived from the echo impulse response, taking account of the sampling frequency. The filter must be adaptive because the transmission line characteristics may evolve over time. Regarding input signals, the context is favorable, because the filter input signal is the transmitted data sequence $x_A(n)$, which is generally uncorrelated, unit power, and has the AC matrix $R_N = I_N$. Then, the performance of the gradient algorithm, LMS, is equivalent to that of the recursive least squares algorithm (RLS). The adaptation step $\delta$ is bounded by $2/N$ and the time constant is $\tau = 1/\delta$. In the learning phase, the mean output error power is deduced from (14.28) as:

$$E_r = ||H_{\text{opt}}||_2^2 \ (1-\delta)^{2n} \tag{17.7}$$

The norm $L_2$ of the echo coefficient vector $||H_{\text{opt}}||_2^2$ reflects the power of the echo signal.

During bidirectional transmission, the useful signal $y_B(n)$ in the reference is smaller than the echo $r_A(n)$ and the echo attenuation $A_e$ must satisfy the inequality:

$$A_e > A_s + \text{SNR (dB)} \tag{17.8}$$

where $A_s$ is the echo-to-useful signal ratio. For example, SNR = 40 dB and $A_s$ = 20 dB leads to $A_e$ = 60 dB, which implies that the residual error after convergence is very small.

In the adaptation process, the useful signal impacts the coefficients, and the consequence is an increase of the output residual error. The filter coefficient variance after convergence is $\sigma_y^2\,\delta/2$, and the residual error is $N$ times more, $N\sigma_y^2\,\delta/2$. The term $\sigma_y^2$ is the power of the useful signal – that is, the received data. The SNR objective can be reached if the following inequality is satisfied:

$$N\frac{\delta}{2} < \frac{1}{\text{SNR}} \tag{17.9}$$

In the above derivations, the output error power is assumed to be close to the useful signal power. As an illustration, take SNR = $10^{-4}$ (40 dB), $N = 60$, then $\delta < 3.3 \times 10^{-6}$, which is a very small value and makes the learning phase very long. The impact on the coefficient accuracy is worth pointing out. Referring to Section 14.5, a simplified expression for the number of bits of the coefficients is obtained as:

$$b_c = \log_2\left(\frac{1}{\delta}\right) + \frac{1}{2}\log_2(A_e) \tag{17.10}$$

With the above figures, $b_c = 29$. In practice, there is no need to perform multiplications with such a high degree of accuracy; it is needed only in the coefficient updating operations.

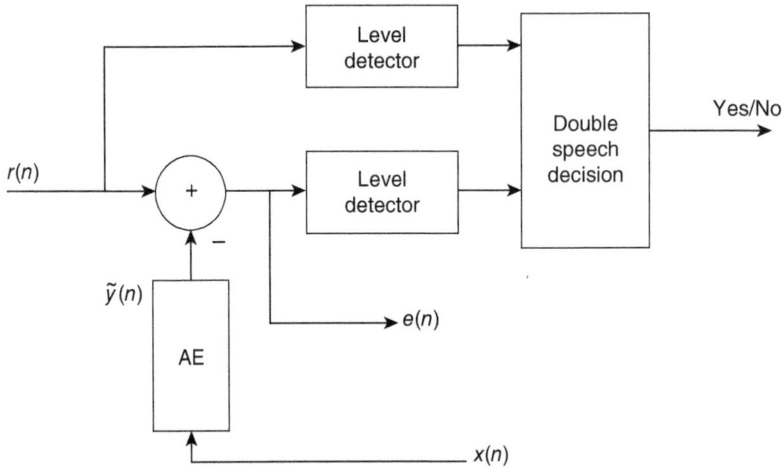


**Figure 17.9** Double speech detection.

### 17.5.2 Acoustic Echo Canceler

Acoustic echo cancelation leads to very long filters. Taking $v = 330\,\text{m/s}$ as the propagation speed of acoustic waves in the air, sampling frequency $f_s = 8\,\text{kHz}$, and distance $2D = 100\,\text{m}$, we obtain $N = 2Df_s/v = 2400$. This case is encountered in audioconference rooms.

An additional adverse issue in speech echo cancelation is double speech. Double speech is produced when the speech of the local subscriber is superimposed on the echo in the reference signal. To cancel the echo, this disturbing signal entails a change in the coefficients which must be stopped as rapidly as possible. Adaptation is therefore stopped as soon as detection has occurred. Detection can simply take place by comparing the levels of the received signal $r(n)$ and the signal $e(n)$ after the synthetic echo has been subtracted. The scheme is represented in Figure 17.9.

In the absence of a distant speech signal in $r(n)$, provided cancelation works properly, levels are very different. In contrast, during double speech, levels come close, and the information can be exploited to determine and stop the coefficient change.

Note that the parameters of level detection and decision must be chosen with care, to avoid false decisions and excessive delays. Level detection may be based on power or amplitude estimations.

## 17.6 Television Image Processing

The transmission of image signals on the communication network requires very high bit rates. Therefore, bit rate reduction techniques, based on digital processing, are crucial – particularly for television signals.

Overall, a television image is a function of four variables $s(x, y, t, \lambda)$. There are two spatial variables: time and wavelength. For transmission purposes, this signal is transformed into a one-dimensional signal.

The wavelength variable can be dropped by considering that the human visual system basically consists of three types of receivers, which perform filtering and produce three signals associated with the primary colors – namely red, green, and blue ($R$, $G$, $B$).

The television scanning process converts these three-dimensional signals into a one-dimensional signal. The images are scanned 25 times per second, with 625 lines per image. In fact, the odd- and even-numbered lines form two consecutive frames which are multiplexed in time. Hence the recurrence of 50 interleaved frames per second.

For transmission, the primary components $R$, $G$, and $B$ are replaced by linear combinations called luminance $Y$ and color differences, or chrominance, $U$ and $V$.

$$Y = 0.30R + 0.59G + 0.11B$$

$$U = R - Y = 0.70R - 0.59G - 0.11B$$

$$V = B - Y = -0.30R - 0.59G + 0.89B$$

Digitization is performed with a frequency of 13.5 MHz for the luminance and 6.75 MHz for the chrominance signals. Since the analog-to-digital conversion is 8 bits, the corresponding data rate rises to 216 Mbit/s. This format corresponds to Recommendation CCIR 601 of the ITU and is described as Type 422. It leads to images presented in the form of tables of 8-bit numbers containing 720 points per line and 576 useful lines, in the case of 625-line scanning. Thus, one image corresponds to 414 720 bytes for the luminance and 207 360 bytes for each chrominance component.

Bit rate reduction techniques rely on the fact that good modeling is provided by the output of a first-order IIR filter to which Gaussian white noise is applied. The corresponding two-dimensional autocorrelation function can be written as:

$$V(x, y) = r_0 e^{-(\alpha x + \beta y)}$$

where $\alpha$ and $\beta$ are positive constants. For the associated spectrum, this gives:

$$S(\omega_1, \omega_2) = r_0 \left( \frac{4\alpha\beta}{\left(\alpha^2 + \omega_1^2\right)\left(\beta^2 + \omega_2^2\right)} \right) \tag{17.11}$$

The greatest compression in the representation of a signal is obtained with a transformation based on the eigenvectors of the autocorrelation matrix. In the case of first-order signals, this transformation is well approximated by the discrete cosine or sine transformation, presented in Sections 3.3.3 and 3.3.4. In the image compression standards, it is the DCT applied to blocks of $8 \times 8$ picture element points, or pixels, which has been retained. The standards formulated for videophones, image storage, and digital television use the following three techniques [8]:

(1) Motion estimation in order to be able to minimize the difference between the current image and the preceding one.
(2) The discrete cosine transform to minimize spatial redundancy.
(3) Variable-length statistical coding (VLC).

The general structure of an image encoder is shown in Figure 17.10. The digitizer $Q$ operates above thresholds which can be set by a control device, allowing a constant bit rate to be obtained with the help of a buffer memory. For commercial-quality television, the bit rate can be reduced to around 4 Mbit/s, which represents a compression factor of the order of 50 [9].

Digital filters are used for interpolation and subsampling operations during changes in image format or movement estimation. These are separable filters. Digital compression of multimedia signals, speech, image, and sound allows for considerable reduction of the bit rates required for broadcasting programs, and in combination with the techniques of digital transmission, it offers the possibility of high spectral efficiency by transmitting several programs in the channels formerly used for a single analog program. The resulting saving may be quite significant, as is the case in satellite television broadcasting.

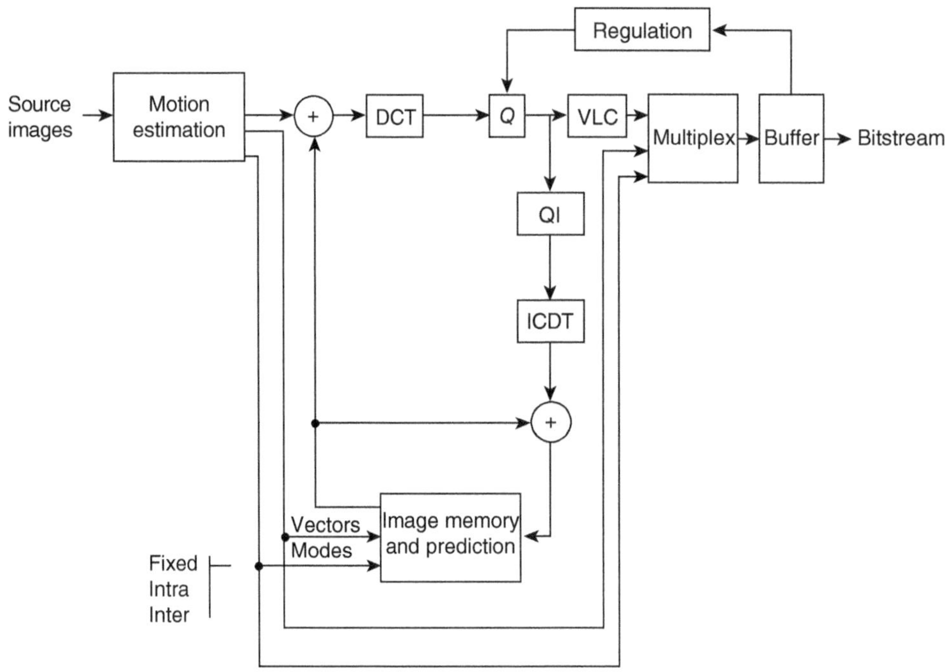


**Figure 17.10** General structure of a moving image encoder.

Techniques with high spectral efficiency make intensive use of digital processing and they make the best use of the characteristics of the channels. In this way, multicarrier techniques can lead to capacities of several bit/s per hertz on channels which are of limited quality or susceptible to interference.

## 17.7 Multicarrier Transmission – OFDM

The objective of multicarrier transmission techniques is to approach the theoretical capacity of a channel, firstly by limiting the effect of distortion and secondly by adjusting the data rate in accordance with the spectral density of the noise. In fact, by dividing a given channel into tens, hundreds, or thousands of subchannels, the effect of distortion on each subchannel is made negligible and the data rate it is able to support can be allocated to each subchannel. One simple and effective approach to implementing this procedure involves the fast Fourier transform; known as OFDM (orthogonal frequency-division multiplexing), its principles are illustrated in Figure 17.11. The stream of data to be transmitted is converted into $N$ elementary streams, each $N$ times smaller, which are applied to the input of an inverse DFT computer. In accordance with the definition of the inverse DFT given in Chapter 2, this operation corresponds to a modulation of $N$ carriers by the elementary streams, at frequency multiples of $f_s/N$, and the addition of the set of signals modulated in this way. Thus, the rate of OFDM symbols is $f_s/N$. On reception, after passing through the channel, a direct DFT performs the set of demodulations and reconstructs the original data, which merely needs to be serialized to recover the initial stream.

This simple principle is a direct illustration of the definition of the DFT and its inverse. However, to operate correctly, it requires several precautions and adaptations [10].

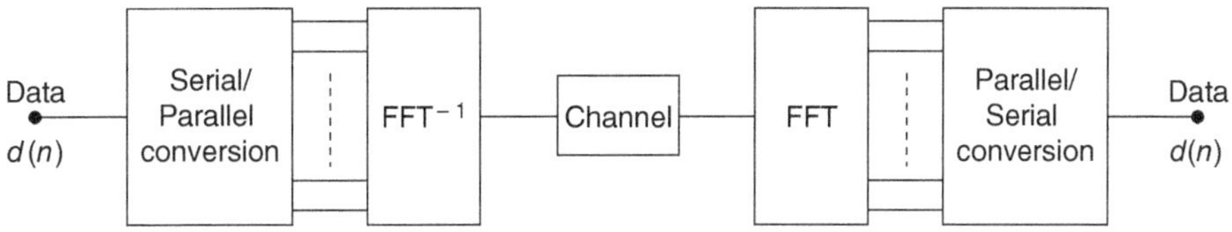


**Figure 17.11** The principle of OFDM transmission.

Referring to Section 2.4, and in particular to Figure 2.8, notice that orthogonality of the signals is valid only for frequencies which are at the center of the interval of length $fs/2N$ allocated to each subchannel; and notice how the subchannels have an area of overlap and that the amplitude of overlap reduces with increasing frequency difference. On the edges of the transmission channel, the frequency responses of the subchannels are not symmetric, which can lead to interference. It is therefore necessary to avoid using extreme subchannels and to provide a margin of at least a few subchannels on each side of the chosen frequency band.

In the time domain, a practical transmission channel has an impulse response of duration $\tau$. To avoid superposition of two consecutive OFDM symbols on reception, the symbols must be separated by sufficient time; that means it is necessary to introduce a guard interval $T_g > \tau$. During this guard interval, it is necessary to prolong the OFDM symbol, to introduce the circular convolution mentioned in Section 2.1, and hence to avoid interference between the subchannels. In practice, the receiver operation is facilitated by the end of the symbol being reproduced at the start after a time $T_g$ (Figure 17.12).

With this device, the received signals are simply multiplied by the DFT of the channel – an effect which can be compensated for by an equalization in amplitude and phase in each subchannel. To show this, the Z-transfer function of the channel which contains $P \leq Ng$ coefficients is defined as $C(Z)$:

$$C(Z) = \sum_{p=0}^{P} C_p Z^{-p} \tag{17.12}$$

If $x(n)$ is the transmitted signal, the received signal $y(n)$ can be written:

$$y(n) = \sum_{p=0}^{P} C_p x(n-p)$$

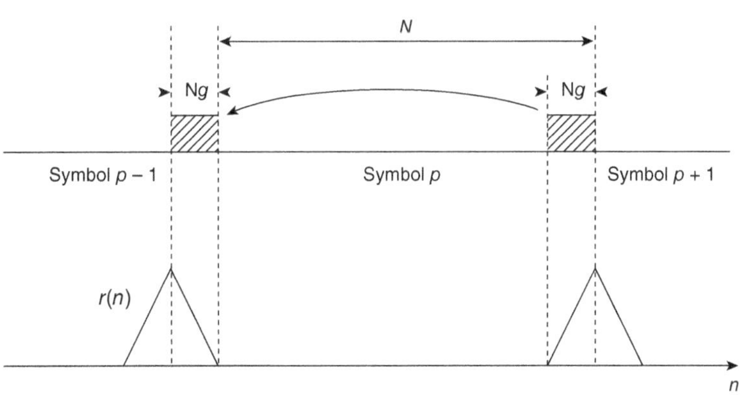


**Figure 17.12** Introducing a guard interval and the correlation function.

As $x(n)$ is expressed in terms of the data $d_k$ by:

$$x(n) = \sum_{k=0}^{N-1} d_k e^{j(2\pi/N)kn} \tag{17.13}$$

The double summation is obtained for $y(n)$:

$$y(n) = \sum_{p=0}^{P} C_p \sum_{k=0}^{N-1} d_k e^{j(2\pi/N)k(n-p)} \tag{17.14}$$

Setting:

$$H_k = \sum_{p=0}^{P} C_p e^{-j(2\pi/N)kp}$$

finally gives:

$$y(n) = \sum_{k=0}^{N-1} (d_k H_k) e^{j(2\pi/N)kn} \tag{17.15}$$

The circular convolution property of the DFT is again found, and the receiver provides the transmitted data multiplied by the channel spectrum $H_k$.

The redundancy of the transmitted signals can be exploited by the receiver for synchronization. In fact, by calculating the following correlation function:

$$r(n) = \sum_{i=n-N_g+1}^{n} y(i)y^*(i-N) \tag{17.16}$$

peaks appear as shown in Figure 17.12 which characterize the start of each symbol and allow the temporal analysis window of the receiver to be adjusted and a contribution made to synchronization of the clocks.

Synchronization in time and frequency is a major problem in systems with a large number of carriers, and special reference symbols are introduced or some subchannels are reserved for fixed signals called pilots.

Figure 17.13 shows the block diagram of a digital television receiver for terrestrial broadcasts [11]. The analog interfaces carry the signal in the band 0.76–8.37 MHz and the analog-to-digital conversion is performed at $fs = 18.28$ MHz. Then a real-to-complex conversion using a quadrature filter is performed and the sampling frequency is reduced to 9.14 MHz. Before calculation of the FFT at $N = 8192$ points, a complex multiplier performs frequency adjustment on the spectrum of the signal. Temporal synchronization controls the positioning of the window of the FFT. The transmitted signal contains 6817 active carriers, and 177 of them are dedicated to pilot signals which allow for exact synchronization of the receiver, an estimate of the frequency response of the channel for equalization, and a measure of the distortion in each subchannel. The guard interval can reach 20% of the symbol duration. This system must facilitate transmission rates up to 32 Mbit/s in a channel with 8 MHz spacing, or 4 bit/s/Hz.

Alternative multicarrier transmission techniques require filter banks as described in Chapter 12. At the cost of an increase in computational complexity, they allow the cyclic prefix to be avoided, and deliver a high level of out-of-band signal rejection, which may be critical for coexistence in networks [12].

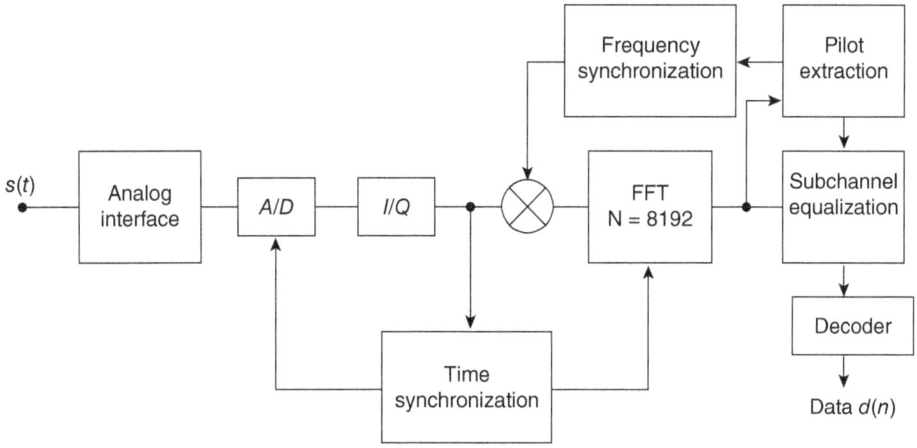


**Figure 17.13** Receiver for terrestrial digital television.

## 17.8 Mobile Radiocommunications

The basic mobile radio channel, called Rayleigh, corresponds to the non-line-of-sight (NLOS) transmission between transmitter and receiver. It is represented by the sum of a large set of independent paths of equivalent propagation delays, with the addition of Gaussian white noise. As a result, useful signals can be viewed as being multiplied by a complex coefficient whose real and imaginary parts are independent, centered, Gaussian random variables.

In practice, transmission may happen with multiple reflections and diffractions, with differing propagation delays. Then, the channel is characterized by multipaths with fading, and it is said to be doubly dispersive, because it exhibits dispersions in both the time and frequency domains. These obstacles can be overcome by the multicarrier technique OFDM, which is able to exploit the channel with high bit rates [13].

Processing design and performance evaluations rely on channel models which have been defined for three multipath configurations:

Extended pedestrian A (EPA)

| **Delay (ns)** | **0** | **30** | **70** | **90** | **110** | **190** | **410** |
|---|---|---|---|---|---|---|---|
| Power (dB) | 0 | −1 | −2 | −3 | −8 | −17.2 | −20.8 |

Extended vehicular A (EVA)

| **Delay (ns)** | **0** | **30** | **150** | **310** | **370** | **710** | **1090** | **1730** | **2510** |
|---|---|---|---|---|---|---|---|---|---|
| Power (dB) | 0 | −1.5 | −1.4 | −3.6 | −0.6 | −9.1 | −7.0 | −12.0 | −16.9 |

Extended typical urban (ETU)

| Delay (ns) | 0 | 50 | 120 | 200 | 230 | 500 | 1600 | 2300 | 5000 |
|---|---|---|---|---|---|---|---|---|---|
| Power (dB) | −1 | −1 | −1 | 0 | 0 | 0 | −3 | −5 | −7 |

For each incident wave, and for each of the multipath channels, mobility entails a frequency shift due to the Doppler effect given by:

$$f = \frac{v}{c} f_0 \cos(\theta) \tag{17.17}$$

where $v$ is the speed of the mobile, $c$ is the celerity of light, $f_0$ is the carrier frequency, and $\theta$ is the angle of incidence. As a consequence, the signal associated with a given path undergoes a slow evolution, whose Fourier transform, called the Doppler spectrum, is generally represented by the following theoretical formula:

$$S(f) = \frac{1}{\left(1 - \left(\frac{f}{f_D}\right)^2\right)^{0.5}}; \quad -f_D < f < f_D \tag{17.18}$$

where $f_D$ represents the maximum frequency.

Two approaches can be contemplated to model this Doppler spectrum [14].

- Purely recursive second-order section. The approximation applies whenever $f_D$ is much smaller than the frequency step – that is, the product $f_D T$ is small ($T$: symbol period). The complex coefficients of the path are determined by:

$$C(n) = b_1 C(n-1) - b_2 C(n-2) + u(n); \quad b_1 = 2r_d \cos\left(2\pi \frac{f_D}{\sqrt{2}} T\right); \quad b_2 = r_d^2 \tag{17.19}$$

  The input $u(n)$ is zero-mean Gaussian white noise. The pole is very close to the unit circle – for example: $r_d = 0.999 - 0.1\ 2\pi f_D\ T$.
- Sum of $N$ cissoids uniformly spread on the unit circle with random initial phases $\theta(k)$. The cissoid with index $k$ undergoes the Doppler shift: $\Delta f(k) = f_D \cos\left(\frac{2\pi k}{N}\right)$ and the coefficient of the path is expressed by:

$$C(n) = \frac{1}{\sqrt{N}} \sum_{k=1}^{N} e^{j(2\pi \Delta f(k)\, n + \theta(k))} \tag{17.20}$$

Next, the amplitudes defined in the tables of the multipath models are multiplied by $C(n)$ to get the coefficients of the radio channel.

The mobile radio channel so defined entails a set of constraints for OFDM. However, thanks to its richness in multipaths, it offers noteworthy potential gains in capacity and flexibility. In particular, the MIMO technique described in Section 13.9 allows for multiuser transmission. The principle is shown in Figure 17.14 for four antennas at the transmitter, two users and two antennas per user at the receiver.

The principle consists of decomposing the radio channel into signal and null spaces, so that each user transmits their data in the other user's null space. Accordingly, the $2 \times 4$ channel matrix

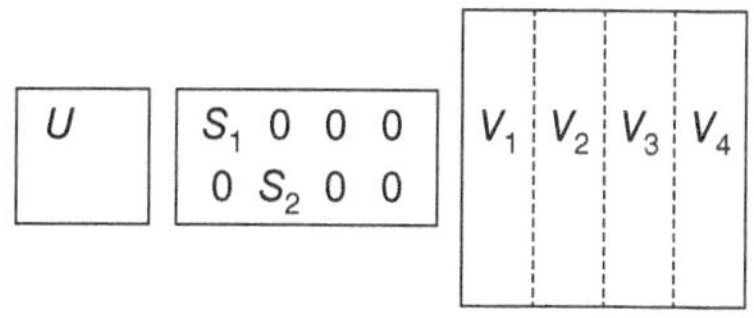


**Figure 17.14** Decomposition of the channel matrix of user 1.

between the transmitter and a user, $H_1$, is decomposed into singular values to yield the matrices $U$, $S$, and $V$, as in Figure 17.14. $V$ is the modulation matrix in the transmitter, $S$ is the singular value matrix, and $U^t$ is the demodulation matrix.

The channel matrix decomposition is written as: $H_1 = U\ S\ V^t$ . Vectors $V_1$ and $V_2$ define the signal space of user 1, while $V_3$ and $V_4$ define the null space. If user 1 is alone, the data couples to be transmitted are applied to the $4 \times 2$ matrix $[V_1\ V_2]$, whose outputs are connected to four antennas through four OFDM interfaces. Now, if user 2 is active, it must use the null space and its transmission matrix becomes $H_2\ [V_3\ V_4]$. This $2 \times 2$ matrix must be decomposed to get the modulation matrix applied to the data couples of user 2. Then, the same procedure is enforced on user 1, starting from the decomposition of $H_2$ and performing the transmission in the corresponding null space. Finally, with the same set of antennas, the radio base station is able to target a set of different users [15].

Multiantenna systems, with four and eight antennas, are included in the basic mobile radio standards.

The constraints imposed by the channel on OFDM are as follows, in particular:

- The subcarrier spacing must account for the Doppler frequency of the fastest mobile.
- Transmission by blocks of a limited number of multicarrier symbols.
- Estimation of the channel through preamble and/or scattered pilot data.
- Some measures to counter fading and noise.

A typical configuration is as follows:

- Bandwidth: 10 MHz.
- Sampling frequency: 15.36 MHz.
- FFT size: 1024; cyclic prefix: 72.
- Subcarrier spacing: 15 kHz.
- Number of symbols per block: 14.

Assuming the carrier frequency is 4 GHz and mobile speed is 100 km/h, the maximum Doppler frequency is 370 Hz, and it has to be compensated for with every transmitted symbol.

Leveraging signal processing techniques and making use of current computation capabilities, the new generations of wireless and mobile radio systems are able to reach communication quality levels and bit rates that come close to those of hardwired systems and they contribute significantly to developing the flexibility of digital networks.

## References


**1** U. Mengali and A.N. d'Andrea, *Synchronization Techniques for Digital Receivers*, Plenum Press, 1997.

**2** H. Meyr, M. Moeneclay and S.A. Fechtel, *Digital Communication Receivers: Synchronization, channel estimation and Signal Processing*, Wiley, Chichester, 1998.

**3** J.D. Gibson, Adaptive Prediction in Speech Differential Encoding Systems. *Proceedings of the IEEE*, 68, 1980.

**4** CCITT, *Digital Networks*, Transmission Systems and Multiplexing Equipment Book III.3, Geneva, 1985.

**5** ISO-IEC 13818, *Information Technology*, Coding of Moving Pictures and Audio, Geneva, 1996.

**6** ISO-IEC 14496, *Information Technology*, Generic Coding of Audio-Visual Objects, Geneva, 1998.

**7** C. Breining, et al., Acoustic echo control: application of very-high-order adaptive filters, *IEEE Signal Processing Magazine*, 16(4), 42-69, 1999.

**8** J. Chen, U. Koc and K .J. Ray Liu, *Design of Digital Video Coding Systems*, Marcel Dekker Inc., New York, 2002.

**9** G. Sullivan and T. Wiegand, Video Compression—from concepts to the H264/AVC standard. *Proceedings of the IEEE*, 93(1), 18-31, 2005.

**10** T. de Couasnon, R. Monnier and J. B. Rault, *OFDM for Digital TV Broadcasting, Signal Processing*, Vol. 39, 1994, 1-32.

**11** U. Ladebusch and C.A. Liss, Terrestrial DVB—a broadcast technology for stationary, portable and mobile use, *Proceedings of the IEEE*, 94(1), 183-193, 2006.

**12** M. Renfors, X. Mestre, E. Kofidis and F. Bader, *Orthogonal Waveforms and Filter Banks for Future Communication Systems*, Academic Press, Cambridge, MA, USA, 2017.

**13** Document: 3rd Generation Partnership Project; *Technical Specification Group Radio Access Network; Evolved Universal Terrestrial Radio Access (E-UTRA); Base Station (BS)* Radio Transmission and Reception (Release 15), 3 GPP TS 36.104 v15.4.0, 2018.

**14** F. Pérez Fontan and P. Marino Espineira, *Modeling the Wireless Propagation Channel*, Wiley, Chichester, 2008.

**15** E. Dahlman, S. Parkvall and J. Skold, *5G-NR: the Next Generation Wireless Access Technology*, Academic Press, Cambridge, MA, USA, 2018.

# Exercises: Solutions and Hints

## Chapter 1

**1.1** $I_L(t) = \frac{1}{2} + \frac{2}{\pi}\sum_{p=1}^{4}\frac{(-1)^{p+1}}{2p-1}\cos 2\pi(2p-1)\frac{t}{T}$

**1.2** $s(nT) = \sin(n\pi + \varphi) = (-1)^n \sin\varphi$.
The possibility of reconstruction depends on $\varphi$. $\left(\varphi = \frac{\pi}{2}\text{ yes; } \varphi = 0\text{ no}\right)$.

**1.3** $H\left(\frac{f_s}{2}\right) = \frac{2\sqrt{2}}{\pi}(0.92\text{ dB})$.

**1.4** $f_2 < f_s < 2f_1$.

**1.5** $s(nT) = s_r(nT) + js_i(nT) = e^{j(\pi/2)n}\frac{\sin\frac{3\pi}{8}n}{\sin\left(\frac{\pi}{8}n\right)}$.

**1.6** Maximum value of $s(n) = 8$;

$$s(n) = 0 \quad\text{for}\quad \varphi_k = -2\pi\frac{k}{8}n + k\pi$$

**1.7** $f_s = 2$ MHz; $\Delta f = 1$ kHz.

**1.8** $p(1) = \frac{1}{\pi}\frac{1}{\sqrt{A^2-s^2}}$; $r(\tau) = 2(f_2 - f_1)\frac{\sin\pi(f_2-f_1)\tau}{\pi(f_2-f_1)\tau}\cos\pi(f_2+f_1)\tau$.

**1.9** Periodic part; Fourier coefficient: $C_n = \frac{p\sin\frac{\pi}{2}n}{\pi n}$.
Non-periodic part; spectrum: $S_2(f) = p(1-p)T\frac{1-\cos\pi fT}{\pi^2 f^2 T^2}$.

**1.10** Signal-to-noise ratio in the band: 300–500 Hz = 75 dB ($f_s = 16$ kHz; gain 3 dB).

**1.11** Quantizing distortion: line at $\frac{3}{8}f_s$; power: $0.0195^2$.

**1.12** If the characteristic is centered:

$$a_1 = 0\text{ for }0 \le |\alpha| \le \frac{1}{2};\quad a_1 = \frac{4q}{\pi}\sqrt{1-\frac{1}{4\alpha^2}}\text{ for }\frac{1}{2} \le |\alpha| \le 1$$

Centering at $\frac{q}{2}$ : $a_1 = \frac{2q}{\pi}$ for $0 \le |\alpha| \le 1$.

**1.13** Without clipping (peak factor): 10 bits; with clipping at 1% = 9 bits.

**1.14** Linear coding $(S/N)_{max} = 50$ dB; with nonlinear coding, it varies from 35 to 38 dB when the signal varies from −36 dB to 0 dB.

**1.15** Optimal values: $x_0 = 0$; $x_1 = 0.9816$: $y_1 = 0.4528$; $y_2 = 1.510$.

## Chapter 2

**2.1** The DFT of the second set is related to the DFT of the first by $X'(k) = e^{-jk(\pi/4)}X(k)$

**2.2** Real multiplications: $M_R = 28$: real additions: $A_R = 84$.

**2.3** The small differences come from aliasing and decrease when $N$ increases.

**2.4** Number of complex multiplications: 160, 96, 72. Additions: 384.

**2.5** Maximum noise-power at any output: 28. $q^2/12$. With quantization at eight bits of the coefficients: $|\varepsilon(i,k)| \leq 0.003$.

**2.6** Recursion: $X_0 = x(N-1)$; $X_m = x(N-1-m) + WX_m - 1$ for $1 \leq m \leq N-1$
$N-1$ complex multiplications are needed.

**2.7** Total roundoff noise: $N\frac{q^2}{12} + Nq^2$; signal-to-noise ratio degradation: $\Delta SNR = 11.5$ dB (input noise $q^2/12$).

**2.8** Recording: 20 000 samples; memory 160 kbits; cycle time per multiplication: 1 µs.

**2.9** Cosine, Hamming, and Blackman windows attenuate the secondary lobes but do not allow for the detection of weak components.

## Chapter 3

**3.1** It is sufficient to verify that the products $I_3 \times A$ and $A \times I_3$ are different.

**3.2** It is sufficient to verify the relations 3.3, 3.4, 3.5, and 3.6.

**3.3** Number of real multiplications in radices 2, 4, and 8: 384, 284, and 246.

**3.4** The order-12 DFT requires 20 complex multiplications.

**3.5** Use relations (3.18) and (3.21) to get the two factorizations.

**3.6** The complex DFT of order eight leads to 24 real multiplications. The odd transform leads to 26.

**3.7** With that approach, the operations in $\Delta_{12}$ vanish, which reduces the number of complex multiplications to 16.

**3.8** Matrices of the transformation:

$$T = \begin{bmatrix} 1 & 1 & 1 & 1 \\ 1 & 4 & 16 & 13 \\ 1 & 16 & 1 & 16 \\ 1 & 13 & 16 & 4 \end{bmatrix}; \quad T^{-1} = 13 \begin{bmatrix} 1 & 1 & 1 & 1 \\ 1 & 13 & 16 & 4 \\ 1 & 16 & 1 & 16 \\ 1 & 4 & 16 & 13 \end{bmatrix}$$

## Chapter 4

**4.1** Response to the sequence $a^n$ : $y(n) = a^{n-3}\frac{1-a^{9-n}}{1-a}$ for $5 \leq n \leq 8$.

**4.2** Take the derivative, perform series expansion, and an integration

$$\log(z-a) = -\sum_{n=}^{\infty} \frac{1}{na^n} Z^{n;} \quad \frac{Z^{-1}}{(1-aZ^{-1})(1-bZ^{-1})} = \frac{1}{a-b} \sum_{n=0}^{\infty} \left( \frac{1}{b^n} - \frac{1}{a^n} \right) Z^{-n}$$

with $|a| < 1$; $|b| < 1$

**4.3** $H(Z) = \frac{1}{(1-re^{j\theta}Z^{-1})(1-re^{-j\theta}Z^{-1})}$.

**4.4** Output power: 21; $H(\omega) = 4.41 - 1.536 \cos \omega + 0.46 \cos 2\omega$.

**4.5** System response:

$$y(n) = \frac{e^{jn\omega}}{1 - e^{-j\omega} + 0.8e^{-2j\omega}} + r^{n-1} \left[ (a-0.8b)\frac{r \sin(n+1)\theta}{\sin\theta} - 0.8a\frac{\sin(n\theta)}{\sin\theta} \right]$$

## Chapter 5

**5.1** The response is zero for $f = 0.288$; 0.347; 0.408; 0.469; maximal ripple: 0.08.
Zeros of $H(Z)$: 0.606; 1.651; $0.4292 \pm j0.464$; $1.073 \pm j1.161$.

**5.2** Coefficients: −0.012; 0; 0.042; 0; −0.093; 0; 0.314; 0.5. Zeros of $H(Z)$: 0.4816; 2.076; $0.3764 \pm j0.368$; $1.3583 \pm j1.328$; maximal ripple: 0.03.

**5.3** $\delta = 0.017$, which is less than the above values.

**5.4** With the sampling frequency $f_s/2$ at the output, the numbers of memories and multiplications are divided by 2, through interleaving (see Section 10.5).

**5.5** In the complex plane, $H(Z)$ is rotated by $\pi$ and $\pm\pi/2$, which yields a high-pass filter and a low-pass filter.

**5.6** Coefficients: $N = 27$; computation accuracy: $b_c = 12$ bits; $b_i = 20$ bits.

## Chapter 6

**6.1** Follow the derivation in Section 6.1. The difference between filter delay and group delay illustrates the nonlinearity of the phase response.

**6.2** Unit step response: $y(n) = \frac{1}{1.8}(1 - (-0.8)^{n+1}) + (-0.8)^{n+1}y(-1)$.

**6.3** Poles: $P = 0.78 \pm j0.438$. The zeros do not add multiplications to the circuit.

**6.4** $H_m = 85$; $\cos\omega_0 = 0.808$; $\|H\|_2 = 8.53$.
With zeros at $3f_s/8$: $\|H\|_2 = 25.8$.
Limit-cycle frequency close to $f_s/10$; amplitude $\simeq 42q$. Significant amplitude oscillations are possible because (6.56) is not verified.

**6.5** $b_c \geq 13$ bits; infinite attenuation point displacement: $df_i$, $\leq 2.210^{-5}f_s$.

**6.6** $H(Z) = \dfrac{0.796 - 1.42Z^{-1} + Z^{-2}}{1 - 1.42Z^{-1} + 0.796Z^{-2}}$; $\tau_g(\omega)$ calculated by (6.45).
Realization possible with three multiplications.

## Chapter 7

**7.1** First-order section: $|H(\omega)|^2 = \frac{1.49+1.4\cos\omega}{1.81-1.8\cos\omega}$

$$\varphi(\omega) = \tan^{-1}\frac{1.6\sin\omega}{1.63 - 0.2\cos\omega};\quad \tau(\omega) = \frac{1.6(0.37\cos\omega - 0.2)}{(0.37\cos\omega - 0.2)^2 + 2.66\sin^2\omega}$$

**7.2** Characteristic frequencies: 0.162; 0.231; 0.538; 0.736.

**7.3** Transfer function: $H(Z) = 0.094\frac{(1+Z^{-1})^4}{(1+0.039Z^{-2})(1+0.447Z^{-2})}$.

**7.4** Transformation: $Z^{-1} = -\frac{Z^{-1}-\alpha}{1-\alpha Z^{-1}}$; $\alpha = \frac{\sin\pi(f_1-f_1')}{\sin\pi(f_1+f_1')}$
with $f_1 = 0.1725$; $f'_1 = 0.1$; $\alpha = 0.3$.

**7.5** Scale factors: $a_0^0 = 2^{-6}$; $a_0^1 = 2^{+2}$; $a_0^2 = 2$.
For $H(j) = 1$, we get: $a_0 = 0.515$ and $a_0^3 = 4.12$; accuracy: $b_i$, $= 16$ bits.

**7.6** Coefficient wordlength: $b_c \simeq 12$ bits. The optimum is obtained through systematic search about rounding. The critical pole $0.9235 \pm j0.189$ means it is not possible to reduce $b_c$ to 11 bits.

**7.7** The filter in Section 7.2.3 can have limit cycles of amplitude less than $3q$ and frequency near $f_s/5$.

**7.8** The IIR filter requires 7 multiplications and 4 memories while the FIR counterpart requires 8 multiplications and 16 memories.

**7.9** Transfer function:

$$H(Z) = 0.0625\frac{1 - 1.165Z^{-1} + Z^{-2}}{1 - 1.404Z^{-1} + 0.84Z^{-2}}\frac{1 - 0.198Z^{-1} + Z^{-2}}{1 - 1.238Z^{-1} + 0.455Z^{-2}}$$

$$f_1 = 4832H_Z; \quad f_2 = 7495H_Z; \quad \Delta f_1 = 4H_Z; \quad \Delta f_2 = -3H_Z.$$

Scale factors: $a_0^0 = 2^{-3}$; $a_0^1 = 2^{-1}$; $a_0^2 = 1$.

**7.10** Theoretic order: $N = 5.19$; for $N = 6$, $\delta_1$ becomes very small. Coefficient wordlength: $b_c \simeq 11$ bits. Difference between input and internal data: 7 bits.

## Chapter 8

**8.1** $S = \frac{1}{z+2}\begin{bmatrix} z & 2 \\ 2 & z \end{bmatrix}$; $t = \begin{bmatrix} 1 - z/2 & z/2 \\ -z/2 & 1 + z/2 \end{bmatrix}$

For LC circuits take, $z = Lp + \frac{1}{Lp}$ or $z = \frac{LC}{Lp + \frac{1}{Cp}}$

**8.2** The diagram is that shown in Figure 8.3, with $N = 6$ and $Y_6 = 0$. For

$$f_s = 40\text{ kHz} : a_1 = a_4 = 0.205; a_2 = a_3 = 0.085;$$

The coefficients are multiplied by four for $f_s = 10\text{ kHz}$.

**8.3** Follow the procedure given at the end of Section 8.2 and show the impact of the sampling frequency. Verify the five-bit curve given in Figure 8.9.

**8.4** Lattice filter zeros: 0.6605; $0.6647 \pm j0.5020$; after rounding the $k_i$ to 5 bits: 0.6661; $0.6377 \pm j0.5002$.

## Chapter 9

**9.1** $$X(f) = \frac{1 - a\cos 2\pi f}{1 + a^2 - 2a\cos 2\pi f} + j\frac{-a\sin 2\pi f}{1 + a^2 - 2a\cos 2\pi f}.$$

**9.2** Calculate $X_I(\omega)$ through the Hilbert transform or write:

$$X_R(\omega) = \frac{1}{2}\left[\frac{1}{1 - P} + \frac{1}{1 - P}\right]$$

**9.3** The nonzero terms in the sets $x_R(n)$ and $x_I(n)$ are interleaved. The operation performed is analytic filtering: $y(n) = \frac{1}{2}e^{-jn\frac{\pi}{5}}$.

**9.4** Coefficient wordlength: $b_c \simeq 2 + \frac{1}{2}\log_2\left(\frac{f_e}{24f}\right) + \log_2\left(\frac{1}{\delta}\right)$.

**9.5** Phase shifter order: $N \simeq \log\left(\frac{\pi}{\varepsilon}\right)\log\left(\frac{f_s}{f_1}\frac{f_s}{f_2}\right)$;

coefficients: $b_c \approx \log_2\left(\frac{\pi}{\varepsilon}\right) + \log_2\left(\frac{f_s}{f_1}\right) + \log_2\left(\frac{f_s}{f_2}\right)$.

For the example in Section 9.4: $N = 4.97$; $b_c \simeq 14$ bits.

**9.6** Transfer functions:

$$H_m(Z) = 1 - 2Z^{-1} + 2Z^{-2} - Z^{-3} + 0.25Z^{-4}$$
$$H_L(Z) = 0.5 - 1.5Z^{-1} + 2.25Z^{-2} - 1.5Z^{-1} + 0.5Z^{-4}$$
$$H_M(Z) = 0.25 - Z^{-1} + 2Z^{-2} - 2Z^{-3} + Z^{-4}$$

**9.7** Filter frequency response

$$H(f) = e^{-j2\pi 3f}\,[0.5 + 0.592\cos 2\pi f - 0.1012\cos 6\pi f]$$

$$H(0) = H\left(\frac{f_s}{8}\right) = 0.989;\ \delta_1 = \delta_2 = 0.011;\ \Delta f = f_s/4$$

Output of IQ modulator

$$y(n) = 0.4945\, e^{-j(n-5)\pi/4} + 0.0055\, e^{j(n-5)\pi/4}$$

After undersampling, setting $n = 2p+1$

$$y(p) = 0.4945\, e^{-jp\pi/2} - 0.0055\, e^{jp\pi/2}$$

Two components at frequencies $f_s/4$ and $-f_s/4 = 3f_s/4$.

**9.8** Pass-band width: [0; 0.25]; ripple: $4\times10^{-2}$; delay: $2T$.

**9.9** Filter impulse response:

$$h(n) = \frac{2}{N}\,\frac{\sin\left(\frac{n\pi}{2}\right)}{\sin\left(\frac{\pi n}{N}\right)}\, e^{j\left(\frac{N}{2}-1\right)n\frac{\pi}{N}}$$

With respect to (9.24), the DFT introduces double periodicity in time and frequency. The values of $x_c(n)$ in the vicinity of $n = 8$ are approximately restored from the first 3 or 7 values of $h(n)$. Clearly, the DFT is efficient but the accuracy is limited.

**9.10**

$$\begin{bmatrix} x(5) \\ x(6) \end{bmatrix} = \begin{bmatrix} -1 - 0.866j \\ 0.612 - 1.319j \end{bmatrix} = -\begin{bmatrix} 1 & j \\ -1 & e^{-j3\pi/4} \end{bmatrix}^{-1} T_{21}X_1$$

## Chapter 10

**10.1** Coefficient wordlengths: $b_c = 1, 2, 5, 6, 9, 10, 10, 11, 14$. For the half-band filter:

$$b_c \approx 2\log_2(1/\delta_m - \delta_0).$$

**10.2** Filters in the cascade of 3 filters: $\Delta f = 0.4$ with $M = 2$; $\Delta f = 0.15$ with $M = 3$; $\Delta f = 0.025$ with $M = 8$. Roundoff noise at the output of a half-band filter: $2M\frac{q^2}{12}$.
After three filters: $P_N = 20\frac{q^2}{12}$.

**10.3** The function can be carried out with a half-band filter ($M = 3$) and a low-pass filter with 54 coefficients – hence the computation rate of 264 kmult/s. A direct realization with 100 coefficients leads to 400 kmult/s.

**10.4** The odd DFT corresponds to a frequency shift of: $f_s/2N$.

**10.5** Polyphase network functions:

$$D(Z)(1 - 0.1354Z^{-1} + 0.069Z^{-2})(1 + 0.98Z^{-1} + 0.51Z^{-2})$$
$$N_1(Z) = 1 + 7.806Z^{-1} + 9.718Z^{-2} + 3.773Z^{-3} + 0.1883Z^{-4}$$
$$N_2(Z) = 3.713(1 + 2.908Z^{-1} + 2.035Z^{-2} + 0.317Z^{-3})$$

The diagram is that given in Figure 10.10; multiplication rate: $8f_s$.

## Chapter 11

**11.1** Frequency response:

$$H(f) = e^{-j2\pi f2.5}\,[0.904\cos\pi f + 0.234\cos 3\pi f - 0.1\cos 5\pi f]$$

Output signal:

$$y(n) = 0.963\cos(n - 2.5)\pi/4$$

Output of analysis filters, after downsampling:

$$u_1(n) = \frac{0.963\cos(n - 2.5)\pi}{4} + 0.963\cos(n - 2.5)3\pi/4$$
$$u_2(n) = \frac{0.037\cos(n - 2.5)\pi}{4} + 0.037\cos(n - 2.5)3\pi/4$$

Total transfer function:

$$T(Z) = Z^{-1}\,[-0.023 + 0.121\,Z^{-2} + 0.882Z^{-4} + 0.121Z^{-6} - 0.023Z^{-8}]$$

Reconstructed signal: $x'(n) = 0.929\cos(n - 5)\pi/4$

**11.2** Transfer functions of the two factors:

$$H_0(Z) = (1 + Z^{-1})^3/4;\ H_1(Z) = (-1 - 3Z^{-1} + 3\,Z^{-2} + Z^{-3})/4$$

Amplification factor: $20/16 = 1.25$.

**11.3** Use the procedure described in Section 11.3, without double zero at $-1$ for $H_1(-Z)$.

**11.4** Output signals: $y_1(n) = 0.951\cos(n - 4)\pi/4$

$$y_2(n) = 0.15\cos(n - 3)\pi/4$$

Downsampling introduces image components at frequency 3/8.
Cancelling of the image components at reconstruction can be verified.

**11.5** The reconstruction error is bounded by the quantization step multiplied by twice the sum of the absolute values of the coefficients.

## Chapter 12

**12.1** The transfer function of the first branch of the polyphase network is:

$$B1(Z) = Z^{-1/2}[-0.0218 + 0.0621\,Z^{-4} + 0.1996\,Z^{-8} + 0.0160\,Z^{-12}]$$

To simplify, let

$$H1(Z) = [-0.0218 + 0.0621\,Z^{-1} + 0.1996\,Z^{-2} + 0.0160\,Z^{-3}]$$

The frequency response is analyzed, taking $Z = e^{j2\pi f}$.
A delay $\tau$ corresponds to the factor $e^{-j2\pi f\tau}$; the delays in the branches are as follows:

$$\tau = [1.8875 \;\; 1.5863 \;\; 1.3198 \;\; 1.0188].$$

These delays compensate the factors due to interleaving: (1/2, 3/2, 5/2, 7/2).

**12.2** Convolution in time is multiplication in the frequency domain.
The coefficients are obtained by calculating the inverse Fourier transform of the prototype filter impulse response. They are applied at FFT output in the analysis part and at inverse FFT input in the synthesis part.
The coefficients are given by (12.40) and (12.42). For the first filter of the analysis bank after the basis filter, the outputs 1 to 7 must be used and the multiplication results are summed. For the synthesis, the signal intended for the first filter must be applied to inputs 1 to 7 of the iFFT, after multiplication by the coefficients.

**12.3** Assuming the length of the prototype filter impulse response is unity, with a continuous-time half-sinusoid, we get:

$$H(f) = \frac{2}{\pi}\frac{\cos(\pi f)}{1 - 4f^2}$$

To be compared with the FFT response in similar conditions (duration: 0.5):

$$H_{\text{FFT}}(f) = \frac{\sin\left(\frac{\pi f}{2}\right)}{\pi f/2}$$

The frequency response $H(f)$ decreases with the square of the frequency.
In each branch of the polyphase network, the coefficient values are $h(n)$ and $h(n + N)$.
Realization by $2N$-FFT – it suffices to add two adjacent outputs. More multiplications are needed.

## Chapter 13

**13.1** AC function:

$$r_1(p) = \frac{N-p}{N}\cos(2\pi pf) + \frac{N-p}{N}\left[\cos(2\pi(N-1)f)\frac{\sin(2\pi(N-p)f)}{(N-p)\sin(2\pi f)}\right]$$

$$r_1(1) = \frac{N-1}{N}\left[\cos(2\pi f) + \frac{\sin(2\pi 2(N-1)f)}{2(N-1)\sin(2\pi f)}\right]$$

For $f = 1/8$ and $N = 16$, we get $r_1 = 0.618$.

Cancelling $y(n)$ for $n > N-1$ corresponds to applying a quadrature filter which, due to the ripple, introduces a deviation for the frequency values which are non-multiples of $1/2N$ – hence the difference between the estimated and actual frequencies (1.4%). The bound is $CR = 3.7 \times 10^{-6}$.

**13.2** Time constant $\tau = 1/\delta$. For $y(n)$ to approach $m$ within 1% on average, 23 samples are needed. After the transition phase, the quadratic residual error is given by (14.39). Recursive and non-recursive estimators are equivalent for $n \simeq 2/\delta$.

**13.3** AC function: $r(0) = 1$; $r(1) = 0.707$; $r(2) = 0$.
Eigenvalues of $R_3$: [2, 1, 0].

**13.4** Output power: $P_s = \left(1 + a_1^2 + a_2^2\right)\sigma_b^2 + |1 - a_1 \quad e^{-j\omega} - a_2 e^{-2j\omega}|^2$
Setting the derivatives to zero, we obtain the expressions of the coefficients.

**13.5** AC function:

$$r(0) = 1; \quad r(1) = \frac{\left(\cos\frac{\pi}{4} + \cos\frac{\pi}{3}\right)}{2}; \quad r(2) = \frac{1}{4}\cos\frac{2\pi}{3}$$

We verify that the zeros of the predictor are situated in the Z-plane between $e^{j\pi/4}$ and $e^{j\pi/3}$.

**13.6** The roots of the polynomials sit on the unit circle, and they verify the alternation principle.

**13.7** Covariance matrix: $R3 = [2\ 1.866\ 1.5;\ 1.866\ 2\ 1.866;\ 1.5\ 1.866\ 2]$
Weighting coefficients: $A = [3.73\ \ -6.46\ \ 3.73]$

**13.8** Eigenvalues: $\lambda_1 = 1.06 + 0.37j$; $\lambda_2 = 0.30 - 0.60j$
Signal-to-noise ratio in each channel: $|\lambda_i|^2/\sigma_b^2$; theoretical capacity: 6.2 bit/s/Hz.

## Chapter 14

**14.1** Impulse response: $H(Z) = 0.5(1 + 1.5\ Z^{-1}(1 + 0.5\ Z^{-1} + 0.5^2\ Z^{-2} + \ldots))$
Prediction coefficients: $a_1 = 1.07$; $a_2 = -0.43$.
Output power: $Emin = 0.357 = 1/Gp$

$$H(Z)P(Z) = 0.5\,(1 + 0.430\,Z^{-1} - 0.425Z^{-2} + 0.220Z^{-3} + \ldots)$$

$$\delta_m = 1; \quad Er = 1.25\,E\min; \quad \tau = \frac{1}{\delta} = 4$$

**14.2** The prediction filter exhibits infinite attenuation at frequency $f_s/8$. Hence: $a_1 = \sqrt{2}$; $a_2 = -1$.
With the noise $\sigma^2$, one gets:

$$a_1 = \sqrt{2}\frac{1 + 2\sigma^2}{1 + 8\sigma^2 + 8\sigma^4} \simeq \sqrt{2}(1 - 6\sigma^2); \quad a_2 = -\frac{1}{1 + 8\sigma^2 + 8\sigma^4} \approx -(1 - 8\sigma^2)$$

**14.3** Signal AC function:

$$r_0 = P_x = \frac{1}{3}; \quad r_1 = \frac{1}{6}; \quad r_2 = \frac{1}{12}$$

Optimum coefficient values: $H^t_{\text{opt}} = [2; -1; 0]$
With noise: $H^t_{\text{opt}} = [1.35; -0.5; -0.07\,]$
Noise amplification factor: $\|H_{\text{opt}}\|_2^2 = 2.09$
The residual interference power is derived from the total response (channel + equalizer).
Minimum eigenvalue: $\lambda_{\min} = 0.235$. Simulations confirm that the time constant is in agreement with the estimation based on $\lambda_{\min}$.

**14.4** Give the expression of the output error and search for the coefficients which minimize its power.
To begin with, in the input–output relation of filter $H(Z)$, substitute the reference $y(n)$ for the output $\tilde{y}(n)$ and compute the optimal coefficient values in that case.

## Chapter 15

**15.1** Equation of a separation line: $-x_1 + x_2 = 0$;

$$Distances: A1: 1; \quad A2: 2; \quad A3: 2; \quad B1: -1; \quad B2: -1; \quad B3: -4$$

Iterative method: The adaptation step in (15.2) is set to ½ and inputs are used in the following order: $A1, B1, A2, B2, A3, B3$. Normalizing $h_2$ to unity, we get:
$H(1) = [0 - 2\ 1]$ (line 0 – A1); $H2 = [0.2 - 1.2\ 1]$; .......; $H6 = [-0.45 - 1.55\ 1]$

**15.2** Equations of two separation lines: $1 - 3x_1 + x_2 = 0; -5 + \frac{5}{7}x_1 + x_2 = 0$
Iterative procedure: In a two-step process, for example, a line separating $A$ from $B + C$ is determined first, and, then, $B$ is separated from $C$.

**15.3** Optimum coefficient values: $h_{1\text{opt}} = 1.4$; $h_{2\text{opt}} = 1$. Error power: $E_0 = 0.023$. With respect to the estimations given in Section 15.4, the degradation is due to the nonlinearity, which is not taken into account.
The step $\delta = 0.1$ entails an additional error power of 5%.

**15.4** Figure 15.6 is supplemented with the set of coefficients $h_{ij2}$. The coefficients of the last stage become $h_{j13}$ $(j = 1, 2)$. As estimations, we get $a = 1.19$ and $b = 2.19$.
With coefficients $h_{11j} = [1.23\ 1.20\ 1.44]$, the two nonlinearities are accounted for to determine $c$, which yields $c = 1.5$. In the presence of noise, the drift of the coefficients is reduced.

**15.5** The circuit leads to 1.535 for the slope of the curve at the origin instead of $\pi/2$. This result is justified by the series expansion of the sine function.

**15.6** Coefficient values: $\pm 1/(0.7 \times 15)$.
Total number of coefficients: $64 \times 32 + 32 \times 15 + 15 \times 10 = 2678$.
Updating for one digit: $32 \times 64 + 15 \times 32 + 15 = 2543$.
With the $8 \times 8$ grid, one might be able to handle 4 (64/16) different writings of each digit.

**15.7** The equations of the circuits are as set forth in Section 15.6. Parameter C determines the coefficients of the recursive section.
Coefficients $h_{11,1}$ and $h_{11,3}$ must, on average, compensate for the attenuation brought by $f(a)$. Coefficients $h_{ij,2}$ must compensate for the attenuation of the signals they multiply. Approximate values can be obtained by simple calculations based on development (15.28).

**15.8** Values obtained, with $d_k(n)$ reference values:

$$x_{i1} = \pm 0.76; \;\; x_{i2} = \pm 0.64; \;\; e_{i2} \approx 0.02 * 0.64 * \sum_{k=1}^{16} d_k(1)d_k(2); \;\; e_{i1} = 0.59 * e_{i2}$$

Reducing the initial values entails a reduction of the errors involved in the initial coefficient updating and an increase of the system time constant.
As for stability issues and time constant, see (14.16), (15.22), and (6.18).

**15.9** The diagram is extended to a third hidden layer by repeating the first-layer circuit.
The slopes of the lines that bear the segments are expressed by: $8\left[\sin\left(\frac{k\pi}{8}\right) - \sin\left(\frac{(k-1)\pi}{8}\right)\right]$; $1 \leq k \leq 8$. The coefficient values follow. To verify, take $x = \frac{k-1}{8} + \varepsilon$ and compute the output $y$.
Maximum error estimation: $\sin\left(\frac{\pi}{16}\right) - \frac{\sin\left(\frac{\pi}{8}\right)}{2} = 0.0037$.

## Chapter 16

**16.1** Prediction coefficients: $a_1 = 1.85j$; $a_2 = 1$.
Signal to be subtracted in the frequency domain: $E(6) = 1$; $E(10) = 2j$.

**16.2** Splitting matrix T8 into four appropriate blocks A, B, C, and D, we find: $Sx = -C^{-1}DX$.
Spurious impulse: amplitude 2 and index 3.
Between 2 index values, the impulse impacts the whole useful signal, and it is necessary to subtract the signal provided by linear prediction.

**16.3** For a single error, $N1 = 7$ and for a double error, $N2 = 6$. Coding gain: 3 (4.7 dB)

**16.4** Error-free transmission of binary data requires at least SNR = 3, or 4.77 dB.
If the system signal-to-noise ratio is 8 dB, the code must bring a gain of 7 dB.
A code with length $L = 9$ provides the gain 7 dB for bit error rate $10^{-8}$.

**16.5** Free distance: 3; decoded sequence (systematic coding): 10010101
Recursive code: rate $R = 3/8$; free distance: 3.

# Index

Note: *Italicized* and **bold** page numbers refer to figures and tables, respectively.



*Digital Signal Processing: Theory and Practice*, Tenth Edition. Maurice Bellanger.